NANOMATERIALS FOR DIRECT ALCOHOL FUEL CELLS

NANOMATERIALS FOR DIRECT ALCOHOL FUEL CELLS

Characterization, Design, and Electrocatalysis

Edited by

FATIH ŞEN

Şen Research Group, Department of Biochemistry, Faculty of Science, Dumlupinar University, Kütahya, Turkey

ELSEVIER

Elsevier
Radarweg 29, PO Box 211, 1000 AE Amsterdam, Netherlands
The Boulevard, Langford Lane, Kidlington, Oxford OX5 1GB, United Kingdom
50 Hampshire Street, 5th Floor, Cambridge, MA 02139, United States

Copyright © 2021 Elsevier Inc. All rights reserved.

No part of this publication may be reproduced or transmitted in any form or by any means, electronic or mechanical, including photocopying, recording, or any information storage and retrieval system, without permission in writing from the publisher. Details on how to seek permission, further information about the Publisher's permissions policies and our arrangements with organizations such as the Copyright Clearance Center and the Copyright Licensing Agency, can be found at our website: www.elsevier.com/permissions.

This book and the individual contributions contained in it are protected under copyright by the Publisher (other than as may be noted herein).

Notices
Knowledge and best practice in this field are constantly changing. As new research and experience broaden our understanding, changes in research methods, professional practices, or medical treatment may become necessary.

Practitioners and researchers must always rely on their own experience and knowledge in evaluating and using any information, methods, compounds, or experiments described herein. In using such information or methods they should be mindful of their own safety and the safety of others, including parties for whom they have a professional responsibility.

To the fullest extent of the law, neither the Publisher nor the authors, contributors, or editors, assume any liability for any injury and/or damage to persons or property as a matter of products liability, negligence or otherwise, or from any use or operation of any methods, products, instructions, or ideas contained in the material herein.

Library of Congress Cataloging-in-Publication Data
A catalog record for this book is available from the Library of Congress

British Library Cataloguing-in-Publication Data
A catalogue record for this book is available from the British Library

ISBN: 978-0-12-821713-9

For information on all Elsevier publications
visit our website at https://www.elsevier.com/books-and-journals

Publisher: Matthew Deans
Acquisitions Editor: Simon Holt
Editorial Project Manager: Chiara Giglio
Production Project Manager: Nirmala Arumugam
Cover Designer: Greg Harris

Typeset by STRAIVE, India

Contents

Contributors xv

Chapter 1 Fundamentals of electrochemistry **1**
*Muhammed Bekmezci, Elif Esra Altuner, Vildan Erduran,
Ramazan Bayat, Iskender Isik, and Fatih Şen*

1 Introduction 1
2 Conclusion 12
References 13

Chapter 2 Nanomaterials and their classification **17**
İsmail Mert Alkaç, Burcu Çerçi, Cisil Timuralp, and Fatih Şen

1 Introduction 17
2 Classification of nanomaterials 18
3 Conclusion 29
References 30

**Chapter 3 The electrochemical mechanism and transport
phenomenon of liquid fuel cells** **35**
Gamze Karanfil, Ramazan Bayat, and Fatih Şen

1 Introduction 35
2 Direct methanol fuel cells 40
3 Direct ethanol fuel cells 43
4 Direct ethylene glycol fuel cells 45
5 Conclusion 48
References 49

Chapter 4 The material development and characterization of direct alcohol fuel cells 53

Saadet Güler, Ahmet Yavaş, Sibel Demiroglu Mustafov, and Fatih Şen

1 Introduction 53
2 Fuel cells 54
3 Catalysts 62
4 Characterization methods of catalysts 66
5 Conclusion 70
References 70

Chapter 5 Fundamentals of alcohol fuel cells 75

Merve Akin, Vildan Erduran, Elif Esra Altuner, Cisil Timuralp, Iskender Isik, and Fatih Şen

1 Introduction 75
2 Fuel cells 76
3 Fuel cell history 77
4 Use of alcohols in fuel cells 77
5 Alcohol fuel cell thermodynamics 79
6 Direct methanol fuel cells 83
7 Direct ethanol fuel cell 86
8 Conclusion 89
References 89

Chapter 6 The electrocatalysts with pH of the electrolyte for the complete pathways of the oxidation reactions 95

Muhammed Bekmezci, Ramazan Bayat, Merve Akin, Hakan Burhan, Iskender Isik, and Fatih Şen

1 Introduction 95
2 General considerations about fuel cells 96
3 Conclusions 101
References 102

Contents **vii**

Chapter 7 Pt-based catalysts for alcohol oxidation **109**

Hakan Burhan, Kubilay Arıkan, Muhammed Bekmezci, Tugba Gur,
and Fatih Şen

1 Introduction 109

2 Frequently used electrochemical techniques for fuel cells 114

3 Single-cell tests 116

4 Conclusions 122

References 123

Chapter 8 Monometallic nanomaterials for direct alcohol fuel cells . **129**

Ramazan Bayat, Vildan Erduran, Muhammed Bekmezci,
Iskender Isik, and Fatih Şen

1 Introduction 129

2 Nanostructured materials 131

3 Catalyst layer 134

4 Conclusion 138

References 139

Chapter 9 Bimetallic nanomaterials for direct alcohol fuel cells . . . **145**

Haydar Goksu, Muhammed Bekmezci, Vildan Erduran, and Fatih Şen

1 Introduction 145

2 Monometallic nanomaterials 146

3 Bimetallic nanomaterials 147

4 Conclusions 152

References 153

Chapter 10 Ternary/quaternary nanomaterials for direct alcohol
fuel cells **157**

Elif Esra Altuner, Tugba Gur, and Fatih Şen

1 Introduction 157

2 Fuel cells 158

3 Conclusion 168

References 168

viii Contents

Chapter 11 Catalysts for high-temperature fuel cells operated by alcohol fuels 173

Ali Cherif, Nimeti Doner, and Fatih Şen

1 Introduction 173
2 High-temperature alcohol fuel cells 175
3 Operating catalyst 177
4 Conclusion 180
References 181

Chapter 12 Porous metal materials for polymer electrolyte membrane fuel cells 187

Fatma Aydın Ünal, Cisil Timuralp, Vildan Erduran, and Fatih Şen

1 Introduction 187
2 Polymer exchange membrane fuel cells (PEMFCs) 188
3 Fundamental components and materials of PEMFCs 189
References 203

Chapter 13 Novel materials structures and compositions for alcohol oxidation reaction 209

Vildan Erduran, Muhammed Bekmezci, Merve Akin, Ramazan Bayat, Iskender Isik, and Fatih Şen

1 Introduction 209
2 Pt-based electrocatalysts 211
3 Pt-free electrocatalysts 217
4 Supporting materials 220
5 Challenges and future perspectives 237
6 Conclusions 238
References 238

Chapter 14 Synthesis and characterization of nanocomposite membranes for high-temperature polymer electrolyte membranes (PEM) methanol fuel cells **251**

Fatma Aydın Ünal, Vildan Erduran, Ramazan Bayat, Sadin Ozdemir, and Fatih Şen

1 Introduction 251
2 Why high-temperature polymer electrolyte membranes? 252
3 Opportunities and challenges for high-temperature PEMFCs 253
4 Properties of composite membranes 253
5 Synthesis and characterization 254
6 Synthesis and characterization of nanocomposite membranes via the sol-gel method 261
7 Characterization and synthesis of the polybenzimidazole-base membranes 269
8 Conclusion 278
References 278

Chapter 15 Fabrication and properties of polymer electrolyte membranes (PEM) for direct methanol fuel cell application **283**

Fatma Aydın Ünal, Vildan Erduran, Cisil Timuralp, and Fatih Şen

1 Introduction 283
2 Basics of the DMFC 286
3 Requirements for DMFC membranes 287
4 Materials and properties 287
5 Membrane and its properties 289
6 Fabrication methods of nanocomposite membranes 297
7 Conclusions 299
References 299

x Contents

Chapter 16 Carbonaceous nanomaterials (carbon nanotubes, fullerenes, and nanofibers) for alcohol fuel cells 303

Vildan Erduran, Muhammed Bekmezci, Ramazan Bayat, Iskender Isik, and Fatih Şen

1 Introduction 303
2 Carbon nanomaterials 305
3 Conclusion 312
References 312

Chapter 17 Carbon-based nanomaterials for alcohol fuel cells 319

Merve Akin, Ramazan Bayat, Vildan Erduran, Muhammed Bekmezci, Iskender Isik, and Fatih Şen

1 Introduction 319
2 Fuel cells 320
3 Alcohol fuel cells 323
4 Nanomaterials 326
5 Conclusions 330
References 330

Chapter 18 Dendrimer-based nanocomposites for alcohol fuel cells . 337

Elif Esra Altuner, Muhammed Bekmezci, Ramazan Bayat, Merve Akin, Iskender Isik, and Fatih Şen

1 Introduction 337
2 Alcohol fuel cells 338
3 Nanocomposites types and applications of direct alcohol fuel cells 343
4 Conclusions 347
References 348

Chapter 19 Metal organic framework-based nanocomposites for alcohol fuel cells 353

Bahar Yilmaz, Ramazan Bayat, Muhammed Bekmezci, and Fatih Şen

1 Introduction 353
2 Fuel cells 354
3 Nanocomposites 358
4 Metal organic frameworks 359
5 Conclusion 365
References 365

Chapter 20 Carbon-polymer hybrid-supported nanomaterials for alcohol fuel cells 371

Ramazan Bayat, Nimeti Doner, and Fatih Şen

1 Introduction 371
2 Fuel cells 372
3 Carbon nanomaterials 375
4 Polymer materials 378
5 Conclusion 380
References 380

Chapter 21 Polymer-based nanocatalyts for alcohol fuel cells 389

Ilyas Ilker Isler, Haydar Goksu, Vildan Erduran, Iskender Isik, and Fatih Şen

1 Introduction 389
2 Alcohol fuel cells 390
3 Using nanocatalyst polymer fuel cells, which reduces costs and increases conductivity 397
4 Conclusion 398
References 399

xii Contents

Chapter 22 Different synthesis methods of nanomaterials for direct alcohol fuel cells **405**

Vildan Erduran, Muhammed Bekmezci, Iskender Isik, and Fatih Şen

1 Introduction 405
2 Direct alcohol fuel cells 410
3 Studies related to nanocatalyst synthesis 415
4 Nanomaterial synthesis 420
5 Conclusion 425
References 425

Chapter 23 The synthesis and characterization of size-controlled bimetallic nanoparticles **433**

Haydar Goksu, Muhammed Bekmezci, Ramazan Bayat, Elif Esra Altuner, and Fatih Şen

1 Introduction 433
2 Synthesis of bimetallic and monometallic nanoparticles and their differences 434
3 Using electrical current for fabrication of metallic nanoparticles . . 437
4 Porous form and surface area of bimetallic nanoparticles 439
5 Important metals used for the preparation bimetallic nanoparticles 441
6 Conclusion 442
References 443

Chapter 24 The synthesis and characterization of size-controlled monometallic nanoparticles **449**

Muhammed Bekmezci, Vildan Erduran, Mustafa Ucar, and Fatih Şen

1 Introduction 449
2 Direct alcohol fuel cells 450
3 Conclusions 458
References 458

Chapter 25 Topics on the fundamentals of the alcohol oxidation reactions in acid and alkaline electrolytes 465

Vildan Erduran, Merve Akin, Hakan Burhan, Iskender Isik, and Fatih Şen

1 Introduction 465
2 Alcohol oxidation reactions in acidic media 468
3 Alcohol oxidation reaction in alkaline media 470
4 Catalysts for alcohol oxidation reactions 473
5 Conclusion 475
References 475

Chapter 26 Direct alcohol-fed solid oxide fuel cells 481

Hakan Burhan, Kubilay Arıkan, Sadin Ozdemir, Iskender Isik, and Fatih Şen

1 Introduction 481
2 Direct alcohol-fed, low-temperature solid oxide fuel cells 483
3 Membranes 486
4 Electrochemical performance 501
5 Conclusion 502
References 502

Chapter 27 Commercial aspects of direct alcohol fuel cells 511

Elif Esra Altuner, Kubilay Arıkan, Hakan Burhan, Sadin Ozdemir, and Fatih Şen

1 Introduction 511
2 Fuel cells 512
3 Development of direct alcohol fuel cells 516
4 Process of direct alcohol fuel cells 516
5 The commercial market of direct alcohol fuel cells 518
6 Conclusion 520
References 520

Index 525

Contributors

Merve Akin Şen Research Group, Department of Biochemistry; Department of Materials Science and Engineering, Faculty of Engineering, Dumlupinar University, Kütahya, Turkey

İsmail Mert Alkaç Şen Research Group, Department of Biochemistry, University of Dumlupinar, Kütahya; Department of Materials Science and Engineering, Faculty of Engineering, Dumlupinar University, Kütahya, Turkey

Elif Esra Altuner Şen Research Group, Department of Biochemistry, Dumlupinar University, Kütahya, Turkey

Kubilay Arıkan Şen Research Group, Department of Biochemistry, Dumlupinar University, Kütahya, Turkey

Fatma Aydın Ünal Metallurgical and Materials Engineering Department, Faculty of Engineering, Alanya Alaaddin Keykubat University, Alanya/Antalya, Turkey

Ramazan Bayat Şen Research Group, Department of Biochemistry; Department of Materials Science and Engineering, Faculty of Engineering, Dumlupinar University, Kütahya, Turkey

Muhammed Bekmezci Şen Research Group, Department of Biochemistry; Department of Materials Science and Engineering, Faculty of Engineering, Dumlupinar University, Kütahya, Turkey

Hakan Burhan Şen Research Group, Department of Biochemistry, Dumlupinar University, Kütahya, Turkey

Burcu Çerçi Faculty of Medicine, Department of Medical Biology and Genetics, Izmir Katip Celebi University, Izmir, Turkey

Ali Cherif Şen Research Group, Department of Biochemistry, Dumlupinar University, Kütahya; Department of Mechanical Engineering, Faculty of Engineering, Gazi University, Ankara, Turkey

Nimeti Doner Department of Mechanical Engineering, Faculty of Engineering, Gazi University, Ankara, Turkey

Vildan Erduran Şen Research Group, Department of Biochemistry; Department of Materials Science and Engineering, Faculty of Engineering, Dumlupinar University, Kütahya, Turkey

Haydar Goksu Kaynasli Vocational College, Duzce University, Duzce, Turkey

Saadet Güler Department of Material Science and Engineering, Izmir Katip Çelebi University; Department of Metallurgical and Materials Engineering, The Graduate School of Natural and Applied Sciences, Dokuz Eylul University, Izmir, Turkey

Tugba Gur Vocational School of Health Services, Van Yuzuncu Yıl University, Van, Turkey

Iskender Isik Department of Materials Science and Engineering, Faculty of Engineering, Dumlupinar University, Kütahya, Turkey

Ilyas Ilker Isler Kaynasli Vocational College, Duzce University, Duzce, Turkey

Gamze Karanfil Karamanoglu Mehmetbey University, Faculty of Engineering, Department of Energy Systems Engineering, Karaman, Turkey

Sibel Demiroglu Mustafov Department of Metallurgical and Materials Engineering, Dokuz Eylul University, Izmir, Turkey

Sadin Ozdemir Mersin University, Food Processing Programme, Technical Science Vocational School, Mersin, Turkey

Fatih Şen Şen Research Group, Department of Biochemistry, Dumlupinar University, Kütahya, Turkey

Cisil Timuralp Mechanical Engineering Department, Eskisehir Osmangazi University, Eskişehir, Turkey

Mustafa Ucar Faculty of Science and Arts, Department of Chemistry, Afyon Kocatepe University, Afyonkarahisar, Turkey

Ahmet Yavaş Department of Material Science and Engineering; Graduate Program of Materials Science and Engineering, Graduate School of Natural and Applied Science, Izmir Katip Çelebi University, Izmir, Turkey

Bahar Yilmaz Karamanoglu Mehmetbey University, Faculty of Engineering, Department of Bioengineering, Karaman, Turkey

1

Fundamentals of electrochemistry

Muhammed Bekmezci[a,b], Elif Esra Altuner[a], Vildan Erduran[a,b], Ramazan Bayat[a,b], Iskender Isik[b], and Fatih Şen[a]

[a]*Şen Research Group, Department of Biochemistry, Dumlupinar University, Kütahya, Turkey.* [b]*Department of Materials Science and Engineering, Faculty of Engineering, Dumlupinar University, Kütahya, Turkey*

1 Introduction

Electrochemistry deals with the interactions between electrical and chemical effects. The history of electrochemistry began in 1800 when Alessandro Volta challenged the ideas of Luigi Galvani, who claimed that animals produce electricity. Volta announced the invention of the voltaic battery as the first modern electric battery. Thus, new fields of study have emerged concerning the chemical production of electricity and the effects of electricity on chemicals [1]. At the heart of electrochemistry lies the study of the electrical energy produced by chemical reactions and the chemical reactions produced by the transition of an electric current [2].

While the fundamental concepts of electrochemistry apply to both of these, the primary focus of this chapter is the application of electrochemical methods to the activity of chemical structures. Scientists conduct electrochemical measurements on chemical systems for a number of purposes. Doing so allows electrochemical measurements to be taken on chemical structures for a number of purposes, such as obtaining thermodynamic reaction results, creating an unstable intermediate such as a radical ion, testing its degradation rate or spectroscopic properties, and evaluating a large quantity of metal ions or a solution for organic species. These examples use electrochemical methods as instruments for spectroscopic methods in the analysis of chemical structures. For these approaches, an understanding of the basic concepts of electrode reactions and the electrical properties of the electrode solution interface is needed [3, 4].

Nanomaterials for Direct Alcohol Fuel Cells. https://doi.org/10.1016/B978-0-12-821713-9.00023-8
Copyright © 2021 Elsevier Inc. All rights reserved.

1.1 Oxidation–reduction redox reactions

The term *electricity* refers to several events related to the presence and flow of an electrical charge carried by electrons or ions. In electrochemistry, it is first necessary to understand oxidation and reduction reactions [5]. In an oxidation and reduction (redox) reaction, electrons are transported between particles or change the amount of oxidation of atoms. In addition, the total number of electrons with oxidation should be equal to the total number incorporated with reduction. A half-reaction is one of two parts of the reaction, some of which contain electron loss (oxidation), and the other include electron gain (reduction) [6]:

$$\text{Oxidation}: Zn^{2+}_{(qe)} + 2e^- \rightarrow Zn_{(s)} \tag{1}$$

$$\text{Reduction}: Cu^{2+}_{(qe)} + 2e^- \rightarrow Cu_{(s)} \tag{2}$$

$$\text{Overall}: Zn_{(s)} + Cu^{2+}_{(qe)} \rightarrow Zn^{2+}_{(qe)} + Cu_{(s)} \tag{3}$$

In the half-reactions (1 and 2), zinc oxidation produces electricity, while the reduction of copper ions is occurring. By placing an electrical circuit between these two reactions, electricity is flowing from the produced place to the consumed place, thereby the electrode potential determines whether a half-cell acts as an electron generator or as an electron consumer. In the half-cell with the smallest (i.e., least negative or lowest positive) electrode potential, the reaction proceeds in the opposite direction, and the oxidized element in this half-cell lose electrons, allowing it to flow around the circuits. The Zn^{2+}/Zn half-cell has lower electrode potential, and it will be the anode terminal of the cell. Here, we can see that the Cu^{2+}/Cu half-cell has more positive electrode potential, so that will be the cathode terminal, and the reaction will go forward. As reactions continue, both half-cells result in an imbalance between the positive and negative ions. With the aid of a salt bridge, it allows ions to move from one half-cell to another to correct this imbalance; otherwise, the electrons may stop flowing in the outer circuit [7].

1.2 Electrochemical cells

There are two types of electrical cells in electrochemistry:
- Galvanic cells (Fig. 1), which automatically consume electrical energy
- Electrolytic cells (Fig. 2), which consume electrical energy

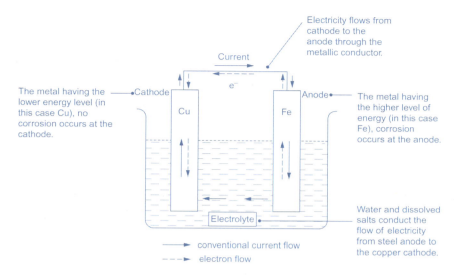

Fig. 1 Schematic presentation of a galvanic cell [8].

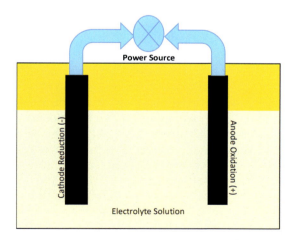

Fig. 2 Schematic presentation of an electrolytic cell [9].

- Galvanic cells (voltaics) convert chemical energy into electrical energy and generate electricity from the resulting chemical reaction.
- Redox reaction in galvanic cells is spontaneous.
- In galvanic cells, the anode is the negative electrode and the cathode is the positive electrode.

4 Chapter 1 Fundamentals of electrochemistry

- Galvanic cells are more commonly used as an electrical current source, such as a battery or accumulator.
- Electrolytic cells convert electrical energy into chemical energy, thereby causing a chemical reaction from electrical energy.
- In electrolytic cells, the redox reaction is induced by an external source of electrical energy.
- In electrolytic cells, the anode is the positive electrode and the cathode is the negative electrode.
- Some electrolytic cells are for commercial and industrial applications. There are several practical uses, such as hydrogen and oxygen gas production, electroplated coating, and alloying of pure metals.

In electrochemical systems, the processes that affect the transport of charge across the interface between an ionic conductor and an electronic conductor, the outer circuit, and the membrane system, take place in the designs called "electrochemical cells." These processes consist of the movement of electrons and ions. These systems are characterized as two electrodes that are separated by at least one electrolyte step. The transition region between the two phases consists of a charge imbalance region (electrical double layer) that forms in the inner monomolecular layer of adsorbed water molecules and ions [10].

The same is true for most chemical and biological fuel cells [1,11,12]. In the case of a metal immersed in water, it reflects the distribution of unequal charges (ions) at the liquid-solid interface and consists of two layers surrounding the object. The first layer is the surface charge layer, which forms two thin positive and negative charge planes that adsorb to the surface of the electron liquid polar water molecules in the metal. The inner layer closest to the electrodes is called the "compact Helmholtz layer (inner Helmholtz layer (IHL) and outer Helmholtz layer (OHL))." The second layer, known as the "diffuse layer," consists of free ions that are brought to the surface by electrical attraction and thermal movement. In this second layer, if water-soluble ions are present, some of the larger (and more polarizable) anions are loosely bound to the metal (chemisorb) to form a negative inner layer corresponding to the excess cations in the outer layer [13]. This equilibrium double layer acts as a capacitor, producing half-cell potential, also called "electrode solution potential" (Fig. 3) [14,15].

Also, Faraday's law applies if the current is carried by electron transfer at the electrode/solution interface and causes a reduction or oxidation. The transport of current by electron transfer occurs at the electrode/solution interface and causing a reduction or oxidation applied. The level of chemical alteration in the electrodes is

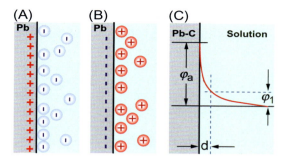

Fig. 3 An electric charge and electrical field in the electrical double layer on the electrode surface: (A) when positively charged and (B) when negatively charged. (C) Electrochemical electrode systems [15].

commensurate with how much energy is transmitted. These activities are termed "Faradan operations." Electrodes in which operations occur are often referred to as "charge transfer electrodes." Often, the potential spectrum is such that there is no thermodynamic or kinetic charge transfer within that range. This means that alternating negative and positive ions are attracted to the electrode to produce alternating current, the direction of which changes constantly during this period. However, processes such as adsorption and desorption may occur, and the configuration of the electrode solution interface may differ with various potentials or compositions of the solution. These phenomena are considered as non-Faradaic systems. In general, both Faradaic and non-Faradaic currents are found in electrode reactions [16].

The interface potential differences in electrode solution systems are only a few volts. In the case of an electrode immersed in a solution, this difference corresponds to layers of thin water molecules attached to the electrode surface that are only a few atomic diameters. That is, a very large potential gradient can be generated with a very small amount of voltage. In these electrochemical cells, the difference in electrical potential between electrodes can be determined. With a high impedance voltmeter, this calculation is usually performed: A calculation of the available energy (1 V = 1 joule/coulomb (J/C)) is the externally determined cell potential in Volts (V) to increase the voltage between the electrodes [17,18].

6 Chapter 1 Fundamentals of electrochemistry

1.3 Electrode and cell potentials

The cell potential is an indication of the progression of cell response to reference electrode.

The half-cell potentials are then measured according to the potential of the reference cell [19]. Thus, the standard cell potentials (E_{cell}) of any cell are determined [20]. The reaction reportedly takes place if the probability is positive and the opposite reaction is allegedly bad.

Because the cell potential value, as well as the free energy sign, assert whether the reactions are spontaneous, there is a relationship between these two terms. The shift in free energy ($\Delta G°$) is also an indicator of how much work can be achieved during the chemical process (4):

$$\Delta G° = -nFE_{cell} \tag{4}$$

where n indicates the electrons exchanged between the electrodes and F is an electron charge per electron mole, called a "Faraday constant," the number of electrons passed between the electrodes. The Faraday constant is equal to 96,485 C mol^{-1} [21].

1.3.1 Thermodynamics

Thermodynamics, a set of mathematical functions that govern all forms of energy and their transformations, examines the transformation of energy flow, particularly the form of heat and its transformation from one form to another (such as the transformation of chemical energy into heat energy). Energy storage and power properties of electrochemical energy conversion systems, with the knowledge of cell potentials obtained from two half-cell potential measurements following the determination of other half cell potentials and reactions, measure cell potential as a function of temperature and follow from thermodynamic formulations. The basic thermodynamic equations for a reversible electrochemical conversion are given as follows (5);

$$\Delta G = \Delta H - T\Delta S \tag{5}$$

where ΔH enthalpy, or the energy released by the reaction, ΔS is entropy, and T is the absolute temperature. Enthalpy refers to the temperature changes in a system, and if it absorbs heat during the process, it is said to be "endothermic," and the change in enthalpy (ΔH) is said to be greater than zero ($\Delta H > 0$). However, if the system emits heat, the process is called exothermic and the change is less than zero ($\Delta H < 0$). ΔS entropy is a measure of irregularity or randomness in a system; the lower the entropy, the lower the irregularity [22].

1.3.2 Nernst equation

Electrochemical reactions are generally carried out at a standard temperature (298.15 K). However, the value of electrode potential depends on the electrode material, the solvent medium, and the temperature and concentration of the ion medium. The Nernst equation allows for the determination of cell potential under unusual conditions, and is obtained by combining Eqs. (6)–(8). It compares the measured cell potential with the reaction coefficient and provides an accurate determination of equilibrium constants (including solubility constants). The actual free energy change ΔG for a reaction under unusual conditions is:

$$\Delta G = \Delta G^{\circ} + RTlnQ \tag{6}$$

$$-nFE_{cell} = -nFE^{\circ}{}_{cell} + RTlnQ \tag{7}$$

$$E_{cell} = E^{\circ}{}_{cell} - \left(\frac{RT}{nF}\right) lnQ \tag{8}$$

In these equations; $\triangle$G is the standard Gibbs free energy change, R is the ideal gas constant (8314 J/mol · K), T is temperature (K), and Q is a concentration coefficient used in place of equilibrium concentrations [23].

1.4 Corrosion

Corrosion is a chemical or electrochemical oxidation process in which electrons are passed into the environment and a positive value is changed. Liquid, gas, or a compound can affect the corrosion process. Because they have their own conductivity for electron transfer, these environments are called "electrolytes."

The green covering (known as a "patina") is a typical image and a further example of corrosion, which is common in bronze sculptures. The green color is caused by the degradation of bronze copper (II) compounds. Iron and its alloys can also experience corrosion. Rust is a blend of oxide forms of hydrated iron (III), as defined by Eqs. (9)–(11):

$$Fe_{(s)} \rightarrow Fe^{2+}_{(aq)} + 2e^- \qquad\qquad E^{\circ} = +0.44\ V \tag{9}$$

$$O_{2\ (g)} + 4H^+_{(aq)} + 4e^- \rightarrow 2H_2O_{(l)} \qquad\qquad E^{\circ} = +1.229\ V \tag{10}$$

$$4Fe^{2+}_{(aq)} + O_{2\ (g)} + (4 + 2x) \cdot H_2O_{(l)}$$

$$\rightarrow 2Fe_2O_3 \cdot xH_2O + 8H^+_{(aq)} \quad E^\circ = +1.229 \, V \tag{11}$$

Corrosion, which can be chemical or electrochemical in nature due to current flow, requires a specific environment and at least two reactions.

When it comes to the application theory, the ions on the surface of the metal migrate through the voltaic cell into water. The acidity of the aqueous solution affects the basic cathode reaction ability. The agglomeration of electrolytes in an aqueous solution influences the rate of metal corrosion. Due to high salt, the corrosion and degradation of metals occurred much faster in coastal regions. Atmospheric sulfur oxides break into acidic solutions, which improve corrosion by metal.

Besides the unsightly appearance of corrosion, certain corrosion processes can damage metal structures. Also, the cost of repairing the metal parts in such structures as bridges is high. Several important studies have looked into these topics [10,24,25].

1.4.1 Protection from corrosion

Stopping corrosion is important not only to prevent economic loss, but also damage to the environment, loss of yield, failure of key components due to corrosion, deformation, and collapse of structures, and other negative effects. Simple ways can be used to increase the corrosion resistance of metals. While there are many convenient methods of preventing corrosion, the simplest of these is to prevent the surface of the metal object from coming into contact with the atmosphere. The paint coating is among these techniques. This will be cost-effective in preventing corrosion. As a precaution, it can minimize unwanted reactions by ensuring environmental interaction control. This may include measures to reduce exposure to water or control the amount of sulfur, chlorine, or oxygen in the surrounding area. The most common of these methods is cathodic protection, an electrochemical technique in which a more easily oxidized metal is used. The sacrificial coating method features coating with additional sacrificial electrode types, which are more likely to oxidize. Anodizing protection involves coating iron alloy steel with a less active metal such as tin. The tin will not rust as a result of the coating process, so the steel will be protected so long as the tin remains coated and does not wear off. This method makes a steel anode of the electrochemical cell (hence its name). Corrosion inhibitors, passivation, and design modifications are other methods that help reduce corrosion [26,27].

1.5 Battery and classifications

There is a need to limit the use of fossil fuels such as coal, oil, and gas, which are the main energy source in the world and the main cause of the greenhouse gas effect released by combustion. In recent years, scientists have been researching designs that are intended to be used as energy sources without posing a threat to the environment. One of the oldest and most important applications of electrochemistry is the conversion and storage of energy. While galvanic cells convert chemical energy into work, electrolytic cells convert chemical work into chemical free energy [28]. Electrochemical oxidation-reduction (redox) reactions of chemical energy directly into electrical energy materials are called "batteries." Batteries are closed systems and anode and cathode charge transfer media, and they play an active role as an active mass in the redox reaction. In other words, energy storage and conversion take place in the same tank. In principle, a galvanic cell can work as a battery [29].

1.5.1 Primary batteries

Primary batteries are single-use, which means that they cannot be easily or efficiently charged with electricity. Such systems containing an adsorbing agent or separating material (i.e., no free or liquid electrolyte) as the electrolyte are referred to as "dry cells."

A Leclanché cell contains the reactions shown in Eqs. (12)–(14):

$$\text{Anode}: Zn_{(g)} + 2OH^-_{(aq)} \rightarrow ZnO_{(g)} + H_2O_{(l)} + 2e^- \quad E^\circ = -1.28\,\text{V} \tag{12}$$

$$\text{Cathode}: 2MnO_{2\,(g)} + H_2O_{(l)} + 2e^-$$
$$\rightarrow 2Mn_2O_{3\,(g)} + 2OH^-_{(aq)} \quad E^\circ = +0.15\,\text{V} \tag{13}$$

$$\text{Overall}: Zn_{(g)} + 2MnO_{2\,(g)} \rightarrow ZnO_{(g)} + Mn_2O_{3\,(g)} \quad E^\circ = +1.43\,\text{V} \tag{14}$$

The primary battery is suitable for portable electronic and electrical devices and is generally a cheap and lightweight package of energy [30].

1.5.2 Secondary or rechargeable cells or batteries

Electrode reactions in a secondary or rechargeable battery can proceed in either direction, allowing the battery to be charged. During charging, in order to force the reaction in a nonspontaneous direction; current is passed through the circuit in the opposite direction to the current on the cell to provide the required free energy. These devices, also known as "storage batteries" or "accumulators," are used to store electrical energy in devices such as rechargeable and smartphones, electronic tablets, and automobiles [29].

The lithium ion battery presented in Fig. 4 is the most popular rechargeable battery and is used in many portable electronic devices.

1.5.3 Reserve batteries

In a cell such as a reserve battery, a basic component may include the separation of the remaining components just before activation. The most frequently isolated component is electrolytes. In this case, the battery is capable of long-term storage because chemical degradation or self-discharge is largely eliminated. In this structure, the electrolyte is active until it reaches

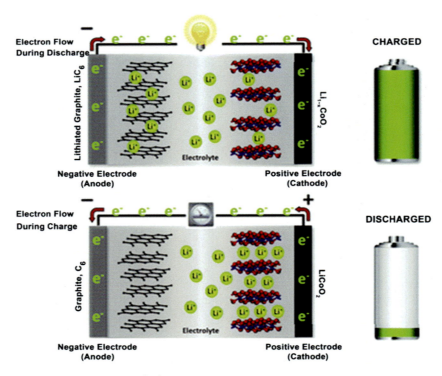

Fig. 4 Illustration of the lithium ion battery [31]. Reprinted (adapted) with permission from T. Chen, Y. Jin, H. Lv, A. Yang, M. Liu, B. Chen, et al, Applications of lithium-ion batteries in grid-scale energy storage systems, Trans. Tianjin Univ. 26 (2020) 208–217, doi:10.1007/s12209-020-00236-w. Creative Commons CC-BY license.

the melting point and allows ionic conduction, thus activating the battery. Most spare batteries are used only once and then discarded [32].

1.5.4 Fuel cells

Fuel cells are galvanic cell-type devices that continuously convert the chemical energy of the fuel into electrical energy. Like batteries, fuel cells convert this energy electrochemically and are not subject to the Carnot cycle limitation of thermal motors; thus, they offer the potential for high-efficiency conversion. There is a difference between fuel cells and batteries in terms of providing energy. Power in the fuel cells is continuously supplied from an external source if necessary. If the limiting reagent that disrupts the generation of electrical energy is consumed, it must be replaced or recharged. Unlike metal anodes used in most fuel cells, anode active substances are generally gas or liquid fuels such as hydrogen, methanol, hydrocarbons, or natural gas [30]. The main parts of the fuel cell are classified as follows:

- The anode (fuel electrode) should catalyze the oxidation reaction of the fuel, provide a collective interface for the fuel and electrolyte, and transmit electrons from the reaction zone to the external circuit catalyze.
- The cathode (oxygen electrode) oxygen electrolyte, oxygen recovery, and transfer reaction are the main parts of the fuel cell.
- The electrolyte should carry one of the ionic species involved in fuel and oxygen electrode reaction while preventing the transmission of electrons. In addition, the role of gas release in practical cells is generally provided by the electrolyte system [33].

1.5.4.1 Types of fuel cells

Solid oxide fuel cells (SOFCs): SOFCs are suitable for generating electricity from hydrocarbon fuels, working where some oxidic electrolytes become highly conductive oxygen ions. They can operate at a temperature of about 1000 °C with 60% efficiency [34].

Molten carbonate fuel cells (MCFCs): The electrolyte of MCFCs is generally a molten carbonate of lithium and potassium (Li_2CO_3/K_2CO_3) or an alkali metal such as lithium or sodium. The operating temperature of the electrolyte is generally between 600 and 650 °C and its efficiency is about 50%. It has a special working system, and it is involved in the carbon dioxide electrode reactions consumed in the cathode produced in the anode. MCFCs have important competencies, such as the use of carbon monoxide as fuel [35].

Phosphoric acid fuel cells (PAFCs): PAFCs are the first commercially available fuel cells; platinum or platinum alloys are used as catalysts in both electrodes. These cells have been designed with a cell structure retained in the silicon carbide matrix; they use phosphoric acid as the electrolyte and have a working temperature in the range of 150–220 °C [36].

Proton exchange membrane fuel cells (PEMFCs): These cells, with 60% efficiency and a low-temperature operating range, can be one of the most promising possibilities as a clean energy source due their low emissions [37].

Microbial fuel cells (MFCs): In MFCs, live microbes are used as catalysts. In this design cell, organic fuels can be produced by converting the energy stored in chemical bonds to produce an electrical current using the biocatalytic properties of living microorganisms [38].

Enzymatic fuel cells (EFCs): EFCs work on similar principles as all fuel cells, but what makes them different from conventional fuel cells is the catalysts they use and the fuels they accept. While most fuel cells use metals such as platinum and nickel as catalysts, enzymes derived from living EFCs are used [39].

Alkaline electrolyte fuel cells (AFCs): The electrolyte of AFCs is an alkaline aqueous solution, such as potassium or sodium hydroxide, which is an inexpensive electrolyte material that is contained in a porous matrix or continuously recirculated. Electrodes are usually made with nonprecious metals such as porous carbon or nickel [39].

Direct methanol fuel cells (DMFCs): DMFCs are powered by the conversion of liquid methanol (without the need for hydrogen) directly to hydrogen ions on the anode side of the fuel cell. DMFCs, which allow the direct use of methanol, constitute a type of PEMFC that uses methanol as fuel instead of hydrogen or hydrogen-rich gas. Platinum is used as an anode catalyst in DMFCs [40,41].

2 Conclusion

As a result, it can be said that electrochemistry is established with a simple mathematical relation. It is important to understand the ideas behind the laws of electrochemistry. There are a wide variety of concepts for electrochemistry in the literature, and these concepts facilitate research by helping to explain equations that are difficult to understand.

This chapter discusses the development of electrochemistry throughout history, a process that still needs to be explored today. Examples of the electrochemical-energy generation process focused on fuel cells and energy conversion systems are briefly presented. In recent years, as a result of increasing energy consumption, various energy-saving methods have been researched and developed. It is necessary to emphasize the functionality of fuel cells by continuing to study their applications, which will contribute to the creation of energy-efficient and environmentally favorable industrial processes.

References

[1] A.J. McEvoy, Fundamentals and applications of electrochemistry, EPJ Web Conf. 54 (2013), https://doi.org/10.1051/EPJCONF/20135401018, 01018.

[2] A.J. Bard, L.R. Faulkner, J. Wiley, Electrochemical Methods Fundamentals and Applications, 2001.

[3] C.G. Zoski, Handbook of Electrochemistry, Elsevier, 2007, https://doi.org/10.1016/B978-0-444-51958-0.X5000-9.

[4] E.L. Hopley, S. Salmasi, D.M. Kalaskar, A.M. Seifalian, Carbon nanotubes leading the way forward in new generation 3D tissue engineering, Biotechnol. Adv. 32 (2014) 1000–1014, https://doi.org/10.1016/j.biotechadv.2014.05.003.

[5] P.S. Poskozim, General chemistry, principles and modern applications, seventh edition (Petrucci, Ralph H.; Harwood, William S.) and general chemistry, fifth edition (Whitten, Kenneth W.; Davis, Raymond E.; Peck, M. Larry), J. Chem. Educ. 74 (1997) 491, https://doi.org/10.1021/ed074p491.

[6] R.H. Petrucci, J.D. Madura, C. Bissonnette, General Chemistry: Principles and Modern Applications, Pearson Canada, Toronto, 2011.

[7] C. Breitkopf, K. Swider-Lyons, Electrochemical science—historical review, in: Springer Handbooks, Springer, 2017, pp. 1–9, https://doi.org/10.1007/978-3-662-46657-5_1.

[8] A. Ahmad, Basic concepts in corrosion. Principles of Corrosion Engineering and corrosion control, in: Principles of Corrosion Engineering and Corrosion Control, 2006, pp. 9–56, https://doi.org/10.1016/b978-075065924-6/50003-9.

[9] A. Escapa, R. Mateos, E.J. Martínez, J. Blanes, Microbial electrolysis cells: an emerging technology for wastewater treatment and energy recovery. From laboratory to pilot plant and beyond, Renew. Sustain. Energy Rev. 55 (2016) 942–956, https://doi.org/10.1016/j.rser.2015.11.029.

[10] N. Perez, Electrochemistry and Corrosion Science, Kluwer Academic, 2004.

[11] F. Şen, Mesoporous materials in biofuel cells, in: Mater. Res. Found., Materials Research Forum LLC, 2019, pp. 157–172, https://doi.org/10.21741/9781644900079-7.

[12] F.A. Unal, M.H. Calimli, H. Burhan, F. Sismanoglu, B. Yalcın, F. Şen, Microbial fuel cells characterization, in: Materials Research Foundations, 2019, pp. 75–100, https://doi.org/10.21741/9781644900116-4.

[13] L.A. Jurado, R.M. Espinosa-Marzal, Insight into the electrical double layer of an ionic liquid on graphene, Sci. Rep. 7 (2017) 1–12, https://doi.org/10.1038/s41598-017-04576-x.

[14] R. Norsworthy, Understanding corrosion in underground pipelines: basic principles, in: Underground Pipeline Corrosion: Detection, Analysis and

14 Chapter 1 Fundamentals of electrochemistry

Prevention, Elsevier Ltd., 2014, pp. 3–34, https://doi.org/10.1533/9780857099266.1.3.

[15] D. Pavlov, in: Lead–Carbon Electrodes, 2017, pp. 621–662, https://doi.org/10.1016/B978-0-444-59552-2.00015-8.

[16] P.M. Biesheuvel, J.E. Dykstra, The Difference Between Faradaic and Nonfaradaic Processes in Electrochemistry, 2018.

[17] K.H.J. Buschow, Encyclopedia of Materials : Science and Technology, Elsevier, Amsterdam ;New York, 2001.

[18] S. Chen, Practical electrochemical cells, in: The Handbook of Electrochemistry, Elsevier, 2007, pp. 33–56, https://doi.org/10.1016/B978-044451958-0.50003-3.

[19] M. Ciobanu, J.P. Wilburn, M.L. Krim, D.E. Cliffel, Fundamentals, in: The Handbook of Electrochemistry, Elsevier, 2007, pp. 3–29, https://doi.org/10.1016/B978-044451958-0.50002-1.

[20] V. Tripkovic, M.E. Björketun, E. Skúlason, J. Rossmeisl, Standard hydrogen electrode and potential of zero charge in density functional calculations, Phys. Rev. B: Condens. Matter Mater. Phys. 84 (2011) 115452, https://doi.org/10.1103/PhysRevB.84.115452.

[21] D. Reger, S. Goode, D. Ball, Chemistry: Principles and Practice, 2009.

[22] D.A. Jones, Principles and Prevention of Corrosion, Prentice Hall, 1996.

[23] X. Yi, W. Dong, X. Zhang, J. Xie, Y. Huang, MIL-53(Fe) MOF-mediated catalytic chemiluminescence for sensitive detection of glucose, Anal. Bioanal. Chem. 408 (2016) 8805–8812, https://doi.org/10.1007/s00216-016-9681-y.

[24] H. Tamura, The role of rusts in corrosion and corrosion protection of iron and steel, Corros. Sci. 50 (2008) 1872–1883, https://doi.org/10.1016/j.corsci.2008.03.008.

[25] Y. Liu, X. Liang, L. Gu, Y. Zhang, G.D. Li, X. Zou, J.S. Chen, Corrosion engineering towards efficient oxygen evolution electrodes with stable catalytic activity for over 6000 hours, Nat. Commun. 9 (2018) 2609, https://doi.org/10.1038/s41467-018-05019-5.

[26] D.A. Snow, Plant Engineer's Reference Book, Elsevier, 2002, https://doi.org/10.1016/B978-075064452-5/50103-6.

[27] P.A. Schweitzer, Fundamentals of Corrosion : Mechanisms, Causes, and Preventative Methods, CRC Press, 2010.

[28] T.M. Letcher, Future Energy: Improved, Sustainable and Clean Options for Our Planet, Elsevier, 2008.

[29] D.L. Thomas Reddy, Linden's Handbook of Batteries, 4/e (SET 2), McGraw-Hill Education Ltd, 2011.

[30] M. Winter, R.J. Brodd, What are batteries, fuel cells, and supercapacitors? Chem. Rev. 104 (2004) 4245–4270, https://doi.org/10.1021/cr020730k.

[31] T. Chen, Y. Jin, H. Lv, A. Yang, M. Liu, B. Chen, Y. Xie, Q. Chen, Applications of lithium-ion batteries in grid-scale energy storage systems, Trans. Tianjin Univ. 26 (2020) 208–217, https://doi.org/10.1007/s12209-020-00236-w.

[32] A.G. Ritchie, N.E. Bagshaw, Military applications of reserve batteries, Philos. Trans. R. Soc. A Math. Phys. Eng. Sci. 354 (1996) 1643–1652, https://doi.org/10.1098/rsta.1996.0070.

[33] D. Linden, Handbook of Batteries, McGraw-Hill, 1995.

[34] A.B. Stambouli, E. Traversa, Solid oxide fuel cells (SOFCs): a review of an environmentally clean and efficient source of energy, Renew. Sustain. Energy Rev. 6 (2002) 433–455, https://doi.org/10.1016/S1364-0321(02)00014-X.

[35] P.E.V. de Miranda, Science and Engineering of Hydrogen-Based Energy Technologies : Hydrogen Production and Practical Applications in Energy Generation, 2018.

[36] N. Fourati, N. Blel, Y. Lattach, N. Ktari, C. Zerrouki, Reference Module in Materials Science and Materials Engineering, Elsevier, 2016, https://doi.org/10.1016/B978-0-12-803581-8.01733-1.

[37] R.E. Yonoff, G.V. Ochoa, Y. Cardenas-Escorcia, J.I. Silva-Ortega, L. Meriño-Stand, Research trends in proton exchange membrane fuel cells during 2008–2018: a bibliometric analysis, Heliyon 5 (2019), https://doi.org/10.1016/j.heliyon.2019.e01724.

[38] A.J. Slate, K.A. Whitehead, D.A.C. Brownson, C.E. Banks, Microbial fuel cells: an overview of current technology, Renew. Sustain. Energy Rev. 101 (2019) 60–81, https://doi.org/10.1016/j.rser.2018.09.044.

[39] A. Coralli, B.J.M. Sarruf, P.E.V. de Miranda, L. Osmieri, S. Specchia, N.Q. Minh, Chapter 2—Fuel Cells, Elsevier Inc., 2019, https://doi.org/10.1016/B978-0-12-814251-6.00002-2.

[40] A. Demirbas, Direct use of methanol in fuel cells, Energ. Sources Part A 30 (2008) 529–535, https://doi.org/10.1080/15567030600817159.

[41] U. Desideri, K.-L. Hsueh, R.K. Shah, A.V. Vikar, Research opportunities and challenges, in: Fuel Cell Science and Engineering, The 3rd South East Europe Conference on Sustainable Development of Energy, Water and Environment Systems-3rd SEE SDEWES Novi Sad View project, The 13th Conference on Sustainable Development of Energy, Water and Environment Systems-SDEWES2018 View project Research Opportunities and Challenges in Fuel Cell Science and Engineering, Y.Y, 2018.

2

Nanomaterials and their classification

İsmail Mert Alkaç[a,b], Burcu Çerçi[c], Cisil Timuralp[d], and Fatih Şen[a]

[a]Şen Research Group, Department of Biochemistry, Dumlupinar University, Kütahya, Turkey. [b]Department of Materials Science and Engineering, Faculty of Engineering, Dumlupinar University, Kütahya, Turkey. [c]Faculty of Medicine, Department of Medical Biology and Genetics, Izmir Katip Celebi University, Izmir, Turkey. [d]Mechanical Engineering Department, Eskisehir Osmangazi University, Eskişehir, Turkey

1 Introduction

A "nanomaterial (NM)" can be defined as a material with a nanoscale internal structure or with nanoscale external dimensions or surface structure. Most of the materials in the world qualify as NMs according to this definition because the structure of NMs is arranged at the nanoscale [1].

In general, materials can be divided into two classes as NMs and bulk materials. Nanomaterials are produced at nanometer scale between 1 and 100 nm. Bulk materials with a size greater than 100 nm are known as "particles." Physical properties in bulk materials are independent of size, while their size and shape can depend on various physical properties [2].

There are different definitions of NMs internationally, which are accepted by different organizations [3]. According to the British Standards Institution, the specific terms have been used in scientific applications as shown in Table 1 [3].

NMs are produced using a broad range of chemical components, including carbon, semiconductors, polymers, metals, and metal oxides. They are intended for unique features and can be coated or treated on the surface. They come in a wide range of shapes, including balls, sticks, cables, tubes, fibers, coatings, rings, needles, shells, and plates [4].

Nanomaterials for Direct Alcohol Fuel Cells. https://doi.org/10.1016/B978-0-12-821713-9.00011-1
Copyright © 2021 Elsevier Inc. All rights reserved.

18 Chapter 2 Nanomaterials and their classification

Table 1 Terminology for NMs, British Standards Institution. PAS 136 2007.

Term	Definition
Nanoscale	Size ranges from approximately 1 to 100 nm
NMs	Materials with one or more external dimensions in the nanoscale or nanostructure
Nanostructure	A structure consisting of contiguous elements with one or more nanoscale dimensions, but excluding any primary atomic or molecular structure
Nanoparticle	A nanoobject with all three external dimensions in the nanoscale
Nanoobject	A discrete piece of material with one or more external dimensions in the nanoscale
Nanotechnology	A technology that pervades virtually every aspect of life and will enable dramatic advances to be realized in most areas of communication, manufacturing, health, materials, and knowledge-based technologies
Nanofiber	A flexible nanorod
Nanorod	A nanoobject with two similar external dimensions in the nanoscale and the third dimension being significantly larger than the other two external dimensions

2 Classification of nanomaterials

The interest in NMs is growing because of their unique physicochemical properties. However, the classification of the NMs started to be ambiguous in research studies. Therefore, the classification of NMs is important for the various utilization areas [5]. NMs can be divided into groups based on their morphology, size, shape, chemical composition, state, and dimensionality, as shown in Fig. 1 [6]. In this chapter, the classification of the NMs is discussed according to the features of the materials.

2.1 Dimensionality

According to their shape and dimension, NMs can be classified into four types according to dimensionality: zero-dimensional (0D), one-dimensional (1D), two-dimensional (2D), and three-dimensional (3D), as shown in Fig. 2. NMs can be produced from elementary blocks in 0D, 1D, and 2D, while 3D NMs are formed in a more complex way [8]. It must be noted that NMs can be harmful to human health because of the substrats that they are fixed on or because of nanopores. Likewise, free nanoparticles can cause respiratory problems when they become airborne [1].

Chapter 2 Nanomaterials and their classification   19

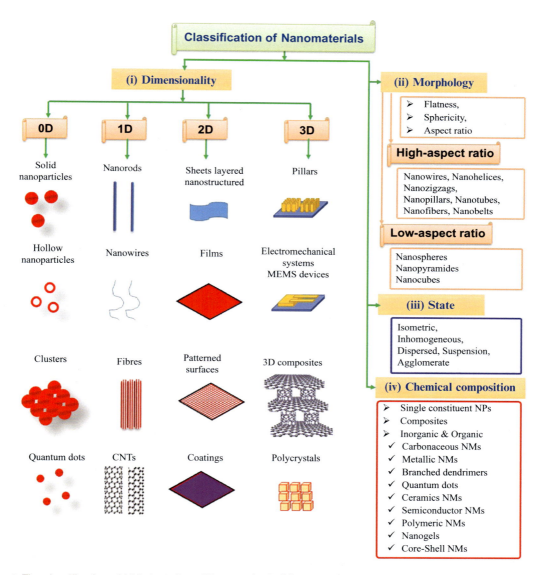

Fig. 1 The classification of NMs based on different criteria [2].

2.1.1 0D nanomaterials

The 0D NMs contain several types of nanoparticles, including dendrimers, branched macromolecules, palladium, anatase titanium dioxide, carbon (fullerenes), gold or tungsten, zinc oxide, vanadium oxide, magnesium oxide, molybdenum disulfide, and silicon carbide nanoparticles. All these substances have their

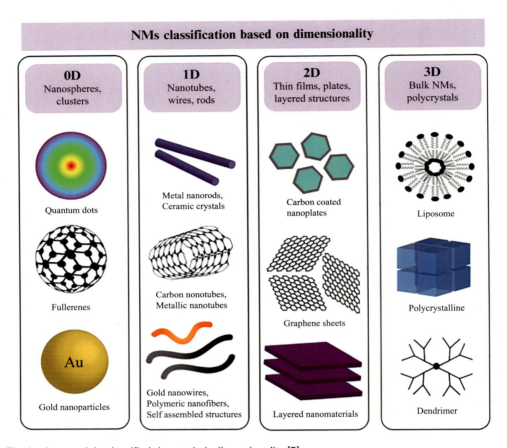

Fig. 2 The basic materials classified due to their dimensionality [7].

dimensions in nanoscales lower than 100 nm [9, 10]. These NMs can be produced for several purposes, such as cell markers [11], emulsifiers [12], or reinforcement fillers [13]. One of the best-known 0D materials is the quantum dot (QD), whose dimensions are below 10 nm. One of the most important properties of these materials is that they are semiconductors and therefore they have important roles in electronics to confine electrons and holes [14].

2.1.2 1D nanomaterials

1D NMs consist of metallic, nanotube, polymeric, nanowire, ceramic, nanorod filament, or fiber substances, and nanofibers with 2D are in nanoscale; however, one of the dimensions is not in nanoscale. Such materials can used as a thin film for surface coatings attached to a substrate [10].

Nanotubes have a peculiarly crystalline, cylindrical form, shaped like pentagons, hexagons, and heptagons [15]. The best-known nanotubes are made of carbon, but molybdenum, boron nitride, copper sulfides or tungsten, and various halides such as cadmium chloride, cadmium iodine and nickel chloride can also be manufactured with nanotubes [15].

Nanowires, classified as 1D NMs with length of 1–1000 nm in at least one dimension have multiple unique features for photoelectrochemical applications that exhibit better performance than their bulk counterparts [16]. They can be produced with semiconductor compounds, metals, and electrical insulators such as silicon, nickel, and titanium dioxide [15].

Nanorods are positioned with an aspect ratio of usually between 3:1 and 2:1 at the border between 0D and 1D NMs [17]. Some nanorods have the potential to alter their reflectivity when an electric field modifies their orientation. They can also emit heat when they adsorb infrared radiation [9].

Nanofibers are widely used NMs that can be developed using inorganic materials, such as carbon, nitride, platinum, and titanium; and from various polymers including nylon, polyurethane, and polyvinyl alcohol. The establishment of the nanofibers can be like a web that can be used in filtration applications or formed yarn [18].

2.1.3   2D nanomaterials

2D NMs have a structure with one dimension in nanoscale while the others are not. These 2D materials comprise amorphous or crystalline nanocoating, thin films, nanoplates, and single-layered and multilayered [10].

Ceramic and metal coatings consist of thin films [19]. In physics and electronics, they are often used to modify optical reflectivity of surfaces or to create electronic components with conductive or insulating surface properties. Several materials, including polymers and composites, can also be used to manufacture nanocoatings [20]. In that case, the aim would be to increase hardness, corrosion, or resistance to abrasion, or to have an insulating layer. Nanoplates can be produced as smectic clay or developed as nanoplates of graphene, gold, silver, bismuth telluride, or bismuth selenide. Nanoplates are usually a number of nanometers thick, and their length and width are up to 600 nm for nanoribbons of graphene [21, 22], and 70–150 nm for nanoclay [21]. Nanoplates can be used to improve the properties of their mechanical, thermal, and diffusion barriers as modules in electronics and as composites, they respond favorably [9].

2.1.4 3D nanomaterials

3D NMs exhibit internal nanoscale characteristics, but at the nanoscale, there is no external dimension [23]. Therefore, it can be noted that 3D materials have dimensions over 100 nm size [8, 24]. Nanotubes, carbon nanobuds, foams, fibers, polycrystals, pillars, fullerenes, layer skeletons, and honeycombs are the types of 3D NMs that can be combined in different directions [24, 25]. Also, nanocomposites and nanostructured materials are the best-known 3D NMs.

To define nanofillers distributed in a bulk matrix, the term "nanocomposite" is commonly used. Matrices can be polymers, ceramics, or metals, while nanofillers may consist of 0D, 1D, or 2D NMs. The final material can be a film, fiber, or volume. Bones are one of the best-known natural nanocomposites, and it consists of a collagen matrix and calcium hydroxyapatite nanocrystals distributed within it.

Owing to the high surface-to-volume ratio of nanofilling, there is a difference between nanocomposite and composite materials. Due to their strength and rigidity, polymer matrix nanocomposites have been identified for various applications [26], superhydrophobic coatings with nanoprotrusions [27], block copolymers [28], and nanostructured alloys and metals.

2.2 Classification of nanomaterials by morphology and state

NPs can be classified as inhomogeneous and isometric or agglomerate and dispersed, based on uniformity. Sphericity and flatness, as well as aspect ratio, comprise the morphological nature of NPs. On the other hand, nanoparticles with a decreased aspect ratio display various shapes such as cubes, helical, spherical, pillarlike, and pyramidal. The electromagnetic properties of NPs such as magnetism and surface charge rely on this stack. In addition, the agglomeration of NPs in a fluid depends on functionalization and morphology, leading to hydrophilicity or hydrophobicity. NPs are nanotubes, nanostars, nanohooks, nanorods, nanocubes, nanohelices, nanozigzags, nanobelts, nanopillars, and nanoplates with various morphologies [2, 10].

The properties of NPs, such as durability, are provided by nanocomposites that integrate NPs into a standard material matrix. Metals such as zinc, iron, alumina, silver, titanium, silica, and copper are made from these metallic NMs. Another type of NMs is the linked dendrimer, which has a branchlike nanoscale structure [2].

NPs can be categorized according to their chemical composition, such as monocomposite or nanocomposites. Carbonaceous NMs are mainly derived from carbon, such as carbon nanotubes (CNTs), graphenes and fullerenes. NMs may occur in scattered shapes, colloids, and suspensions or in an agglomerated state according to their chemical and electromagnetic properties. For example, magnetic NPs appear to be clustered into an agglomerate if their surface is not functionalized [2].

2.3 Classification of NMs by chemical composition

NMs can be divided into two groups as organic and inorganic materials [29]. Moreover, NMs can be produced using two or more materials. For example, they can be encapsulated or coated [10].

2.3.1 Inorganic-based NMs

These NMs can be developed by using metals, metal oxides, and semiconductors such as Au, TiO_2, and silicon, respectively. On the other hand, Ag, ZnO, and ceramics can be used to synthesize inorganic-based NMs [3].

2.3.1.1 Carbon-based NMs

As the name implies, carbon is the main component of carbon-based NMs, and they can be developed in different morphologies, such as hollow tubes, ellipsoids, and spheres. These NMs can be used in different fields according to their structure and properties that depend on the unique structure of carbon. Carbon can be found in various forms, such as diamond and graphite [30]. Carbon-based NMs include such materials as carbon nanofibers, CNTs, carbon onions, carbon black, fullerenes (C60), and graphene. Production methods for carbon-based NMs include arc discharge, laser ablation, and chemical vapor deposition (CVD) [31]. These NMs contain hybridized sp^2 carbon atoms, which can be produced in several dimensions (Fig. 3) [30, 32, 33].

Carbon-based NMs can be used in many applications. Therefore, these materials become attractive to many scientists and producers according to their chemical and physical properties that include conductivity, thermal properties, and other characteristics. Carbon-based NMs and carbon can be divided into four groups: fullerene, CNTs, graphene sheets, and graphite and nanodiamonds.

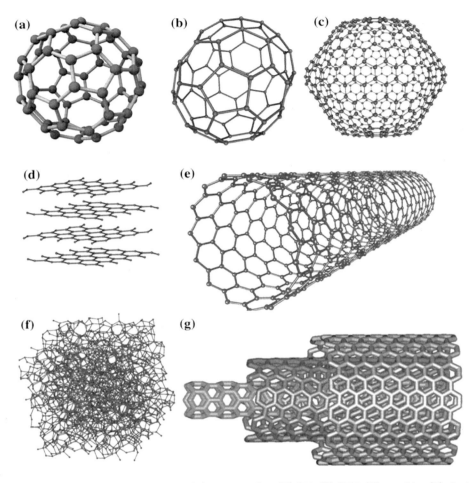

Fig. 3 Carbon allotrope structures in nanoform: (A) fullerene C60; (B) C70; (C) C540; (D) graphite; (E) single-walled carbon nanotube (SWCNT); (F) amorphous carbon; (G) multiwalled carbon nanotube (MWCNT) [10].

(1) *Fullerene (0D):* Derivatives are an allotrope of carbon consisting of 60 °C atoms that can neutralize oxygen and nitrogen [34].

(2) *CNTs (1D):* Hexagonal structures consisting of carbon atoms that linked to form a hexagonal structure. Chemical and physical modifications of CNTs have varying characteristics based on the materials used. It can be synthesized directly using the CVD process [35].

(3) *Graphene sheets (2D):* Performed by graphene and carbon atoms that bond covalently to each other to form a honeycomb or hexagonal lattice. Graphene sheets have several

structural properties, including good stability, chemical reactivity, high electrical conductivity, and large surface area. Graphite is the source of graphene in which mechanical and chemical methods can be used for the extraction of graphene [36].

(4) *Graphite and nanodiamonds (3D):* Basically used for semiconductors, coating, and abrasives. Graphite is formed by sp^2 carbon atoms that are arranged in a hexagonal shape. On the other hand, nanodiamonds have a layered, spherelike shape, and their optical and magnetic features distinguish them from the other materials [37, 38].

2.3.1.2 Metallic nanomaterials

Gold, silver, and iron are metal NMs that have unique chemical, electrical, and optical properties [24]. Metallic NMs can be produced by ultrasonic vibration and sonoelectrochemical [25]. Designation processes can affect the size of metallic nanoparticles. For example, if you reduce the size of a metal particle in a nanometer range, with a significant number of atoms appearing on the top by different electronic quantities, the electronic build band gap can vary. These surface atoms can become more effective by reducing the size of nanometers due to the growing distance between unsaturated sites and atomic coordinates. Au NPs, for example, can be prepared and synthesized by mixing ultrasonic vibration and sonoelectrochemical [25]. Metal NPs can typically absorb light through in-band (such as Cu, Au, Al, and Ag) and interband (such as Ni, Pt, Ru, and Pd) transitions. Earlier studies have found that light irradiation can increase the catalytic activity of metal NPs [39].

2.3.1.2.1 Metal oxide nanomaterials Metal oxides form another chemical group of NMs. Applications have been reported for titanium dioxide as an ultraviolet (UV) filter, a catalyst in chemical reactions, and an antibacterial agent that can be used in cosmetics, cleaning products and wall paint, and filtration systems, respectively. Moreover, titanium dioxide is also important in the production of solar cells [40].

Zinc oxide is also used as an antistatic agent, with fascinating antibacterial and catalytic properties [9]. Nanosilica is used as a hollow carrier for drug delivery because of its chemical properties [41]. Nanoalumina is used as a chemical catalyst, a mechanical, optical reinforcement, and a thermally conductor for abrasion-resistant coatings [9]. Iron oxides can be used as ion exchangers and pigment flocculants in electrical, water purification, and biomedical applications due to their excellent magnetic properties [42]. Cerium oxide is used as a vehicle exhaust catalyst, a chemical polishing agent for silicone wafers, a fuel cell, and a UV absorber used to coat wood [9].

Several kinds of metal oxides, such as Fe_2O_3, TiO_2, ZnO, SiO_2, and Al_2O_3, have been synthesized through hydrothermal or sol-gel reactions. A metal oxide has a considerable number of uses, such as catalysts, semiconductors, and chemical sensors, because of differentiation of its surface properties that affects the bandgap energy of the material [43]. The biocompatibility of these materials with a highly active surface area is the reason why these materials have a more important property. Several responses, such as adding a polymer chain, doping metal ions, or a binding agent, can easily change the surface of NPs [44, 45].

To modify the structure of NPs, many organic compounds are used, such as amines [46], epoxies [47], anionic compounds [48], and thiols [49]. Modification of the structure of NPs, especially surface modification, provides them with unique chemical and physical properties [2].

2.3.1.3 Quantum dots

QDs are ultrasmall semiconductor nanocrystals composed of 100–10,000 atoms in the 1.5–10-nm size range [50]. They are nanoparticles of semiconductors that exhibit optical and electronic (optoelectronic) properties depending on their size and composition. Recently, QD nanotechnology has been used in various electronic and biomedical sectors successfully [50].

Due to quantum mechanics, QDs are small semiconductor particles with optical and electronic properties that differ from those of larger particles. They exhibit distinctive electronic and optical properties by absorbing UV or white light and reemitting it at a different wavelength [2].

Initially used in determining surface kinetics, QDs are now successfully used as drug delivery agents and diagnostic markers. They are nanocrystals made up of a core semiconductor material further enclosed within another semiconductor material with a wider spectral band gap [51]. Besides their nanosize, QDs have an application potential in nanodiagnostics, imaging, targeted drug delivery, and photodynamic therapy due to their specific characteristics, such as wide-range excitation, high fluorescence, size-tunable narrow emission spectra, and high photostability [51].

2.3.1.4 Semiconductor nanomaterials

Semiconductor NMs were characterized with various morphologies and compositions that can be used easily in several applications [52]. They are controlled by quantum mechanics [53]. Nanosilicon is one of the most usable semiconductor NMs, which can play a role in solar cells, rechargeable batteries, electronics, and solid-state lighting. Gallium nitride is used to produce transistors for electric vehicles, light-emitting diodes (LEDs), and photonic and optoelectronic devices, which are versatile and water resistant. Indium gallium nitride can be developed as LED nano-rings, photovoltaic nanocolumns, and optoelectronic nanoflowers. Gallium arsenide, a semiconductor NM, can be used as a optomechanics and photodetector in solar cells [9].

2.3.1.5 Ceramics nanomaterials

Ceramic nanomaterial is defined as inorganic, heat-resistant metal compounds that are less than 100 nm size and nonmetallic solids [6, 8].

Several chemical and physical methods have been explored and reported for preparation of ceramic NMs. These materials have been found to show improved superconductive, electrooptical, structural, ferroelectric, and ferromagnetic properties [54]. Similarly, by changing the concentration of doping, the physical and structural properties of nanoceramics can be changed, which could reduce oxygen vacancy and induce structural material distortion [55].

Calcium hydroxyapatite (CaHAP) porous nanoceramics have been produced by high-purity methylcellulose and can be prepared by the precipitation of hydroxyapatite (HAP). The porosity of CaHAP and its exact surface area depend on the sintering process temperature. Their capacity to activate antimicrobials is impaired by interactions between bacteria and their surface charges. Al_2O_3-ZrO_2 nanoceramics are generated in low-temperature and pressureless conditions by synthesizing nanoceramics. The results of this analysis showed that the grain size of NPs and homogeneous mixture were successfully formed with amorphous materials at low temperatures [56].

2.3.1.6 Polymeric nanomaterials

Polymeric NMs are dense particles composed of synthetic or natural polymers that are nanosized. In medical and pharmaceutical applications, these materials are commonly used as drug-release controllers in the human body [57]. NMs based on the polymer include the following:

(1) Polymeric NPs, which typically consist of polymers that are biodegradable and biocompatible with a size of approximately between 10 and 1000 nm. They are commonly used to distribute medications to particular targets [58].

(2) Polymeric nanocomposites, a mixture of polymers and other nanofillers used to develop [59].

(3) Polymeric micelles, produced in a specified solvent through the self-assembly of amphiphilic block copolymers. Due to their special characteristics, they can be used in drug delivery [57].

(4) Dendrimers, with 3D macromolecules that are less than 15 nm in size. These materials are commonly produced for pharmaceutical and medical applications due to their characteristics [60].

To render nanoobjects, nanospheres (including nanoporous membranes and nanofibers), thermosets, thermoplastics, and elastomers can be used. Reinforcement for structural composite materials, barrier membranes, fuel cell membranes, filtration membranes, fire-resistant optical components, versatile electrical elements, and antibacterial textiles are used in their applications [61].

2.3.1.7 Core-shell nanomaterials

Single-component NMs have been widely studied, and various properties have been discovered, including easy processing and higher surface atom ratio. Researchers have concluded that the combination of multicomponent NMs could achieve much better properties with the rapid evaluation of characterization and synthesis techniques, and the diversity of their structure and composition could enrich their applications in various fields. Thus, standardized core-shell NMs have become a prominent research subject in recent years [62]. Several metals can be used for the core particles, such as Au, Ag, Pt, Pd, Zn, and Ni, while TiO_2, ZnO, Cu_2O, SiO_2, Fe_2O_3, and SnO_2 metal oxide semiconductors can be developed as shells. They can play a role in catalysis, photocatalysis, and solar photovoltaics [63].

NPs formed by encapsulation of shell material can be investigated mainly in two categories: (1) inorganic-organic, organic-inorganic, and inorganic-inorganic types; and (2) chemical coating and physical coating of the shell are used to improve the activity of crust and maintain the chemical integrity of the core. Coated main particles are designed to have magnetic, optical, and catalytic properties [64].

2.3.2 Organic-based NMs

Organic materials can be used to develop organic-based NMs. By using noncovalent interactions, organic NMs can be conerted into liposomes, micelles, polymers and dendrimers [3].

2.3.3 Composite-based NMs

Composite NMs can consist of the mixture of different NPs, or they can be more complex, like organic metal frames. A composite can be a combination of metal, organic NMs or carbon. Composite-based NMs are synthesized in a number of morphologies based on the properties needed for their desired uses [3].

Composites are solid materials composed of multiple phases with dimensions less than 100 nm in a structure or phase with a nanoscale repetition gap between phases [65, 66]. In the process of constructing composite structures, physical measurements on the nanometer scale range are often used. The combination of several composite-forming materials can produce various properties, such as optical properties, flexural strength, wear of gloss, and water sorption and retention [67–69]. Composite materials allow the adsorption potential of the material to be enhanced by multiple types of interactions, such as Yoshida H-bonding and bonding of dipole-dipole hydrogen [70].

2.3.4 Silicates, carbonates, and nitrides

Silicates, carbonates, and nitrides are important NM groups. Clay (magnesium aluminum silicate) is a unique nanosilicate that can take shape of nanoplates and nanotubes [21]. For improvement of mechanical strength, electrical conductivity, and barrier properties, clay is used as a filler in composites. Zeolite can be used as a molecular hydrocarbon sieve, a catalyst, and a germ to produce a separation membrane [71]. In the case of calcium carbonate, if its function as a microscale composite filler is primarily aimed at lowering costs, calcium carbonate increases its resistance to effects, its barrier properties, environmental creep and aging, dimensional stability, surface finish, nanoscale modulus, and thermal conductivity. It also decreases processing contraction [72].

3 Conclusion

Due to advances in technology and discoveries during the last 40 years, a great deal of progress has been made in understanding the production and properties of NMS. They vary in size, natural or anthropogenic origin, chemical and physical properties, and

30 Chapter 2 Nanomaterials and their classification

production techniques such as bottom-up, top-down, or bottom-to-bottom. On a day-to-day basis, physical chemistry features are added to NMs with high diversity and usage areas, and the number of scientific and field areas is steadily increasing. Because of their excellent properties and diversity, NMs are used in many industries. In this respect, the classification of NMs will obviously require revision in the coming years.

We hope that this chapter has given a general overview of the knowledge and perspective on NMs.

References

[1] C. Buzea, I.I. Pacheco, K. Robbie, Nanomaterials and nanoparticles: sources and toxicity, Biointerphases 2 (2007) MR17–MR71.

[2] T.A. Saleh, Nanomaterials: classification, properties, and environmental toxicities, Environ. Technol. Innov. 20 (2020) 101067.

[3] J. Jeevanandam, A. Barhoum, Y.S. Chan, A. Dufresne, M.K. Danquah, Review on nanoparticles and nanostructured materials: history, sources, toxicity and regulations, Beilstein J. Nanotechnol. 9 (2018) 1050–1074.

[4] G. Oberdörster, E. Oberdörster, J. Oberdörster, Nanotoxicology: an emerging discipline evolving from studies of ultrafine particles, Environ. Health Perspect. 113 (2005) 823–839.

[5] A.M. Glezer, Structural Classification of Nanomaterials, Russ. Metall. 2011 (2011) 263–269.

[6] H. Gleiter, Nanostructured materials: basic concepts and microstructure, Acta Mater. 48 (2000) 1–29.

[7] T.Y. Poh, N.A.T.B.M. Ali, M.M. Aogáin, M.H. Kathawala, M.I. Setyawati, K.W. Ng, S.H. Chotirmall, Inhaled nanomaterials and the respiratory microbiome: clinical, immunological and toxicological perspectives, Part. Fibre Toxicol. 15 (2018) 1–16.

[8] V.V. Pokropivny, V.V. Skorokhod, Classification of nanostructures by dimensionality and concept of surface forms engineering in nanomaterial science, Mater. Sci. Eng. C 27 (2007) 990–993.

[9] P.I. Dolez, Nanomaterials definitions, classifications, and applications, Nanoengineering: Global Approaches to Health and Safety, Elsevier, 2015, pp. 3–40.

[10] C. Buzea, I. Pacheco, Nanomaterials and their classification, Advanced Structured Materials, Springer Verlag, 2017, pp. 3–45.

[11] O.V. Salata, Applications of nanoparticles in biology and medicine, J. Nanobiotechnol. 2 (2004) 3.

[12] J. Luo, Q. Zhou, J. Sun, R. Liu, X. Liu, Micelle-assisted synthesis of PANI nanoparticles and application as particulate emulsifier, Colloid Polym. Sci. 292 (2014) 653–660.

[13] F. Gao, Advances in polymer nanocomposites: types and applications, Woodhead, Cambridge, 2012.

[14] A.M. Smith, S. Nie, Semiconductor nanocrystals: structure, properties, and band gap engineering, Acc. Chem. Res. 43 (2010) 190–200.

[15] H. Terrones, M. Terrones, Curved nanostructured materials, New J. Phys. 5 (2003) 126–127.

[16] J. Deng, Y. Su, D. Liu, P. Yang, B. Liu, C. Liu, Nanowire photoelectrochemistry, Chem. Rev. 119 (2019) 9221–9259.

Chapter 2 Nanomaterials and their classification **31**

[17] D. Madhavan, Krishnamurthy, N. Krishnamurthy, P. Vallinayagam, Engineering Chemistry, Prentice-Hall India Pvt. Ltd., 2008.

[18] P.J. Brown, K. Stevens, Nanofibers and Nanotechnology in Textiles, Woodhead, Oxford, 2007.

[19] O. Milton, Materials Science of Thin Films, second ed., Academic Press, Boston, 2001.

[20] N.H. Steven Abbott, Nanocoatings: Principles & Practice—From Research to Production, DEStech Publishing, Lancaster, 2013.

[21] F. Uddin, Clays, nanoclays, and montmorillonite minerals, Metall. Mater. Trans. A 39A (2008) 2804–2814.

[22] J. Baringhaus, M. Ruan, F. Edler, A. Tejeda, M. Sicot, A. Taleb-Ibrahimi, A.P. Li, Z. Jiang, E.H. Conrad, C. Berger, C. Tegenkamp, W.A. De Heer, Exceptional ballistic transport in epitaxial graphene nanoribbons, Nature 506 (2014) 349–354.

[23] P.H.C. Camargo, K.G. Satyanarayana, F. Wypych, Nanocomposites: synthesis, structure, properties and new application opportunities, Mater. Res. 12 (2009) 1–39.

[24] R. Aversa, M.H. Modarres, S. Cozzini, R. Ciancio, A. Chiusole, The first annotated set of scanning electron microscopy images for nanoscience, Sci. Data 5 (2018) 180172.

[25] B.W. Shiau, C.H. Lin, Y.Y. Liao, Y.R. Lee, S.H. Liu, W.C. Ding, J.R. Lee, The characteristics and mechanisms of Au nanoparticles processed by functional centrifugal procedures, J. Phys. Chem. Solid 116 (2018) 161–167.

[26] J. Fricke, A. Emmerling, Aerogels—preparation, properties, applications, Chemistry, Spectroscopy and Applications of Sol-Gel Glasses, 1992, pp. 37–87.

[27] N. Zhao, J. Xu, Q. Xie, L. Weng, X. Guo, X. Zhang, L. Shi, Fabrication of biomimetic superhydrophobic coating with a micro-nano-binary structure, Macromol. Rapid Commun. 26 (2005) 1075–1080.

[28] R.K. O'reilly, C.J. Hawker, K.L. Wooley, Cross-linked block copolymer micelles: functional nanostructures of great potential and versatility, Chem. Soc. Rev. 35 (2006) 1068–1083.

[29] P. Ferreira, M.F. Ashby, D.L. Schodek, in: D.L. Schodek, P. Ferreira, M.F. Ashby (Eds.), Nanomaterials, Nanotechnologies and Design, Elsevier Science, Burlington, 2009. Online by Books.

[30] Z. Li, L. Wang, Y. Li, Y. Feng, W. Feng, Carbon-based functional nanomaterials: preparation, properties and applications, Compos. Sci. Technol. 179 (2019) 10–40.

[31] N. Kumar, S. Kumbhat, Carbon-based nanomaterials, Essentials in Nanoscience and Nanotechnology, Wiley, 2016, pp. 189–236.

[32] F. Xie, M. Yang, M. Jiang, X.J. Huang, W.Q. Liu, P.H. Xie, Carbon-based nanomaterials—a promising electrochemical sensor toward persistent toxic substance, TrAC Trends Anal. Chem. 119 (2019) 115624.

[33] M. Nehra, N. Dilbaghi, A.A. Hassan, S. Kumar, Carbon-Based Nanomaterials for the Development of Sensitive Nanosensor Platforms, Elsevier, 2019, pp. 1–25.

[34] N. Sumi, K.C. Chitra, Fullerene C60 nanomaterial induced oxidative imbalance in gonads of the freshwater fish, Aquat. Toxicol. 210 (2019) 196–206.

[35] S. Battaglia, S. Evangelisti, T. Leininger, F. Pirani, N. Faginas-Lago, A novel intermolecular potential to describe the interaction between the azide anion and carbon nanotubes, Diamond Relat. Mater. 101 (2020) 107533.

[36] A.A. Silva, R.A. Pinheiro, A.C. Rodrigues, M.R. Baldan, V.J. Trava-Airoldi, E.J. Corat, Graphene sheets produced by carbon nanotubes unzipping and their performance as supercapacitor, Appl. Surf. Sci. 446 (2018) 201–208.

32 Chapter 2 Nanomaterials and their classification

[37] P.N. Sudha, K. Sangeetha, K. Vijayalakshmi, A. Barhoum, Emerging Applications of Nanoparticles and Architecture Nanostructures: Current Prospects and Future Trends, Elsevier Inc., 2018, pp. 341–384.

[38] Q.L. Yan, M. Gozin, F.Q. Zhao, A. Cohen, S.P. Pang, Highly energetic compositions based on functionalized carbon nanomaterials, Nanoscale 8 (2016) 4799–4851.

[39] C. Kim, H. Lee, Light-assisted surface reactions on metal nanoparticles, Cat. Sci. Technol. 8 (2018) 3718–3727.

[40] E.V. Varner, K.E.K. Rindfusz, A. Gaglione, EPA/600/R-10/089, Nano Titanium Dioxide Environmental Matters: State of the Science Literature Review, U.S. Environ. Prot. Agency, 2010, p. 486.

[41] I.A. Rahman, V. Padavettan, Synthesis of Silica nanoparticles by Sol-Gel: size-dependent properties, surface modification, and applications in silica-polymer nanocomposites—a review, J. Nanomater. 2012 (2012).

[42] M. Mohapatra, S. Anand, Synthesis and applications of nano-structured iron oxides/hydroxides—a review, Int. J. Eng. Sci. Technol. 2 (2011) 127–146.

[43] T.A. Saleh, G. Fadillah, Recent trends in the design of chemical sensors based on graphene–metal oxide nanocomposites for the analysis of toxic species and biomolecules, TrAC Trends Anal. Chem. 120 (2019) 115660.

[44] L. Das, P. Das, A. Bhowal, C. Bhattachariee, Synthesis of hybrid hydrogel nano-polymer composite using Graphene oxide, Chitosan and PVA and its application in waste water treatment, Environ. Technol. Innov. 18 (2020) 100664.

[45] Y. Qi, J. Ye, S. Zhang, Q. Tian, N. Xu, P. Tian, G. Ning, Controllable synthesis of transition metal ion-doped CeO_2 micro/nanostructures for improving photocatalytic performance, J. Alloys Compd. 782 (2019) 780–788.

[46] P. Gaur, S. Banerjee, C-N cross coupling: novel approach towards effective aryl secondary amines modification on nanodiamond surface, Diamond Relat. Mater. 98 (2019) 107468.

[47] P. Chu, H. Zhang, J. Zhao, F. Gao, Y. Guo, B. Dang, Z. Zhang, On the volume resistivity of silica nanoparticle filled epoxy with different surface modifications, Compos. Part A Appl. Sci. Manuf. 99 (2017) 139–148.

[48] J.U. Hur, J.S. Han, J.R. Shin, H.Y. Park, S.C. Choi, Y.G. Jung, G.S. An, Fabrication of SnO_2-decorated Fe_3O_4 nanoparticles with anionic surface modification, Ceram. Int. 45 (2019) 21395–21400.

[49] T. Zeng, P. Zhang, X. Li, Y. Yin, K. Chen, C. Wang, Facile fabrication of durable superhydrophobic and oleophobic surface on cellulose substrate via thiol-ene click modification, Appl. Surf. Sci. 493 (2019) 1004–1012.

[50] T. Maxwell, M.G. Nogueira Campos, S. Smith, M. Doomra, Z. Thwin, S. Santra, Nanoparticles for Biomedical Applications. Fundamental Concepts, Biological Interactions and Clinical Applications, Elsevier, 2019, pp. 243–265.

[51] R.S. Pawar, P.G. Upadhaya, V.B. Patravale, Handbook of Nanomaterials for Industrial Applications, Elsevier, 2018, pp. 621–637.

[52] K. Qi, R. Selvaraj, L. Wang, Functionalized inorganic semiconductor nanomaterials: characterization, properties, and applications, Front. Chem. 8 (2020) 616728.

[53] T. Ihn, Semiconductor Nanostructures: Quantum States and Electronic Transport, Oxford University Press, Ocford, 2010.

[54] P. Sobierajska, A. Dorotkiewicz-Jach, K. Zawisza, J. Okal, T. Olszak, Z. Drulis-Kawa, R.J. Wiglusz, Preparation and antimicrobial activity of the porous hydroxyapatite nanoceramics, J. Alloys Compd. 748 (2018) 179–187.

[55] Y. Tian, F. Xue, Q. Fu, L. Zhou, C. Wang, H. Gou, M. Zhang, Structural and physical properties of Ti-doped $BiFeO_3$ nanoceramics, Ceram. Int. 44 (2018) 4287–4291.

Chapter 2 Nanomaterials and their classification **33**

[56] X. Xu, Y. Yang, X. Wang, X. Su, J. Liu, Low-temperature preparation of Al_2O_3-ZrO_2 nanoceramics via pressureless sintering assisted by amorphous powders, J. Alloys Compd. 783 (2019) 806–812.

[57] X. Yang, K. Lian, Y. Tan, Y. Zhu, X. Liu, Y. Zeng, T. Yu, T. Meng, H. Yuan, F. Hu, Selective uptake of chitosan polymeric micelles by circulating monocytes for enhanced tumor targeting, Carbohydr. Polym. 229 (2020) 115435.

[58] S. Sur, A. Rathore, V. Dave, K.R. Reddy, R.S. Chouhan, V. Sadhu, Nano-Struct. Nano-Objects 20 (2019) 100397.

[59] S. Fu, Z. Sun, P. Huang, Y. Li, N. Hu, Some basic aspects of polymer nanocomposites: a critical review, Nano Mater. Sci. 1 (2019) 2–30.

[60] B.M. Okrugin, I.M. Neelov, F.A.M. Leermakers, O.V. Borisov, Structure of asymmetrical peptide dendrimers: insights given by self-consistent field theory, Polymer (Guildf) 125 (2017) 292–302.

[61] R. Lee, D. Shenoy, R. Sheel, Micellar Nanoparticles : Applications for Topical and Passive Transdermal Drug Delivery, Elsevier, 2010.

[62] H.P. Feng, L. Tang, G.M. Zeng, J. Zhou, Y.C. Deng, X. Ren, B. Song, C. Liang, M.Y. Wei, J.F. Yu, Core-shell nanomaterials: applications in energy storage and conversion, Adv. Colloid Interface Sci. 267 (2019) 26–46.

[63] K. Mondal, A. Sharma, Recent advances in the synthesis and application of photocatalytic metal-metal oxide core-shell nanoparticles for environmental remediation and their recycling process, RSC Adv. 6 (2016) 83589–83612.

[64] V. Gascón, M.B. Jiménez, R.M. Blanco, M. Sanchez-Sanchez, Semi-crystalline Fe-BTC MOF material as an efficient support for enzyme immobilization, Catal. Today 304 (2018) 119–126.

[65] Y. Oh, J. Lee, M. Lee, Fabrication of Ag-Au bimetallic nanoparticles by laser-induced dewetting of bilayer films, Appl. Surf. Sci. 434 (2018) 1293–1299.

[66] A.S. Lozhkomoev, A.V. Pervikov, A.V. Chumaevsky, E.S. Dvilis, V.D. Paygin, O.L. Khasanov, M.I. Lerner, Fabrication of Fe-Cu composites from electroexplosive bimetallic nanoparticles by spark plasma sintering, Vacuum 170 (2019) 108980.

[67] R.M. Abozaid, Z. Lazarević, I. Radović, M. Gilić, D. Šević, M.S. Rabasović, V. Radojević, Optical properties and fluorescence of quantum dots CdSe/ZnS-PMMA composite films with interface modifications, Opt. Mater. (Amst). 92 (2019) 405–410.

[68] Q. Zhang, Q. Gao, W. Qian, H. Zhang, W. Tian, Z. Li, A C-coated and Sb-doped SnO_2 nanocomposite with high surface area and low charge transfer resistance as ultrahigh capacity lithium ion battery anode, Mater. Today Energy 13 (2019) 93–99.

[69] A. Biswas, I.S. Bayer, A.S. Biris, T. Wang, E. Dervishi, F. Faupel, Advances in top-down and bottom-up surface nanofabrication: techniques, applications & future prospects, Adv. Colloid Interface Sci. 170 (2012) 2–27.

[70] H.L. Parker, A.J. Hunt, V.L. Budarin, P.S. Shuttleworth, K.L. Miller, J.H. Clark, The importance of being porous: polysaccharide-derived mesoporous materials for use in dye adsorption, RSC Adv. 2 (2012) 8992–8997.

[71] T. Tago, T. Masuda, Zeolite Nanocrystals-Synthesis and Applications 191 X Zeolite Nanocrystals-Synthesis and Applications, IntechOpen, 2010.

[72] Y. Boyjoo, V.K. Pareek, J. Liu, Synthesis of micro and nano-sized calcium carbonate particles and their applications, J. Mater. Chem. A 2 (2014) 14270–14288.

3

The electrochemical mechanism and transport phenomenon of liquid fuel cells

Gamze Karanfil[a], Ramazan Bayat[b,c], and Fatih Şen[b]

[a]*Karamanoglu Mehmetbey University, Faculty of Engineering, Department of Energy Systems Engineering, Karaman, Turkey.* [b]*Şen Research Group, Department of Biochemistry, Dumlupinar University, Kütahya, Turkey.* [c]*Department of Materials Science and Engineering, Faculty of Engineering, Dumlupinar University, Kütahya, Turkey*

1 Introduction

Most research and development studies in the field of energy are based on finding efficient, renewable, and environmentally friendly alternative electricity sources. Although many new sources, processes, and technologies have been developed, problems are encountered in the adaptation of these studies to daily life. In particular, the importance of using mobile devices in modern life is increasing day by day, and current technologies offer limited working time in view of energy storage limits [1–3].

Fuel cells have the potential to replace existing batteries and meet the requirements of future sustained power supply. Fuel cells can be classified as electrochemical devices which, without a burning reaction, convert the chemical energy of fuels such as hydrogen, methanol, ethanol, and ethylene glycol into electrical energy. While electrical energy is obtained by oxidizing fuel, heat and water are also produced as by-products. Fuel cells are systems that can produce power with a higher energy conversion yield and are more environmentally friendly than traditional energy conversion systems. Classifying fuel cells is contingent upon a number of parameters, such as the fuel cell construction, operation temperature, electrolyte type, exchange ion, and reactant type. Usually, fuel cells are categorized according to the electrolytes being used [3–5].

Among many types of fuel cells, alkaline fuel cells (AFCs) and proton exchange membrane fuel cells (PEMFCs) are suitable for mobile device adhibition in view of their low operating

---

Nanomaterials for Direct Alcohol Fuel Cells. https://doi.org/10.1016/B978-0-12-821713-9.00018-4
Copyright © 2021 Elsevier Inc. All rights reserved.

temperatures, noncorrosive structure, and flexibility of functional use. The fuel typically used in PEMFCs and compared to its ionization rate at low voltage loss during reactions, AFCs consist of hydrogen because it is the most electrochemically active fuel. Clean energy is produced by hydrogen oxidation because carbon dioxide (CO_2) emissions are not generated. However, the fact that hydrogen is a naturally nonexistent gas eliminates its characteristic of being the primary fuel. Thus, an external reform process is required to get hydrogen. The storage difficulty of hydrogen and the complexity and weight of the required system also increase production costs [6–9].

Alcohols are an alternative fuel used in fuel cells under atmospheric conditions. They are liquid in atmospheric conditions, which makes them easy to use, store, and transport. In addition, because alcohols are liquids at room temperature and 1 atm pressure, they have a greater volumetric energy density than hydrogen. Also, some alcohols are renewable and have low toxicity [10]. The energy density values and conversion efficiencies of divergent alcohols compared with hydrogen as fuel for fuel cells are presented in Table 1 [11]. It can be seen that the energy densities and energy

Table 1 Features and conversion productivities of divergent alcohols noted as fuel for fuel cells compared with hydrogen [11].

Feature	Hydrogen	Methanol	Ethanol	Ethylene glycol
Formula	H_2	CH_3OH	C_2H_5OH	$(CH_2OH)_2$
$-\Delta G^\circ$ (kJ/mol)	237	702	1325	1180
ΔH° (kJ/mol)	286	726	1367	1192
Energy density, LHV (kWh/kg)	33	6.09	8.00	5.29
Energy density, LHV (kWh/L)	2.96×10^{-3}	4.80	6.32	5.80
E°_{cell} (V)	1.23	1.21	1.14	1.22
η_{actual}	2	4	4	8
η_{theo}	2	6	12	10
Stored energy (Ah/kg)	26,802	3350	2330	3458
Stored energy (Ah/L)	2.40	2653	1841	3855
ϵ^{rev}_{cell} (%)	83	97	97	99
ϵ_f (%)	100	67	33	88
$^a\epsilon_v$ (%)	57	41	44	41
$\epsilon_{cell}\left(\epsilon^{rev*}_{cell}\epsilon_f{}^*\epsilon_v\right)$ (%)	54	27	14	36
		40^b	43^b	41^b

[a]Supposing cell operation voltage of 0.7 V for hydrogen and 0.5 V for alcohols.
[b]If an overall electrochemical reaction happens with the participation of all electrons.

storage capacities of alcohols are much higher than hydrogen in liquid form and the overall cell efficiencies are quite high with whole oxidation of alcohols.

The calculations of energy densities and capacities of any fuel and efficiencies of fuel cell systems given in Table 1 can be shown as follows [11]. Calculations of volumetric and specific energy densities of fuels based upon lower heating values (LHVs) are given as follows [11]:

$$\text{Energy density (kWh/L)} = (-\Delta G^{\circ} x \rho)/(3600 \, x \, M_m)$$

$$\text{Energy density (kWh/kg)} = -\Delta G^{\circ}/(3600 \, x \, M_m),$$

where ΔG° is the Gibbs free energy change (at normal pressure and temperature), and M_m and ρ are the molar mass and the density of fuel, respectively.

Calculations of the stored energy in fuel denoted as Ah/L or Ah/kg are shown as follows:

$$\text{Stored energy } (Ah/L) = n * F * \rho/(3600 * M_m)$$

$$\text{Stored energy } (Ah/kg) = n * F/(3600 * M_m)$$

The number of electrons emitted from fuel oxidation is indicated, and F is the Faraday constant (96,487 coulombs/mol equivalent).

The reversible voltage (E_r°) calculation of the fuel cell reactions is determined according to the following equation:

$$E_r^{\circ} = -\Delta G^{\circ}/n*F$$

The ratio of the free energy change from Gibbs, $-\Delta G^{\circ}$ to the enthalpy change of the total fuel oxidation reaction, H°, is defined as the fuel cell's theoretical electrical yield (or reversible cell yield), and can be provided as

$$\varepsilon_{cell}^{rev}(\%) = -(\Delta G^{\circ}/\Delta H^{\circ}) * 100$$

The ratio of the actual number of electrons to the theoretical number of electrons (η_{theo}) involved in the whole fuel oxidation reaction, including the oxidation reaction (η_{actual}), is known as the *faradic* or *current fuel cell reaction yield*:

$$\varepsilon_f (\%) = (\eta_{actual}/\eta_{theo}) * 100$$

The output voltage of the cell is defined as the rate of the working voltage of the cell to the reversible voltage of the cell, as seen here:

$$\varepsilon_v(\%) = (E_{op}/E_r^{\circ}) * 100$$

The decrease in cell voltage from reversible voltage to working voltage is the product of electrode polarization losses (activation and polarization of the concentration) and resistance losses between many cell constituents such as electrolytes, interconnectors, and touch resistance.

The overall yieldance of the cell is a consequence of reversible cell yieldance, Faradic yieldance, and voltage yieldance and is expressed as

$$\varepsilon_{cell}(\%) = \varepsilon_{cell}^{rev} * \varepsilon_f * \varepsilon_v * 10^{-4}$$

Instead of gaseous hydrogen, direct liquid fuel cells (DLFCs) have made tremendous progress using liquid alcohols and soluble organic substances as fuel. These have been declared one of the most promising new power sources, particularly for mobile electronics. In addition to the advantages mentioned previously, DLFCs have a wide range of other benefits, including improved manufacturing, easy transportation, and comfortable use of liquid fuels compared to gaseous hydrogen [12–15].

Direct alcohol fuel cells (DAFCs) have several attractive features, such as low startup time, use of waste resources as a fuel source (i.e., methanol or ethanol that can be obtained from waste), high energy density, easy handling and use of the fuel, and finally cost-effectiveness [3]. The current DAFCs, containing direct ethanol fuel cells (DEFCs), direct methanol fuel cells (DMFCs), and direct ethylene glycol fuel cells (DEGFCs), typify rewarding liquid fuel resources for DAFCs. Nevertheless, there are limitations related to the application of DAFCs, such as the high price of the final product, low electrochemical reaction rates, and poor durability [16].

DMFCs, as the first work of DAFC, have been the object of research due to properties such as high energy density (6.09 kWh/kg), high energy conversion yieldance, their lack of need of electrical energy for charging, having a fast fuel filling system, giving off a low level of pollutant emissions, having a longer cell life owing to outstanding energy density, cheap fuel, and ease of transportation and storage. On the other hand, because methanol is volatile and toxic, is not renewable, requires an expensive platinum (Pt) catalyst, and has a high rate of methanol passage through the polymeric Nafion membrane in its structure, the current and cathode yield decreases. Thus, DMFCs have many disadvantages [17–21].

Ethanol is an interesting and promising alcohol available at DAFC (Direct Alcohol Fuel Cells). Compared to other fuels, such as hydrogen, methanol, and ethylene glycol, ethanol has a higher energy density (8 kWh/kg). In addition, it is nontoxic, in that it does not cause environmental pollution. It is also known that the direct oxidation productivity of ethanol in fuel cells is better than that of methanol owing to the lower transition rate of the fuel, which reduces its impact on cathode productivity. Besides these advantages, ethanol is sustainable because of it occurs naturally and is easy to find, transport, and store, with fast fuel filling. Further, it is recognized as a renewable energy source because it can be generated collectively by agrarian bioprocesses through fermentation of biomass from husbandry, forestry, and civic waste. Therefore, DEFCs seem to be a perfect alternating power source for mobile devices owing to their high energy density (energy per unit volume), nontoxicity, and easily accessible fuel [22–25].

Ethylene glycol is an attractive fuel, especially for mobile electronic devices, owing to its electron transfer rate of up to 80% and an energy density of 5.29 kWh/kg [11, 12]. The most remarkable feature of ethylene glycol is that the entity of an OH group on every carbon atom does it more efficient than ethanol against oxidation at the anode. Also, compared to both methanol and ethanol, ethylene glycol has a relatively low vapor pressure and high boiling point, which is very important for handheld devices. This is why DEGFCs have become one of the important research points in DLFCs. For safety reasons, ethylene glycol is comparatively more volatile than ethanol; however, it is nevertheless conceived to have low toxicity compared to methanol.—In addition, alike to ethanol oxidation, ethylene glycol oxidation meets problems with the slow kinetics of C–C bond break and catalytic empoisonment by intermediate species [26–29].

In DAFCs, acidic proton exchange membranes (PEMs) are often used. With a highly acidic membrane, it is simple to eliminate CO_2 created by anodic reaction. One of the most important factors causing the low yield of DAFCs is the slow kinetics of alcohol oxidation reactions in acidic environments. Alcohol oxidation kinetics can be improved by replacing the acidic medium with the alkaline medium. In addition, ionic conduction in alkali fuel cells occurs via the transmission of hydroxide ions from the cathode to the anode in a way that is opposite to the proton conduction. As a result, the alcohol permeability rate is reduced by reversing the direction of electro-osmic drag. For the reasons already stated, it is important to clarify the reaction kinetics and ion transport mechanisms of DAFCs in alkaline environments [10, 30].

As mentioned previously, the use of alcohol in fuel cells has significant advantages, but also some disadvantages. To overcome the disadvantages, the electrochemical oxidation reaction mechanisms of alcohols should be well understood. In addition, it is important to explain the transport phenomenon in DLFCs in detail and to determine the design parameters in accordance with this phenomenon. In the next section, the oxidation mechanisms of methanol, ethanol, and ethylene glycol, the three most commonly used alcohols in DLFCs, in acidic and alkaline environments will be explained comprehensively.

2 Direct methanol fuel cells

2.1 In acidic media

Like a traditional fuel cell, the DMFC has seven components: cathode diffusion layer, anode catalyst layer, cathode catalyst layer, anode diffusion layer, electrolyte (membrane), anode flow channel, and cathode flow channel. Through the use of the anode flow channel, a methanol-water solution is dispersed through the anode diffusion layer to reach the anode catalyst layer, where methanol is catalytically oxidized to emit protons, electrons, and CO_2. The protons are carried to the cathode catalyst layer via the membrane, while electrons are transferred by an external circuit. At the cathode side, oxygen/air is supplied to the cathode flow channel and transfers through the cathode diffusion layer to the cathode catalyst layer, where oxygen reacts with protons and electrons to form water. The gas CO_2 produced on the anode side and the liquid water produced on the cathode side are then eliminated from the cell. The electrochemical reactions at anode/electrolyte and cathode/electrolyte interfaces and the total oxidation responses of the acidic medium of methanol are illustrated in the following equations [31–34] and schematically shown in Fig. 1:

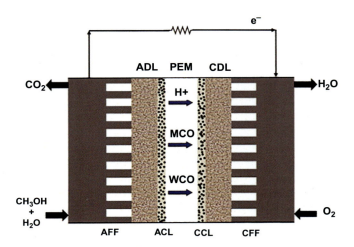

Fig. 1 Schematic representation of a DMFC [35].

$$\text{Anode reaction}: CH_3OH + H_2O \rightarrow 6H^+ + 6e^- + CO_2$$

$$\text{Cathode reaction}: {}^3/_2 O_2 + 6H^+ + 6e^- \rightarrow 3H_2O$$

$$\text{Overall reaction}: CH_3OH + {}^3/_2 O_2 \rightarrow 3H_2O + CO_2$$

Oxidation of methanol occurring in DMFCs at the anode involves many complex intermediate reaction steps. It was observed that methanol was converted to formaldehyde (CH_2O) by giving $2e^-$, then formaldehyde to formic acid (HCOOH) by giving $2e^-$, and finally formic acid to carbon dioxide by giving $2e^-$.

Methanol conversion to formaldehyde is given as follows:

$$CH_3OH \rightarrow HCOH + 2H^+ + 2e^-$$

Methanol conversion to formic acid is given as follows:

$$HCOH + H_2O \rightarrow HCOOH + 2H^+ + 2e^-$$

Formic acid to CO_2 conversion is given as follows:

$$HCOOH \rightarrow CO_2 + 2H^+ + 2e^-$$

Bimetallic and trimetallic metal mixtures of Pt used as catalysts in DMFC. The dissociation of water fed with methanol on Pt (0.6 V) makes methanol oxidation difficult with pure Pt. This problem is alleviated by placing ruthenium (Ru) next to Pt because water prefers lower-potential Ru (0.2 V). The Pt—Ru catalyst reaction used on the anode side takes place as follows:

$$CH_3OH + xPt \rightarrow Pt_xCH_2OH + H^+ + e^-$$

$$CHOH + xPt \rightarrow PtxCO + 2H^+ + e^- + xPt$$

PtCOH and PtCO components are platinum poison. As a result of previous studies, it was understood that the presence of ruthenium (Ru) prevents platinum poisoning [36]:

$$Ru + H_2O \rightarrow RuOH + H^+ + e^-$$

$$Pt_xCHOH + PtOH \rightarrow HCOOH + H^+ + e^- + Pt$$

$$Pt_xCO + RuOH \rightarrow CO_2 + H^+ + e^- + xPt + Ru$$

Fig. 2 shows the proposed mechanism for a methanol oxidation reaction (MOR) [37].

2.2 In alkaline media

DMFC is an alternative energy conversion system that uses liquid methanol, thus enabling easy fuel transportation and storage. Two kinds of electrolyte membrane are applied in these

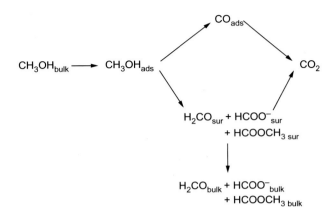

Fig. 2 The proposed mechanism for an MOR [37].

cells: anion exchange membranes (AEMs) and PEMs. PEM materials such as Nafion are widely applied in hydrogen fuel cells owing to their high productivity. However, the Nafion membrane in DMFCs exhibits high methanol permeability, which leads to fuel passage even at low temperatures, and thus to a reduce in cell productivity.

The physicochemical properties of AEMs are slightly different from PEMs and can be applied as an alternative electrolyte membrane in DMFCs. A DMFC with AEM works in alkaline conditions. Studies suggest that methanol's electrooxidation kinetics in an alkaline environment are faster than in an acidic medium. Due to its elementary molecular form, methanol in a DMFC is simpler to break down than other aliphatic alcohols. By reacting with hydroxide ions, methanol is electrochemically oxidized at the anode, and carbon dioxide and water are formed as a result. When an oxidant (air or oxygen) reacts with water, hydroxide (OH^-) ions are formed in the cathode. The anode, cathode, and total reactions occurring in an alkaline medium are given as follows [38–41]:

$$\text{Anode reaction}: CH_3OH + 6OH^- \rightarrow 5H_2O + CO_2 + 6e^-$$

$$\text{Cathode reaction}: {}^3/_2 O_2 + 3H_2O + 6e^- \rightarrow 6OH^-$$

$$\text{Overall reaction}: CH_3OH + {}^3/_2 O_2 \rightarrow 2H_2O + CO_2$$

Alkaline electrolyte conditions provide a suitable environment for reducing fuel passage in DMFCs. The direction of passage of hydroxide ions across the membrane is from the cathode to the anode, which resists the movement of alcohol to the cathode side, thereby reducing the crossover of alcohol. Non-noble metals can be applied as electrode catalysts in alkali electrolytes, which are more cost-effective than PEMs (which require the use of platinum catalyst in acidic electrolytes), and also makes it possible to reduce the catalyst loading because the reaction kinetics are faster in an alkaline medium than in an acidic environment. In spite of these advantages, the cell productivity of AEMs is much lower than PEMs owing to their lower ionic conductivity and weaker electrochemical, thermal, and mechanical stability [42].

3 Direct ethanol fuel cells

3.1 In acidic media

Because of the high energy density, bioavailability, low toxicity, sustainability, low expense, storage, and transport ease of ethanol used as fuel, DEFCs aim to be a safe and fertile energy conversion technology.

The schematic representation of a DEFC is given in Fig. 3 [11]. Air is procured to the cathode side, while anhydrous or diluted ethanol is fed to the anode side. Ethanol is oxidized at the electrode/electrolyte interface (in the catalyst layer) to its protons, electrons, and carbon dioxide, together with intermediate outputs/partly oxidized ethanol outputs such as acetic acid and acetaldehyde. Electrons pass through the electrical circuit and protons move through the membrane. The moved protons react with oxygen in the air at the cathode to generate water. The electrochemical reactions occurring at the anode and cathode and the

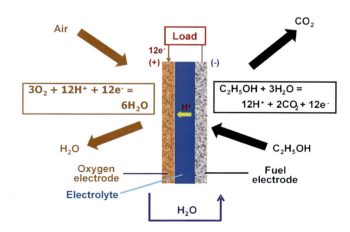

Fig. 3 Schematic representation of a DEFC [11].

complete ethanol oxidation reaction (EOR) can be given as follows [11,43]:

$$\text{Anode reaction}: C_2H_5OH + 3H_2O \rightarrow 12H^+ + 12e^- + 2CO_2$$

$$\text{Cathode reaction}: 3O_2 + 12H^+ + 12e^- \rightarrow 6H_2O$$

$$\text{Overall reaction}: C_2H_5OH + 3O_2 \rightarrow 2CO_2 + 3H_2O$$

In a DEFC, if ethanol ever specifically participates in an electrochemical reaction, due to the 12-electron transmission that takes place in complete oxidation, the reaction kinetics are very stagnant. Nevertheless, as mentioned before, an EOR can continue via conversion to acetic acid or acetaldehyde containing two or four electrons (Fig. 4), respectively [24].

3.2 In alkaline media

In traditional DEFCs, PEMs and precious metal catalysts are used. The need for advances in DEFCs was created by the sluggish kinetics of the EOR in acidic media and the need to break C—C bonds to ensure the entire oxidation of ethanol to CO_2. It has been observed that when acidic media are replaced by alkaline media, DEFC productivity can be improved by increasing the EOR kinetics, even with inexpensive materials such as AEMs and nonprecious metal catalysts [44].

Typical alkaline DEFC structure, in which the membrane electrode assembly (MEA) consists of the anode and cathode electrodes separated by an anion exchange membrane (AEM). At the anode, the ethanol- and water-containing fuel solution flows through the anode flow channel and travels to the anode catalyst layer via the anode diffusion layer, where ethanol is potentially oxidized to release electrons, CO_2 and water. The water is passed through the membrane in the anode and reaches the cathode,

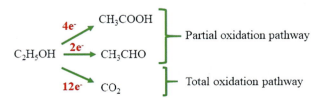

Fig. 4 Schematic diagram of parallel paths during an EOR in an acidic medium [24]. Reprinted (adapted) with permission from Y. Zheng, X. Wan, X. Cheng, K. Cheng, Z. Liu, Z. Dai, Advanced catalytic materials for ethanol oxidation in direct ethanol fuel cells, Catalysts. 10 (2020), https://doi.org/10.3390/catal10020166. Copyright (2020) Catalysts (MDPI).

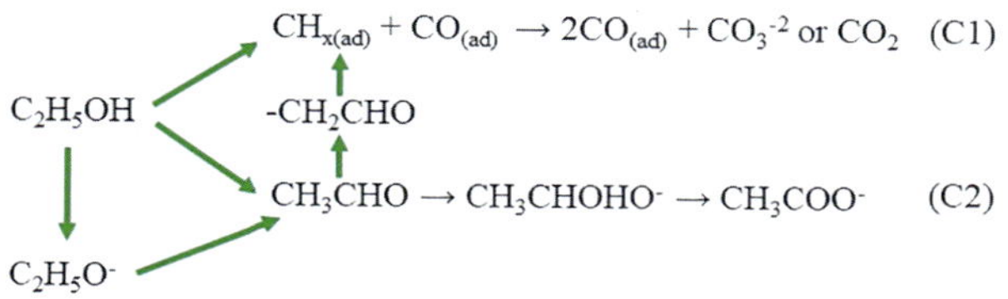

$$CH_{x(ad)} + CO_{(ad)} \rightarrow 2CO_{(ad)} + CO_3^{-2} \text{ or } CO_2 \quad (C1)$$

$$-CH_2CHO$$

$$CH_3CHO \rightarrow CH_3CHOHO^- \rightarrow CH_3COO^- \quad (C2)$$

$$C_2H_5OH$$

$$C_2H_5O^-$$

Fig. 5 Schematic diagram of parallel pathways during an EOR in an alkaline medium [48].

while the emitted electrons are transferred through the external circuit to the cathode. In the cathode, oxygen passes to the cathode catalyst layer via the cathode diffusion layer, where oxygen reacts to form hydroxide ions with electrons and water. At the cathode, the hydroxide ions pass through the AEM to the anode for the EOR [45–47]. The electrodes and total DEFC reactions in alkaline media are seen in the following equations:

$$\text{Anode reaction}: C_2H_5OH + 12OH^- \rightarrow 9H_2O + 2CO_2 + 12e^-$$

$$\text{Cathode reaction}: 3O_2 + 6H_2O + 12e^- \rightarrow 12OH^-$$

$$\text{Overall reaction}: C_2H_5OH + 3O_2 \rightarrow 3H_2O + 2CO_2$$

The main challenge of using DEFCs in alkaline media, however, is completing the oxidation of ethanol to utilize the total chemical energy stored in ethanol molecules in electricity. If the C—C bond breaks (Fig. 5, path C1), ethanol can convert to CO_2 (or carbonate in alkaline solutions) with 100% yield, but there is no catalyst that can perform this path owing to mechanical and/or sterical effects at low temperatures. It is assumed that the EOR takes place in parallel paths (Fig. 5, path C2) with the catalysts currently used. Although the C—C bond is not dissolved in this parallel way, ethanol is first oxidized into acetaldehyde, which is then oxidized into acetic acid (or acetate in alkaline solution) [48,49].

4 Direct ethylene glycol fuel cells

4.1 In acidic media

Because of its outstanding physicochemical properties, various organic molecules such as methanol, ethanol, and ethylene glycol have been found to be the most powerful fuel for DAFCs. Methanol, however, is poisonous, is highly flammable, and appears to flow through the membrane of the fuel cell. The predominant final oxidation result of acetic acid is ethanol. The electron transfer rate is low when the device is working at a temperature below 60 °C (33%). The fuel to be used in DAFC must have properties such as low toxicity, high energy density, and safety of use. It should also be a commercially available and inexpensive fuel with a high boiling point, and fully oxidized to CO_2 with little or no by-products. Recently, due to its excellent properties such as a potential energy ability of 5.8 kWh/L, a boiling point of 198°C, and an electron transfer rate of 80%, another fuel alternative called ethylene glycol has attracted considerable interest [50,51]. As shown in Fig. 6 [52], DEGFC, like all other DAFCs, is symmetrically formed, consisting of a diffusion layer and a catalytic layer in both anodine and cathode, and a PEM between the two electrodes. CO_2 oxidation produces $10e^-$ per ethylene glycol molecule; anode, cathode, and overall reactions occurring at DEGFC in the acidic medium are given as follows. [51]:

Anode reaction : $(CH_2OH)_2 + 2H_2O \rightarrow 10H^+ + 10e^- + 2CO_2$

Cathode reaction : $^5/_2 O_2 + 10H^+ + 10e^- \rightarrow 5H_2O$

Overall reaction : $(CH_2OH)_2 + ^5/_2 O_2 \rightarrow 2CO_2 + 3H_2O$

However, studies have concluded that ethylene glycol is partially oxidized and produces a range of intermediate substances, including glycoxal, glycolaldehyde, glyoxylic acid, glycolic acid, and oxalic acid. The direct oxidation yield of ethylene glycol to CO_2 at room temperature is insignificant, not exceeding more than a few percentage points. This problem makes it extremely

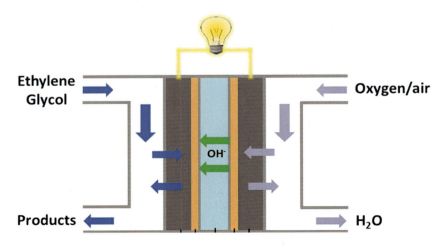

Fig. 6 Schematic representation of a DEGFC [52].

important to develop new catalysts for the entire ethylene glycol oxidation reaction (EGOR). General oxidation reactions of ethylene glycol to intermediates can be written as follows [27]:

$$(CH_2 - OH)_2 \rightarrow OH - C - CH_2 - OH$$

$$OH - C - CH_2 - OH \rightarrow OH - C - C - HO$$

$$OH - C - CH_2 - OH \rightarrow HOO - C - CH_2 - OH$$

$$OH - C - C - HO \rightarrow OH - C - C - OOH$$

$$HOO - C - CH_2 - OH \rightarrow OH - C - C - OOH$$

$$OH - C - C - OOH \rightarrow HOO - C - C - OOH$$

4.2 In alkaline media

In terms of the membrane used, DEGFCs are characteristically split into two groups: AEM- and PEM-dependent-based. Although PEM-based DEGFCs have been shown to perform well at high temperatures (i.e., >110°C), due to the slow kinetics associated with acidic membranes and the mixed potential caused by the passing of ethylene glycol, they usually perform poorly at room temperature. Recently, it has been shown that when the acidic membrane is replaced with an alkaline one, cell productivity can be greatly improved by improved kinetics of the oxidation reaction of ethylene glycol and the oxygen reduction reaction (ORR). In addition, the direction of electro-osmotic drag is from cathode to anode in AEM-based DEGFCs, which can increase the cell efficiency by decreasing the rate of fuel transfer from anode to cathode. In particular, ethylene glycol increases cell productivity by showing the ultimate reactivity in alkaline solutions and much less electrode poisoning in alkaline solutions. Besides, the cost of AEMs is much less than PEMs. The mentioned advantages make AEM-based DEGFCs more attractive than PEM-based DEGFCs [28].

An anode diffusion layer, an anode catalyst layer, a PEM between the anode and the cathode, a cathode catalyst layer, and an MEA-forming cathode diffusion layer constitute the fuel cell. In the flow channel, the fuel solution flows and the reactants pass from the flow field via the anode diffusion layer to the catalyst layer of the anode. Then the EGOR occurs in the anode catalyst layer, where ethylene glycol is oxidized to release electrons, CO_2, and water in the presence of hydroxide ions. Oxygen/air then is transferred into the cathode catalyst layer on the cathode, where the ORR occurs. The produced hydroxide ions move through the membrane to the anode for the EGOR. The EGOR and ORR results in an overall reaction given by the following equations [52]:

$$Anode\ reaction: (CH_2OH)_2 + 10OH^- \rightarrow 8H_2O + 2CO_2 + 10e^-$$

$$Cathode\ reaction: {}^5/_2 O_2 + 5H_2O + 10e^- \rightarrow 10OH^-$$

$$Overall\ reaction: (CH_2OH)_2 + {}^5/_2 O_2 \rightarrow 3H_2O + 2CO_2$$

Oxalate may be the primary product of the EGOR. In these grounds, for the processing of electrons, water and oxalate, ethylene glycol is partly oxidized according to

$$(CH_2OH)_2 + 10OH^- \rightarrow (COO^-)_2 + 8H_2O + 8e^-$$

In a cathode, the emitted electrons migrate from the outside circuit and enter the cathode catalytic sheet, engaging in the hydrogen peroxide reduction response (HPRR), in order to react to water accordingly with hydrogen peroxide and protons:

$$4H_2O_2 + 8H^+ + 8e^- \rightarrow 8H_2O$$

As a result, the total reaction of this fuel cell is accomplished by integrating the EGOR and the HPRR as follows [12]:

$$4H_2O_2 + 8H^+ + 8e^- \rightarrow 8H_2O$$

5 Conclusion

Liquid alcohols such as methanol, ethanol, and ethylene glycol attract attention as an alternative energy source for fuel cells due to their compatibility with existing storage and transportation infrastructures. Direct liquid/alcohol fuel cells become advantageous compared to hydrogen fuel cells due to their high energy conversion efficiency, more convenient use of liquid fuels, and safety of use. Studies have shown that fuel cells using AEMs have greater efficiency than conventional PEM fuel cells due to their fast reaction kinetics. In addition to the fast reaction kinetics of alcohols in the alkaline environment, less costly catalyst requirements have increased interest in studying these substances. In this chapter, the electrochemical reactions and reaction kinetics of various alcohols used in DLFCs in acidic and alkaline environments have been examined.

Nomenclature

AEM	**Anion exchange membrane**
$\Delta G°$	Gibbs free energy
$\Delta H°$	Enthalpy
$E_{cell}°$	Reversible cell voltage
ε_{cell}^{rev}	Theoretical electric yieldance (or reversible cell yieldance)
ε_f	Faradic or current yieldance
ε_v	Voltage yieldance of the cell
ε_{cell}	The overall yieldance of the cell
η_{actual}	Actual number of electrons involved in the oxidation reaction
η_{theo}	Theoretical number of electrons involved in the entire fuel oxidation reaction

References

[1] S.O. Ganiyu, C.A. Martínez-Huitle, The use of renewable energies driving electrochemical technologies for environmental applications, Curr. Opin. Electrochem. 22 (2020) 211–220. https://doi.org/10.1016/j.coelec.2020.07.007.

[2] X. Chen, W. Cao, Q. Zhang, S. Hu, J. Zhang, Artificial intelligence-aided model predictive control for a grid-tied wind-hydrogen-fuel cell system, IEEE Access. 8 (2020) 92418–92430. https://doi.org/10.1109/ACCESS.2020.2994577.

[3] M.A. Abdelkareem, K. Elsaid, T. Wilberforce, M. Kamil, E.T. Sayed, A. Olabi, Environmental aspects of fuel cells: a review, Sci. Total Environ. 752 (2021) 141803. https://doi.org/10.1016/j.scitotenv.2020.141803.

[4] N. Sazali, W.N.W. Salleh, A.S. Jamaludin, M.N.M. Razali, New perspectives on fuel cell technology: a brief review, Membranes (Basel) 10 (2020). https://doi.org/10.3390/membranes10050099.

[5] A. Kirubakaran, S. Jain, R.K. Nema, A review on fuel cell technologies and power electronic interface, Renew. Sust. Energ. Rev. 13 (2009) 2430–2440. https://doi.org/10.1016/j.rser.2009.04.004.

[6] S. Song, P. Tsiakaras, Recent progress in direct ethanol proton exchange membrane fuel cells (DE-PEMFCs), Appl. Catal. B Environ. 63 (2006) 187–193. https://doi.org/10.1016/j.apcatb.2005.09.018.

[7] S. Shamim, K. Sudhakar, B. Choudhary, J. Anwar, A Review on Recent Advances in Proton Exchange Membrane Fuel Cells: Materials, Technology and Applications, (n.d.). www.pelagiaresearchlibrary.com (accessed February 4, 2021).

[8] A. Albarbar, M. Alrweq, A. Albarbar, M. Alrweq, Proton exchange membrane fuel cells: review, in: Proton Exchange Membrane Fuel Cells, Springer International Publishing, 2018, pp. 9–29. https://doi.org/10.1007/978-3-319-70727-3_2.

[9] B. Şen, B. Demirkan, A. Savk, R. Kartop, M.S. Nas, M.H. Alma, S. Sürdem, F. Şen, High-performance graphite-supported ruthenium nanocatalyst for hydrogen evolution reaction, J. Mol. Liq. 268 (2018) 807–812. https://doi.org/10.1016/j.molliq.2018.07.117.

[10] E. Antolini, E.R. Gonzalez, Alkaline direct alcohol fuel cells, J. Power Sources 195 (2010) 3431–3450. https://doi.org/10.1016/j.jpowsour.2009.11.145.

[11] S.P.S. Badwal, S. Giddey, A. Kulkarni, J. Goel, S. Basu, Direct ethanol fuel cells for transport and stationary applications—a comprehensive review, Appl. Energy 145 (2015) 80–103. https://doi.org/10.1016/j.apenergy.2015.02.002.

[12] Z. Pan, Y. Bi, L. An, A cost-effective and chemically stable electrode binder for alkaline-acid direct ethylene glycol fuel cells, Appl. Energy 258 (2020) 114060. https://doi.org/10.1016/j.apenergy.2019.114060.

[13] S.S. Siwal, S. Thakur, Q.B. Zhang, V.K. Thakur, Electrocatalysts for electrooxidation of direct alcohol fuel cell: chemistry and applications, Mater. Today Chem. 14 (2019) 100182. https://doi.org/10.1016/j.mtchem.2019.06.004.

[14] T.S. Zhao, Z.X. Liang, J.B. Xu, Fuel cells—direct alcohol fuel cells | overview, in: Encyclopedia of Electrochemical Power Sources, Elsevier, 2009, pp. 362–369. https://doi.org/10.1016/B978-044452745-5.00240-9.

[15] A. Brouzgou, F. Tzorbatzoglou, P. Tsiakaras, Direct alcohol fuel cells: challenges and future trends, Proc. 2011 3rd Int. Youth Conf. Energ. IYCE 2011 (2011) 1–6.

[16] H. An, L. Pan, H. Cui, B. Li, D. Zhou, J. Zhai, Q. Li, Synthesis and performance of palladium-based catalysts for methanol and ethanol oxidation in alkaline fuel cells, Electrochim. Acta 102 (2013) 79–87. https://doi.org/10.1016/j.electacta.2013.03.142.

[17] Z. Zakaria, S.K. Kamarudin, A review of quaternized polyvinyl alcohol as an alternative polymeric membrane in DMFCs and DEFCs, Int. J. Energy Res. 44 (2020) 6223–6239. https://doi.org/10.1002/er.5314.

[18] F. Şen, G. Gökağaç, Different sized platinum nanoparticles supported on carbon: An XPS study on these methanol oxidation catalysts, J. Phys. Chem. C 111 (2007) 5715–5720. https://doi.org/10.1021/jp068381b.

[19] Ö. Karatepe, Y. Yildiz, H. Pamuk, S. Eris, Z. Dasdelen, F. Sen, Enhanced electrocatalytic activity and durability of highly monodisperse Pt@PPy-PANI nanocomposites as a novel catalyst for the electro-oxidation of methanol, RSC Adv. 6 (2016) 50851–50857. https://doi.org/10.1039/c6ra06210e.

[20] Y. Yıldız, S. Kuzu, B. Sen, A. Savk, S. Akocak, F. Şen, Different ligand based monodispersed Pt nanoparticles decorated with rGO as highly active and reusable catalysts for the methanol oxidation, Int. J. Hydrog. Energy 42 (2017) 13061–13069. https://doi.org/10.1016/j.ijhydene.2017.03.230.

[21] H. Burhan, K. Cellat, G. Yılmaz, F. Şen, Direct methanol fuel cells (DMFCs), in: Direct Liquid Fuel Cells, Elsevier, 2021, pp. 71–94. https://doi.org/10.1016/B978-0-12-818624-4.00003-0.

[22] N. Shaari, S.K. Kamarudin, Z. Zakaria, Enhanced alkaline stability and performance of alkali-doped quaternized poly(vinyl alcohol) membranes for passive direct ethanol fuel cell, Int. J. Energy Res. 43 (2019) 5252–5265. https://doi.org/10.1002/er.4513.

[23] M.Z.F. Kamarudin, S.K. Kamarudin, M.S. Masdar, W.R.W. Daud, Review: direct ethanol fuel cells, Int. J. Hydrog. Energy 38 (2013) 9438–9453. https://doi.org/10.1016/j.ijhydene.2012.07.059.

[24] Y. Zheng, X. Wan, X. Cheng, K. Cheng, Z. Liu, Z. Dai, Advanced catalytic materials for ethanol oxidation in direct ethanol fuel cells, Catalysts 10 (2020). https://doi.org/10.3390/catal10020166.

[25] H. Burhan, M. Yılmaz, K. Cellat, A. Zeytun, G. Yılmaz, F. Şen, Direct ethanol fuel cells (DEFCs), in: Direct Liquid Fuel Cells, Elsevier, 2021, pp. 95–113. https://doi.org/10.1016/b978-0-12-818624-4.00004-2.

[26] D.M. Fadzillah, S.K. Kamarudin, M.A. Zainoodin, M.S. Masdar, Critical challenges in the system development of direct alcohol fuel cells as portable power supplies: An overview, Int. J. Hydrog. Energy 44 (2019) 3031–3054. https://doi.org/10.1016/j.ijhydene.2018.11.089.

Chapter 3 The electrochemical mechanism and transport phenomenon of liquid fuel cells **51**

[27] A. Serov, C. Kwak, Recent achievements in direct ethylene glycol fuel cells (DEGFC), Appl. Catal. B Environ. 97 (2010) 1–12. https://doi.org/10.1016/j.apcatb.2010.04.011.

[28] L. An, T.S. Zhao, S.Y. Shen, Q.X. Wu, R. Chen, Performance of a direct ethylene glycol fuel cell with an anion-exchange membrane, Int. J. Hydrog. Energy 35 (2010) 4329–4335. https://doi.org/10.1016/j.ijhydene.2010.02.009.

[29] Z. Pan, B. Huang, L. An, Performance of a hybrid direct ethylene glycol fuel cell, Int. J. Energy Res. 43 (2019) 2583–2591. https://doi.org/10.1002/er.4176.

[30] G. Merle, M. Wessling, K. Nijmeijer, Anion exchange membranes for alkaline fuel cells: a review, J. Memb. Sci. 377 (2011) 1–35. https://doi.org/10.1016/j.memsci.2011.04.043.

[31] T.S. Zhao, C. Xu, R. Chen, W.W. Yang, Mass transport phenomena in direct methanol fuel cells, Prog. Energy Combust. Sci. 35 (2009) 275–292. https://doi.org/10.1016/j.pecs.2009.01.001.

[32] K.Z. Yao, K. Koran, K.B. McAuley, P. Oosthuizen, B. Peppley, T. Xie, A review of mathematical models for hydrogen and direct methanol polymer electrolyte membrane fuel cells, Fuel Cells 4 (2004) 3–29. https://doi.org/10.1002/fuce.200300004.

[33] N. Kakati, J. Maiti, S.H. Lee, S.H. Jee, B. Viswanathan, Y.S. Yoon, Anode catalysts for direct methanol fuel cells in acidic media: do we have any alternative for Pt or Pt-Ru? Chem. Rev. 114 (2014) 12397–12429. https://doi.org/10.1021/cr400389f.

[34] S. Eris, Z. Daşdelen, Y. Yıldız, F. Sen, Nanostructured Polyaniline-rGO decorated platinum catalyst with enhanced activity and durability for methanol oxidation, Int. J. Hydrog. Energy 43 (2018) 1337–1343. https://doi.org/10.1016/j.ijhydene.2017.11.051.

[35] N.K. Shrivastava, T.A.L. Harris, Direct methanol fuel cells, in: Encyclopedia of Sustainable Technologies, Elsevier, 2017, pp. 343–357. https://doi.org/10.1016/B978-0-12-409548-9.10121-6.

[36] M. Gülcan, Development of Inorganic/Organic Hybride Proton Exchange Membranes for Direct Methanol Fuel Cells (DMFC), Istanbul University, 2014.

[37] M. Kübler, T. Jurzinsky, D. Ziegenbalg, C. Cremers, Methanol oxidation reaction on core-shell structured ruthenium-palladium nanoparticles: relationship between structure and electrochemical behavior, J. Power Sources 375 (2018) 320–334. https://doi.org/10.1016/j.jpowsour.2017.07.114.

[38] U.K. Gupta, H. Pramanik, Physically crosslinked KOH impregnated polyvinyl alcohol based alkaline membrane for direct methanol fuel cell, Can. J. Chem. Eng. 96 (2018) 1888–1895. https://doi.org/10.1002/cjce.23233.

[39] E.H. Yu, U. Krewer, K. Scott, Principles and materials aspects of direct alkaline alcohol fuel cells, Energies 3 (2010) 1499–1528. https://doi.org/10.3390/en3081499.

[40] C.C. Yang, S.J. Chiu, W.C. Chien, Development of alkaline direct methanol fuel cells based on crosslinked PVA polymer membranes, J. Power Sources 162 (2006) 21–29. https://doi.org/10.1016/j.jpowsour.2006.06.065.

[41] K. Scott, E. Yu, G. Vlachogiannopoulos, M. Shivare, N. Duteanu, Performance of a direct methanol alkaline membrane fuel cell, J. Power Sources 175 (2008) 452–457. https://doi.org/10.1016/j.jpowsour.2007.09.027.

[42] J. Ryu, J.Y. Seo, B.N. Choi, W.J. Kim, C.H. Chung, Quaternized chitosan-based anion exchange membrane for alkaline direct methanol fuel cells, J. Ind. Eng. Chem. 73 (2019) 254–259. https://doi.org/10.1016/j.jiec.2019.01.033.

[43] Y.S. Li, T.S. Zhao, Z.X. Liang, Performance of alkaline electrolyte-membrane-based direct ethanol fuel cells, J. Power Sources 187 (2009) 387–392. https://doi.org/10.1016/j.jpowsour.2008.10.132.

[44] L. An, T.S. Zhao, Transport phenomena in alkaline direct ethanol fuel cells for sustainable energy production, J. Power Sources 341 (2017) 199–211. https://doi.org/10.1016/j.jpowsour.2016.11.117.

[45] I. Grimmer, P. Zorn, S. Weinberger, C. Grimmer, B. Pichler, B. Cermenek, F. Gebetsroither, A. Schenk, F.A. Mautner, B. Bitschnau, V. Hacker, Ethanol tolerant precious metal free cathode catalyst for alkaline direct ethanol fuel cells, Electrochim. Acta 228 (2017) 325–331. https://doi.org/10.1016/j.electacta.2017.01.087.

[46] B. Cermenek, B. Genorio, T. Winter, S. Wolf, J.G. Connell, M. Roschger, I. Letofsky-Papst, N. Kienzl, B. Bitschnau, V. Hacker, Alkaline ethanol oxidation reaction on carbon supported ternary PdNiBi Nanocatalyst using modified instant reduction synthesis method, Electrocatalysis 11 (2020) 203–214. https://doi.org/10.1007/s12678-019-00577-8.

[47] Y. Li, System design and performance in alkaline direct ethanol fuel cells, in: Lecture Notes in Energy, Springer Verlag, 2018, pp. 217–247. https://doi.org/10.1007/978-3-319-71371-7_7.

[48] J. Guo, R. Chen, F.C. Zhu, S.G. Sun, H.M. Villullas, New understandings of ethanol oxidation reaction mechanism on Pd/C and Pd2Ru/C catalysts in alkaline direct ethanol fuel cells, Appl. Catal. B Environ. 224 (2018) 602–611. https://doi.org/10.1016/j.apcatb.2017.10.037.

[49] B. Cermenek, J. Ranninger, B. Feketeföldi, I. Letofsky-Papst, N. Kienzl, B. Bitschnau, V. Hacker, Novel highly active carbon supported ternary PdNiBi nanoparticles as anode catalyst for the alkaline direct ethanol fuel cell, Nano Res. 12 (2019) 683–693. https://doi.org/10.1007/s12274-019-2277-z.

[50] V. Livshits, E. Peled, Progress in the development of a high-power, direct ethylene glycol fuel cell (DEGFC), J. Power Sources 161 (2006) 1187–1191. https://doi.org/10.1016/j.jpowsour.2006.04.141.

[51] Z. Pan, Y. Bi, L. An, Performance characteristics of a passive direct ethylene glycol fuel cell with hydrogen peroxide as oxidant, Appl. Energy 250 (2019) 846–854. https://doi.org/10.1016/j.apenergy.2019.05.072.

[52] L. An, R. Chen, Recent progress in alkaline direct ethylene glycol fuel cells for sustainable energy production, J. Power Sources 329 (2016) 484–501. https://doi.org/10.1016/j.jpowsour.2016.08.105.

4

The material development and characterization of direct alcohol fuel cells

Saadet Güler[a,b], Ahmet Yavaş[a,c], Sibel Demiroglu Mustafov[d], and Fatih Şen[e]

[a]Department of Material Science and Engineering, Izmir Katip Çelebi University, Izmir, Turkey. [b]Department of Metallurgical and Materials Engineering, The Graduate School of Natural and Applied Sciences, Dokuz Eylul University, Izmir, Turkey. [c]Graduate Program of Materials Science and Engineering, Graduate School of Natural and Applied Science, Izmir Katip Çelebi University, Izmir, Turkey. [d]Department of Metallurgical and Materials Engineering, Dokuz Eylul University, Izmir, Turkey. [e]Şen Research Group, Department of Biochemistry, Dumlupinar University, Kütahya, Turkey

1 Introduction

The world's energy consumption levels have been increasing daily, particularly throughout the 20th and 21st centuries thus far. In the last decade, a large part (87%) of the energy used was generated by burning fossil fuels (oil 33.7%, coal 29.7%, and natural gas 23.6%) [1–3]. However, the use of fossil fuels, which are not renewable and lead to high levels of carbon dioxide emissions, as conventional energy sources is seen as one of the major causes of global warming [4,5]. Further, significant increases in the prices of fossil fuels and in energy requirements of late have made it vital to develop new energy resources [5–9]. Accordingly, it has been considered that energy production from fuel cells, which are renewable and highly efficient sources, could meet a significant part of the world's energy requirements and substantially reduce the emission of greenhouse gases and other pollutants [1–3, 10–12].

The fuel cell is an electrochemical device that converts chemical energy into electrical energy utilizing hydrogen or hydrocarbon and oxygen as input fuel, and the fuel cell yields electricity and water as outputs [13–17]. Also, they have many types of classification according to the electrolyte type, temperature range, or combination of fuels and oxidants. To give an example, in the

Nanomaterials for Direct Alcohol Fuel Cells. https://doi.org/10.1016/B978-0-12-821713-9.00002-0
Copyright © 2021 Elsevier Inc. All rights reserved.

classification made in terms of temperature range, proton exchange membrane fuel cells (PEMFCs) are room-temperature fuel cells, molten carbonate fuel cells (MCFCs) are intermediate temperature fuel cells, while solid oxide fuel cells (SOFCs) are high-temperature fuel cells [11, 18]. Direct alcohol fuel cells (DAFCs) use liquid and renewable alcohol fuels (e.g., methanol and ethanol) with unique features including simple storage and handling, wide availability, and high energy density, all of which make them a promising option for both fuel utilization and feed strategies [1, 19].

The materials used in these fuel cells are very demanding. Each type of fuel cell may offer different needs for material combinations and operating conditions, as well as different restrictions that may be required depending on the application. For the purposes of this discussion, catalysts are essential materials for fuel cells, not only in terms of performance but cost as well. As is well known, noble metal catalysts such as platinum (Pt) and palladium (Pd) provide the best catalytic activity in fuel cell reactions. Nevertheless, the wide utilization of these metals is cost-prohibitive. For this reason, material research and development for fuel cells is done mostly to reduce their cost and increase their durability [11, 20].

2 Fuel cells

Fuel cells, environmentally friendly energy conversion technologies discovered a long time ago, are widely used today and likely to be used even more in the future. Although fuel cells are one of the earliest developments in the field of energy conversion, the history of their invention is not clear. The U.S. Department of Energy stated that in January 1839, Christian Friedrich Schönbein published the first study on the concept of fuel cells in the *Philosophical Journal*. In the meantime, a month later, William Grove published the findings of a similar study and devised the fuel cell in the same journal [2, 21].

Fuel cells are electrochemical tools that transform the chemical energy of a reaction occurring through fuels and oxidants like methanol, ethanol, and natural gas directly into electrical energy. For a variety of applications, including distributed power generation, transport, and portable electronics, fuel cells are seen as viable power sources. There is a wide range of types of fuel cells, grouped by the utilization of various classifications based on the combination of fuel and oxidant type, whether the fuel is operated inside or outside the fuel cell, the type of electrolyte, the operating temperature, and whether the reactants are supplied by inner or external manifolds to the cell. The most widely recognized fuel cell classification is based on the electrolyte type of material (i.e., a substance serving as a bridge of the ion exchange so as to create electric current between the anode and cathode) that is utilized [11, 12, 22, 23].

These categories include the following:
- PEMFCs, often identified as polymer electrolyte fuel cells (PEFCs)
- Alkaline fuel cells (AFCs)
- Phosphoric acid fuel cells (PAFCs)
- Molten carbonate fuel cells (MCFCs)
- Solid oxide fuel cells (SOFCs)

Methanol can also be utilized as a fuel in PEMFCs, so these fuel cells are frequently termed "direct methanol fuel cells (DMFCs)" [11, 22, 24].

2.1 Proton exchange membrane fuel cells

As mentioned before, after the German chemist Christian Friedrich Schönbein first proposed the basic principle of fuel cells in 1839, many kinds of fuel cells have been developed, including PEMFCs, AFCs, PAFCs, MCFCs, and SOFCs. Among them, PEMFCs have attracted a lot of interest; they are considered up-and-coming power sources for electric vehicles and mobile phones because of their numerous advantages, which include high energy transformation efficiency, great power density, and cleanliness because they emit no pollutants [24–26].

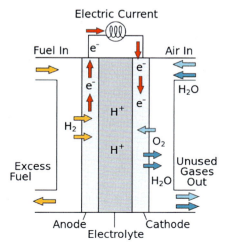

Fig. 1 PEMFC schematic diagram [27].

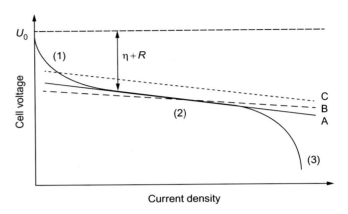

Fig. 2 A fuel cell's characteristic curve (U-i curve) [11].

The main principle of a PEMFC (Fig. 1) is that the two electrodes are split through an electrically insulating, yet ionically conductive polymer electrolyte (membrane), in line with other types of fuel cells. The hydrogen is oxidized to form protons in a PEMFC as an anodic reactant; oxygen is reduced in the cathode and creates water with protons that are brought along the membrane. The total reaction is

$$H_2 + {}^1/_2 O_2 \rightarrow H_2O \tag{1}$$

In Fig. 2, the typical curve of a PEMFC is given. Through the curve, the various contributions making up the cell's potential can be clarified. The thermodynamic reversible potential (also known as the "remaining potential"), called U_0, is the potential that can be determined in the absence of current in the setup. For PEMFCs, the thermodynamic reversible potential corresponds to 1.23 V. In the first region of the curve indicated by (1), the initial exponential decay of the voltage through growing current density is triggered by weak electrode kinetics. This first region can be enhanced by the features of the electrocatalytic layer. There is a linear relationship in the second part (2), known as the "ohmic-loss field" owing to the ohmic resistance of the cell constituents, which is underlined by the straight-line A. Consequently, deviations from the thermodynamic reversible potential U_0 are owing to ohmic contributions (R) and overpotential (η) contributions (signified by the total R + η). Eventually, the constraints on mass transport in (3) make way for a dramatic reduction in potential with growing current densities. Numerous factors can lead to these transport restrictions [11].

Fuel cell membrane materials must be able to conduct protons. Accordingly, the material selection relies on the fuel cells' service temperature range. Fuel cells working at temperatures of less than 100 °C (i.e., 60–80 °C) are called "low-temperature PEMFCs (LT-PEMFCs)." Cation-exchange components like sulfonated polymers are utilized in them. They need liquid water so as to dissociate the proton from the sulfonic acid group (with a commercial name of Nafion) linked to the polymer backbone. High-temperature PEMFCs (HT-PEMFCs), on the other hand, have an operating temperature range between 120 and 00 °C, in which the partial pressure of water (H_2O) diminishes continuously with temperature. Therefore, amphoteric proton self-conductive substances like phosphoric acid/phosphonic acid, heterocycles, and acidic phosphates, which are able to donate and accept protons without the need for "vehicle molecule" water, have to be utilized as proton-conducting electrolytes among the fuel cell's electrodes [11, 12, 22, 28].

2.2 Direct methanol fuel cells

As alternative energy systems, DMFCs have attracted great attention over the years. These cells have been proved feasible for a wide range of application areas, including portable electronic devices and field equipment, automotive, and stationary power plants owing to their good theoretical energy density, simplicity, simple fuel recharge, rapid start-up, and low environmental impact [11, 12, 29, 30]. Of these fuel cells, DMFCs are the only kind in which one of the reactants is supplied in liquid phase rather than the types of gas-fed cells. There are many advantages to utilizing methanol as fuel compared to hydrogen: it is an inexpensive liquid fuel that is simply processed, transported, and stored. Moreover, liquid methanol–fed fuel cells also have easy fuel process systems and moderate operating circumstances. After all, having such advantages makes DMFCs a strong rival for use in sustainable energy conversion and storage equipment [11, 29, 31]. The basic concept of DMFCs depends on the electrochemical reactions of two reactants, liquid methanol, and oxygen, in the anode and cathode, respectively, as shown in the following equations [29–31]:

Reaction of anode:

$$CH_3OH + H_2O \rightarrow CO_2 + 6H^+ + 6e^- \tag{2}$$

Reaction of cathode:

$$^3/_2O_2 + 6H^+ + 6e^- \rightarrow 3H_2O \tag{3}$$

Overall reaction:

$$CH_3OH + {}^3/_2O_2 \rightarrow CO_2 + 2H_2O \tag{4}$$

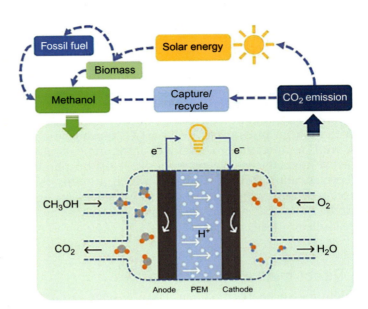

Fig. 3 An illustration of the working principle [29].

The thermodynamic potential of DMFCs at 25 °C is 1.21 V, and the theoretical thermodynamic conversion efficiency of methanol could reach up to 96.4% by way of the electrochemical reactions in the previous expressions. The operating basis of a single DMFC is demonstrated in Fig. 3. Owing to the polarization loss caused by multiple resistances in electrochemical reactions, the practical operating potential for a single DMFC is less than the theoretical amount [29, 30].

2.3 Phosphoric acid fuel cells

PAFCs were the initial fuel cells to be marketed through the United Technologies Corporation (UTC) among fuel cell types in which a 4-kW PAFC was developed in 1968. PAFCs have since been scaled up and field tested, and their performance, durability, and price have been considerably enhanced [32].

A technological wonder, PEMFCs depend on perfluorosulfonic acid membranes like Nafion combining good conductivity with great chemical and structural stability. Nevertheless, due to the robust proton conductivity dependency of the Nafion membranes on relative humidity, the working temperature is restricted to a maximum of 80 °C to provide high water content in the membrane. Additionally, the necessity of the high-purity H_2 for Nafion membrane-based PEMFCs emerges as another critical issue because the unfavorable potential loss is grounded by several fuel contaminants, especially CO, which poisons Pt electrocatalysts through robust adsorption [33].

In comparison to traditional PEMFCs operating at temperatures below 80 °C, the HT-PEMFCs, also called PAFCs, work at higher temperatures (typically between 120 and 200 °C). They use Pt as the catalyst and phosphoric acid as the proton carrier. HT-PEMFCs are a significant fuel cell technology owing to their simpler water and thermal management, high catalytic efficiency, rapid reaction kinetics, and improved antipoisoning capability of Pt catalysts (high tolerance to carbon monoxide), which can utilize direct methanol or formic acid-transformed gases with no impurity eliminations as fuel. Additionally, when PAFC employs at an elevated working temperature above 150 °C, the cathode water management will be easier compared to an LT-PEMFC concerning water vapor only. Eventually, it can be easily stated that the water and thermal management of the PAFC power system are much simpler than for the LT-PEMFC. The electrode kinetics at these elevated operating temperatures are also anticipated to be superior [34–37].

2.4 Alkaline fuel cells

The first kind of fuel cell to be put through usable service, which occurred at the start of the twentieth century, was the AFC, which is also the oldest fuel cell using concentrated potassium hydroxide (KOH) as a liquid electrolyte. This fuel cell, which is also identified as the "Bacon fuel cell" after the major contributions made by the British inventor F. T. Bacon, is one of the most advanced fuel cell systems, and it is called "the cell that flew Man to the Moon" [11]. Since the mid-1960s, the National Aeronautics and Space Administration (NASA) has employed AFCs on flights in the *Apollo* series of missions and on the space shuttle. This technology is still used in shuttle missions today. Later, many research groups continued to focus on these fuel cells for use in other fields of study. By the 1970s, Kordesh built a car powered by AFCs integrated with a lead-acid battery. In light of these developments and the findings of many studies by Kordesh, it has been found possible to commercialize these fuel cells, particularly for automotive applications, since the 1970s [11, 21, 38].

Having the best electrical performance of all fuel cells, AFCs generally utilize very pure gases and potassium hydroxide solution as a liquid electrolyte because that is the most conductive of all alkaline hydroxides. The anode-charged hydrogen reacts with water- and electron-producing hydroxyl anions (recombination). Electrons are passed to the cathode via an external circuit where oxygen reacts with water to form hydroxyl ions (reduction of oxygen). The following equations give the total reactions [11, 21]:

Reaction of anode:

$$2H_2 + 4OH^- \rightarrow 4H_2O + 4e^- \tag{5}$$

Reaction of cathode:

$$O_2 + 2H_2O + 4e^- \rightarrow 4OH^- \tag{6}$$

Reaction of total cell:

$$2H_2 + O_2 \rightarrow 2H_2O + \text{electrical energy} + \text{heat} \tag{7}$$

AFCs have several benefits over other fuel cells, including that they are simpler to manage because the working temperature is comparatively low (about 23–70 °C) [21]. They are very convenient for dynamic operational modes and can be developed for both small compact systems and big power plants. Another benefit is that the reaction kinetics at the electrodes are faster than in acidic environments of PEMFCs (to cite one example), resulting in greater cell voltages. This great electrical productivity makes it possible to use a lower amount of noble metal catalysts such as Pt, which is quite expensive [11, 21].

Even though there are these advantages, AFCs also have a few drawbacks, such as that liquid electrolytes must be used. The solution of KOH is very sensitive to the existence of CO_2. Therefore, an important working limitation is a need for low CO_2 levels in the oxidant stream. Hydroxyl ions can react with carbon dioxide in the air when using air instead of oxygen, and they form potassium carbonate (K_2CO_3) as per the reaction given here:

$$CO_2 + 2OH^- \rightarrow CO_3^- + H_2O \tag{8}$$

and/or

$$CO_2 + 2KOH \rightarrow K_2CO_3 + H_2O \tag{9}$$

The major reason for the reduced carbonate creation is the precipitation of great metal carbonate crystals like K_2CO_3 (as per the equation given here). This reaction initially reduces the amount of hydroxyl ions that exist at the anode for the reaction. Also, it changes the electrolyte composition and herewith decreases the ionic conductivity. For these fuel cells, the excess or lack of electrolytes constitutes another disadvantage that leads to electrode overflow and drying. AFCs have different forms. The materials employed in an AFC are highly dependent on the form of that fuel cell. Therefore, the materials for a conventional AFC differ meaningfully from those in a direct methanol AFC or anion exchange membrane [11, 21].

2.5 Solid oxide fuel cells

As solid electrochemical devices converting the chemical potential of fuels into electrical power through electrochemical reactions, SOFCs have attracted extensive attention owing to their great conversion efficiency and low pollution emissions [39–41]. Typically, a SOFC is made of three components: an ionic-conducting solid oxide electrolyte (generally made of yttria-stabilized zirconia [42], and two porous electrodes, the anode and the cathode [41, 43].

Unlike other types of fuel cells, in SOFCs, oxygen ions flow through the electrolyte from the cathode to the anode, while hydrogen or carbon monoxide reacts at the anode with the oxygen. To be more precise, in the course of the operation of an SOFC, oxygen molecules receive electrons from the external circuit for becoming oxygen ions at the cathode, as seen in Eq. (10). Then, oxygen ions pass from the electrolyte to the anode, oxidize the fuel like hydrogen, and donate electrons (Eq. 11). Thus, SOFCs can repeatedly produce electricity on the condition that the fuel and oxidant are consecutively provided at the anode and cathode, respectively [39, 43].

Cathode reaction:

$$O_2 + 4e^- \rightarrow 2O^{2-} \tag{10}$$

Anode reaction:

$$2O^{2-} + 2H_2 \rightarrow 2H_2O + 4e^- \tag{11}$$

Overall cell reaction:

$$O_2 + 2H_2 \rightarrow 2H_2O \tag{12}$$

Compared to the traditional technologies of electrical power production from fuels (e.g., combustion transformation), SOFCs have a number of benefits. Normally, SOFCs work at comparatively high temperatures (600–1000 °C) and possess rapid reaction kinetics. Hence, noble metal catalysts are not needed and this property carries a much lower cost. Also, because the conversion efficiency of SOFCs is good, there is no thermal-mechanical transformation associated with the process. In addition, the fuels are oxidized through oxygen ions instead of air comprising a great quantity of nitrogen, so little or no NO_x emissions from the SOFCs occur throughout their process. Furthermore, one of the unique advantages of SOFCs is that there is no maintenance of electrolyte loss for SOFCs, as they are a nonliquid-based fuel cell. They also eliminate the probability of electrode corrosion and liquid leakage [39, 40, 43].

3 Catalysts

In several applications, fuel cells are devices used to directly generate electrical energy from the combustion of a chemical product. Catalysts have been used to investigate various catalytic performances in fuel cells. Pt-based materials in these catalysts showed superior performance [10, 12, 44].

Various types of catalysts are used for fuel cells. Several studies have been carried out on the effective use of Pt catalysts, which accelerate the electrochemical reactions taking place in a fuel cell, in small amounts. The effectiveness of Pt catalysts decreases with increasing current density. Therefore, lower loads of Pt are suitable for high current densities of the practical current without adversely affecting the cell's performance. Various materials are often used to enhance the surface area of Pt. A large surface area and high electrical conductivity carbon materials are preferred. Pt is utilized as a catalyst both for reactions occurring at the anode and reactions occurring at the cathode. Usually, the Pt catalyst consists of small particles on the surface of slightly larger carbon particles used as support materials. Electrocatalysts have been used to improve the efficiency of fuel cells in the development of DAFCs for practical applications including electric vehicles and portable electronics [30, 45].

Due to the limited traditional energy resources and high energy requirements in the world, cost-effective and environmentally friendly energy resources have attracted attention. Accordingly, the DAFC has been the center of attention owing to its simple configuration system, weight, and high-power generation efficiency [2, 10].

3.1 Platinum-based catalysts

To date, Pt has been the most effective metal catalyst to alcohol oxidation in acid media However, adsorption of CO, one of the intermediates in alcohol electro-oxidation in pure Pt catalysts, can cause slow reaction kinetics. To solve this problem, Pt-based alloys were made with a second and third metal as the alloy compound. Thanks to these Pt-based alloys, the catalyst performance has increased. Comparing pure Pt to Pt-based alloys for alcohol oxidation, the increased activity of the latter has been connected to a bifunctional mechanism and a ligand effect (i.e., electronic effect). Generally, it is widely accepted in bifunctional mechanisms that oxygen-containing species contain adsorption to various metal atoms at lower potentials. Various factors affecting the catalytic activity of Pt-based alloy catalysts have been seen. It has been determined that these are mostly dependent on the composition, structure, morphology, particle size, and content of the alloy. Metal such as Ni, Co, Ru, Bi, Mo, Pd, Cu, and Sn is used as alloying metal in Pt alloys [20, 46, 47]. This not only reduces the use of Pt by adding a second metal to the catalysts in fuel cells, but also increases the activity of catalysts for alcohol oxidation.

Different effects of these binary Pt-based alloy catalysts have been seen. While Pt-Ru alloys are considered the most adapted for methanol oxidation, Pt-Sn alloys have been used in ethanol electro-oxidation. For example, looking at the intrinsic mechanism, it has been detected that the existence of Ru/Sn in the structure changes Pt's electronic structure and, consequently, the adsorption of oxygen-containing species. Another study found that Pt-Ru bimetallic catalysts could activate H_2O at a lower potential than Pt. It has been observed that the adsorbed OH^- bonds on the ruthenium element oxidize the CO on the surface of the Pt atoms into CO_2 [1, 10, 30, 44].

Wang et al. showed that the concave dendritic $PtCu_2/C$ nanocrystals exhibited enhanced electrocatalytic activity in the methanol oxidation reaction compared to other concave $PtCu/C$, $PtCu_3/C$, Pt_3Cu/C, and commercial Pt/C electrocatalysts. This improved efficiency toward methanol oxidation could be due to the concave structure and potential synergetic effect of Pt and Cu components (Fig. 4) [48].

Variant catalysts are utilized in AFCs. Metal catalysts in the hydrogen oxidation reaction involve carbon-supported Pt and Pt-Pd. A porous nickel arranged from Raney nickel is the most ordinarily utilized catalyst of hydrogen oxidation in AFCs. Several materials for cathode catalysts were used. Noble metal catalysts such as Pt and gold were primarily utilized in fuel cells concerning space implementations. As cathode catalysts, oxides are often utilized. For example, in hydrocele AFCs, perovskite (e.g., $La_{0.1}Ca_{0.9}MnO_3$) or spinel ($MnCo_2O_4$) catalysts are utilized [10–12].

3.2 Other catalysts

Due to the universally high price of the Pt element and the global scarcity of Pt, alternative metals have been investigated. In this regard, many attempts have been made to develop material catalysts in DAFCs. It has been determined in research

64 Chapter 4 The material development and characterization of direct alcohol fuel cells

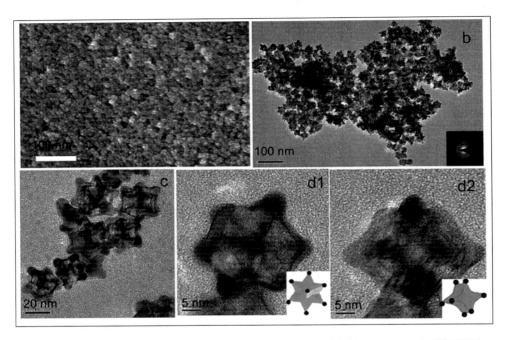

Fig. 4 (A) The images of FESEM, (B) The images of TEM, SAED pattern of PtCu$_2$ nanocrystals, (C) TEM images of concaved dendritic PtCu$_2$ nanocrystals, (D1, D2) HR-TEM images of PtCu$_2$, the insets being the geometric models [48–50].

studies that Pd can be an alternative to Pt catalysts. The Pd catalyst was found to provide greater electrocatalytic activity for alcohol electro-oxidation in an alkaline medium. Moreover, Pd is found in much greater quantities in the world. It was therefore thought that the catalyst cost could be significantly reduced. It has been developed in addition to noble metals such as cobalt (Co) and nickel (Ni) as electrocatalysts for the electrocatalytic oxidation of soft organic molecules in alkaline media. Many studies have been done on the modification of Ni-based catalysts to increase catalytic strength. Zhang et al. worked to make an Ni-phosphorus alloy nanowire network embedded in solvothermal reduced graphene oxide (RGO). The hybrid electrode of the Ni-P nanowire/RGO composite has been shown to provide superior electrocatalytic activity toward the oxidation of methanol in an alkaline solution. Chen et al. (2007) used the element Au as a base catalyst in ethanol oxidation, and it turned out that Au has properties similar to Pd. It has been observed that Au is inert in a low-pH environment but active in a high-pH environment. The oxidation ability of Au has also been compared with other fuels like methanol and ethylene glycol. Many studies have been conducted to examine the catalytic activity of cheap metals such as Ir-based materials, Ni-based materials, and W-carbide materials [1, 2, 11, 27, 51].

3.3 Synthesis of catalysts

The methods used in the preparation of catalysts depend on the desired physical and chemical characteristics and what the goal is. According to literature studies, only chemical composition, it is not a criterion for the designed catalytic activity. Properties such as support type, surface area, and pore size affect the catalytic activity of the produced catalyst. At the same time, many preparation processes and methods have been developed to prepare active catalysts [49, 50]. Impregnation, coprecipitation, and the sol-gel method are the main catalyst synthesis methods. In choosing the catalyst synthesis methods, the properties required of the catalyst, and the advantages and disadvantages of the method used should be taken into account. In the presence of a low amount of CO in fuel cells, poisoning on the electrode surfaces and a decrease in performance occur. The solution to this problem is the development of new catalysts for H_2-CO mixtures that will not affect CO but will give good results in the operating conditions of the fuel cell, and nanotechnology provides an advantage in this regard. Reduction, microwave, and hydrothermal methods are also used in catalyst synthesis. In these methods, carbon-based nanomaterials are generally used as support, and they provide an advantage in reducing carbon monoxide poisoning and increasing the efficiency of the catalyst [52].

3.3.1 Use in catalyst development of nanostructures

Nanotechnology seems to have the technological key to enable fuel cells to enter daily life both safely and cheaply. It is now known that catalysts that can be produced at nanoscale play a significant role in reaction kinetics. Nanosized materials have the advantages of having most of their active centers on the surface and having high catalytic activity with their large surface area, as well as great thermal, chemical, and mechanical resistance [53, 54]. Recently, the effect of nanostructures on fuel cell catalysts has been great. Recent studies have suggested that sequential mesoporous carbon, carbon aero- and xero-gels, carbon nanofibers (CNFs), and carbon nanotubes (CNTs) increase the efficiency of electrocatalysts [1, 12, 52].

Catalysts made of tiny, metallic particles, like ruthenium or Pt, assisted by nanocarbons or metal oxides, are generally utilized in DAFCs. The predominant types of CNF structures are fishbone (herringbone), thrombocyte, ribbon, and tubular structures. Dissimilar traditional graphite substances and nanotubes in which the basal plane is visible are revealed in CNFs only at the edges. This characteristic permits CNFs to be utilized as catalyst support in fuel cells.

4 Characterization methods of catalysts

The catalytic properties of the catalyst surface can be determined by examining its composition and structure on the atomic scale. The surface shape of heterogeneous catalysts is related to the chemical composition and atomic-scale structure of the catalyst. For this reason, the basis of the characterization method is to examine the atoms formed under the reaction conditions where the catalyst is located.

Catalyst characterization studies are carried out to develop new catalysts with activity, selectivity, stability, and mechanically more durable for industrial purposes. For this reason, characterization techniques are developed to produce effective materials by defining certain properties of materials. In principle, all spectroscopic methods serve this purpose, while focusing on particle size and shape, pore structure, pore size and pore size distribution, catalytic performance, and composition analysis to develop more effective catalysts [49, 50].

Löffler et al. [55] performed a study in that they examined a new preparation technique for the catalyst structures utilized in the PEM. The peaks obtained as a result of electrochemical deposition are shown in Fig. 5. When the X-ray diffraction (XRD) analysis was examined, it was found that the negative peaks in the structure were caused by the pure state of Pt. The approximation of the Lorentzian function determined the matching peak. Using the Debby-Scherrer formula, the average particle size was determined as approximately 9 nm.

In addition, particle size and size distribution are analyzed utilizing HR-TEM. In the study illustrated in Fig. 6, Pt particles were shown as dark spots (gray background) on the carbon support after electrochemical deposition. Moreover, particle size analysis is shown in the histogram chart.

When the produced Pt nanocatalysts were examined, it was confirmed that the particles were in nanocrystalline form. Preliminary findings indicate that the application of pulsed electrodeposition enables an increase in catalytic activity [55].

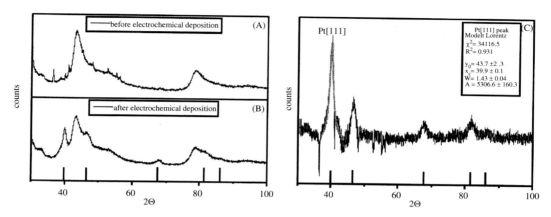

Fig. 5 XRD pattern of electrode (A) before and (B) after electrochemical deposition. (C) The difference diffractogram displays the diffraction pattern of the Pt NPs [55].

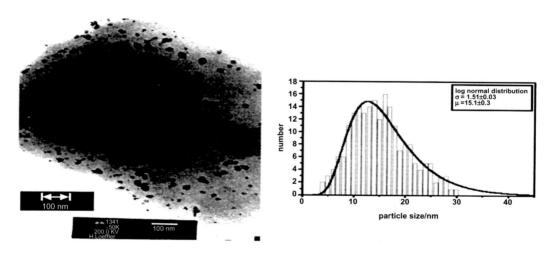

Fig. 6 HR-TEM image of an electrode after electrochemical deposition and size distribution of Pt particles [55].

Heping Xie et al. found that the anode utilized in the fuel cell is infiltrated through silver (Ag) NPs as an efficient catalyst to encourage the carbon electro-oxidation. Thanks to silver nanoparticles, maximum power density has been achieved in the fuel cell produced compared to coal-based fuel cells. As shown in Fig. 7A, XRD was utilized to study the structure and mineral phase of the fuel samples. As a result of this analysis, it was determined that there are three types of minerals; ankerite ($Ca_{0.997}(Mg_{0.273}Fe_{0.676}Mn_{0.054})(CO_3)_2$), calcite ($CaCO_3$), and kaolinite $Al_2Si_2O_5(OH)_4$) [56].

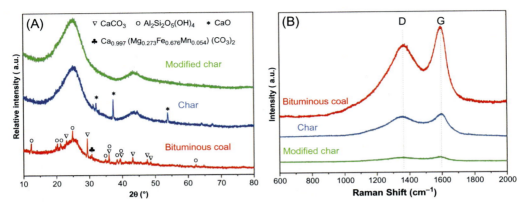

Fig. 7 (A) XRD patterns and (B) Raman spectra of modified char, char, and bituminous coal [56].

To form oxides, the minerals have been pyrolyzed after thermal treatment. Consequently, only calcium oxide (CaO) phases were detected for coal. Two basic diffraction peaks of carbon crystalline structures existed in all samples. The first peak is attributed to the (002) graphite carbon diffraction peak, and the second to the (100) amorphous carbon diffraction peak. In two peaks following pyrolysis, the detected enlargement of these shows that the crystalline structures of carbon determined more disorganized. Raman analysis was also performed to examine the crystal structure of the samples in detail. Two characteristic peaks are visible in the spectrum, as shown in Fig. 7B: the G (graphite) band and the D (irregular) band. The $I_{(D)}/I_{(G)}$ numerical quantity of bituminous coal, coal, and modified coal samples are roughly 0.850, 0.921, and 0.946. The modified coal has the largest amount of unstructured carbon and carbon electrochemical oxidation activity. The particle size of the amorphous carbon caused a rapid increase in Raman analysis as the carbon material was annealed to a particular temperature [56].

The chemical bonds of carbon fuels are analyzed using Fourier transform infrared spectra (FTIR). When the analysis result shown in Fig. 8 is examined, chemical bond groups in bituminous coal emerged largely as a result of the removal of volatile organic products after coal burning. The modification with acetic acid managed to boost the hydroxyl functional group. Carbon oxidation reactions are considered to be more acceptable for carbon atoms associated with functional groups containing oxygen. Moreover, thermogravimetric (TGA) analysis, another catalyst characterization method, was performed to examine the reactivity of modified coal and the effect of carbonates on reactivity (Fig. 9) [56].

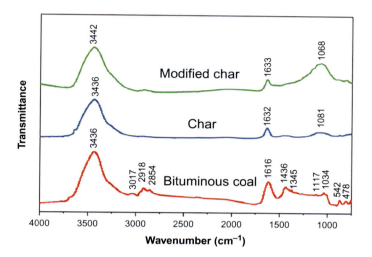

Fig. 8 The result of the FTIR analyses made by bituminous coal, char, and modified char [56].

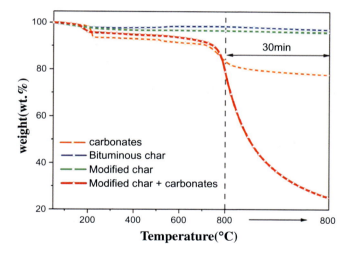

Fig. 9 TGA results of char and qualified char-carbonate mixture below CO_2 atmosphere [56].

X-ray photoelectron spectroscopy is a common technique employed for catalyst characterization. It is utilized to get information about the surfaces of catalysts. An X-ray beam is sent to the solid sample and stimulates the catalyst surface, causing photoelectron scattering. While the atoms form a peak with binding energy, the metal amounts and the valence of the metal atoms are determined. While the atoms form a peak with binding energy, the metal amounts and the valence of the metal atoms are determined by this analysis method [57].

5 Conclusion

Significant progress has been made in the production and optimization of new catalysts for DMFCs in recent years. Consequently, various studies on catalysts to increase electrocatalytic activity and efficiency have been presented. In addition to conventional catalysts, some new carbon materials, such as nanostructured carbons, have been shown as high potential catalyst support materials. Research and development in fuel cell materials frequently aim at developing and finding materials that will minimize the cost or enhance the durability of an individual subcomponent. This chapter covers the important aspects of the material development and characterization methods for DAFCs.

References

[1] Y. Wang, Nanomaterials for Direct Alcohol Fuel Cell, Pan Stanford Publishing, Singapore, 2016.

[2] H.R. Corti, E.R. Gonzalez, Direct Alcohol Fuel Cells: Materials, Performance, Durability and Applications, Springer, 2014, ISBN: 97894-007-7708-8.

[3] N. Jafri, W.Y. Wong, V. Doshi, L.W. Yoon, K.H. Cheah, A review on production and characterization of biochars for application in direct carbon fuel cells, Process. Saf. Environ. Prot. 118 (2018) 152–166, https://doi.org/10.1016/j.psep.2018.06.036.

[4] Z. Daşdelen, Y. Yıldız, S. Eriş, F. Şen, Enhanced electrocatalytic activity and durability of Pt nanoparticles decorated on GO-PVP hybride material for methanol oxidation reaction, Appl. Catal. B Environ. 219 (2017) 511–516, https://doi.org/10.1016/j.apcatb.2017.08.014.

[5] B. Çelik, Y. Yildiz, H. Sert, E. Erken, Y. Koşkun, F. Şen, Monodispersed palladium-cobalt alloy nanoparticles assembled on poly(N-vinyl-pyrrolidone) (PVP) as a highly effective catalyst for dimethylamine borane (DMAB) dehydrocoupling, RSC Adv. 6 (29) (2016) 24097–24102, https://doi.org/10.1039/c6ra00536e.

[6] B. Sen, S. Kuzu, E. Demir, S. Akocak, F. Sen, Highly monodisperse RuCo nanoparticles decorated on functionalized multiwalled carbon nanotube with the highest observed catalytic activity in the dehydrogenation of dimethylamine–borane, Int. J. Hydrog. Energy 42 (36) (2017) 23292–23298, https://doi.org/10.1016/j.ijhydene.2017.06.032.

[7] B. Sen, B. Demirkan, B. Şimşek, A. Savk, F. Sen, Monodisperse palladium nanocatalysts for dehydrocoupling of dimethylamineborane, Nano-Struct. Nano-Objects 16 (2018) 209–214, https://doi.org/10.1016/j.nanoso.2018.07.008.

[8] B. Şen, B. Demirkan, M. Levent, A. Şavk, F. Şen, Silica-based monodisperse PdCo nanohybrids as highly efficient and stable nanocatalyst for hydrogen evolution reaction, Int. J. Hydrog. Energy 43 (44) (2018) 20234–20242, https://doi.org/10.1016/j.ijhydene.2018.07.080.

[9] B. Şen, et al., High-performance graphite-supported ruthenium nanocatalyst for hydrogen evolution reaction, J. Mol. Liq. 268 (2018) 807–812, https://doi.org/10.1016/j.molliq.2018.07.117.

[10] S.S. Siwal, S. Thakur, Q.B. Zhang, V.K. Thakur, Electrocatalysts for electrooxidation of direct alcohol fuel cell: chemistry and applications, Mater. Today Chem. 14 (2019) 100182, https://doi.org/10.1016/j.mtchem.2019.06.004.

Chapter 4 The material development and characterization of direct alcohol fuel cells **71**

[11] M. Gasik, Materials for Fuel Cells, 2008.

[12] J. San Ping, S. Pei Kang, Nanostructured and Advanced Materials for Fuel Cells, CRC Press, 2013.

[13] S. Eris, Z. Daşdelen, F. Sen, Investigation of electrocatalytic activity and stability of Pt@f-VC catalyst prepared by in-situ synthesis for methanol electrooxidation, Int. J. Hydrog. Energy 43 (1) (2018) 385–390, https://doi.org/10.1016/j.ijhydene.2017.11.063.

[14] Ö. Karatepe, Y. Yildiz, H. Pamuk, S. Eris, Z. Dasdelen, F. Sen, Enhanced electrocatalytic activity and durability of highly monodisperse Pt@PPy-PANI nanocomposites as a novel catalyst for the electro-oxidation of methanol, RSC Adv. 6 (56) (2016) 50851–50857, https://doi.org/10.1039/c6ra06210e.

[15] Y. Yildiz, E. Erken, H. Pamuk, H. Sert, F. Şen, Monodisperse Pt nanoparticles assembled on reduced graphene oxide: highly efficient and reusable catalyst for methanol oxidation and dehydrocoupling of dimethylamine-borane (DMAB), J. Nanosci. Nanotechnol. 16 (6) (2016) 5951–5958, https://doi.org/10.1166/jnn.2016.11710.

[16] Y. Yıldız, S. Kuzu, B. Sen, A. Savk, S. Akocak, F. Şen, Different ligand based monodispersed Pt nanoparticles decorated with rGO as highly active and reusable catalysts for the methanol oxidation, Int. J. Hydrog. Energy 42 (18) (2017) 13061–13069, https://doi.org/10.1016/j.ijhydene.2017.03.230.

[17] F. Şen, S. Şen, G. Gökağaç, Efficiency enhancement of methanol/ethanol oxidation reactions on Pt nanoparticles prepared using a new surfactant, 1,1-dimethyl heptanethiol, Phys. Chem. Chem. Phys. 13 (4) (2011) 1676–1684, https://doi.org/10.1039/c0cp01212b.

[18] R. Kaur, A. Marwaha, V.A. Chhabra, K.H. Kim, S.K. Tripathi, Recent developments on functional nanomaterial-based electrodes for microbial fuel cells, Renew. Sustain. Energy Rev. 119 (July) (2020), https://doi.org/10.1016/j.rser.2019.109551, 109551.

[19] E.H. Yu, U. Krewer, K. Scott, Principles and materials aspects of direct alkaline alcohol fuel cells, Energies 3 (8) (2010) 1499–1528, https://doi.org/10.3390/en3081499.

[20] E. Kuyuldar, S.S. Polat, H. Burhan, S.D. Mustafov, A. Iyidogan, F. Sen, Monodisperse thiourea functionalized graphene oxide-based PtRu nanocatalysts for alcohol oxidation, Sci. Rep. 10 (1) (2020) 7811, https://doi.org/10.1038/s41598-020-64885-6.

[21] G. Merle, M. Wessling, K. Nijmeijer, Anion exchange membranes for alkaline fuel cells: a review, J. Membr. Sci. 377 (1–2) (2011) 1–35, https://doi.org/10.1016/j.memsci.2011.04.043.

[22] M. Mohsin, R. Raza, M. Mohsin-ul-Mulk, A. Yousaf, V. Hacker, Electrochemical characterization of polymer electrolyte membrane fuel cells and polarization curve analysis, Int. J. Hydrog. Energy (2019), https://doi.org/10.1016/j.ijhydene.2019.08.246.

[23] M.A. Abdelkareem, A. Allagui, E.T. Sayed, M. El Haj Assad, Z. Said, K. Elsaid, Comparative analysis of liquid versus vapor-feed passive direct methanol fuel cells, Renew. Energy 131 (2019) 563–584, https://doi.org/10.1016/j.renene.2018.07.055.

[24] S. Zhu, J. Ge, C. Liu, W. Xing, Atomic-level dispersed catalysts for PEMFCs: progress and future prospects, Energy Chem. 1 (3) (2019), https://doi.org/10.1016/j.enchem.2019.100018, 100018.

[25] S. Shahgaldi, J. Hamelin, Improved carbon nanostructures as a novel catalyst support in the cathode side of PEMFC: a critical review, Carbon 94 (July) (2015) 705–728, https://doi.org/10.1016/j.carbon.2015.07.055.

[26] L. Wang, X. Wan, S. Liu, L. Xu, J. Shui, Fe-N-C catalysts for PEMFC: progress towards the commercial application under DOE reference, J. Energy Chem. 39 (2019) 77–87, https://doi.org/10.1016/j.jechem.2018.12.019.

72 Chapter 4 The material development and characterization of direct alcohol fuel cells

[27] M.A.F. Akhairi, S.K. Kamarudin, Catalysts in direct ethanol fuel cell (DEFC): an overview, Int. J. Hydrog. Energy 41 (7) (2016) 4214–4228, https://doi.org/10.1016/j.ijhydene.2015.12.145.

[28] H.Q. Nguyen, B. Shabani, Proton exchange membrane fuel cells heat recovery opportunities for combined heating/cooling and power applications, Energy Convers. Manag. 204 (November) (2020), https://doi.org/10.1016/j.enconman.2019.112328, 112328.

[29] Z. Xia, X. Zhang, H. Sun, S. Wang, G. Sun, Recent advances in multi-scale design and construction of materials for direct methanol fuel cells, Nano Energy 65 (August) (2019), https://doi.org/10.1016/j.nanoen.2019.104048, 104048.

[30] M. Wang, X. Wang, M. Chen, Z. Yang, C. Dong, Nanostructured electrocatalytic materials and porous electrodes for direct methanol fuel cells, Cuihua Xuebao/Chin. J. Catal. 37 (7) (2016) 1037–1048, https://doi.org/10.1016/S1872-2067(16)62477-4.

[31] E. Akbari, Z. Buntat, A. Nikoukar, A. Kheirandish, M. Khaledian, A. Afroozeh, Sensor application in direct methanol fuel cells (DMFCs), Renew. Sust. Energ. Rev. 60 (2016) 1125–1139, https://doi.org/10.1016/j.rser.2016.02.001.

[32] H. Ito, Economic and environmental assessment of phosphoric acid fuel cell-based combined heat and power system for an apartment complex, Int. J. Hydrog. Energy 42 (23) (2017) 15449–15463, https://doi.org/10.1016/j.ijhydene.2017.05.038.

[33] Y. Cheng, et al., High CO tolerance of new SiO_2 doped phosphoric acid/polybenzimidazole polymer electrolyte membrane fuel cells at high temperatures of 200–250 °C, Int. J. Hydrog. Energy (2018) 22487–22499, https://doi.org/10.1016/j.ijhydene.2018.10.036.

[34] C.L. Lu, et al., High-performance and low-leakage phosphoric acid fuel cell with synergic composite membrane stacking of micro glass microfiber and nano PTFE, Renew. Energy 134 (2019) 982–988, https://doi.org/10.1016/j.renene.2018.11.011.

[35] J. Jang, et al., Phosphoric acid doped triazole-containing cross-linked polymer electrolytes with enhanced stability for high-temperature proton exchange membrane fuel cells, J. Membr. Sci. 595 (2020) 117508, https://doi.org/10.1016/j.memsci.2019.117508.

[36] H. Bai, et al., Poly(arylene piperidine)s with phosphoric acid doping as high temperature polymer electrolyte membrane for durable, high-performance fuel cells, J. Power Sources 443 (July) (2019), https://doi.org/10.1016/j.jpowsour.2019.227219, 227219.

[37] S. Martin, J.O. Jensen, Q. Li, P.L. Garcia-Ybarra, J.L. Castillo, Feasibility of ultra-low Pt loading electrodes for high temperature proton exchange membrane fuel cells based in phosphoric acid-doped membrane, Int. J. Hydrog. Energy 44 (52) (2019) 28273–28282, https://doi.org/10.1016/j.ijhydene.2019.09.073.

[38] G. Couture, A. Alaaeddine, F. Boschet, B. Ameduri, Polymeric materials as anion-exchange membranes for alkaline fuel cells, Prog. Polym. Sci. 36 (11) (2011) 1521–1557, https://doi.org/10.1016/j.progpolymsci.2011.04.004.

[39] A.J. Abd Aziz, N.A. Baharuddin, M.R. Somalu, A. Muchtar, Review of composite cathodes for intermediate-temperature solid oxide fuel cell applications, Ceram. Int. (2020), https://doi.org/10.1016/j.ceramint.2020.06.176.

[40] H. Li, Z. Lü, A highly stable cobalt-free LaBa0.5Sr0.5Fe2O6-δ oxide as a high performance cathode material for solid oxide fuel cells, Int. J. Hydrog. Energy (2020) 8, https://doi.org/10.1016/j.ijhydene.2020.05.115.

[41] S. Zhou, Y. Yang, H. Chen, Y. Ling, In situ exsolved Co–Fe nanoparticles on the Ruddlesden-Popper-type symmetric electrodes for intermediate temperature solid oxide fuel cells, Ceram. Int. 46 (11) (2020) 18331–18338, https://doi.org/10.1016/j.ceramint.2020.05.057.

Chapter 4 The material development and characterization of direct alcohol fuel cells **73**

[42] G. Brus, P.F. Raczkowski, M. Kishimoto, H. Iwai, J.S. Szmyd, A microstructure-oriented mathematical model of a direct internal reforming solid oxide fuel cell, Energy Convers. Manag. 213 (March) (2020), https://doi.org/10.1016/j.enconman.2020.112826, 112826.

[43] M. Zhou, X. Wang, Y. Zhang, Q. Qiu, M. Liu, J. Liu, Effect of counter diffusion of CO and CO_2 between carbon and anode on the performance of direct carbon solid oxide fuel cells, Solid State Ionics 343 (October) (2019), https://doi.org/10.1016/j.ssi.2019.115127, 115127.

[44] E. Antolini, Catalysts for direct ethanol fuel cells, J. Power Sources 170 (1) (2007) 1–12, https://doi.org/10.1016/j.jpowsour.2007.04.009.

[45] Z. Liu, L.M. Gan, L. Hong, W. Chen, J.Y. Lee, Carbon-supported Pt nanoparticles as catalysts for proton exchange membrane fuel cells, J. Power Sources 139 (1–2) (2005) 73–78, https://doi.org/10.1016/j.jpowsour.2004.07.012.

[46] H. Göksu, H. Burhan, S.D. Mustafov, F. Şen, Oxidation of benzyl alcohol compounds in the presence of carbon hybrid supported platinum nanoparticles (Pt@CHs) in oxygen atmosphere, Sci. Rep. 10 (1) (2020) 5439, https://doi.org/10.1038/s41598-020-62400-5.

[47] M.S. Nas, M.H. Calimli, H. Burhan, M. Yılmaz, S.D. Mustafov, F. Sen, Synthesis, characterization, kinetics and adsorption properties of Pt-Co@GO nano-adsorbent for methylene blue removal in the aquatic mediums using ultrasonic process systems, J. Mol. Liq. 296 (2019), https://doi.org/10.1016/j.molliq.2019.112100, 112100.

[48] Y.-X. Wang, H.-J. Zhou, P.-C. Sun, T.-H. Chen, Exceptional methanol electro-oxidation activity by bimetallic concave and dendritic Pt–Cu nanocrystals catalysts, J. Power Sources 245 (2014) 663–670, https://doi.org/10.1016/j.jpowsour.2013.07.015.

[49] H.O. Folkins, E. Miller, Preparation and properties of catalysts, Ind. Eng. Chem. 49 (2) (1957) 241–244, https://doi.org/10.1021/ie50566a037.

[50] R.L. Moss, Preparation and characterization of supported metal catalysts, in: Experimental Methods in Catalytic Research, Elsevier, 1976, pp. 43–94.

[51] L. An, R. Chen, Recent progress in alkaline direct ethylene glycol fuel cells for sustainable energy production, J. Power Sources 329 (2016) 484–501, https://doi.org/10.1016/j.jpowsour.2016.08.105.

[52] S. Basri, S.K. Kamarudin, W.R.W. Daud, Z. Yaakub, Nanocatalyst for direct methanol fuel cell (DMFC), Int. J. Hydrog. Energy 35 (15) (2010) 7957–7970, https://doi.org/10.1016/j.ijhydene.2010.05.111.

[53] S. Cavaliere, S. Subianto, I. Savych, D.J. Jones, J. Rozière, Electrospinning: designed architectures for energy conversion and storage devices, Energy Environ. Sci. 4 (12) (2011) 4761, https://doi.org/10.1039/c1ee02201f.

[54] N. Abdullah, S.K. Kamarudin, L.K. Shyuan, N.A. Karim, Fabrication and characterization of new composite TiO_2 carbon nanofiber anodic catalyst support for direct methanol fuel cell via electrospinning method, Nanoscale Res. Lett. 12 (1) (2017) 613, https://doi.org/10.1186/s11671-017-2379-z.

[55] M.S. Löffler, B. Groß, H. Natter, R. Hempelmann, T. Krajewski, J. Divisek, New preparation technique and characterisation of nanostructured catalysts for polymer membrane fuel cells, Scr. Mater. 44 (8–9) (2001) 2253–2257, https://doi.org/10.1016/S1359-6462(01)00760-6.

[56] H. Xie, et al., Coal pretreatment and Ag-infiltrated anode for high-performance hybrid direct coal fuel cell, Appl. Energy 260 (November) (2020), https://doi.org/10.1016/j.apenergy.2019.114197, 114197.

[57] J.F. Watts, J. Wolstenholme, An Introduction to Surface Analysis by XPS and AES, John Wiley & Sons, Ltd, Chichester, UK, 2003.

5

Fundamentals of alcohol fuel cells

Merve Akin[a,b], Vildan Erduran[a,b], Elif Esra Altuner[a], Cisil Timuralp[c], Iskender Isik[b], and Fatih Şen[a]

[a]*Şen Research Group, Department of Biochemistry, Dumlupinar University, Kütahya, Turkey.* [b]*Department of Materials Science and Engineering, Faculty of Engineering, Dumlupinar University, Kütahya, Turkey.* [c]*Mechanical Engineering Department, Eskisehir Osmangazi University, Eskişehir, Turkey*

1 Introduction

Increasing demand for portable power devices, increasing need for energy, and depletion of resources have led to the exploration into and discovery of new energy alternatives. Research and development on green energy resources such as fuel cells began to be carried out. In these studies, it is desired that the products are affordable, environmentally friendly, and high in energy density. Fuel cells are electrochemical cells, often known as an environmentally friendly and effective alternative for generating electricity with low toxic emissions, low operating temperature, and easy storage [1]. It is remarkable that alcohol fuel cells are more suitable and do not harm the environment, alcohol is easily stored and transportable, and alcohol shows high energy density. In many studies, methanol and ethanol fuel cells have been focused on, and it has been revealed that they have a very high potential for energy production. Many studies are conducted to reduce the carbon monoxide poisoning in alcohol fuel cells and to eliminate their disadvantages, such as the slow oxidation kinetics of the fuel. Pt catalysts are used to activate methanol and ethanol in alcohol fuel cells. However, the high cost of Pt causes disadvantages in fuel cells. Carbon-derived materials are promising as cathode catalysts due to their stability in acid and basic environments, high surface areas, good electrical conductivity, and low cost [2]. It is remarkable that alcohol fuel cells exhibit superior characteristics for portable applications.

Nanomaterials for Direct Alcohol Fuel Cells. https://doi.org/10.1016/B978-0-12-821713-9.00010-X
Copyright © 2021 Elsevier Inc. All rights reserved.

In this chapter, the components and history of fuel cells and the use of alcohols in fuel cells are examined. To understand the performance of alcohol fuel cells, thermodynamics laws are mentioned; methanol and ethanol fuel cells, which are the most used fuels in alcoholic fuel cells, are examined; and their applications were mentioned.

2 Fuel cells

Fossil fuels used today lead to many environmental problems. At the same time, the need for energy increases with increases in population. At this point, fuel cells attract attention thanks to features such as high efficiency, low emission, environmental friendliness, zero noise pollution, and increased energy security [3]. Fuel cells are devices that convert chemical energy directly into electrical energy. Fuel cells, which exhibit superior properties such as efficiency, lower amounts of noise, and being free of waste, are considered the best choice for electricity generation [4]. In short, fuel cells provide a cleaner, more efficient, and possibly more flexible chemical for electrical energy conversion [5]. The fuel cell is formed by the electrode, the cathode where oxidant reduction takes place, and the anode where the fuel oxidation takes place [6]. So long as fuel and oxidant are provided, fuel cells do not run out; unlike batteries, they do not need to be charged [7]. Fuel cell components consist of the membrane, gas diffusion layers, bipolar plates, and a catalyst. These components are important for fuel cell performance.

Membranes are layers sandwiched between anode and cathode. These layers are ion-permeable and do not allow electrons to pass through them.

Gas diffusion layers (GDL) are compressed between the catalyst and the gas flow channel and provide support to the catalyst [8]. The amount of compression in the gas diffusion layers plays an important role in fuel cell performance by affecting the fraction of porosity occupied by liquid water and contact resistance [9]. GDL is effective in diffusing reactants into the catalyst [10].

Bipolar plates transmit one side and the other negatively charged electrical current from cell to cell and distribute fuel gas and air homogeneously [11]. They also provide channels for reactive gases by removing reaction products [12]. BPs must be electrically and thermally conductive and chemically very resistant [13]. Plates made of graphite or composite material make up most of the fuel cell volume [14].

Catalysts are components in fuel cells that affect the reaction rate. They are very effective for fuel cell performance. For the development of fuel cells, catalysts should exhibit high performance, high durability, and low cost [15]. It is stated in many studies that Pt is used as a catalyst. However, due to its high cost, studies for either improved Pt-based catalyst design or the replacement of Pt altogether are desirable [16].

3 Fuel cell history

Fuel cells are devices that can convert chemical energy directly into electrical energy and generate electricity by providing oxidation-reduction reactions. The development of the fuel cell took place over many years. The decomposition of water into hydrogen and oxygen using electricity was discovered in 1800 by the British scientists Sir Anthony Carlisle and William Nicholson [17]. In 1839, Sir William Grove, the pioneer of fuel cell technology, revealed that electricity could be produced by reversing the electrolysis of water [18]. In 1889, Ludwig Mond and his assistant described gas-powered battery experiments using Mond gas derived from coal and called their system a *fuel cell* [19]. In 1896, the first direct carbon fuel cell (DCFC), with alkaline electrodes with a power of 1.5 kW and consisting of 100 separate cells, was developed by William W. Jacques [20]. In subsequent years, the alkaline fuel cell has been investigated.

In the late 1950s, the space race started and at the same time, interest in fuel cells increased and accelerated as a result of concerns about energy resources and the environment [21]. In 1990, the Jet Propulsion Laboratory at the National Aeronautics and Space Administration (NASA), together with the University of Southern California, developed the first methanol fuel cell [17]. In summary, a diagram showing the important milestones in fuel cell development is shown in Fig. 1. Studies for fuel cell development dating to the 1800s continue today.

4 Use of alcohols in fuel cells

Today, the need and demand for portable technological products are increasing. Alcohols are thought to be an alternative fuel in fuel cell development, and the use of alcohols is being investigated as well.

Alcohols are compounds bearing the hydroxyl group ($-OH$) attached to the carbon atom. Alcohols have a very high potential for use as energy fuel, especially in transportation [22]. Alcohols

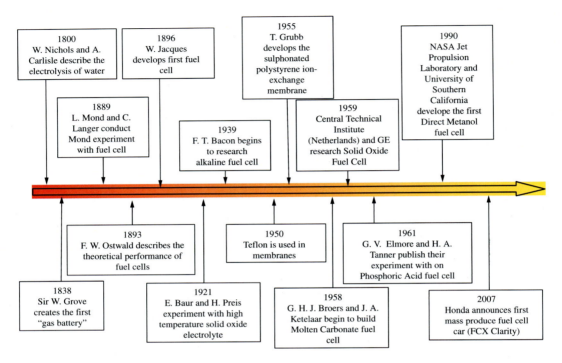

Fig. 1 The history of fuel cells [17].

are easily oxidizable hydrocarbons in fuel cell anodes and the net reaction of an alcohol fuel cell is as shown in Fig. 2 [23].

alcohol + oxygen → carbon dioxide + water + electric energy + heat.

Especially low molecular weight alcohols (methanol, ethanol, ethylene glycol, and glycerol) have high advantages over hydrogen [24]. The most preferred fuel in fuel cell technology is hydrogen [25]. However, the difficulties in the storage and distribution of hydrogen led to the investigation of different fuels. The most commonly used DAFC for fuel cells is methanol [26]. Light alcohols such as methanol and ethanol can electro-oxidize at temperatures below 90 °C and have a greater energy density than hydrogen [27]. Direct alcohol fuel cells for portable power applications attract attention because alcohol provides higher power density and comfortable workable properties compared to hydrogen-fed and polymer electrolyte membrane fuel cells [28]. The advantages offered by alcohol-based fuel cells attract attention as an alternative source. In addition to being easily storable, the energy density of these alcohols (5–8 kW h kg^{-1}) is close to the energy density of gasoline (12 kW h kg^{-1}) [24]. DAFCs are suitable for portable

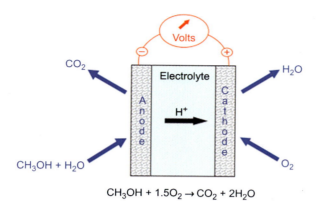

Fig. 2 Alcohol fuel cell general reaction scheme [23].

power applications (e.g., laptops, battery chargers, auxiliary power units) [27]. To ensure that alcohols can be used as fuel, it is necessary to design the engine to properly use the fuel properties or to add additives to ethanol and methanol [29].

5 Alcohol fuel cell thermodynamics

Thermodynamics consists of two parts: "Thermo" means heat and "dynamic" means power. In short, it is the science that deals with the conversion of heat to work [30]. It can also be defined as the study of change from a thermodynamic energy state to a different energy state. In fuel cells, there is a direct change from chemical energy to electrical energy. Thermodynamic equations can be used to understand the performance of a fuel cell [31]. For this, it is necessary first to understand the laws of thermodynamics. There are four laws of thermodynamics, which will be explained briefly next.

5.1 Zeroth law of thermodynamics

In the zeroth law of thermodynamics, it is stated that the heat exchange occurs as a result of the contact of two substances, and they have the same temperature after a certain period of contact. The zeroth law was formulated by R. H. Fowler half a century after the first and second laws of thermodynamics, but because it had to be formulated before these laws, it became known as the zeroth law [32].

5.2 The first law of thermodynamics

The first law of thermodynamics states that energy cannot be destroyed or created; rather, energy is conserved. At this point, there are concepts that we encounter in thermodynamics. System and environment concepts are important for thermodynamics. The system can be called the population of substances on which chemical events take place, have boundaries, and are examined. The part outside the system boundaries is called the "perimeter." There is no substance exchange between the environment and the system in closed systems, but the systems allow energy exchange. To explain the transfer of this energy with an equation, the change in its internal energy ($\triangle U$) is equal to the heat transferred to the system (Q) and the work done by the system (W), as shown in Eq. (1) [33]. In open systems, substance exchange occurs between the environment and the system. At this point, energy is transferred to the system from the environment, as shown in Eq. (2). Considering the equation of this law:

$$\Delta U = Q - W \tag{1}$$

$$\Delta U = Q + W \tag{2}$$

5.3 Second law of thermodynamics

The second law of thermodynamics says that part of the energy conversion will turn into a form other than the desired form. The second law assumes that it is impossible for a device in a cyclic state to completely convert heat to work [34]. The second law, in short, distinguishes between the change in the entropy of the universe and the irreversible and reversible processes [35]. The concept of entropy, the measure of the amount of disturbance in a system, is important for the second law. The higher the entropy, the less information is known about the system [36]. In entropy, which is an important thermodynamic property, the less energy that can be used, the higher the entropy [37]. The second law of thermodynamics explains the nature of chemical reactions and processes as processes, that the entropy change in the universe occurs spontaneously only if, by its processes, it is greater than or equal to zero [38].

The change in entropy in a closed system can be described by Eq. (3). It is equal to the inversely added heat divided by the absolute temperature of the system (3) [39]:

$$dS = \frac{\delta Qrev}{T} \tag{3}$$

Entropy is simply a measure of disorder. In the transformation from point A to point B, the new equation becomes

$$S_B - S_A = \int \frac{dQAB}{T} \geq 0 \tag{4}$$

The probabilities of Eq. (4) are as follows:

If $\Delta S = 0$, the process is reversible; if $\Delta S > 0$, the process cannot be undone; and if $\Delta S < 0$, the process is impossible [34].

5.4 Third law of thermodynamics

The third law of thermodynamics says that as the temperature of any system approaches 0 K, the entropy of the system assumes its lowest value when the system is in its lowest energy state [40]. In short, it can be said that the entropy change approaches zero as the system approaches absolute zero. The impossibility of a system to be absolute zero in the third law is the most important consequence, and the closer it is to absolute zero, the harder it is to lower the temperature of the system [41]. This difficulty makes it impossible for the system to be 0 K. However, in this law, the entropy of the system approaches a phase-independent constant because the absolute temperature tends to zero, and this constant is taken as zero [42].

5.5 Gibbs free energy

Gibbs free energy (G) is a measure of the maximum available work that can be derived from any system under conditions of pressure (P) and constant temperature (T) [43]. Gibbs free energy, also called "free enthalpy" can be expressed in G, which is the maximum work or maximum available energy, as shown in Eq. (5) [44]:

$$G = H - TS \qquad (5)$$

In this equation, G is the Gibbs energy, H is the enthalpy, T is the temperature in Kelvin, and S is the entropy value. As can be understood from Eq. (5), Gibbs energy is obtained by subtracting entropy and absolute temperature from enthalpy. Gibbs energy change is expressed by Eq. (6):

$$\Delta G = \Delta H - T\Delta S \qquad (6)$$

Energy efficiency in many systems can be calculated according to enthalpy, Gibbs energy. In fuel cells, the reactions taking place on the anode and cathode sides have enthalpy and Gibbs free energy values. The theoretical efficiency of a fuel cell is the ratio between energy output and energy input, and the ratio of energy input and output of the fuel cell can be expressed in Eq. (7) [45]:

$$\eta = \Delta G / \Delta H \qquad (7)$$

According to thermodynamics, the maximum chemical energy amount of the system, which can be converted into energy forms such as high-quality electricity, is also given by G (free energy) [46].

Table 1 shows Gibbs energy change, enthalpy, energy density, and theoretical voltage values. To give an example of Eq. (7) by using the data in Table 1, the $\Delta G^{0\ 0}$ value of ethanol is given as $-1325\,kJ/mol^{-1}$ and the value of ΔH^0 as $-1367\,kJ/mol$.

The theoretical energy efficiency of ethanol fuel cell can be calculated as follows:

$$\Delta G^0/\Delta H^0 = 1325/1367 = 0.97.$$

The theoretical efficiency in this case has been shown to be % 97.

The current efficiency of a fuel cell is equal to the ratio of the number of electrons (n_{actual}), during oxidation to the theoretical number of electrons (n_{teo}) in the complete oxidation reaction (8) [47].

$$\varepsilon f\,(\%) = (n_{actual}/n_{teo}) \times 100 \tag{8}$$

This equation can be exemplified using Table 1 as follows: the n_{actual} value for methanol is 4, and the n_{teo} value is 6.

As a result, the current efficiency for methanol will be found to be approximately % 67.

Table 1 Properties of different alcohols, conversion efficiencies, and comparison with hydrogen [47].

Property	Hydrogen	Methanol	Ethanol	Ethylene glycol
Formula	H_2	CH_3OH	C_2H_5OH	$(CH_2OH)_2$
$-\Delta G°$ (kJ/mol)	237	702	1325	1180
$-\Delta H$ (kJ/mol)	286	726	1367	1192
Energy density, LHV (kW h kg^{-1})	33	6.09	8.00	5.29
Energy density, LHV (kW h/L)	2.96×10^{-3}	4.80	6.32	5.80
$E°_{cell}$ (V)	1.23	1.21	1.14	1.22
n_{exp}	2	4	4	8
n_{theo}	2	6	12	10
Energy stored (Ah/kg/Ah/L)	26,802/2.40	3350/2653	2330/1841	3458/3855
ε_{cell}^{rev}(%)	83	97	97	99
ε_f(%)	100	67	33	88

We can obtain information about fuel cell efficiencies using thermodynamics. Many studies are carried out to increase fuel cell efficiency and to ensure complete oxidation. In one study, thermodynamic calculations of ethanol at different temperatures are shown in Fig. 3. In this calculation for steam reformation, it is observed that as the temperature increases, the ethanol conversion increases, and it is highest for complete oxidation to carbon dioxide (CO_2) after 150 °C [48].

6 Direct methanol fuel cells

Direct methanol fuel cells (DMFCs) are considered suitable for portable applications due to their high energy density, simple structure, easy storage, and low operating temperature [49]. One of the important differences that distinguish a DMFC from other fuel cells is that the fuel is liquid and can react directly at the cell electrode [50]. Although there are many advantages of DMFCs, some disadvantages of liquid methanol are found in fuel cells. In DMFCs, liquid methanol is fed to the anode and the water passage from the anode to the cathode causes water loss and cathode overflow [51]. Preventing this situation, which is one of the major problems of methanol fuel cells, will increase the performance of these cells and support the proliferation of application areas.

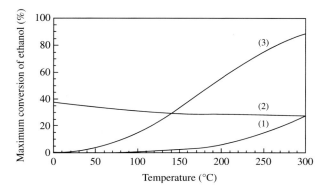

Fig. 3 Maximum conversion for ethanol oxidation at certain temperatures [48].
(1) $CH_3CH_2OH \rightarrow CH_3CHO + H_2$,
(2) $CH_3CH_2OH + H_2O \rightarrow CH_3COOH + 2H_2$,
(3) $CH_3CH_2OH + 3H_2O \rightarrow 2CO_2 + 6H_2$ [48].

6.1 DMFC working principles

Unlike other fuel cells, DMFCs do not require a separate production system [52]. They are devices that generate electrical energy using methanol oxygen. The anode fuel is dilute methanol in water, and CO_2 formation is observed as the methanol is consumed [23]. A DMFC consists of an anode, a cathode, and an electrolyte. As shown in Fig. 4, a proton, an electron, and CO_2 are formed by reacting with a methanol-water mixture at the anode. Protons pass through the electrolyte and reach the cathode. Electrons, on the other hand, reach the cathode by following a different route, and electricity generation take place. The reason that electrons take a different path is that the electrolyte is permeable only to ions. Electrons and protons reaching the cathode react with oxygen and release water. The water and carbon dioxide that is formed is discharged from the cell.

If the anode and cathode equations of the methanol fuel cell are examined directly, the DMFC consists of an anode (9), where methanol is electro-oxidized to CO_2 through the reaction; and a cathode, where oxygen is reduced to water or steam (10) [53]. In the general reaction, it is seen that as a result of the reaction of methanol with oxygen, CO_2 and water are released (11).

$$\text{Anode}: CH_3OH + H_2O \rightarrow CO_2 + 6H^+ + 6e^- \quad (9)$$

$$\text{Cathode}: 3/2O_2 + 6H^+ + 6e^- \rightarrow 3H_2O \quad (10)$$

$$\text{General response}: CH_3OH + 3/2O_2 \rightarrow CO_2 + 2H_2O \quad (11)$$

To increase the performance of DMFCs, there has been a focus on developing these systems by producing materials with electrocatalytic activity and stability [54]. Pt-based electrocatalysts are used

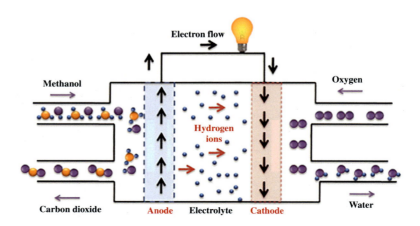

Fig. 4 A DMFC [52].

in many studies at this point, but different support materials are investigated due to the high cost of Pt and CO poisoning. It is believed that better results will be obtained with nanoparticle-supported electrocatalysts. In DMFC technology, as in other fuel cells, membrane and electrocatalysts affect performance. At this point, it is desired to overcome slow anode kinetics by developing new membranes in DMFCs and by developing new anode catalysts and methanol transition [55]. In DMFCs, the slow reaction of the anode at the cathode and methanol cross-reactions cause problems in energy conversion [56]. In these cells, it is important that methanol is adsorbed to the electrode surface [57].

6.2   Methanol fuel cell advantages

DMFCs offer expandable advantages, including smaller system sizes, weights, and fermentation of agricultural products [58]. Thus, methanol has a more affordable cost. Advantages of using dilute methanol in DMFCs include high energy density (such as 6000 Wh kg^{-1}) and longer life than lithium-ion batteries [59]. In addition, the fact that methanol is liquid makes it much easier to carry and store than hydrogen.

6.3   Methanol fuel cell disadvantages

One of the disadvantages in methanol fuel cells is the crossover phenomenon, in which methanol molecules diffuse through the membrane and are oxidized by the oxygen in the cathode, which reduces the fuel energy density [60]. Methanol oxidation kinetics are heavy, and methanol passes through the membrane, preventing oxygen reduction at the cathode [61]. Fuel loss remains a major challenge with methanol passing through the membrane [62]. This causes a decrease in fuel cell system efficiency as the crossed methanol is converted to CO_2 at the cathode. Due to the increase in gas diffusion limitation in DMFCs, as water accumulates on the cathode, performance degradation occurs [63]. Fig. 5 shows the performance of a DMFC that works stably for 16 h as an example of this situation. Studies have been carried out to look at these problems.

6.4   Some applications made to increase DMFC performance and reduce costs

Heinzel and Barragan revealed that there are five parameters that prevent methanol cross-over: membrane thickness, methanol concentration, pressure, operating temperature, and catalyst morphology [64].

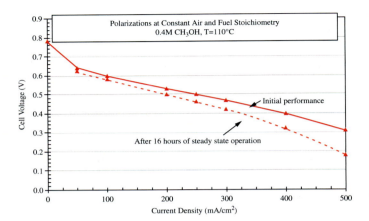

Fig. 5 Stable working DMFC performance for 16 h [63].

Studies have been carried out to reduce the disadvantage of methanol transition in DMFCs. In a study conducted, it has been observed that increasing the thickness of the cathode catalyst layer (CCL) increases fuel cell performance [65]. The use of a thicker CCL is bound to the oxidation of methanol in part of the catalyst layer, and it has been observed that the effects of mixed potentials decrease [66].

The use of catalysts in DMFCs affects performance, as in other fuel cells. Increased catalyst loading has been shown to improve performance, and catalysts made of platinum-ruthenium materials are actively used today [67]. Because the applications of these materials increase the cost of the fuel cell, studies are carried out to reduce the cost by supporting them with different materials. Abdullah et al. reported the synthesis of nanofibers accumulating titanium dioxide–carbon nanofibers on platinum-ruthenium (PtRu) catalysts for DMFC application in a catalyst study [68]. As a result of this study, it has been shown that the catalyst in DMFC applications is promising [68].

7 Direct ethanol fuel cell

Ethanol (ethyl alcohol) is alcohol with the chemical formula C_2H_5OH. It is low cost and can be easily produced by the fermentation of agricultural products. Liquid fuel ethanol is less toxic than hydrocarbons and methanol. Pure gasoline has two-thirds of its energy density, as ethanol has a short-chain carbon compound [69]. It is believed that ethanol fuel cell hydrogen will solve the storage and transportation problems. Ethanol as a fuel will not change the balance of CO_2 in the atmosphere as fossil fuels do, with the advantage of being nontoxic [70]. The slow oxidation reaction of ethanol in ethanol fuel cells prevents these cells from commercializing [71].

7.1 Ethanol fuel cell working principles

The reactions and working principles that occur in an ethanol fuel cell are shown in Fig. 6. Ethanol fuel cells consist of anode, cathode, and membrane. The DEFC converts Gibbs ethanol combustion energy directly into electricity without a fuel processor [73]. The ethanol-water mixture circulates through the anode, producing 12 protons and 12 electrons simultaneously per CO_2 and ethanol molecule [72]. Protons form to pass through the membrane to reach the cathode, and here it reacts with oxygen and releases water. Because the membrane does not allow the passage of electrons, they reach the cathode by following a different path. As a result, electricity is generated.

The reaction equations for ethanol oxidation in fuel cells are as follows: Because ethanol contains two bonded carbon atoms, it must break the C–C bond to generate electrical energy (12) [70]. It is seen in Eq. (12) that 12 electrons and 12 protons are produced at the anode. Eq. (13) shows that protons reacting with oxygen at the cathode expose water. In the general reaction, it is observed that carbon dioxide and water are released (14).

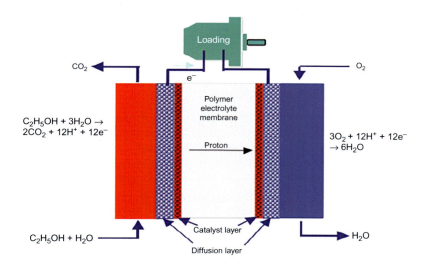

Fig. 6 Ethanol fuel cell working principles [72].

$$\text{Anode}: CH_3CH_2OH + 3H_2O \rightarrow 2CO_2 + 12H^+ + 12e^- \qquad (12)$$

$$\text{Cathode}: 3O_2 + 12H^+ + 12e^- \rightarrow 6H_2O \qquad (13)$$

$$\text{General reaction}: CH_3CH_2OH + 3O_2 \rightarrow 2CO_2 + 3H_2O \qquad (14)$$

Song et al. showed that due to the larger molecular size of ethanol, the membrane permeability rate is lower than for methanol, and that there is a slower oxidation process at the anode due to its low electrochemical activity [74].

7.2 Ethanol fuel cell advantages

Ethanol fuel is nontoxic, renewable, and has a high power density, which are all excellent characteristics [75]. As a result of the use of ethanol as a fuel, only the release of CO_2 and water does not harm the environment. It is affordable and easy to carry. Ethanol can be easily produced by fermentation. Ethanol (8.0 kWh kg^{-1}) has greater specific energy than methanol (6.1 kWh/kg) [76].

7.3 Ethanol fuel cell disadvantages

For the complete oxidation of ethanol in ethanol fuel cells, it is necessary to break the C–C bond at low temperatures, which requires an effective nanoanode catalyst [77]. A carbon-supported Pt–Sn electrocatalyst is known to be the most efficient for the oxidation of ethanol in ethanol fuel cells [78]. Ethanol is more difficult to oxidize than methanol because breaking the C–C bonds in ethanol is not easy. At this point, it is important to increase the electroactivity of methanol, and full oxidation to carbon dioxide is a difficult challenge as well [79].

7.4 Applications made to improve the performance of ethanol fuel cells

Problems such as the slowness of ethanol reaction kinetics at low temperatures, catalyst poisoning at the cathode in the fuel passage, and cathode overflow preventing oxygen transport reduce the fuel cell performance of direct ethanol fuel cells [80].

The ethanol concentration in the anode catalyst layer affects the ethanol transition rate [81]. Andreadis et al. found that ethanol transition rate is influenced by ethanol feed concentration, current density, and cell temperature [82].

If it can efficiently carry out 12-electron oxidation of ethanol to water and CO_2, DEFC offers a high-density energy source [83]. At this point, it has been found that composite nanoparticle catalysts containing Pt, tin oxide, or Pt indium tin oxide allow partial transformation [83].

Dutta et al. addressed the challenge of designing an economically viable electrocatalyst for ethanol oxidation reaction in DEFCs and the controlled synthesis of the non-Pt-bonded Au nanostructure as a catalyst group [84]. As a result of this study, it has been observed that PdAu nanocatalyst has high activity in an ethanol oxidation reaction, and it is an effective catalyst in a high-pH environment [84].

For the first time in the world, a vehicle using direct ethanol fuel cells was presented by Shell's University of Applied Sciences in France [85]. At this point, it is understood that DEFC systems are effective and important for portable applications.

8 Conclusion

It is seen that alcohol fuel cells are easy to store, portable, cost-effective, and have a high energy density, which is very intriguing for their use in fuel cells. Many studies are being done about membranes, catalysts, and the determination of suitable operating temperatures for alcohol fuel cells. It is understood that more cost-effective electrocatalysts must be developed for the electro-oxidation of ethanol and methanol. When appropriate conditions are provided, the potential of using alcohol fuel cells is very high, especially in portable electronic devices.

References

[1] Y. Yıldız, S. Kuzu, B. Sen, A. Savk, S. Akocak, F. Şen, Different ligand based monodispersed Pt nanoparticles decorated with rGO as highly active and reusable catalysts for the methanol oxidation, Int. J. Hydrog. Energy 42 (18) (2017) 13061–13069, https://doi.org/10.1016/j.ijhydene.2017.03.230.

[2] S. Eris, Z. Daşdelen, F. Sen, Enhanced electrocatalytic activity and stability of monodisperse Pt nanocomposites for direct methanol fuel cells, J. Colloid Interface Sci. 513 (2018) 767–773, https://doi.org/10.1016/j.jcis.2017.11.085.

[3] S.M. Haile, Fuel cell materials and components, Acta Mater. 51 (19) (2003) 5981–6000, https://doi.org/10.1016/j.actamat.2003.08.004.

[4] B. Çelik, S. Kuzu, E. Erken, H. Sert, Y. Koşkun, F. Şen, Nearly monodisperse carbon nanotube furnished nanocatalysts as highly efficient and reusable catalyst for dehydrocoupling of DMAB and C1 to C3 alcohol oxidation, Int. J. Hydrog. Energy 41 (4) (2016) 3093–3101, https://doi.org/10.1016/j.ijhydene.2015.12.138.

[5] O.Z. Sharaf, M.F. Orhan, An overview of fuel cell technology: fundamentals and applications, Renew. Sust. Energ. Rev. 32 (2014) 810–853, https://doi.org/10.1016/j.rser.2014.01.012.

90 Chapter 5 Fundamentals of alcohol fuel cells

[6] A. Coralli, B.J.M. Sarruf, P.E.V. De Miranda, L. Osmieri, S. Specchia, N.Q. Minh, Fuel cells, in: Science and Engineering of Hydrogen-Based Energy Technologies: Hydrogen Production and Practical Applications in Energy Generation, Elsevier, 2018, pp. 39–122.

[7] M.C. Williams, Fuel cells, in: Fuel Cells: Technologies for Fuel Processing, Elsevier, 2011, pp. 11–27.

[8] S. Park, J.-W. Lee, B.N. Popov, A review of gas diffusion layer in PEM fuel cells: materials and designs, Int. J. Hydrog. Energy 37 (7) (2012) 5850–5865, https://doi.org/10.1016/j.ijhydene.2011.12.148.

[9] J. Ge, A. Higier, H. Liu, Effect of gas diffusion layer compression on PEM fuel cell performance, J. Power Sources 159 (2) (2006) 922–927, https://doi.org/10.1016/j.jpowsour.2005.11.069.

[10] B.G. Pollet, A.A. Franco, H. Su, H. Liang, S. Pasupathi, Proton exchange membrane fuel cells, in: Compendium of Hydrogen Energy, Elsevier, 2016, pp. 3–56.

[11] A. Hermann, T. Chaudhuri, P. Spagnol, Bipolar plates for PEM fuel cells: a review, Int. J. Hydrog. Energy 30 (12) (2005) 1297–1302, https://doi.org/10.1016/j.ijhydene.2005.04.016.

[12] H. Tawfik, Y. Hung, D. Mahajan, Bipolar plate durability and challenges, in: Polymer Electrolyte Fuel Cell Degradation, Elsevier Inc., 2012, pp. 249–291.

[13] S. Pasupathi, J.C. Calderon Gomez, H. Su, H. Reddy, P. Bujlo, C. Sita, HT-PEMFC modeling and design, in: Recent Advances in High-Temperature PEM Fuel Cells, Elsevier, 2016, pp. 32–54.

[14] B. Bhattacharyya, Microdevices fabrication for microelectromechanical systems and other microengineering applications, in: Electrochemical Micromachining for Nanofabrication, MEMS and Nanotechnology, Elsevier, 2015, pp. 185–204.

[15] E. Carcadea, M. Varlam, A. Marinoiu, M. Raceanu, M.S. Ismail, D.B. Ingham, Influence of catalyst structure on PEM fuel cell performance—a numerical investigation, Int. J. Hydrog. Energy 44 (25) (2019) 12829–12841, https://doi.org/10.1016/j.ijhydene.2018.12.155.

[16] S. Holdcroft, Fuel cell catalyst layers: a polymer science perspective, Chem. Mater. 26 (1) (2013) 381–393, https://doi.org/10.1021/cm401445h.

[17] J.M. Andújar, F. Segura, Fuel cells: history and updating. A walk along two centuries, Renew. Sust. Energ. Rev. 13 (9) (2009) 2309–2322, https://doi.org/10.1016/j.rser.2009.03.015.

[18] S.S. Siwal, S. Thakur, Q.B. Zhang, V.K. Thakur, Electrocatalysts for electrooxidation of direct alcohol fuel cell: chemistry and applications, Mater. Today Chem. 14 (2019) 100182, https://doi.org/10.1016/j.mtchem.2019.06.004.

[19] E.I. Ortiz-Rivera, A.L. Reyes-Hernandez, R.A. Febo, Understanding the history of fuel cells, in: 2007 IEEE Conference on the History of Electric Power, HEP 2007, 2007, pp. 117–122, https://doi.org/10.1109/HEP.2007.4510259.

[20] D. Cao, Y. Sun, G. Wang, Direct carbon fuel cell: fundamentals and recent developments, J. Power Sources 167 (2) (2007) 250–257, https://doi.org/10.1016/j.jpowsour.2007.02.034.

[21] M.L. Perry, T.F. Fuller, A historical perspective of fuel cell technology in the 20th century, J. Electrochem. Soc. 149 (7) (2002) S59, https://doi.org/10.1149/1.1488651.

[22] G.A. Mills, E.E. Ecklund, Alcohols as components of transportation fuels, Annu. Rev. Energy 12 (1) (1987) 47–80, https://doi.org/10.1146/annurev.eg.12.110187.000403.

[23] D. Gervasio, Fuel cells—direct alcohol fuel cells | new materials, in: Encyclopedia of Electrochemical Power Sources, Elsevier, 2009, pp. 420–427.

[24] K.I. Ozoemena, Nanostructured platinum-free electrocatalysts in alkaline direct alcohol fuel cells: catalyst design, principles and applications, RSC Adv. 6 (92) (2016) 89523–89550, https://doi.org/10.1039/c6ra15057h.

Chapter 5 Fundamentals of alcohol fuel cells **91**

[25] A. Brouzgou, F. Tzorbatzoglou, P. Tsiakaras, Direct alcohol fuel cells: challenges and future trends, in: Proceedings of the 2011 3rd International Youth Conference on Energetics (IYCE), IEEE Conference Publication, 2011, pp. 1–6. (Online). Available from: https://ieeexplore.ieee.org/document/6028346/metrics#metrics. (Accessed 19 January 2021).

[26] A. Santasalo-Aarnio, S. Tuomi, K. Jalkanen, K. Kontturi, T. Kallio, The correlation of electrochemical and fuel cell results for alcohol oxidation in acidic and alkaline media, Electrochim. Acta 87 (2013) 730–738, https://doi.org/10.1016/j.electacta.2012.09.100.

[27] J. Sánchez-Monreal, M. Vera, P.A. García-Salaberri, Fundamentals of electrochemistry with application to direct alcohol fuel cell modeling, in: Proton Exchange Membrane Fuel Cell, InTech, 2018.

[28] D.M. Fadzillah, S.K. Kamarudin, M.A. Zainoodin, M.S. Masdar, Critical challenges in the system development of direct alcohol fuel cells as portable power supplies: an overview, Int. J. Hydrog. Energy 44 (5) (2019) 3031–3054, https://doi.org/10.1016/j.ijhydene.2018.11.089.

[29] V.R. Surisetty, A.K. Dalai, J. Kozinski, Alcohols as alternative fuels: an overview, Appl. Catal. A Gen. 404 (1–2) (2011) 1–11, https://doi.org/10.1016/j.apcata.2011.07.021.

[30] M. Ghassemi, A. Shahidian, Thermodynamics, in: Nano and Bio Heat Transfer and Fluid Flow, Elsevier, 2017, pp. 9–30.

[31] L. Khotseng, Fuel cell thermodynamics, in: Thermodynamics and Energy Engineering, IntechOpen, 2019.

[32] I. Dincer, Exergy, in: Comprehensive Energy Systems, vol. 1–5, Elsevier Inc., 2018, pp. 212–264.

[33] R. O'Hayre, S.W. Cha, W. Colella, F. Prinz, Fuel Cell Fundamentals, Wiley, 2016. https://books.google.com.tr/books/about/Fuel_Cell_Fundamentals.html?id=O2JYCwAAQBAJ&printsec=frontcover&source=kp_read_button&redir_esc=y#v=onepage&q&f=false. (Accessed 24 January 2021).

[34] A.C. Dimian, C.S. Bildea, A.A. Kiss, Generalised computational methods in thermodynamics, in: Computer Aided Chemical Engineering, vol. 35, Elsevier B.V., 2014, pp. 157–200.

[35] K.S. Schmitz, Gibbs free energy, work, and equilibrium, in: Physical Chemistry, Elsevier, 2017, pp. 99–157.

[36] D.C. Marinescu, G.M. Marinescu, Classical and quantum information theory, in: Classical and Quantum Information, Elsevier, 2012, pp. 221–344.

[37] A.M.Y. Razak, Thermodynamics of gas turbine cycles, in: Industrial Gas Turbines, Elsevier, 2007, pp. 13–59.

[38] G. Hanrahan, Introduction to environmental chemistry, in: Key Concepts in Environmental Chemistry, Elsevier, 2012, pp. 3–30.

[39] I. Tosun, Review of the first and second laws of thermodynamics, in: The Thermodynamics of Phase and Reaction Equilibria, Elsevier, 2013, pp. 1–12.

[40] J.M. Honig, Fundamentals, in: Thermodynamics, Elsevier, 2007, pp. 1–110.

[41] J. Ge, H. Liu, Experimental studies of a direct methanol fuel cell, J. Power Sources 142 (1–2) (2005) 56–69, https://doi.org/10.1016/j.jpowsour.2004.11.022.

[42] R.F. Sekerka, Third law of thermodynamics, in: Thermal Physics, Elsevier, 2015, pp. 49–52.

[43] C.J. Cleveland, C. Morris, Dictionary of Energy, Elsevier, 2015, pp. 247–273.

[44] R.A. Cottis, L.L. Shreir, G.T. Burstein, Chemical thermodynamics, in: Shreir's Corrosion, Elsevier, 2010, pp. 1–12.

[45] F. Barbir, PEM Fuel Cells: Theory and Practice, vol. 2, Academic Press, Elsevier, Waltham, San Diego, 2012.

[46] Fuel cells, in: Hydrogen and Fuel Cells, Elsevier, 2012, pp. 95–200.

92 Chapter 5 Fundamentals of alcohol fuel cells

[47] S.P.S. Badwal, S. Giddey, A. Kulkarni, J. Goel, S. Basu, Direct ethanol fuel cells for transport and stationary applications—a comprehensive review, Appl. Energy 145 (2015) 80–103, https://doi.org/10.1016/j.apenergy.2015.02.002.

[48] S. Song, Y. Wang, P. Shen, Thermodynamic and kinetic considerations for ethanol electrooxidation in direct ethanol fuel cells, Chin. J. Catal. 28 (9) (2007) 752–754, https://doi.org/10.1016/S1872-2067(07)60063-1.

[49] N.K. Shrivastava, T.A.L. Harris, Direct methanol fuel cells, in: Encyclopedia of Sustainable Technologies, Elsevier, 2017, pp. 343–357.

[50] A.S. Arico, V. Baglio, V. Antonucci, Direct Methanol Fuel Cells, Nova Science Publishers, Inc., 2010.

[51] C. Xu, T.S. Zhao, W.W. Yang, Modeling of water transport through the membrane electrode assembly for direct methanol fuel cells, J. Power Sources 178 (1) (2008) 291–308, https://doi.org/10.1016/j.jpowsour.2007.11.098.

[52] O.K. Simya, P. Radhakrishnan, A. Ashok, K. Kavitha, R. Althaf, Engineered nanomaterials for energy applications, in: Handbook of Nanomaterials for Industrial Applications, Elsevier, 2018, pp. 751–767.

[53] S.K. Kamarudin, F. Achmad, W.R.W. Daud, Overview on the application of direct methanol fuel cell (DMFC) for portable electronic devices, Int. J. Hydrog. Energy 34 (16) (2009) 6902–6916, https://doi.org/10.1016/j.ijhydene.2009.06.013.

[54] Ö. Karatepe, Y. Yildiz, H. Pamuk, S. Eris, Z. Dasdelen, F. Sen, Enhanced electrocatalytic activity and durability of highly monodisperse Pt@PPy-PANI nanocomposites as a novel catalyst for the electro-oxidation of methanol, RSC Adv. 6 (56) (2016) 50851–50857, https://doi.org/10.1039/c6ra06210e.

[55] H. Liu, C. Song, L. Zhang, J. Zhang, H. Wang, D.P. Wilkinson, A review of anode catalysis in the direct methanol fuel cell, J. Power Sources 155 (2) (2006) 95–110, https://doi.org/10.1016/j.jpowsour.2006.01.030.

[56] Y. Qiao, C.M. Li, Nanostructured catalysts in fuel cells, J. Mater. Chem. (2010), https://doi.org/10.1039/c0jm02871a.

[57] M.D. Esrafili, R. Nurazar, Potential of C-doped boron nitride fullerene as a catalyst for methanol dehydrogenation, Comput. Mater. Sci. 92 (2014) 172–177, https://doi.org/10.1016/j.commatsci.2014.05.043.

[58] R. Dillon, S. Srinivasan, A.S. Aricò, V. Antonucci, International activities in DMFC R&D: status of technologies and potential applications, J. Power Sources 127 (1–2) (2004) 112–126, https://doi.org/10.1016/j.jpowsour.2003.09.032.

[59] V. Parthiban, S. Akula, A.K. Sahu, Surfactant templated nanoporous carbon-Nafion hybrid membranes for direct methanol fuel cells with reduced methanol crossover, J. Membr. Sci. 541 (2017) 127–136, https://doi.org/10.1016/j.memsci.2017.06.081.

[60] M. Broussely, G. Archdale, Li-ion batteries and portable power source prospects for the next 5–10 years, J. Power Sources 136 (2 Special Issue) (2004) 386–394, https://doi.org/10.1016/j.jpowsour.2004.03.031.

[61] B. Gurau, E.S. Smotkin, Methanol crossover in direct methanol fuel cells: a link between power and energy density, J. Power Sources 112 (2) (2002) 339–352, https://doi.org/10.1016/S0378-7753(02)00445-7.

[62] J.Y. Park, J.H. Lee, S.K. Kang, J.H. Sauk, I. Song, Mass balance research for high electrochemical performance direct methanol fuel cells with reduced methanol crossover at various operating conditions, J. Power Sources 178 (1) (2008) 181–187, https://doi.org/10.1016/j.jpowsour.2007.12.021.

[63] S.D. Knights, K.M. Colbow, J. St-Pierre, D.P. Wilkinson, Aging mechanisms and lifetime of PEFC and DMFC, J. Power Sources 127 (1–2) (2004) 127–134, https://doi.org/10.1016/j.jpowsour.2003.09.033.

Chapter 5 Fundamentals of alcohol fuel cells **93**

[64] H. Junoh, et al., Performance of polymer electrolyte membrane for direct methanol fuel cell application: perspective on morphological structure, Membranes 10 (3) (2020) 34, https://doi.org/10.3390/membranes10030034.

[65] S. Matar, J. Ge, H. Liu, Modeling the cathode catalyst layer of a direct methanol fuel cell, J. Power Sources 243 (2013) 195–202, https://doi.org/10.1016/j.jpowsour.2013.05.122.

[66] S. Matar, H. Liu, Effect of cathode catalyst layer thickness on methanol crossover in a DMFC, Electrochim. Acta 56 (1) (2010) 600–606, https://doi.org/10.1016/j.electacta.2010.09.001.

[67] A. Heinzel, V.M. Barragán, Review of the state-of-the-art of the methanol crossover in direct methanol fuel cells, J. Power Sources 84 (1) (1999) 70–74, https://doi.org/10.1016/S0378-7753(99)00302-X.

[68] N. Abdullah, S.K. Kamarudin, L.K. Shyuan, Novel anodic catalyst support for direct methanol fuel cell: characterizations and single-cell performances, Nanoscale Res. Lett. 13 (1) (2018) 1–13, https://doi.org/10.1186/s11671-018-2498-1.

[69] M.A.F. Akhairi, S.K. Kamarudin, Catalysts in direct ethanol fuel cell (DEFC): an overview, Int. J. Hydrog. Energy 41 (7) (2016) 4214–4228, https://doi.org/10.1016/j.ijhydene.2015.12.145.

[70] S. Rousseau, C. Coutanceau, C. Lamy, J.-M. Léger, Direct ethanol fuel cell (DEFC): electrical performances and reaction products distribution under operating conditions with different platinum-based anodes, J. Power Sources 158 (1) (2006) 18–24, https://doi.org/10.1016/j.jpowsour.2005.08.027.

[71] S. Abdullah, S.K. Kamarudin, U.A. Hasran, M.S. Masdar, W.R.W. Daud, Development of a conceptual design model of a direct ethanol fuel cell (DEFC), Int. J. Hydrog. Energy 40 (35) (2015) 11943–11948, https://doi.org/10.1016/j.ijhydene.2015.06.070.

[72] L. Jiang, G. Sun, Fuel cells—direct alcohol fuel cells | direct ethanol fuel cells, in: Encyclopedia of Electrochemical Power Sources, Elsevier, 2009, pp. 390–401.

[73] C. Lamy, C. Coutanceau, J.M. Leger, The direct ethanol fuel cell: a challenge to convert bioethanol cleanly into electric energy, in: Catalysis for Sustainable Energy Production, Wiley-VCH Verlag GmbH & Co. KGaA, Weinheim, Germany, 2009, pp. 1–46.

[74] B.C. Ong, S.K. Kamarudin, S. Basri, Direct liquid fuel cells: a review, Int. J. Hydrog. Energy 42 (15) (2017) 10142–10157, https://doi.org/10.1016/j.ijhydene.2017.01.117.

[75] S. Song, P. Tsiakaras, Recent progress in direct ethanol proton exchange membrane fuel cells (DE-PEMFCs), Appl. Catal. B Environ. 63 (3–4) (2006) 187–193, https://doi.org/10.1016/j.apcatb.2005.09.018.

[76] P. Saisirirat, B. Joommanee, Study on the performance of the micro direct ethanol fuel cell (Micro-DEFC) for applying with the portable electronic devices, Energy Procedia 138 (2017) 187–192, https://doi.org/10.1016/j.egypro.2017.10.148.

[77] S. Şen, F. Şen, G. Gökağaç, Preparation and characterization of nano-sized Pt–Ru/C catalysts and their superior catalytic activities for methanol and ethanol oxidation, Phys. Chem. Chem. Phys. 13 (15) (2011) 6784, https://doi.org/10.1039/c1cp20064j.

[78] F. Colmati, M.M. Magalhães, R. Sousa, E.G. Ciapina, E.R. Gonzalez, Direct ethanol fuel cells: the influence of structural and electronic effects on Pt–Sn/C electrocatalysts, Int. J. Hydrog. Energy 44 (54) (2019) 28812–28820, https://doi.org/10.1016/j.ijhydene.2019.09.056.

[79] Z.B. Wang, G.P. Yin, J. Zhang, Y.C. Sun, P.F. Shi, Investigation of ethanol electrooxidation on a Pt-Ru-Ni/C catalyst for a direct ethanol fuel cell, J. Power Sources 160 (1) (2006) 37–43, https://doi.org/10.1016/j.jpowsour.2006.01.021.

[80] S. Abdullah, S.K. Kamarudin, U.A. Hasran, M.S. Masdar, W.R.W. Daud, Modeling and simulation of a direct ethanol fuel cell: an overview, J. Power Sources 262 (2014) 401–406, https://doi.org/10.1016/j.jpowsour.2014.03.105.

[81] M.Z.F. Kamarudin, S.K. Kamarudin, M.S. Masdar, W.R.W. Daud, Review: direct ethanol fuel cells, Int. J. Hydrog. Energy 38 (22) (2013) 9438–9453, https://doi.org/10.1016/j.ijhydene.2012.07.059.

[82] G. Andreadis, S. Song, P. Tsiakaras, Direct ethanol fuel cell anode simulation model, J. Power Sources 157 (2) (2006) 657–665, https://doi.org/10.1016/j.jpowsour.2005.12.040.

[83] J. Mann, N. Yao, A.B. Bocarsly, Characterization and analysis of new catalysts for a direct ethanol fuel cell, Langmuir 22 (25) (2006) 10432–10436, https://doi.org/10.1021/la061200c.

[84] A. Dutta, A. Mondal, P. Broekmann, J. Datta, Optimal level of Au nanoparticles on Pd nanostructures providing remarkable electro-catalysis in direct ethanol fuel cell, J. Power Sources 361 (2017) 276–284, https://doi.org/10.1016/j.jpowsour.2017.06.063.

[85] A. Kirubakaran, S. Jain, R.K. Nema, A review on fuel cell technologies and power electronic interface, Renew. Sust. Energ. Rev. 13 (9) (2009) 2430–2440, https://doi.org/10.1016/j.rser.2009.04.004.

6

The electrocatalysts with pH of the electrolyte for the complete pathways of the oxidation reactions

Muhammed Bekmezci[a,b], Ramazan Bayat[a,b], Merve Akin[a,b], Hakan Burhan[a], Iskender Isik[b], and Fatih Şen[a]

[a]Şen Research Group, Department of Biochemistry, Dumlupinar University, Kütahya, Turkey. [b]Department of Materials Science and Engineering, Faculty of Engineering, Dumlupinar University, Kütahya, Turkey

1 Introduction

Humanity survives on earth with energy. All kinds of economic and financial activities are in this systemic order. It is very important to control the energy in the most appropriate way and apply it to ideal systems. It is now more urgent to popularize renewable energy systems that cause the least harm to the environment.

A nation's energy needs are directly proportional to its development level [1]. The energy requirements of high technological systems are also increasing. Much research is being done to fulfill these needs. There is a trend toward energy types that reduce the damage to nature and provide convenience in storage applications. Energy storage is very important, especially for smartphones and automobiles working with energy storage systems. The energy sector affects the relations of various sectors and nations with each other. However, the focus on renewable energy has reached a very valuable point [2, 3].

Society, driven by the depletion of fossil fuels and the threat of global warming, is considering the use of renewably produced hydrogen as an alternative clean fuel for transportation [4, 5]. Direct electrochemical separation of water is a nonpolluting way of producing pure hydrogen [5]. Many electrocatalyst systems have been developed. These systems are detailed next.

Nanomaterials for Direct Alcohol Fuel Cells. https://doi.org/10.1016/B978-0-12-821713-9.00020-2
Copyright © 2021 Elsevier Inc. All rights reserved.

2 General considerations about fuel cells

It arose from the need for sustainable energy systems of direct liquid fuel cells (DLFCs) [6]. Electrode reactions have slow kinetics. They need precious metals on the cathode surface. Materials such as Nafion are used as membranes in these systems. Due to these situations, the cost is very high in DLFCs [6]. Besides, many studies are carried out to improve the electrode reaction kinetics. In terms of this approach, membrane development studies are also carried out to work in alkaline environments [6, 7].

DLFCs are classified into several subgroups, such as direct ethanol fuel cells (DEFCs), which use ethanol as an anode fuel, and oxygen in the cathode. These gadgets are suitable for compact use, such as electronic chargers. It functions in environmental temperatures [8, 9]. Water is not only a solvent during reactions, but it usually decomposes into the species H^{+*} and OH^{-*}, and this interacts more through proton-coupled electron transfer with other reactive intermediates [10–13].

However, the two main types, proton exchange membrane (PEM) and liquid alkaline systems, have some limitations. For example, PEM systems requires costly platinum group metal (PGM) catalysts [14]. The cost is a very important factor in industrial work [15, 16]. There is a need for low-cost transition metal as catalysts. A highly efficient and low cost alkaline polymer system design based on anion exchange membrane (AEM) or using a membrane-electrode assembly (MEA) may be required [14, 17–20]. Fig. 1 indicates the reaction mechanism of ethanol electro-oxidation in Pt electrodes. As shown in this figure, some of the intermediate species such as adsorbed species (CO, CHx, CH_3CHOH), etc. can be formed during this electrooxidation process. This system performance depends on the effectiveness of PGM catalysts. [14, 15, 21–25]. Fig. 1 shows the detailed electro-oxidation steps.

2.1 Ethanol oxidation reactions

Activity, selectivity, and cost of the catalyst are very important parameters for ethanol oxidation reactions. The extensive studies on ethanol oxidation reactions (EORs) have been performed depending upon the basis of design rules for high efficiency of fuel cells. Pioneering work for the EOR mechanism dates to the 1950s [27].

Different studies of ethanol electro-oxidation have been performed primarily by defining the adsorbed electrode intermediates. However, there have been attempts to understand

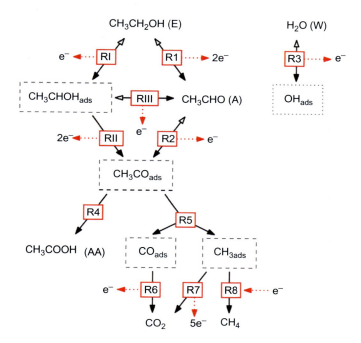

Fig. 1 Schematic of the reaction mechanism of ethanol electro-oxidation in Pt electrodes [26].

reaction concepts, using various techniques such as in situ Fourier transform infrared spectroscopy (FTIR), different electrochemical mass spectrometry (DEMS), etc. A parallel reaction scheme is used to summarize the spherical oxidation process of ethanol in anion-conducting and proton-conducting as shown in Fig. 2 [8, 28, 29]. Generally, the working principle of a DEFC is schematized in Fig. 2. As shown in this figure, DEFC containing anion-conducting (Fig. 2A) and proton-conducting (Fig. 2B) polymer electrolytes are given in detail and total reactions are shown both an alkaline and acidic media. Total reaction are same for both process.

Besides, serious progress has been made in recent years in understanding the ethanol oxidation mechanism as shown in this figure. However, there are still some restrictions. For example, there is some debate as to whether different intermediate species such as acetic acid and/or acetaldehyde are formed in one step or not. These parallel reactions with these intermediate species depending upon pH cause the fuel cell potential to drop significantly [28, 31].

For example, in the first one, reactions take place in an alkaline environment, depending on the critical catalyst choices. Accordingly, we can describe some electrocatalyst systems as follows.

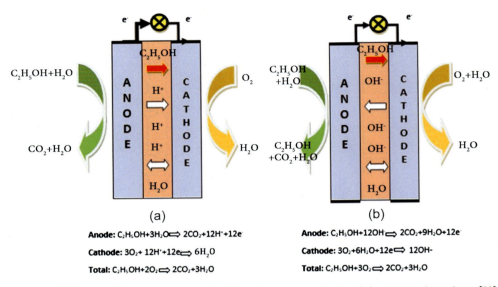

Fig. 2 Study of a DEFC containing anion-conducting (A) and proton-conducting (B) polymer electrolytes [30].

Alkaline electrolyzers (AELs) and proton exchange membrane electrolyzers (PEMELs) are very popular for hydrogen production. AELs are widely used commercially [16, 32], in industry. PEMELs are still in development; for now, it has limited application areas. Compared to AELs, PEMELs have a more compact structure. It is more possible to reach higher current ranges. Recently, however, an alternative technology to PEMELs, the hydroxide exchange membrane electrolyzer (HEMEL), and platinum group metal (PGM)–free electrocatalysts have been developed. In addition, it has become possible to use cheaper membranes, ionomers, and building materials. The other membranes also attract more attention due to their potential for performance equivalence with PEMELs. Here, of course, the current state-of-the-art AELs, PEMELs, and HEMELs have both positive and negative aspects [33].

On the other hand, owing to their high specific surface area and diverse active fields, nanomaterials are more effective and cheaper compared to the others. Therefore, a primary research idea at present is the creation of more active, nanostructured electrocatalysts [13, 34].

Some special applications can also be listed here. Among them, 5-hydroxymethylfurfural (HMF) and 2,5-furandicarboxylic acid (FDCA) electrocatalytic oxidation have been studied extensively.

Chapter 6 The electrocatalysts with pH of the electrolyte **99**

However, the short lives of these catalytic electrodes continue, and the high pH of the electrolyte contributes to the denaturation of HMF during the reaction; further, the short lifespan of the electric electrodes remains troublesome [35].

Quantum mechanics comes into play in such processes. Because quantum mechanics gives particles completely new properties, it allows us to deal with many years of unsolved problems [36].

For example, quantum mechanics analysis guidelines are taken into account in analysis applications in energy storage systems. Water electrolysis uses hydrogen as an energy vector to store large amounts of energy and typically operates on either acidic or alkaline electrolytes [37].

Chemical conversion and storage of energy can be achieved effectively by linking or combining anodic, cathodic, and catalytic subunits in a single device in several ways. The water electrolysis cells are combined with subunits, and the end units are connected in series to provide the electrodes with adequate potential to conduct the water-splitting operation. Electrodes constructed from different types of materials coated with catalysts are based on another form of electrochemical water separation cells. In this case, an electrode (whether anode or cathode) is immersed into an electrolyte and attached to an electrode counter that is also very effective for the ideal water-division reaction half-cell. Both of these systems are usually more efficient than photocatalytic water separation systems that are fully integrated. Every facility, however, has comprehensive studies, and the advantages of viable new construction techniques are immense [38].

In clean energy technology, including fuel cells and electrolyzers, oxygen electrochemistry plays a key role, but the sluggish kinetics of oxygen reactions (OERs) limits the efficiency and marketing of such products [39]. Fuel cells are electrochemical devices that convert chemical energy directly into electrical energy. Fuel cells are generally classified as polymer electrolyte membrane fuel cells (PEMFCs), alkaline fuel cells (AFC), molten carbonate fuel cell, solid oxide fuel cell, phosphoric acid fuel cell, and direct methanol fuel cell (DMFCs) depending on the type of electrolyte used in the cell. Besides, Direct Borohydride Fuel Cells (DBHFCs), which are similar to DMFCs due to the use of membrane electrolyte, cathode properties the alkaline environment and liquid fuel feeding to fuel cells, does not have a problem with hydrogen storage compared to other fuel cells. It is safe due to the use of liquid fuel, and has a number of advantages, such as high battery voltage (1.64 V) and power density (9.3 kWh/kg). Because of these advantages, DBHFCs are considered a potential candidate for small and portable applications. Theoretically, each BH_4 ion can produce eight electrons upon its oxidation [40–43].

Anion exchange membrane fuel cells (AEMFCs), as a theoretically lower-cost electric energy conversion system than proton exchange membrane fuel cells (PEMFCs), have gained considerable attention in recent years [44]. As a key component of anion exchange membrane fuel cells (AEMFCs), anion exchange membranes (AEMs) are expected to have high ionic conductivity, low swelling, and excellent mechanical, chemical, and thermal stability [45].

Until now, these high-performing AEMFCs have been obtained with anode-based platinum group metals (PGMs) and cathode catalysts. To realize the ability of AEMFCs, such catalysts must be free of essential raw materials and PGMs. While substantial progress has been made in developing non-PGM catalysts to minimize the oxygen reaction in the base medium, the hydrogen oxidation reaction (HOR) has not been considered as much. The much lower HOR activity of Pt in the base media compared to acids has just recently arisen. Although some PGM-based composites show improved HOR behavior in the basic medium, the HOR kinetics of an ideal nonpolarized electrode remain less than needed. Even when other technical obstacles (e.g., membrane stability) are solved, it would still be a significant impediment to the large-scale application of this technology. Fundamental insights into the HOR framework and the key energy barriers in the underlying environment need to be identified in order to overcome this problem [46, 47].

It is very important to obtain low-cost and competent electrocatalysts. For fuel cell technology to be applied on an industrial scale, important steps must be taken with this technology. In one of the studies, bare nickel foam (NF), which is easily available, was used to generate porous nickel oxide (NiO) with a simple scalable thermal oxidation process as a leading precursor. As a nonbinding electrocatalyst for oxygen production, the resulting NF-supported NiO is used for Oxygen Evolution Reaction (OER). This strong and efficient electrocatalyst can compete with other electrocatalysts, which consist of expensive components and require sophisticated synthesis; NiO electrocatalyst advances toward the replacement of noble metal–based elements of OER [48].

As a general, for metal-based materials, carbon-based materials/particles have been using as a substrate. Carbon nanoparticles are conical nanostructures made of sp^2. Carbon layers form clustered superstructures during their synthesis. They do not need metal catalysts in preparation, are cheaply produced in industrial quantities, and are a suitable candidate for electrocatalytic reactions. There are also many nanomaterial studies that have shown promising results in terms of creating the appropriate catalysis system of these materials and using nanoparticles in ideal catalysis processes [49–68]. Also, important developments in fuel cell studies are expected to occur through nanotechnology. It is to provide a comprehensive overview of the carbon-based materials in the field of electrocatalysis, and in particular of oxygen reduction, methanol oxidation, and hydrogen evolution, as well as the evolution of oxygen resulting from water splitting, and also to highlight and point out areas where significant improvements can be achieved [69]. Sustainable energy production is an essential prerequisite for a projected economy. For this purpose, electrocatalysts are also widely used as cocatalysts to increase the kinetics of reactions [70].

In these catalyst, for instance, a Pt—Ru nanosized materials assisted by poly(3,4-ethylenedioxythiophene) (PEDOT):poly(styrene sulfonic acid) (PSSA) in a solid-polymer electrolyte anode is used as a catalyst and their performance have been compared with traditional Pt-Ru-assisted Vulcan XC-72R for direct methanol fuel cells [71].

There are many studies on oxygen reduction reaction (ORR) activities and their use as electrocatalysts in PEMFCs. For this aim, first, single-layer and multilayer graphene production was carried out via chemical vapor deposition as a substrate. Graphene can be obtained by different optimization conditions. For this purpose, different gas flow rates, temperatures, times, and substrates were performed for graphene production [72].

An anode including graphene and its derivatives needs to have good electrolytic performance and be stable and marketable in the electrolyte environment. For this reason, the range of anodic materials has changed to include low cost transition metals. Among these metals, nickel, cobalt, and their mixed oxides were extensively tested, including nickel, iron-nickel, and iron-nickel alloys [73,74].

In another study, OERs were investigated on nickel electrodes in 15% by weight NaOH solution at 60 °C. Various surface treatments, such as sandblasting and/or chemical pickling, have been applied. Fresh and aged electrodes were subjected to electrochemical measurement to evaluate whether they are in their ideal form [75].

3 Conclusions

As a conclusion, fuel cells are very important electrochemical devices and their importance increase as energy sources, day by day. It is very important to design these systems with increasing application areas in the most ideal conformation. It is also important to develop low-cost products, electrodes, catalysts, membranes, etc., within the scope of research. The use of nanomaterials as catalyst within the scope of some studies is thought to accelerate developments in battery technology. By developing various anode and cathode technologies in battery technology, pH-related restrictions may be avoided. This chapter has provided general information about a number of these types of applications.

References

[1] F.F. Aydin, Enerji Tüketimi Ve Ekonomik Büyüme, May 2015, Accessed: Jan. 30, 2021. [Online]. Available https://dergipark.org.tr/en/pub/erciyesiibd/77939.

[2] M. Tahir, et al., Electrocatalytic oxygen evolution reaction for energy conversion and storage: a comprehensive review, Nano Energy 37 (2017) 136–157, https://doi.org/10.1016/j.nanoen.2017.05.022. Elsevier Ltd.

[3] M.P. Warren, P.L. Forrester, J.S. Hassard, J.W. Cotton, Technological innovation antecedents in the UK ceramics industry, Int. J. Prod. Econ. 65 (1) (2000) 85–98, https://doi.org/10.1016/S0925-5273(99)00092-4.

[4] R.A. Rozendal, H.V.M. Hamelers, G.J.W. Euverink, S.J. Metz, C.J.N. Buisman, Principle and perspectives of hydrogen production through biocatalyzed electrolysis, Int. J. Hydrog. Energy 31 (12) (2006) 1632–1640, https://doi.org/10.1016/j.ijhydene.2005.12.006.

[5] G. Kardaş, R. Solmaz, B. Yazici, Elektroliz Yöntemiyle Hidrojen Gazı Eldesi Electrochemical Synthesis Of Zno Nanotubes View Project Green Inhibitors View project, 1993, Accessed: Jan. 30, 2021. [Online]. Available: https://www.researchgate.net/publication/268205242.

[6] X. Yu, E.J. Pascual, J.C. Wauson, A. Manthiram, A membraneless alkaline direct liquid fuel cell (DLFC) platform developed with a catalyst-selective strategy, J. Power Sources 331 (2016) 340–347, https://doi.org/10.1016/j.jpowsour.2016.09.077.

[7] R. Zeng, et al., Alkaline ionomer with tuneable water uptakes for electrochemical energy technologies, Energy Environ. Sci. 4 (12) (2011) 4925–4928, https://doi.org/10.1039/c1ee02349g.

[8] F.M. Souza, et al., PdxNby electrocatalysts for DEFC in alkaline medium: stability, selectivity and mechanism for EOR, Int. J. Hydrog. Energy 43 (9) (2018) 4505–4516, https://doi.org/10.1016/j.ijhydene.2018.01.058.

[9] B.D. McNicol, D.A.J. Rand, K.R. Williams, Direct methanol-air fuel cells for road transportation, J. Power Sources 83 (1–2) (1999) 15–31, https://doi.org/10.1016/S0378-7753(99)00244-X.

[10] A. Vasileff, C. Xu, Y. Jiao, Y. Zheng, S.Z. Qiao, Surface and Interface engineering in copper-based bimetallic materials for selective CO_2 electroreduction, Chem 4 (8) (2018) 1809–1831, https://doi.org/10.1016/j.chempr.2018.05.001. Elsevier Inc.

[11] J. Deng, J.A. Iñiguez, C. Liu, Electrocatalytic nitrogen reduction at low temperature, Joule 2 (5) (2018) 846–856, https://doi.org/10.1016/j.joule.2018.04.014. Cell Press.

[12] C. Guo, J. Ran, A. Vasileff, S.Z. Qiao, Rational design of electrocatalysts and photo(electro)catalysts for nitrogen reduction to ammonia (NH_3) under

Chapter 6 The electrocatalysts with pH of the electrolyte **103**

ambient conditions, Energy Environ. Sci. 11 (1) (2018) 45–56, https://doi.org/10.1039/c7ee02220d. Royal Society of Chemistry.

[13] X. Wang, C. Xu, M. Jaroniec, Y. Zheng, S.Z. Qiao, Anomalous hydrogen evolution behavior in high-pH environment induced by locally generated hydronium ions, Nat. Commun. 10 (1) (2019) 1–8, https://doi.org/10.1038/s41467-019-12773-7.

[14] C.C. Pavel, et al., Highly efficient platinum group metal free based membrane-electrode assembly for anion exchange membrane water electrolysis, Angew. Chem. Int. Ed. 53 (5) (2014) 1378–1381, https://doi.org/10.1002/anie.201308099.

[15] V. Mehta, J.S. Cooper, Review and analysis of PEM fuel cell design and manufacturing, J. Power Sources 114 (1) (2003) 32–53, https://doi.org/10.1016/S0378-7753(02)00542-6. Elsevier.

[16] K.E. Ayers, et al., Research advances towards low cost, high efficiency PEM electrolysis, ECS Trans. 33 (1) (2019) 3–15, https://doi.org/10.1149/1.3484496.

[17] S.R. Narayanan, et al., Recent advances in PEM liquid-feed direct methanol fuel cells, in: Proceedings of the Annual Battery Conference on Applications and Advances, 1996, pp. 113–122, https://doi.org/10.1109/bcaa.1996.484980.

[18] K. Yamada, et al., Potential application of anion-exchange membrane for hydrazine fuel cell electrolyte, Electrochem. Commun. 5 (10) (2003) 892–896, https://doi.org/10.1016/j.elecom.2003.08.015.

[19] I. Vincent, D. Bessarabov, Low cost hydrogen production by anion exchange membrane electrolysis: a review, Renew. Sustain. Energy Rev. 81 (2018) 1690–1704, https://doi.org/10.1016/j.rser.2017.05.258. Elsevier Ltd.

[20] K. Kouno, Y. Tuchiya, T. Ando, Measurement of soil microbial biomass phosphorus by an anion exchange membrane method, Soil Biol. Biochem. 27 (10) (1995) 1353–1357, https://doi.org/10.1016/0038-0717(95)00057-L.

[21] T.R. Ralph, et al., Low cost electrodes for proton exchange membrane fuel cells: performance in single cells and Ballard stacks, J. Electrochem. Soc. 144 (11) (1997) 3845–3857, https://doi.org/10.1149/1.1838101.

[22] E.J. Taylor, E.B. Anderson, N.R.K. Vilambi, Preparation of high-platinum-utilization gas diffusion electrodes for proton-exchange-membrane fuel cells, J. Electrochem. Soc. 139 (5) (1992) L45–L46, https://doi.org/10.1149/1.2069439.

[23] A. Kumar, R.G. Reddy, Modeling of polymer electrolyte membrane fuel cell with metal foam in the flow-field of the bipolar/end plates, J. Power Sources 114 (1) (2003) 54–62, https://doi.org/10.1016/S0378-7753(02)00540-2.

[24] K.D. Kreuer, On the development of proton conducting materials for technological applications, Solid State Ionics 97 (1–4) (1997) 1–15, https://doi.org/10.1016/s0167-2738(97)00082-9.

[25] P.S. Fedkiw, W. Her, An impregnation-reduction method to prepare electrodes on Nafion SPE, J. Electrochem. Soc. 136 (3) (1989) 899–900, https://doi.org/10.1149/1.2096772.

[26] E. Sitta, R. Nagao, H. Varela, The electro-oxidation of ethylene glycol on platinum over a wide pH range: oscillations and temperature effects, PLoS One 8 (9) (2013) e75086, https://doi.org/10.1371/journal.pone.0075086.

[27] Y. Wang, S. Zou, W. Bin Cai, Recent advances on electro-oxidation of ethanol on Pt- and Pd-based catalysts: from reaction mechanisms to catalytic materials, Catalysts 5 (3) (2015) 1507–1534, https://doi.org/10.3390/catal5031507. MDPI AG.

[28] E. Antolini, Catalysts for direct ethanol fuel cells, J. Power Sources 170 (1) (2007) 1–12, https://doi.org/10.1016/j.jpowsour.2007.04.009. Elsevier.

[29] J. Guo, R. Chen, F.C. Zhu, S.G. Sun, H.M. Villullas, New understandings of ethanol oxidation reaction mechanism on Pd/C and Pd2Ru/C catalysts in

alkaline direct ethanol fuel cells, Appl. Catal. B Environ. 224 (2018) 602–611, https://doi.org/10.1016/j.apcatb.2017.10.037.

[30] M.Z.F. Kamarudin, S.K. Kamarudin, M.S. Masdar, W.R.W. Daud, Review: direct ethanol fuel cells, Int. J. Hydrog. Energy 38 (22) (2013) 9438–9453, https://doi.org/10.1016/j.ijhydene.2012.07.059.

[31] E. Peled, T. Duvdevani, A. Aharon, A. Melman, New fuels as alternatives to methanol for direct oxidation fuel cells, Electrochem. Solid-State Lett. 4 (4) (2001), https://doi.org/10.1149/1.1355036.

[32] A.S. Gago, et al., Degradation of proton exchange membrane (PEM) electrolysis: the influence of current density, ECS Trans. 86 (13) (2018) 695–700, https://doi.org/10.1149/08613.0695ecst.

[33] R. Abbasi, et al., A roadmap to low-cost hydrogen with hydroxide exchange membrane electrolyzers, Adv. Mater. 31 (31) (2019), https://doi.org/10.1002/adma.201805876.

[34] Y. Jiao, Y. Zheng, M. Jaroniec, S.Z. Qiao, Design of electrocatalysts for oxygen- and hydrogen-involving energy conversion reactions, Chem. Soc. Rev. 44 (8) (2015) 2060–2086, https://doi.org/10.1039/c4cs00470a. Royal Society of Chemistry.

[35] M.J. Kang, H.J. Yu, H.S. Kim, H.G. Cha, Deep eutectic solvent stabilised Co-P films for electrocatalytic oxidation of 5-hydroxymethylfurfural into 2, 5-furandicarboxylic acid, New J. Chem. 44 (33) (2020) 14239–14245, https://doi.org/10.1039/d0nj01426e.

[36] R.W. Gurney, The quantum mechanics of electrolysis, Proc. R. Soc. Lond Ser. A 134 (823) (1931) 137–154, https://doi.org/10.1098/rspa.1931.0187.

[37] B.-J. Kim, et al., Oxygen evolution reaction activity and underlying mechanism of perovskite electrocatalysts at different pH, Mater. Adv. (2020), https://doi.org/10.1039/d0ma00661k.

[38] D. Lukács, Ł. Szyrwiel, J.S. Pap, Copper containing molecular systems in electrocatalytic water oxidation—trends and perspectives, Catalysts 9 (1) (2019) 83, https://doi.org/10.3390/catal9010083. MDPI AG.

[39] M. Suszynska, P. Grau, H. Meinhard, M. Szmida, L. Krajczyk, Chemical strengthening of some soda lime silicate glasses, in: Physica Status Solidi C: Conferences, 2005, https://doi.org/10.1002/pssc.200460245.

[40] Ç. Cenk, Doğrudan sodyum borhidrürlü yakıt pilinde proses parametrelerinin verim üzerine etkisinin incelenmesi, Kocaeli Universitesi, Fen Bilimleri Enstitusu, Kocaeli, 2006.

[41] Y. Polat, Nano ve Mikro Yapıdaki (Bi$_2$O$_3$)1-x-y (Sm$_2$O$_3$)x(Ho2O$_3$) y Elektrolit Sistemlerinin Katı Oksit Yakıt Pili için Sentezlenmesi ve Faz Kararlılığı, Elektriksel ve Termal Özelliklerinin İncelenmesi ve Karşılaştırılması, Süleyman Demirel Üniversitesi Fen Bilim. Enstitüsü Derg 21 (3) (2017) 1011, https://doi.org/10.19113/sdufbed.59059.

[42] B.F. Dalğıç, PEM yakıt hücresinde hidrojen gazındaki nem oranına bağlı olarak elektrik üretimindeki değişimin incelenmesi, 2019, Accessed: Jan. 30, 2021. [Online]. Available: http://earsiv.batman.edu.tr/xmlui/handle/20.500.12402/2316.

[43] E. Erden, Katı oksit yakıt pilleri için Bi$_2$O$_3$ katkılı katı elektrolit-katot malzemelerin sentezi ve karakterizasyonu, Kütahya Dumlupınar Üniversitesi/Fen Bilimleri Enstitüsü, Jan. 2019. Accessed: Jan. 30, 2021. [Online]. Available: http://openaccess.dpu.edu.tr/xmlui/handle/20.500.12438/7908.

[44] T.J. Omasta, et al., Strategies for reducing the PGM loading in high power AEMFC anodes, J. Electrochem. Soc. 165 (9) (2018) F710–F717, https://doi.org/10.1149/2.1401809jes.

[45] S. Li, X. Zhu, D. Liu, F. Sun, A highly durable long side-chain polybenzimidazole anion exchange membrane for AEMFC, J. Membr. Sci. 546 (2018) 15–21, https://doi.org/10.1016/j.memsci.2017.09.064.

[46] E.S. Davydova, S. Mukerjee, F. Jaouen, D.R. Dekel, Electrocatalysts for hydrogen oxidation reaction in alkaline electrolytes, ACS Catal. 8 (7) (2018) 6665–6690, https://doi.org/10.1021/acscatal.8b00689.

[47] X. Xie, J. Zhou, S. Wu, J.W. Park, K. Jiao, Experimental investigation on the performance and durability of hydrogen AEMFC with electrochemical impedance spectroscopy, Int. J. Energy Res. 43 (14) (2019) 8522–8535, https://doi.org/10.1002/er.4851.

[48] P.T. Babar, A.C. Lokhande, M.G. Gang, B.S. Pawar, S.M. Pawar, J.H. Kim, Thermally oxidized porous NiO as an efficient oxygen evolution reaction (OER) electrocatalyst for electrochemical water splitting application, J. Ind. Eng. Chem. 60 (2018) 493–497, https://doi.org/10.1016/j.jiec.2017.11.037.

[49] A. Brouzgou, A. Podias, P. Tsiakaras, PEMFCs and AEMFCs directly fed with ethanol: a current status comparative review, J. Appl. Electrochem. 43 (2) (2013) 119–136, https://doi.org/10.1007/s10800-012-0513-2.

[50] F. Şen, G. Gökagaç, Activity of carbon-supported platinum nanoparticles toward methanol oxidation reaction: role of metal precursor and a new surfactant, tert-octanethiol, J. Phys. Chem. C 111 (3) (2007) 1467–1473, https://doi.org/10.1021/jp065809y.

[51] F. Sen, Y. Karatas, M. Gulcan, M. Zahmakiran, Amylamine stabilized platinum (0) nanoparticles: active and reusable nanocatalyst in the room temperature dehydrogenation of dimethylamine-borane, RSC Adv. 4 (4) (2014) 1526–1531, https://doi.org/10.1039/c3ra43701a.

[52] S. Şen, F. Şen, G. Gökağaç, Preparation and characterization of nano-sized Pt-Ru/C catalysts and their superior catalytic activities for methanol and ethanol oxidation, Phys. Chem. Chem. Phys. 13 (15) (2011) 6784–6792, https://doi.org/10.1039/c1cp20064j.

[53] Z. Daşdelen, Y. Yıldız, S. Eriş, F. Şen, Enhanced electrocatalytic activity and durability of Pt nanoparticles decorated on GO-PVP hybride material for methanol oxidation reaction, Appl. Catal. B Environ. 219 (2017) 511–516, https://doi.org/10.1016/j.apcatb.2017.08.014.

[54] B. Sen, B. Demirkan, B. Şimşek, A. Savk, F. Sen, Monodisperse palladium nanocatalysts for dehydrocoupling of dimethylamineborane, Nano-Struct. Nano-Objects 16 (2018) 209–214, https://doi.org/10.1016/j.nanoso.2018.07.008.

[55] B. Sen, E. Kuyuldar, A. Şavk, H. Calimli, S. Duman, F. Sen, Monodisperse ruthenium–copper alloy nanoparticles decorated on reduced graphene oxide for dehydrogenation of DMAB, Int. J. Hydrog. Energy 44 (21) (2019) 10744–10751, https://doi.org/10.1016/j.ijhydene.2019.02.176.

[56] B. Şen, B. Demirkan, M. Levent, A. Şavk, F. Şen, Silica-based monodisperse PdCo nanohybrids as highly efficient and stable nanocatalyst for hydrogen evolution reaction, Int. J. Hydrog. Energy 43 (44) (2018) 20234–20242, https://doi.org/10.1016/j.ijhydene.2018.07.080.

[57] B. Sen, S. Kuzu, E. Demir, S. Akocak, F. Sen, Highly monodisperse RuCo nanoparticles decorated on functionalized multiwalled carbon nanotube with the highest observed catalytic activity in the dehydrogenation of dimethylamine–borane, Int. J. Hydrog. Energy 42 (36) (2017) 23292–23298, https://doi.org/10.1016/j.ijhydene.2017.06.032.

[58] B. Şen, et al., High-performance graphite-supported ruthenium nanocatalyst for hydrogen evolution reaction, J. Mol. Liq. 268 (2018) 807–812, https://doi.org/10.1016/j.molliq.2018.07.117.

[59] S. Eris, Z. Daşdelen, F. Sen, Investigation of electrocatalytic activity and stability of Pt@f-VC catalyst prepared by in-situ synthesis for methanol electrooxidation, Int. J. Hydrog. Energy 43 (1) (2018) 385–390, https://doi.org/10.1016/j.ijhydene.2017.11.063.

[60] S. Eris, Z. Daşdelen, F. Sen, Enhanced electrocatalytic activity and stability of monodisperse Pt nanocomposites for direct methanol fuel cells, J. Colloid Interface Sci. 513 (2018) 767–773, https://doi.org/10.1016/j.jcis.2017.11.085.

[61] Y. Yildiz, E. Erken, H. Pamuk, H. Sert, F. Şen, Monodisperse Pt nanoparticles assembled on reduced graphene oxide: highly efficient and reusable catalyst for methanol oxidation and dehydrocoupling of dimethylamine-borane (DMAB), J. Nanosci. Nanotechnol. 16 (6) (2016) 5951–5958, https://doi.org/10.1166/jnn.2016.11710.

[62] F. Şen, S. Şen, G. Gökağaç, Efficiency enhancement of methanol/ethanol oxidation reactions on Pt nanoparticles prepared using a new surfactant, 1,1-dimethyl heptanethiol, Phys. Chem. Chem. Phys. 13 (4) (2011) 1676–1684, https://doi.org/10.1039/c0cp01212b.

[63] S. Eris, Z. Daşdelen, Y. Yıldız, F. Sen, Nanostructured polyaniline-rGO decorated platinum catalyst with enhanced activity and durability for methanol oxidation, Int. J. Hydrog. Energy 43 (3) (2018) 1337–1343, https://doi.org/10.1016/j.ijhydene.2017.11.051.

[64] Ö. Karatepe, Y. Yildiz, H. Pamuk, S. Eris, Z. Dasdelen, F. Sen, Enhanced electrocatalytic activity and durability of highly monodisperse Pt@PPy-PANI nanocomposites as a novel catalyst for the electro-oxidation of methanol, RSC Adv. 6 (56) (2016) 50851–50857, https://doi.org/10.1039/c6ra06210e.

[65] F. Gulbagca, S. Ozdemir, M. Gulcan, F. Sen, Synthesis and characterization of Rosa canina-mediated biogenic silver nanoparticles for anti-oxidant, antibacterial, antifungal, and DNA cleavage activities, Heliyon 5 (12) (2019) e02980, https://doi.org/10.1016/j.heliyon.2019.e02980.

[66] F. Göl, A. Aygün, A. Seyrankaya, T. Gür, C. Yenikaya, F. Şen, Green synthesis and characterization of *Camellia sinensis* mediated silver nanoparticles for antibacterial ceramic applications, Mater. Chem. Phys. 250 (2020), https://doi.org/10.1016/j.matchemphys.2020.123037.

[67] F.M. Ertosun, K. Cellat, O. Eren, Ş. Gül, E. Kuşvuran, F. Şen, Comparison of nanoscale zero-valent iron, fenton, and photo-fenton processes for degradation of pesticide 2,4-dichlorophenoxyacetic acid in aqueous solution, SN Appl. Sci. (2019), https://doi.org/10.1007/s42452-019-1554-5.

[68] F. Diler, et al., Efficient preparation and application of monodisperse palladium loaded graphene oxide as a reusable and effective heterogeneous catalyst for suzuki cross-coupling reaction, J. Mol. Liq. 298 (2020) 111967, https://doi.org/10.1016/j.molliq.2019.111967.

[69] A. Kagkoura, N. Tagmatarchis, Carbon nanohorn-based electrocatalysts for energy conversion, Nanomaterials 10 (7) (2020) 1407, https://doi.org/10.3390/nano10071407.

[70] S. Chen, S.S. Thind, A. Chen, Nanostructured materials for water splitting—state of the art and future needs: a mini-review, Electrochem. Commun. 63 (2016) 10–17, https://doi.org/10.1016/j.elecom.2015.12.003. Elsevier Inc.

[71] K.K. Tintula, S. Pitchumani, P. Sridhar, A.K. Shukla, A solid-polymer-electrolyte direct methanol fuel cell (DMFC) with Pt-Ru nanoparticles supported onto poly(3,4-ethylenedioxythiophene) and polystyrene sulphonic acid polymer composite as anode, J. Chem. Sci. 122 (3) (2010) 381–389, https://doi.org/10.1007/s12039-010-0043-6.

[72] M.A. Azder, Çok tabakalı grafenin azot katkılanması, transferi ve PEM yakıt pilinde kullanılmasının çalışılması, Recep Tayyip Erdoğan Üniversitesi/Fen Bilimleri Enstitüsü/Fizik Anabilim Dalı, Rize, 2017.

[73] A.K.M. Fazle Kibria, S.A. Tarafdar, Electrochemical studies of a nickel-copper electrode for the oxygen evolution reaction (OER), Int. J. Hydrog. Energy 27 (9) (2002) 879–884, https://doi.org/10.1016/S0360-3199(01)00185-9.

[74] I. Arulraj, D.C. Trivedi, Characterization of nickel oxyhydroxide based anodes for alkaline water electrolysers, Int. J. Hydrog. Energy 14 (12) (1989) 893–898, https://doi.org/10.1016/0360-3199(89)90076-1.

[75] C. Bocca, A. Barbucci, G. Cerisola, The influence of surface finishing on the electrocatalytic properties of nickel for the oxygen evolution reaction (OER) in alkaline solution, Int. J. Hydrog. Energy 23 (4) (1998) 247–252, https://doi.org/10.1016/s0360-3199(97)00049-9.

7

Pt-based catalysts for alcohol oxidation

Hakan Burhan[a], Kubilay Arıkan[a], Muhammed Bekmezci[a,b], Tugba Gur[c], and Fatih Şen[a]

[a]Şen Research Group, Department of Biochemistry, Dumlupinar University, Kütahya, Turkey. [b]Department of Materials Science and Engineering, Faculty of Engineering, Dumlupinar University, Kütahya, Turkey. [c]Vocational School of Health Services, Van Yuzuncu Yıl University, Van, Turkey

1 Introduction

Energy consumption is increasing rapidly all over the world. Scientific circles and energy producers have focused on alternative energy sources. Fossil energy sources such as coal, oil, and natural gas are limited, and nonrenewable energy resources may face the risk of depletion over time. For these reasons, fuel cells are one of the most important energy sources of the future [1–4].

Despite their high-tech applications in recent years, fuel cells have actually been known to scientists for over 150 years. They were first developed in the late 19th century. In 1839, Sir William Grove produced hydrogen and oxygen in a system consisting of two platinum electrodes immersed in a diluted sulfuric acid solution. The term "fuel cell" was first introduced in 1889 by Ludwing Mond and Charles Langer, who repeated Grove's work. Mond and Langer developed a fuel cell that produced 1.5 watts of power and had a 50% working efficiency, using air as an oxygen source and industrial coal gas as a hydrogen source. In 1894, Wilheam Oswalt made an electrochemical cell by working with coal-derived fuels. In 1932, Francis T. Bacon developed the successful another fuel cell. In 1952, Bacon and his friends built a fuel cell that produced 5 kW of power. At the end of the same year, Harry Karl Ihring designed a 20-hp, fuel cell–driven tractor. Ihring's fuel cell was the start of today's modern fuel cell machines. Druing World War II, fuel cells gained great importance. Intensive research and development of fuel cells were done

Nanomaterials for Direct Alcohol Fuel Cells. https://doi.org/10.1016/B978-0-12-821713-9.00014-7
Copyright © 2021 Elsevier Inc. All rights reserved.

beginning in that period and continuing afterward. In the 1940s, government-sponsored fuel cell research was initiated in the United States; the Los Alamos National Laboratory and the Brookhaven National Laboratory were established and the National Aeronautics and Space Administration (NASA) invested heavily in fuel cell technology in 1960s. Because fuel cells are light and produce water as a by-product, they are being considered for space applications. The use of fuel cells in space studies provides a number of advantages, such as high efficiency, low noise and flicker, and high energy density. Proton exchange membrane (PEM) fuel cells produced by General Electric were used in the *Gemini* spacecraft. In the 1970s, General Motors developed a fuel cell vehicle called Electrovan. The use of fuel cells in space projects continues up to the present day [5–8].

Fuel cells are components that transform chemical energy directly into electrical energy by electrochemical reactions. The fuels are delivered to the fuel cells directly or indirectly. The fuel cells can be supplied directly with hydrogen, methane, steam, air gases, liquefied petroleum gas (LPG), and hydrazine [8–12]. From this point of view, many nanomaterials can be used for fuel cells. Many studies have researched different applications such as fuel cells, hydrogen synthesis, and electrochemical sensor applications [13–27].

1.1 Advantage and disadvantages of fuel cells

Some of the advantages of fuel cells, which provide a clean and renewable energy source that is intended to be used in the future, are the following:
- Minimal emissions, as well as being environmentally friendly in other ways.
- Fuel flexibility (high fuel alternatives such as natural gas, LPG, methanol, and ethanol).
- Direct energy conversion (without burning).
- Lack of noise due to the lack of moving parts.
- High efficiency at low temperatures.
- No solid-waste problems.
- Portable and can be customized depending on user needs.
- Dimensions/size flexibility.
  Fuel cells also have some disadvantages:
- Because this is a new technology, fuel cell utilization is not common, and these cells are expensive to manufacture.
- The storage of hydrogen is problematic.
- Some fuel cell types operate at very high temperatures [4–8].

1.2 Solvent environmental effects on the catalyst effectives

Research studies have found the following about preparation procedures of catalysts: Catalysts are generally reduced in the distilled water in the temperature range of 0–5°C in an inert gas atmosphere (this gas is usually N_2). It is then allowed to dry again in an inert gas atmosphere. Recently, as a novel method, the catalytic activity of the catalyst has been increased by changing the solvent medium during synthesis. Solvents such as methanol, ethanol, and propyl alcohol can be used for this purpose [9, 10].

Among those solvents, it was determined that the catalyst activity order was methanol > pure water > ethanol > propyl alcohol. Besides, another study investigated the catalytic activity of the Co-Cu-B catalyst in the methanolic and pure water medium, and the results indicated that the catalyst activity in the methanolic medium was better than in pure water. Similarly, the catalytic activity of the Co-B-TiO$_2$ catalyst in pure water and methanol was investigated, and the catalysts synthesized in the methanol medium were found to perform better than in pure water [11, 12].

Patel et al. synthesized Co—B, Co—B, TiO$_2$ and Ni—B catalysts in pure water, methanol, and ethanol media. They observed that the synthesis of catalysts in the ethanolic and methanolic media showed better performance than in pure water [28].

1.3 Plasma effects on catalyst efficiency

Besides the catalyst synthesis, new applications have been utilized to increase the catalytic effect of the catalysts, such as sol-gel coating, calcination, microwave, and plasma. Plasma treatment is a method that significantly increases catalyst efficiency [29–31]. In this method, it was observed that the catalytic activity of Co-Cu-B catalyst synthesized in the plasma treatment in Ar, O_2, and N_2 gases was increased. It has been found that the best result is with the plasma treatment under N_2 gas. Sahin et al. synthesized the Co-B-TiO$_2$ catalyst and exposed it to plasma in several gas conditions [30–32]. When the time of the activity was considered, the best efficiency was determined to be 10-min plasma treatment.

The results show that the catalytic activities can increase upon synthesis. Due to the reaction in the anode, 12 electrons per mole of ethanol are released. As a result, the energy density of the cell is higher than the energy density of the methanol fuel cell (M) and

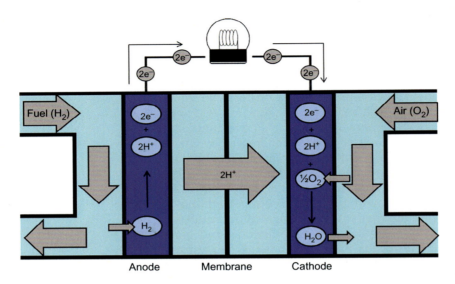

Fig. 1 PEM fuel cell operation [33].

the hydrogen fuel cell (Fig. 1). The electromotive force of the cell is 1.15 V.

The total reaction in the cell is like a combustion reaction (Fig. 2). Although the main products are CO_2 and water, intermediates such as acetaldehyde, acetic acid, and CO also occur [34]. There are so many types of fuel cells, such as methanol, ethanol, and propanol. For instance, direct ethanol fuel cells use ethanol as the fuel.

Recently, studies on direct ethanol fuel cells have increased considerably. This is because these fuel cells have the following advantages:
- High hydrogen content.
- High energy density.
- Easy to obtain.
- Less toxic liquid than methanol.
- Easy to store and transport.
- Can produce more energy with less fuel.
- Does not cause environmental pollution.

However, ethanol fuel cells do have some disadvantages:
- High cost of electrocatalysts and fuel cells.
- CO-like poisonous intermediate species that results in poisoning of the electrocatalyst and decreasing the anode performance.
- During the oxidation of ethanol, the molecules that adhere to the surface cannot be fully determined [35–37].

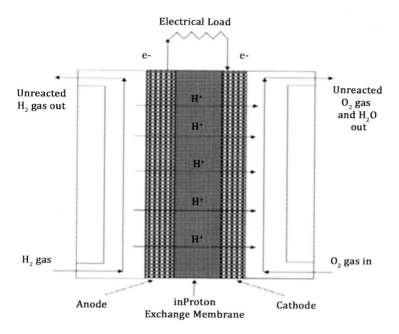

Fig. 2 Reactions in the PEM fuel cell [33].

Among these electrocatalyst, although the use of platinum (Pt) catalyst directly in ethanol fuel cells is costly, it is widely used in fuel cells due to the superior properties of Pt. The Pt catalyst accelerates the ethanol oxidation reaction and increases cell performance. Furthermore, the addition of a second transition element such as tin (Sn) to the platinum catalyst has the positive effect of reducing the formation of intermediate products during ethanol degradation. The effect of Sn during ethanol oxidation is thought to take two forms. The conversion of by-products of ethanol oxidation with the bifunctional mechanism to CO_2 by oxidized Sn weakens the Pt-CO bond with Pt near Sn atoms with electronic impact and Pt surface cleaning [21, 27, 38–40].

By the way, a strong electro-catalyst not only enables the oxidation of ethanol to CO_2 but also avoids the contamination of the surface with CO. Further, it facilitates the elimination of CO from the surface by accelerating the electrochemical oxidation of water:

$$Pt + H_2O \rightarrow Pt - OH + H + + e-$$

$$Pt - CO + Pt - OH \rightarrow 2Pt + CO_2 + H + + e-$$

In recent years, studies on direct ethanol fuel cells have increased. Significant performance differences are noticeable when low-temperature direct ethanol fuel cells are compared with high-temperature solid oxide fuel cells. The current densities of the solid oxide fuel cells are eight times higher than the polymer electrolyte fuel cells. This comparison gives an idea of the way in which PEM fuel cells should be used [41].

Another study emphasized that ethanol transfer to the cathode electrode is less in direct ethanol fuel cells than in direct methanol fuel cells, and to further highlight this advantage, the researchers proposed using the PtCo (3:1)/C catalyst, which has an increased ethanol tolerance, to accelerate the oxygen reduction reaction (ORR) in the cathode compared to Pt/C. As a result of this study, Pt—Co (3:1)/C showed higher performance compared to Pt/C at 60–100°C [42].

Generally, Pt has also been used in direct methanol fuel cells as anode catalysts because of its good catalytic properties in the synthesis of small organic molecules. However, the poisoning of the intermediate species provided by Pt in low-temperature fuel cells diminishes the methanol fuel cell's operations, and the catalyst is totally poisoned and has no activity.

In a Pt-based catalyst that consists of only platinum, it is frequently seen that the catalyst is supported with other substances such as supporting agents and a second metal. Two-metal catalysts show high activity for alcohol oxidation. These catalysts are typically supplemented after reactions with carbon-carbon decoupling by spraying, pressing, or binding of the electrode surface. It is very important to use such catalysts on the anode surface. A parameter that directly influences the strength and efficacy of a fuel cell is its catalytic ability to convert to hydrogen. Methanol electro-oxidation behavior improved as the volume of the second metal was increased, but several experiments have shown an optimal ratio that varies based on operational conditions [15, 19, 20, 43, 44].

2 Frequently used electrochemical techniques for fuel cells

2.1 Cyclic voltammetry

The cyclic voltammetry (CV) technique uses a working electrode coated with a thin-film catalyst added from outside with a given potential cycle. The potential loop started by a given voltage is introduced by increasing the potential at a constant scanning speed and then returning to the original value, and repeating until the optimal upper value is reached. Reduction or oxidation reactions are conducted based on the externally applied surface potential of the electrode, which is stored in an electrolyte solution under PEM fuel cell operating conditions and is covered with a catalytic film. The active catalyst surface depends on the extent to which the metal surface has been hydrogen adsorbed. Hydrogen adsorption or desorption can entail a hydrogen molar–adsorbed volume from the current amount corresponding to the adsorption of hydrogen. Desorption peaks are used as a base. The CV technique measures the mass behavior and particular activity of the catalysts [45].

Hydrogen (H^+) is taken from the Pt surface by the desorption and the electrochemical active surface (EAS) area is calculated, with the amount of charge in the potential region corresponding to -0.2 V to 0.1 V taken as the basis. For a hydrogen atom desorbed (or adsorbed) per each Pt atom, a charge exchange of 210 qC/cm^2 is accepted. The current is observed with the termination of the hydrogen desorption peak ($\sim$0.1 V), although the intensity of this flow is slightly increased with the potential. The resulting current is known as "double layer current" because it originates from the desorption of the second hydrogen layer and is not considered in its active account. The total load is included in the integration of the load in the region. The electroactive surface area of the catalysts is determined by the total amount of load [46]:

$$EAS = QH/0.21 \, [Pt] \qquad (1)$$

Equation (1) is expressed as Pt loading (mgPt/cm^2) per unit electrode surface; 0.21 is a charge exchange with a single hydrogen atom removed from the platinum surface. In the presence of mass activity, current density (mA/cm^2), Pt—O reduction and accumulation zone corresponding to 0.65 V are used in the region where Pt—O oxides begin to be reduced [47].

2.2 Hydrodynamic voltammetric methods

The general name of the voltammetry tests in electrolyte solutions carried out during continuous mixing is "hydrodynamic methods." The action of the electrolyte solution decreases the concentration limits on the surface of the catalyst and emphasizes the catalyst activity. By increasing the smooth and axial rotation speed based on the mixing phase, the diffusion limits are supposed to decrease the limit current after a certain potential. The voltammetric rotating disk electrode (RDE) is one of the most general hydrodynamic methods. In oxygen-saturated electrolyte solution, a linear potential scan is conducted and analyzed with fixed rotational velocities and monitored by the current potential. Due to the Koutecky-Levich association between the angular rotation rate of the electrode, the sum of transmitted electrodes and the limit value of the current, the diffusion coefficient of oxygen can be calculated as follows:

$$1/\mathrm{Id} = 1/\mathrm{Ik} + 1/CLV$$

The ID (diffusion limited current) electrode is transmitted by the IK (kinetic current) and the CLV corresponds to the current limit value. In that correspondence, the following is expressed: the number of transmitted electrons, oxygen density, diffusion factor, viscosity, and rotational speed:

$$ISA = 0.62\,\mathrm{n}\,\mathrm{F}\,CO_2\,DO_2\,2/3\,\mathrm{V} - 1/6\mathrm{W}1/2$$

In addition, Tafel ($\Delta \mathrm{V} = a + b\log(\mathrm{I})$) expresses the potential to change the boundary current versus the potential curve, the logarithm of both sides can be linearized where a and b can be found. Important kinetic parameters such as load transfer number and reaction order can be determined experimentally here and can be determined from a and b times [48].

3 Single-cell tests

In PEM fuel cell conditions, individual cell tests will be used for testing catalyst activity. The theoretical shift in current toward the cell generates polarization curves from single-cell experiments. Using the Tafel equation, the behavior of the catalyst can be calculated at low current densities with minimum diffusion limits. An overview of the high-stream polarization curve areas (typically $I > 200$ mA/cm^2) gives details on the shortcomings of other MEA components [49].

3.1 Choosing transition elements

The strength of the metal-adsorbent bond plays an important role during the oxidation of the reaction on the surface of the metal alloy. Reactive and electronic properties on the metallic surface can be changed by forming an alloy with the required metals. Another important aspect in the formation of metal alloys is the CO binding energy of metals. This energy should be low, whereas the C-metal bond should be strong [50].

Generally, Pt metals are ethanol oxidation catalysts that are inaccessible in terms of their performance. PtSn/C, PtRuSn/C, and PtRu/C catalysts has shown the best performance and activity in the research literature. Nevertheless, PtSnNi and PtSnRh triple alloys perform better than PtSn alloys. The reason for this is that the Sn and Ni oxide fragments are close to the Pt nanoparticle in PtSnNi alloys. In a PtSnRh alloy, Rh is added to the PtSn alloy by changing the Pt—Pt bond distance of the triple metal alloy. The bond distance change is caused by the change in the electrochemical structure of the alloy [51].

The reason for considering one of the most suitable alloys in research studies is that FeNi and AgCu alloys are preferred over PtRu or PtSn based on criteria such as surface energy, the radius of Wigner-Seitz, electronegativity, and the d-band centers of binary metal alloys, which do not have high-cost transition elements. FeNi and AgCu binary metal alloys are preferred because they do not contain expensive metals such as Pt, Pd, and Rh [52].

Several studies have been conducted on Pt-based catalysts, supported by materials such as TiO_2, CoO, and $LiCoO_2$. It was found that Pt-$LiCoO_2$ catalysts indicated better catalyst performance than the other options. In one study, Pt-$LiCoO_2$ catalysts tested $NaBH_4$ and H_2 pressure in the presence of water at the stoichiometric ratio, and it was found that the theoretical H_2 yield could be reached almost 0.6 MPa above H_2 pressure. In another study, the Pt-$LiCoO_2$ catalyst was coated on honeycomb like monolithic structures and they developed a hydrogen generator that can produce high-purity hydrogen at a 2-kW scale by working from the continuous system. The resulting system consists of the $NaBH_4$ fuel tank, pump, by-product $NaBO_2$ storage tank, separator, and hydride reactor. The gravimetric and volumetric hydrogen densities of the solution used in the system are set as 2% and 1.5% kg/100 L, respectively, and the maximum hydrogen production rate is 120 mL/min [53]. Another study aimed to see the effects of using various support materials with the support of Pt and Ru on IRA-400 anionic resin and $LiCoO_2$. It was determined that $LiCoO_2$ yields better results when it is used as support material than Pt-Ru nanoparticles produce hydrogen faster than the catalysts prepared by using and at higher concentrations than 10% $NaBH_4$ concentration [54]. An increase in Pt above the optimum value in Ru reduced the methanol electrooxidation activity of the catalyst, according to one study [55].

In another study, different metals and support materials were examined to investigate the effects on hydrogen production. Rh, Pt, Ru, and Pd were used as the metals, while Al_2O_3, C, TiO_2 and $LiCoO_2$ were used as the materials. The researchers determined that the rate of reaction was not affected by the rate of mixing. Hence, the reaction was not affected by external diffusion limitation speeds above 200 rpm. It is also supported for Al_2O_3 and contains 1% Pt; the activation energy is 57 kJ/mol. The activity between the metals used is Rh > Ru > Pt > Pd. The activity of the catalyst according to the supported platinum materials was determined to be $LiCoO_2$ > TiO_2 > C > Al_2O_3 [56]. Ersöz et al. reported that hydrogen production rates increased with the reduction of $NaBH_4$ and NaOH [56, 57].

Xua et al. studied the effects of adding CeO_2 to the Pt—C catalyst upon the oxidation of electrochemical alcohol (methanol, ethanol, glycerol, and ethylene glycol), and reported that Pt-CeO_2 gave the best performance. The highest catalytic activity was Pt-CeO_2, and the weight ratio was 0.3 mg/cm^2 (1.3:1). The high catalytic activity of Pt-CeO_2 occurs due to its tolerance to poisoning and the synergistic effect of Pt-CeO_2 [58].

Another study utilized alcohol reduction technique using a Nafion stabilizer to obtain a more effective platinum electrocatalyst and worked on metal nanoparticle production. A performance comparison of platinum nanoparticles prepared in direct methanol fuel cells, unsupported Pt-black, and carbon-supported Pt electrocatalysts was done in this study. Nafion-platinum nanoparticles showed higher activity for oxygen reduction in the presence and absence of methanol than in platinum black and Pt/C. The activity of Nafion-Pt nanoparticles against methanol oxidation is very low. The electrocatalysts prepared with the naphtha polyelectrolyte have a high methanol tolerance, thus reducing the cross-linking effect of methanol [59].

In another study, starting with PtCr/C and PtV/C, the double alloys of the platinum were made, and then third metals were added. Although most of the studies in which platinum alloys have been investigated have higher activity than pure Pt, these catalysts have a number of disadvantages, such as excessive peroxide formation and unstable structure [60]. In a study conducted with Pd—Co catalysts, the researchers showed that the formation of peroxide was minimal, although it showed lower activity than pure Pt. Thus, they found that the incorporation of Co in certain proportions of the platinum Pd-containing alloys affects the mechanism to promote the electron-based pathway [61]. Also, Pt—Co and Ni with binary alloys on both carbon support and silica support were examined. Where there was no positive effect on the catalyst activity of silica support, it was recorded that alloying with Pt Co in carbon-supported catalysts increased the active surface area by 100%, and its alloying with Ni increased by 200%. It was observed that Co and Ni atoms were dissolved

Chapter 7 Pt-based catalysts for alcohol oxidation **119**

from the catalyst surface during 24 and 170 hours of studies. In addition, XRD and EDS analyses revealed that only the particles remained stable in the crystal lattice, and the Co and Ni atoms outside were dissolved [62].

In one density function theory (DFT) study, the causes of increased activity in platinum binary alloys (PtM, M = Fe, Co, Ni, Cu, and W) were detailed. At the atomic level, geometry optimization and electronic structure calculations showed that the electronic configuration of the Pt atom changed after alloying with nonnoble metals, which could cause an increase in activity [33].

The strength of the oxygen-metal adsorption bond is explained by the increase in the catalytic activity, and the adsorption energy of the oxygen atom is also bound to the d-band (Fermi level) of the metal atom. Considering that the geometric and electronic properties of the platinum with a second metal are changed in PtM/C-type catalysts, it is possible to develop the characteristics of the Pt surface in three-metal systems of type PtM_1M_2/C, with many more combinations. It has been reported that PtCrCo/C triple catalysts provide 20% higher activity than pure Pt. In another study, PtCrCu/C catalysts were found to provide higher activity than Pt/C even after 300 h of run time. In many studies, the relative proportions of the M_1 and M_2 metals in a triple catalyst were discussed. For example, in a PtCoCr (x:1:1) catalyst, it was found that the catalytic activity at $x = 6$ reached its maximum level. In many studies, it has been stated that the molar concentration ratio of the transition metals forming the trimetallic catalyst affects the catalyst activity. The transition metals, which improve the activity of Pt by modifying the electron valence band, affect this activity by closing the Pt surface when used over a certain ratio [63]. There are few studies on quaternary catalysts. In one such study, starting with PtMn/C binary catalysts, the researchers examined the PtMnX/C triple catalysts and expanded their work to include PtMnCuX/C and PtMnMoX/C quaternary catalysts. PtMnCu/C and PtMnCuFe/C and PtMnCuMo/C catalysts were found to provide high activity. Although the percentage of H_2O_2 formation in quaternary catalysts is known to be low enough, the reaction mechanism itself has not been elucidated [64].

PtFM/C (F: Cr, Mn, Co, Ni, Cu), PtCoCr/C, PtCoNi/C, and PtNiFe/C catalysts have been partially successful in some trimetallic catalyst studies using different transition metals [65–67]. It has also been found that Pt also provides high activity with triple catalysts formed with Cu, Ag, and Fe, and the PtAgFe/C catalyst has comparable mass activity with pure Pt. It has also been found that platinum provides high activity with triple catalysts formed with Cu, Ag, and Fe, and the PtAgFe/C catalyst has comparable mass activity with pure Pt [68].

A number of studies have shown that catalyst activity is strongly dependent on catalyst particle size and the properties of support materials. Particle size is affected by the structure of the support materials, and also depends on the processes applied during the catalyst preparation technique. The large catalyst surface and the small particle size are required for high-catalyst activity. It is known that a particle size of 3–4 nm is suitable for the reaction and particles of smaller sizes can be observed in the dissolution. The properties of the support material determine the particle size of the catalyst, as well as the geometry dispersed along the cathode. Furthermore, the support material directly affects catalytic activity by determining the oxygen transport rate to the catalyst surface.

Many Pt-based, bimetallic catalysts are being investigated for methanol oxidation. Conversely, because of its ability to oxidize intermediates, Ru CO is the ingredient most often used in the modification of the Pt catalyst. Many nanocatalysts in Pt—Ru have thus been documented for various methods used to boost the electrooxidation of methanol. The Pt—Ru electrocatalyst studies concentrate on evaluating catalyst efficiency under synthesis conditions, excellent metal compositions, various support materials, and direct fuel cells for methanol. For example, Liu et al. investigated the effect of support materials on fuel cell performance. In this study, involving Pt-Ru/WO_3-C hybrids of various WO_3-C ratios prepared by the microwave-assisted polyol method, Pt-Ru/WO_3-C catalysts with the highest stability and electrocatalytic activity were reached with 5% Pt. It has been reported to show better results than —Ru alloys. This is explained by the fact that WO_3 support increases the number of active sites on Pt [69]. A significant portion of these studies focus on synthesis conditions. For example, Jackson et al. reported the optimum synthesis temperature and metal content of the Pt-Ru/C catalyst prepared by the chemical deposition method as 350 °C and the percent ratio 50:50 (metal to C), respectively. Researchers also emphasized the particle size, active surface area, ruthenium oxidation status, and degree of alloying [70]. Lal et al. investigated the effect of various reducing agents (propylene glycol, glycerol, hydrazine, sodium borohydride, sodium formate, and ethylene glycol) on the performance of the Pt-Ru/C catalyst and reported that the highest power density was achieved with the propylene glycol agent. This high activity is explained by the synthesis of propylene glycol with less than the average particle diameter [71].

Mechanisms of the ORR on the surface of Pt have been proposed for changing conditions and studied by many researchers. Sethuraman et al. investigated peroxide selectivity and mechanism of ORR on platinum-based catalysts. It was estimated that the velocity-determining step was the first electron transfer. The platinum-containing binary and triple alloy catalysts were found to be more selective in the direction of hydrogen peroxide than pure Pt [72]. Cui et al. used dimethyl carbonate (DMC)–assisted PtCo catalysts directly in methanol fuel cells to create methanol oxidation reactions. The methanol oxidation activity of DMC obtained from the aluminum-doped SBA-15 Silica molds was measured by a CV method, and PtCo/DMC and Pt/DMC were found to provide 26% and 97% greater catalytic activity than the commercial Pt/C catalyst, respectively [73]. Banham et al. investigated the effects of pore size, wall thickness, and crystal properties of carbon support materials on ORR. Silicon-doped DMC provides higher ORR activity than carbon black obtained by organic regulation and silicon-doped organic-inorganic regulation. The researchers concluded that this situation had a larger pore structure and thicker pore walls. In addition, it was seen that Pt particles in DMC samples were distributed better than carbon black [74].

Calvillo et al. examined the performance of DMC-supported Pt catalyst in PEM fuel cells and the effect of surface chemistry on the performance of support materials. Nanocast-ordered mesoporous carbon (CMK-3) obtained by means of molding was used as support, and the DMC surface was modified with nitric acid before catalyst loading. Researchers have found that the modification process did not have a positive effect on the success of catalyst loading, as opposed to the methanol oxidation reaction. DMC-supported catalysts were found to provide higher hydrogen oxidation activity than commercial catalysts [75]. Hsueh and others have studied the use of DMC as catalyst support materials in the PEM fuel cell. The large-pore-diameter DMC was prepared by a mixture of organic and organic-inorganic regulation techniques. Powder silica particles were first formed by using tetraethyl orthosilicate (TEOS) as the source of silica; and then in the second step, together with silica, phenolic resins were polymerized and macroporous (with pore diameters above 50 nm) and low-order carbon samples were obtained. The active surface area of carbon supported catalysts was 20% higher than commercial catalysts and was found to be suitable for PEM fuel cells [76].

Viva et al. examined the performance of Pt-loaded mesoporous carbons for PEM fuel cells. Support synthesis was carried out by organic regulation, with resorcinol-formaldehyde as a carbon source and poly-diallyl dimethyl ammonium chloride (PDADMAC) as the surface-active agent. The resulting carbons were found to have a pore diameter of 20 nm. The polarization curves of MEAs prepared with mesoporous carbon-supported catalysts were like those of commercial catalysts, and an anode half-reaction increased power by 8%. The rotating disk electrode analysis showed that three electrons were transferred to the cathode per oxygen molecule in the anode with catalysts supported by MC [77]. Ambrosio and others studied the performance of Pt-loaded DMCs in PEM fuel cells. In the study, the electrocatalytic performance of Pt supported by CMK3 carbon obtained by the molding process against ORR was examined and compared with Pt with Vulcan XC72 with equal metal loading. Considering the polarization curves (current and power curves), the ORR activity of the DMC-supported catalysts was found to be close to the carbon black–supported catalysts. To further increase the activity, it was observed that catalyst particles and naphthyl on DMC were more efficiently distributed [78].

Cellorio et al. studied the PEM fuel cell performance of Pt loaded on C nanostrands that were obtained from the polymerization of formaldehyde and silica solution with resorcinol and nickel and cobalt loading. The resulting catalysts were evaluated in PEM fuel cells catalyzed by the hydrogen oxidation reaction, and single-cell performance tests showed a 4% increase in power compared to commercial catalysts of 0.6 V [79]. Ding and others studied the fuel cell performance of DMC-supported PtRu particles. Pt and PtRu loading by different techniques were obtained by molding. The activities against ORR were measured, and DMC-supported PtRu catalyst was found to be ineffective, whereas DMC-supported platinum was found to be more efficient than commercial catalysts. The transmission electron microscope (TEM) images show that the loaded metal particle size is much larger than the pore size [80].

Zolfaghari and others have synthesized carbon-based materials for the electrocatalysis of the methanol oxidation reaction using a nonionic surfactant. Furfuryl alcohol was used as the carbon source, and two separate agents were used as the surface active agent—one was pluronic F127, and the other was polyethylene glycol. The carbon synthesized using F127 was found to be mesoporous, while the carbons synthesized with polyethylene glycol were microporous. Both carbon samples were loaded with formic acid by a Pt reduction method. When used for the methanol oxidation reaction, the mesoporous carbons produced by F127 provide greater activity than carbon black due to the large pore volume [81].

4 Conclusions

Generally, a Pt-based catalyst increases fuel cell performance by accelerating the oxidation reaction of alcohols. Pt, which has superior properties in fuel cells, is widely used as a catalyst. It is understood that the use of Pt-based nanoparticles in fuel cells increases their efficiency and conformation efficiency coefficients. This chapter has focused on the use of platinum-based catalysts. To increase the efficiency of Pt-based catalysts, the poisoning of the electrodes should be decreased with the help of the addition of other metals. In addition, growing the active surface area of Pt-based catalysts increases the activity of the catalyst and the efficiency of the fuel cells. Further, even though Pt-based catalysts are highly effective and boast good catalytic properties, they are highly expensive, so they should be used with the help of some supporting agents and/or other cost-effective metals.

References

[1] L. Gong, Z. Yang, K. Li, W. Xing, C. Liu, J. Ge, Recent development of methanol electrooxidation catalysts for direct methanol fuel cell, J. Energy Chem. 27 (6) (2018) 1618–1628, https://doi.org/10.1016/j.jechem.2018.01.029. Elsevier B.V.

[2] P. Arku, B. Regmi, A. Dutta, A review of catalytic partial oxidation of fossil fuels and biofuels: recent advances in catalyst development and kinetic modelling, Chem. Eng. Res. Des. 136 (2018) 385–402, https://doi.org/10.1016/j.cherd.2018.05.044. Institution of Chemical Engineers.

[3] H. Kishi, et al., Structure of active sites of Fe-N-C nano-catalysts for alkaline exchange membrane fuel cells, Nanomaterials 8 (12) (2018) 965, https://doi.org/10.3390/nano8120965.

[4] J. Zhang, J.W. Medlin, Catalyst design using an inverse strategy: from mechanistic studies on inverted model catalysts to applications of oxide-coated metal nanoparticles, Surf. Sci. Rep. 73 (4) (2018) 117–152, https://doi.org/10.1016/j.surfrep.2018.06.002. Elsevier B.V.

[5] N. Jung, D.Y. Chung, J. Ryu, S.J. Yoo, Y.E. Sung, Pt-based nanoarchitecture and catalyst design for fuel cell applications, Nano Today 9 (4) (2014) 433–456, https://doi.org/10.1016/j.nantod.2014.06.006. Elsevier B.V.

[6] A. Lolli, V. Maslova, D. Bonincontro, F. Basile, S. Ortelli, S. Albonetti, Selective oxidation of HMF via catalytic and photocatalytic processes using metal-supported catalysts, Molecules 23 (11) (2018), https://doi.org/10.3390/molecules23112792.

[7] G.R.O. Almeida, et al., Methanol electro-oxidation on carbon-supported PtRu nanowires, J. Nanosci. Nanotechnol. 19 (2) (2018) 795–802, https://doi.org/10.1166/jnn.2019.15743.

[8] J. Zhang, et al., Cyclic penta-twinned rhodium nanobranches as superior catalysts for ethanol electro-oxidation, J. Am. Chem. Soc. 140 (36) (2018) 11232–11240, https://doi.org/10.1021/jacs.8b03080.

[9] L. Nan, W. Yue, Exceptional electrocatalytic activity and selectivity of platinum@nitrogen-doped mesoporous carbon nanospheres for alcohol oxidation, ACS Appl. Mater. Interfaces 10 (31) (2018) 26213–26221, https://doi.org/10.1021/acsami.8b06347.

[10] S. Themsirimongkon, K. Ounnunkad, S. Saipanya, Electrocatalytic enhancement of platinum and palladium metal on polydopamine reduced graphene oxide support for alcohol oxidation, J. Colloid Interface Sci. 530 (2018) 98–112, https://doi.org/10.1016/j.jcis.2018.06.072.

[11] J. Georgieva, et al., A simple preparation method and characterization of B and N co-doped TiO_2 nanotube arrays with enhanced photoelectrochemical

124 Chapter 7 Pt-based catalysts for alcohol oxidation

performance, Appl. Surf. Sci. 413 (2017) 284–291, https://doi.org/10.1016/j.apsusc.2017.04.055.

[12] Y.C. Lu, M.S. Chen, Y.W. Chen, Hydrogen generation by sodium borohydride hydrolysis on nanosized CoB catalysts supported on TiO2, Al2O3 and CeO 2, Int. J. Hydrog. Energy 37 (5) (2012) 4254–4258, https://doi.org/10.1016/j.ijhydene.2011.11.105.

[13] F. Şen, G. Gökagaç, Activity of carbon-supported platinum nanoparticles toward methanol oxidation reaction: role of metal precursor and a new surfactant, tert-octanethiol, J. Phys. Chem. C 111 (3) (2007) 1467–1473, https://doi.org/10.1021/jp065809y.

[14] F. Sen, Y. Karatas, M. Gulcan, M. Zahmakiran, Amylamine stabilized platinum (0) nanoparticles: active and reusable nanocatalyst in the room temperature dehydrogenation of dimethylamine-borane, RSC Adv. 4 (4) (2014) 1526–1531, https://doi.org/10.1039/c3ra43701a.

[15] S. Eris, Z. Daşdelen, F. Sen, Enhanced electrocatalytic activity and stability of monodisperse Pt nanocomposites for direct methanol fuel cells, J. Colloid Interface Sci. 513 (2018) 767–773, https://doi.org/10.1016/j.jcis.2017.11.085.

[16] Y. Yildiz, E. Erken, H. Pamuk, H. Sert, F. Şen, Monodisperse Pt nanoparticles assembled on reduced graphene oxide: highly efficient and reusable catalyst for methanol oxidation and dehydrocoupling of dimethylamine-borane (DMAB), J. Nanosci. Nanotechnol. 16 (6) (2016) 5951–5958, https://doi.org/10.1166/jnn.2016.11710.

[17] F. Şen, S. Şen, G. Gökağaç, Efficiency enhancement of methanol/ethanol oxidation reactions on Pt nanoparticles prepared using a new surfactant, 1,1-dimethyl heptanethiol, Phys. Chem. Chem. Phys. 13 (4) (2011) 1676–1684, https://doi.org/10.1039/c0cp01212b.

[18] S. Eris, Z. Daşdelen, Y. Yıldız, F. Sen, Nanostructured polyaniline-rGO decorated platinum catalyst with enhanced activity and durability for methanol oxidation, Int. J. Hydrog. Energy 43 (3) (2018) 1337–1343, https://doi.org/10.1016/j.ijhydene.2017.11.051.

[19] Ö. Karatepe, Y. Yildiz, H. Pamuk, S. Eris, Z. Dasdelen, F. Sen, Enhanced electrocatalytic activity and durability of highly monodisperse Pt@PPy-PANI nanocomposites as a novel catalyst for the electro-oxidation of methanol, RSC Adv. 6 (56) (2016) 50851–50857, https://doi.org/10.1039/c6ra06210e.

[20] S. Şen, F. Şen, G. Gökağaç, Preparation and characterization of nano-sized Pt-Ru/C catalysts and their superior catalytic activities for methanol and ethanol oxidation, Phys. Chem. Chem. Phys. 13 (15) (2011) 6784–6792, https://doi.org/10.1039/c1cp20064j.

[21] Z. Daşdelen, Y. Yıldız, S. Eriş, F. Şen, Enhanced electrocatalytic activity and durability of Pt nanoparticles decorated on GO-PVP hybride material for methanol oxidation reaction, Appl. Catal. B Environ. 219 (2017) 511–516, https://doi.org/10.1016/j.apcatb.2017.08.014.

[22] B. Sen, B. Demirkan, B. Şimşek, A. Savk, F. Sen, Monodisperse palladium nanocatalysts for dehydrocoupling of dimethylamineborane, Nano-Struct. Nano-Objects 16 (2018) 209–214, https://doi.org/10.1016/j.nanoso.2018.07.008.

[23] B. Sen, E. Kuyuldar, A. Şavk, H. Calimli, S. Duman, F. Sen, Monodisperse ruthenium–copper alloy nanoparticles decorated on reduced graphene oxide for dehydrogenation of DMAB, Int. J. Hydrog. Energy 44 (21) (2019) 10744–10751, https://doi.org/10.1016/j.ijhydene.2019.02.176.

[24] B. Şen, B. Demirkan, M. Levent, A. Şavk, F. Şen, Silica-based monodisperse PdCo nanohybrids as highly efficient and stable nanocatalyst for hydrogen evolution reaction, Int. J. Hydrog. Energy 43 (44) (2018) 20234–20242, https://doi.org/10.1016/j.ijhydene.2018.07.080.

Chapter 7 Pt-based catalysts for alcohol oxidation **125**

[25] B. Sen, S. Kuzu, E. Demir, S. Akocak, F. Sen, Highly monodisperse RuCo nanoparticles decorated on functionalized multiwalled carbon nanotube with the highest observed catalytic activity in the dehydrogenation of dimethylamine–borane, Int. J. Hydrog. Energy 42 (36) (2017) 23292–23298, https://doi.org/10.1016/j.ijhydene.2017.06.032.

[26] B. Şen, et al., High-performance graphite-supported ruthenium nanocatalyst for hydrogen evolution reaction, J. Mol. Liq. 268 (2018) 807–812, https://doi.org/10.1016/j.molliq.2018.07.117.

[27] S. Eris, Z. Daşdelen, F. Sen, Investigation of electrocatalytic activity and stability of Pt@f-VC catalyst prepared by in-situ synthesis for methanol electrooxidation, Int. J. Hydrog. Energy 43 (1) (2018) 385–390, https://doi.org/10.1016/j.ijhydene.2017.11.063.

[28] N. Patel, R. Fernandes, A. Miotello, Promoting effect of transition metal-doped Co-B alloy catalysts for hydrogen production by hydrolysis of alkaline NaBH4 solution, J. Catal. 271 (2) (2010) 315–324, https://doi.org/10.1016/j.jcat.2010.02.014.

[29] Y.N. Chun, Y.C. Yang, K. Yoshikawa, Hydrogen generation from biogas reforming using a gliding arc plasma-catalyst reformer, Catal. Today 148 (3–4) (2009) 283–289, https://doi.org/10.1016/j.cattod.2009.09.019.

[30] B. Wang, et al., Effects of dielectric barrier discharge plasma on the catalytic activity of Pt/CeO$_2$ catalysts, Appl. Catal. B Environ. 238 (2018) 328–338, https://doi.org/10.1016/j.apcatb.2018.07.044.

[31] E.C. Neyts, Plasma-surface interactions in plasma catalysis, Plasma Chem. Plasma Process. 36 (1) (2016) 185–212, https://doi.org/10.1007/s11090-015-9662-5.

[32] Z. Wang, et al., Catalyst preparation with plasmas: how does it work? ACS Catal. 8 (3) (2018) 2093–2110, https://doi.org/10.1021/acscatal.7b03723. American Chemical Society.

[33] L. Hupa, et al., Chemical resistance and cleanability of glazed surfaces, Surf. Sci. 584 (1) (2005) 113–118, https://doi.org/10.1016/j.susc.2004.11.048.

[34] H. Xu, P. Song, J. Wang, Y. Du, Shape-controlled synthesis of platinum-copper nanocrystals for efficient liquid fuel electrocatalysis, Langmuir 34 (27) (2018) 7981–7988, https://doi.org/10.1021/acs.langmuir.8b01729.

[35] Y.J. Wang, B. Fang, H. Li, X.T. Bi, H. Wang, Progress in modified carbon support materials for Pt and Pt-alloy cathode catalysts in polymer electrolyte membrane fuel cells, Prog. Mater. Sci. 82 (2016) 445–498, https://doi.org/10.1016/j.pmatsci.2016.06.002. Elsevier Ltd.

[36] M. Zamanzad Ghavidel, M.R. Rahman, E.B. Easton, Fuel cell-based breath alcohol sensors utilizing Pt-alloy electrocatalysts, Sensors Actuators B Chem. 273 (2018) 574–584, https://doi.org/10.1016/j.snb.2018.06.078.

[37] K.I. Ozoemena, et al., Fuel cell-based breath-alcohol sensors: innovation-hungry old electrochemistry, Curr. Opin. Electrochem. 10 (2018) 82–87, https://doi.org/10.1016/j.coelec.2018.05.007. Elsevier B.V.

[38] B. Sen, S. Kuzu, E. Demir, S. Akocak, F. Sen, Polymer-graphene hybride decorated Pt nanoparticles as highly efficient and reusable catalyst for the dehydrogenation of dimethylamine–borane at room temperature, Int. J. Hydrog. Energy 42 (36) (2017) 23284–23291, https://doi.org/10.1016/j.ijhydene.2017.05.112.

[39] E. Erken, Y. Yıldız, B. Kilbaş, F. Şen, Synthesis and characterization of nearly monodisperse Pt nanoparticles for C 1 to C 3 alcohol oxidation and dehydrogenation of dimethylamine-borane (DMAB), J. Nanosci. Nanotechnol. 16 (6) (2016) 5944–5950, https://doi.org/10.1166/jnn.2016.11683.

126 Chapter 7 Pt-based catalysts for alcohol oxidation

[40] B.H. Liu, Z.P. Li, A review: hydrogen generation from borohydride hydrolysis reaction, J. Power Sources 187 (2) (2009) 527–534, https://doi.org/10.1016/j.jpowsour.2008.11.032. Elsevier.

[41] D.R. Dekel, Review of cell performance in anion exchange membrane fuel cells, J. Power Sources 375 (2018) 158–169, https://doi.org/10.1016/j.jpowsour.2017.07.117.

[42] T. Lopes, E. Antolini, F. Colmati, E.R. Gonzalez, Carbon supported Pt–Co (3:1) alloy as improved cathode electrocatalyst for direct ethanol fuel cells, J. Power Sources 164 (1) (2007) 111–114, https://doi.org/10.1016/J.JPOWSOUR.2006.10.052.

[43] Z. Ozturk, F. Sen, S. Sen, G. Gokagac, The preparation and characterization of nano-sized Pt–Pd/C catalysts and comparison of their superior catalytic activities for methanol and ethanol oxidation, J. Mater. Sci. 47 (23) (2012) 8134–8144, https://doi.org/10.1007/s10853-012-6709-3.

[44] S. Ertan, F. Şen, S. Şen, G. Gökağaç, Platinum nanocatalysts prepared with different surfactants for C1–C3 alcohol oxidations and their surface morphologies by AFM, J. Nanopart. Res. 14 (6) (2012) 922, https://doi.org/10.1007/s11051-012-0922-5.

[45] C.G. Zoski, Handbook of Electrochemistry, Elsevier, 2007.

[46] J. Wu, X.Z. Yuan, H. Wang, M. Blanco, J.J. Martin, J. Zhang, Diagnostic tools in PEM fuel cell research: part I electrochemical techniques, Int. J. Hydrogen Energy 33 (6) (Mar. 2008) 1735–1746, https://doi.org/10.1016/j.ijhydene.2008.01.013. Pergamon.

[47] S. Srinivasan, Evolution of electrochemistry, in: Fuel Cells, Springer, US, 2006, pp. 3–25.

[48] S.H. Liu, J.R. Wu, Synthesis and characterization of platinum supported on surface-modified ordered mesoporous carbons by self-assembly and their electrocatalytic performance towards oxygen reduction reaction, Int. J. Hydrog. Energy 37 (22) (2012) 16994–17001, https://doi.org/10.1016/j.ijhydene.2012.08.107.

[49] C. Song, J. Zhang, Electrocatalytic oxygen reduction reaction, in: J. Zhang (Ed.), PEM Fuel Cell Electrocatalysts and Catalyst Layers: Fundamentals and Applications, Springer, London, 2008, pp. 89–134.

[50] E. Christoffersen, P. Liu, A. Ruban, H.L. Skriver, J.K. Nørskov, Anode materials for low-temperature fuel cells: a density functional theory study, J. Catal. 199 (1) (2001) 123–131, https://doi.org/10.1006/jcat.2000.3136.

[51] F. Colmati, E. Antolini, E.R. Gonzalez, Preparation, structural characterization and activity for ethanol oxidation of carbon supported ternary Pt-Sn-Rh catalysts, J. Alloys Compd. 456 (1–2) (2008) 264–270, https://doi.org/10.1016/j.jallcom.2007.02.015.

[52] U.B. Demirci, Theoretical means for searching bimetallic alloys as anode electrocatalysts for direct liquid-feed fuel cells, J. Power Sources 173 (1) (2007) 11–18, https://doi.org/10.1016/j.jpowsour.2007.04.069. Elsevier.

[53] Y. Kojima, et al., Development of 10 kW-scale hydrogen generator using chemical hydride, J. Power Sources 125 (1) (2004) 22–26, https://doi.org/10.1016/S0378-7753(03)00827-9.

[54] P. Krishnan, T.H. Yang, W.Y. Lee, C.S. Kim, PtRu-LiCoO$_2$—an efficient catalyst for hydrogen generation from sodium borohydride solutions, J. Power Sources 143 (1–2) (2005) 17–23, https://doi.org/10.1016/j.jpowsour.2004.12.007.

[55] F. Bensebaa, et al., Microwave synthesis of polymer-embedded Pt-Ru catalyst for direct methanol fuel cell, J. Phys. Chem. B 109 (32) (2005) 15339–15344, https://doi.org/10.1021/jp0519870.

[56] Y. Ersoz, R. Yildirim, A.N. Akin, Development of an active platine-based catalyst for the reaction of H$_2$ production from NaBH$_4$, Chem. Eng. J. 134 (1–3) (2007) 282–287, https://doi.org/10.1016/j.cej.2007.03.059.

Chapter 7 Pt-based catalysts for alcohol oxidation **127**

[57] R.S. Liu, et al., Investigation on mechanism of catalysis by Pt-LiCoO2 for hydrolysis of sodium borohydride using X-ray absorption, J. Phys. Chem. B 112 (16) (2008) 4870–4875, https://doi.org/10.1021/jp075592n.

[58] C. Xu, R. Zeng, P.K. Shen, Z. Wei, Synergistic effect of CeO2 modified Pt/C catalysts on the alcohols oxidation, Electrochim. Acta 51 (6) (2005) 1031–1035, https://doi.org/10.1016/j.electacta.2005.05.041.

[59] Z. Liu, Z.Q. Tian, S.P. Jiang, Synthesis and characterization of Nafion-stabilized Pt nanoparticles for polymer electrolyte fuel cells, Electrochim. Acta 52 (3) (2006) 1213–1220, https://doi.org/10.1016/j.electacta.2006.07.027.

[60] K. Mech, P. Zabiński, R. Kowalik, K. Fitzner, Analysis of Co-Pd alloys deposition from electrolytes based on [Co(NH 3)6]3+ and [Pd(NH3) 4]2+ complexes, Electrochim. Acta 104 (2013) 468–473, https://doi.org/10.1016/j.electacta.2012.12.006.

[61] K. Oishi, O. Savadogo, Electrochemical investigation of Pd-Co thin films binary alloy for the oxygen reduction reaction in acid medium, J. Electroanal. Chem. 703 (2013) 108–116, https://doi.org/10.1016/j.jelechem.2013.04.006.

[62] N. Travitsky, T. Ripenbein, D. Golodnitsky, Y. Rosenberg, L. Burshtein, E. Peled, Pt-, PtNi- and PtCo-supported catalysts for oxygen reduction in PEM fuel cells, J. Power Sources 161 (2) (2006) 782–789, https://doi.org/10.1016/j.jpowsour.2006.05.035.

[63] E. Antolini, Platinum-based ternary catalysts for low temperature fuel cells. Part II. Electrochemical properties, Appl. Catal. B Environ. 74 (3–4) (2007) 337–350, https://doi.org/10.1016/j.apcatb.2007.03.001.

[64] M. Ammam, E.B. Easton, Oxygen reduction activity of binary PtMn/C, ternary PtMnX/C (X = Fe, Co, Ni, Cu, Mo and, Sn) and quaternary PtMnCuX/C (X = Fe, Co, Ni, and Sn) and PtMnMoX/C (X = Fe, Co, Ni, Cu and Sn) alloy catalysts, J. Power Sources 236 (2013) 311–320, https://doi.org/10.1016/j.jpowsour.2013.02.029.

[65] J. Shim, D.Y. Yoo, J.S. Lee, Characteristics for electrocatalytic properties and hydrogen-oxygen adsorption of platinum ternary alloy catalysts in polymer electrolyte fuel cell, Electrochim. Acta 45 (12) (2000) 1943–1951, https://doi.org/10.1016/S0013-4686(99)00414-4.

[66] A.K. Shukla, M. Neergat, P. Bera, V. Jayaram, M.S. Hegde, An XPS study on binary and ternary alloys of transition metals with platinized carbon and its bearing upon oxygen electroreduction in direct methanol fuel cells, J. Electroanal. Chem. 504 (1) (2001) 111–119, https://doi.org/10.1016/S0022-0728(01)00421-1.

[67] J. Luo, et al., Ternary alloy nanoparticles with controllable sizes and composition and electrocatalytic activity, J. Mater. Chem. 16 (17) (2006) 1665–1673, https://doi.org/10.1039/b518287e.

[68] Ç. Güldür, S. Güneş, Performance evaluation of PtAgFe/C as a cathode catalyst in PEM fuel cell, Gazi Univ. J. Sci. 31 (1) (2018) 42–51.

[69] W. Liu, X. Qin, X. Zhang, Z. Shao, B. Yi, Preparation of PtRu/WO3–C by intermittent microwave method with enhanced catalytic activity of methanol oxidation, J. Appl. Electrochem. 46 (8) (2016) 887–893, https://doi.org/10.1007/s10800-016-0953-1.

[70] C. Jackson, O. Conrad, P. Levecque, Systematic study of Pt-Ru/C catalysts prepared by chemical deposition for direct methanol fuel cells, Electrocatalysis 8 (3) (2017) 224–234, https://doi.org/10.1007/s12678-017-0359-9.

[71] O. Sahin, H. Kivrak, A comparative study of electrochemical methods on Pt-Ru DMFC anode catalysts: the effect of Ru addition, Int. J. Hydrog. Energy 38 (2) (2013) 901–909, https://doi.org/10.1016/j.ijhydene.2012.10.066.

[72] V.A. Sethuraman, J.W. Weidner, A.T. Haug, M. Pemberton, L.V. Protsailo, Importance of catalyst stability vis-à-vis hydrogen peroxide formation rates

in PEM fuel cell electrodes, Electrochim. Acta 54 (23) (2009) 5571–5582, https://doi.org/10.1016/j.electacta.2009.04.062.

[73] X. Cui, J. Shi, L. Zhang, M. Ruan, J. Gao, PtCo supported on ordered mesoporous carbon as an electrode catalyst for methanol oxidation, Carbon 47 (1) (2009) 186–194, https://doi.org/10.1016/J.CARBON.2008.09.054.

[74] D. Banham, F. Feng, T. Fürstenhaupt, K. Pei, S. Ye, V. Birss, Effect of Pt-loaded carbon support nanostructure on oxygen reduction catalysis, J. Power Sources 196 (13) (2011) 5438–5445, https://doi.org/10.1016/j.jpowsour.2011.02.034.

[75] L. Calvillo, M. Gangeri, S. Perathoner, G. Centi, R. Moliner, M.J. Lázaro, Synthesis and performance of platinum supported on ordered mesoporous carbons as catalyst for PEM fuel cells: effect of the surface chemistry of the support, Int. J. Hydrog. Energy 36 (16) (2011) 9805–9814, https://doi.org/10.1016/j.ijhydene.2011.03.023.

[76] Y.J. Hsueh, et al., Ordered porous carbon as the catalyst support for proton-exchange membrane fuel cells, Int. J. Hydrogen Energy 38 (25) (2013) 10998–11003, https://doi.org/10.1016/j.ijhydene.2013.01.007.

[77] F.A. Viva, et al., Mesoporous carbon as Pt support for PEM fuel cell, Int. J. Hydrogen Energy 39 (16) (2014) 8821–8826, https://doi.org/10.1016/j.ijhydene.2013.12.027.

[78] E.P. Ambrosio, M.A. Dumitrescu, C. Francia, C. Gerbaldi, P. Spinelli, Ordered mesoporous carbons as catalyst support for PEM fuel cells, Fuel Cells 9 (3) (2009) 197–200, https://doi.org/10.1002/fuce.200800082.

[79] V. Celorrio, J. Flórez-Montaño, R. Moliner, E. Pastor, M.J. Lázaro, Fuel cell performance of Pt electrocatalysts supported on carbon nanocoils, Int. J. Hydrogen Energy 39 (10) (2014) 5371–5377, https://doi.org/10.1016/j.ijhydene.2013.12.198.

[80] J. Ding, K.Y. Chan, J. Ren, F.S. Xiao, Platinum and platinum-ruthenium nanoparticles supported on ordered mesoporous carbon and their electrocatalytic performance for fuel cell reactions, Electrochim. Acta 50 (15) (2005) 3131–3141, https://doi.org/10.1016/j.electacta.2004.11.064.

[81] R. Zolfaghari, F.R. Ahmadun, M.R. Othman, W.R. Wan Daud, M. Ismail, Nonionic surfactant-templated mesoporous carbon as an electrocatalyst support for methanol oxidation, Mater. Chem. Phys. 139 (1) (2013) 262–269, https://doi.org/10.1016/j.matchemphys.2013.01.033.

8

Monometallic nanomaterials for direct alcohol fuel cells

Ramazan Bayat[a,b], Vildan Erduran[a,b], Muhammed Bekmezci[a,b], Iskender Isik[b], and Fatih Şen[a]
[a]Şen Research Group, Department of Biochemistry, Dumlupinar University, Kütahya, Turkey. [b]Department of Materials Science and Engineering, Faculty of Engineering, Dumlupinar University, Kütahya, Turkey

1 Introduction

Today, with the world's increasing population, developing industries, and changing daily habits, human energy needs are increasing day by day. This issue needs to be addressed so that the need for energy does not cause major problems in the future. This increase in energy demand is coupled with the desire for a gradual decrease in the usage of fossil fuels such as oil, natural gas, coal, and other fuels. Because as fossil fuels pollute and damage the environment, the demand for clean and renewable energy sources is increasing day by day [1, 2]. Fuel cells are devices that convert the chemical energy of fuels directly into heat and electrical energy by using fuels such as hydrogen, methanol, natural gas, and petroleum. Today, fuel cells are prominent in alternative energy production due to problems such as cost, resource availability, and continuity of operations in turbine, generator, and renewable energy production systems [3, 4]. Also, fuel cells are similar in electricity generation due to their electrochemical reactions with existing batteries. However, these devices convert previously-stored energy into electrical energy as a result of electrochemical reactions and provide the necessary energy. On the other hand, although the electrical energy is proportional to the energy previously stored in these technologies, so long as fuel cells are fed with air and fuel, they carry out electrochemical reactions and continue to provide energy [5]. The electrical efficiency of fuel

Nanomaterials for Direct Alcohol Fuel Cells. https://doi.org/10.1016/B978-0-12-821713-9.00019-6
Copyright © 2021 Elsevier Inc. All rights reserved.

cells varies between 45% and 60%. Further, as a result of the evaluation of waste heat, the efficiency is up to 70%–80% [6, 7].

Fuel cells, a new technology, have many advantages over other renewable energy systems, but they also have disadvantages.

The important advantages of fuel cells in terms of usability are as follows [8]:
- High conversion efficiency (40%–60%)
- Being able to use hydrocarbon fuels such as methanol, natural gas, and petroleum as fuel
- Environmentally harmless emissions (in high-temperature fuel cell types, waste heat can be used in cogeneration)
- Quiet operation, as they do not contain any mechanical parts

The major disadvantages of fuel cells in terms of usability are as follows [8, 9]:
- Problems with storing hydrogen, which is commonly used as the fuel in fuel cells
- Expensive cost per unit of this energy due to ongoing research studies

Fuel cells are classified according to the type of electrolyte used, as follows:
- Alkaline fuel cells (AFCs)
- Polymer electrolyte membrane fuel cells (PEMFCs)
- Direct methanol fuel cells (DMFCs)
- Molten carbonate fuel cells (MCFCs)
- Solid oxide fuel cells (SOFCs)

The use of fuel cells has become widespread in recent years due to their use in transportation, domestic, military, and mobile areas and because they are environmentally friendly. General application areas of fuel cells are given in Table 1 [10].

Unlike other types of fuels, alcohols can be used directly as fuel in fuel cells, while at the same time releasing environmentally friendly by-products such as water, which do not harm the environment [11]. The ideas of increasing environmental sensitivity, ensuring sustainability, and developing and disseminating fuel systems that do not produce hazardous waste have created a major driving force behind the development of fuel cells by the scientific world. With this growth rate, the portable fuel cell market, which was $80.1 million in 2008, is expected to be $24.8 billion in 2025 [12]. In addition to the details about the structure and classification of the fuel cell mentioned previously, the main purpose of this chapter is to discuss whether direct alcohol fuel cells (DAFCs) allow technology to be used for mobile applications by eliminating the need for slow and expensive fuel converters.

Table 1 Fuel cell types and properties [10].

	Electrolyte	Operating temperature (°C)	Electrical efficiency (%)	Fuel/oxidizer	Usage areas
AFCs	Potassium hydroxide solution	60–250	60–70	H_2/O_2	Space vehicles
PEMFCs	Proton transformation membrane	30–100	40–60	H_2/O_2, air	Automotive and stationary applications
DMFCs	Polymer membrane	25–90	20–30	CH_3OH/O_2, air	Portable devices
MCFCs	Alkali carbonate	600–800	65	H_2/O_2, natural gas, biogas, coal gas, air	Cogenerated fixed applications
SOFC	Molten alkali metal mixture	600–1000	60–65	H_2/O_2, natural gas, biogas, coal gas, air	Cogenerated fixed applications

High-pH intervals are used in alkaline DEFCs. The oxidation reaction energy desired to be obtained in DEFCs can be achieved by the formation of a significant current density in the presence of a catalyst. As is already known, catalysts are materials that reduce the activation energy of the reaction in a chemical reaction and accelerate it. This system also allows the use of cheap metal catalysts such as non-noble metals and transition metals [13].

With the development of nanotechnology, studies in the field of fuel cells are increasing. Nanotechnology provides solutions for issues such as the production, efficiency, and cost of fuel cells.

2 Nanostructured materials

Nanoscience and nanotechnology involve manipulating individual atoms and molecules to produce materials from them for applications that are well below the lower microscopic level [14]. Nanostructured materials have a comprehensive range of applications, such as optical, electronic, magnetic, biological, catalytic, and biomedical materials, and the scope of their chemical synthesis and processes is extensive [15]. Nanoparticles, which are defined as particles with a size of 100 nm or less, are the basis of nanoscale materials, and therefore nanotechnology [16]. These particles generally exhibit properties that are considered to be different and superior to other commercial materials.

The known advantages of nanoparticles include quantum dimensional effects, dimensional dependence of the electronic structure, unique characters of surface atoms, and high surface/volume ratio [17]. Due to the extraordinary features exhibited by these structures, nanoparticle synthesis has paved the way for the preparation of most technological and pharmacological products, such as high-efficiency catalysts, superconductors with specific technological materials for optical applications, antiwear additives, surfactants, and drug release systems. In addition, with the control of materials at the nanoscale level, miniaturized devices can be realized with unique functionality, such as nanocarriers, sensors, nanomachines, and superior-density data storage systems [18–20]. It is clear that the first indispensable step for new developments in nanotechnology, which involves the design, production, and functional use of nanostructured materials and devices, is the production of nanoparticles. Nanoparticles, which form the origin of the production of nanotechnological materials, fall into a wide range of chemicals and morphologies. Today, nanoparticles can be prepared with the desired properties, consisting of a mixture of metal, metal alloy, ceramic, and polymer-based material, or a mixture of different morphologies such as core-shell, additive, sandwich, hollow, spherical, sticklike, and multifaceted [21, 22].

Metal nanoparticles in nanotechnology unlock several features and pave many new roads in nanotechnology. They have received great attention due to their applications in various branches of technology, including fuel cells, pharmaceuticals, electronics, photonics, sensing technologies, therapeutics, and antimicrobial products. The most important feature of metal nanoparticles is the surface area/volume ratio, which allows them to easily interact with other particles. In metal nanoparticles, a high surface area/volume ratio makes diffusion take place faster and can be applied at low temperatures. Metallic nanoparticles are nanosized metals in the size range of 10–100 nm [23]. Metal nanoparticles can be grouped into monometallic, bimetallic, core-shell, cluster-cluster, and alloy according to the number of components that they contain. The use of catalysts prepared with monomeric nanoparticles in DAFCs will be discussed in this section.

Chapter 8 Monometallic nanomaterials for direct alcohol fuel cells   **133**

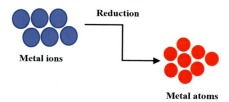

Fig. 1 Reduction pathway for the formation of metal nanoparticles [24]. This image was used from an open-access article distributed under the terms of the Creative Commons Attribution License, which allows unlimited use, distribution, and reproduction in any medium by accurately quoting the original work.

Monometallic nanoparticles occur when metal salt is reduced simultaneously in the presence of reducing reagent, stabilizer, and solvent (Fig. 1) [24].

If a catalyst is able to bind reagents to its surface as strongly as the reaction that takes place, and its products as weakly as they could be separate from the surface, it can be considered as a good catalyst [25, 26]. Some metals such as silver (Ag) catalysts cannot keep the reactive molecules on their surface strong enough. On the other hand, Wolfram (W) holds its reactive molecules on its surface extremely strongly, making it difficult for products to separate from the surface. Catalysts of this type are not effective. In addition, some metal catalysts, such as nickel (Ni) and platinum (Pt), are good catalysts because they can keep their reagents strong enough that the reaction can occur on their surface, and their products weak enough to be separated from the surface [27].

Another important criterion for a good catalyst is that the surface area/volume ratio is large and the homogeneous nanoparticle distribution is excessive. These two effects increase the efficiency of catalysis. This helps to reduce the required amounts of the Pt group and precious metals that are generally used. It also paves the way for the use of Au, which can be an active catalyst only when it is a nanoparticle, and other metals, which are used less often. As nanoparticles decrease in size does their increased surface-to-volume ratio. Catalytic reactivity is related to features such as specific surface area, stabilization of surface area, the topology of nanoparticles (smoothness), support materials and interfaces, as well as other similar features. It has been found that the large surface area and large support porosity result in excellent catalytic efficiency, especially compared to the activity of state-of-the-art catalysts. Noble metal–based and metal oxide catalysts are good examples of this (e.g., CeO_2 catalysts for the automotive industry) [28]. As the surface area of metal nanoparticles increases, so does their biological activity in some practices as a result of the increase in energy on the surface. For instance, Au nanoparticles are used for antimicrobial applications. Furthermore, the increase in the specific surface area decreases the sintering temperature of metal nanoparticles [29, 30].

3 Catalyst layer

The importance of a catalyst is great in fuel cells, as its capacity to convert fuel into hydrogen directly affects the performance and yield of the fuel cell [31–33]. The catalyst layer (CL) is a very thin ($\approx 10\ \mu m$) polymer composite structure containing metal or metal oxides that transmit protons and a stored electron. CLs should be able to perform many tasks [34]:

- It should allow electrochemical reactions.
- It should deliver protons to or from these regions (via polymer material).
- It should transmit electrons (carbon or other elements) to or from these regions (porous media).

Allowing the movement of reactants (hydrogen, oxygen, and water molecules) through the porous structure of the layersis very important. The electrochemical activity should be evaluated to the highest level for minimum catalyst use. In general, catalysts are intended to be long-lasting, highly efficient, and available in the least amount per cell [35]. Active surface area is the most important element in catalysts. The number and distribution of these active surface area is extremely important for catalyst reactions. Catalysts should have a sufficient number of active surface area present in the fuel cell, and their number should reach the maximum. Studies have shown that the best catalysts for activeness and stability are Pt and Pt-containing catalysts [36–40].

Because the most important parameter in electrocatalysts is the catalyst surface, the most effective way to increase the efficiency of an electrocatalyst is to increase the reaction surface at low overvoltages by increasing the active surface area. At this point, the current density is a significant parameter for understanding the electrocatalytic efficiency [41].

The choice of metal suitable for the catalyst task is also very important because this is the catalyst system that forms the electrode. Electrocatalytic properties of metals can be determined by so-called fermi dynamics. "Fermi energy" is the energy of the electrons moving in metals (i.e., the energy of electrons that transfer from metal to ions in solution) [42].

The electrocatalytic properties of the metals used as catalysts are determined by the current density values resulting from

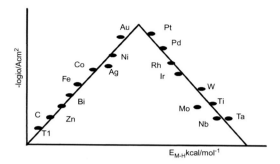

Fig. 2 Volcano curve for a metal electrode [43].

charge changes during reactions. Metals with a high current density show more activity electro catalytically. The curve obtained by plotting the values of current densities that occur on the surface of the metals due to load changes against Metal-H bond energy is called the "volcano curve," as shown in Fig. 2. As metal-H bond energy increases, the current density increases first and then decreases. The metals showing the highest current density on the curve, although they vary according to reaction type and environment, are Pt, Au, Pd, Ni, and Rh [43].

Many transition metals in the volcanic curve called Pt group metals (PGMs) are known to be good catalysts for various chemical and electrochemical reactions in the literature.

The place where hydrogen oxidation reactions (HORs) and oxygen reduction reactions (ORRs) occur is the CL. Here, the following important phases occur for electrochemical reactions: (1) carbon support materials with Pt particles dispersed on the carbon surface, (2) ionomers, and (3) voids. While hydrogen oxidation reactions (HORs) occur in the anode; ORRs occur in the cathode. The most popular catalyst for both types of reactions is Pt, so the catalyst layer creates the bulk of the cost in a fuel cell. Pt shows good catalytic activity [44–52]. The most important factor in catalyst-level development is Pt loading. The U.S. Department of Energy set targeted loading values of 0.3 mg/cm^2 Pt in 2010, 0.2 mg/cm^2 in 2015, and 0.125 mg/cm^2 in 2017 and 2020. Another research topic as opposed to Pt loading reduction is new catalyst substances. There are two main approaches to this subject: (1) the use of cheaper precious metals such as ruthenium or palladium instead of Platin, and (2) the use of nonprecious metal catalysts [53–55].

Wang [55] and Liu [56] published detailed studies on nonprecious metal catalysts. Another issue studied concerning CLs is to prevent significant performance decreases as a result of CO adsorption of the surfaces of Pt particles [57]. In the 1980s and 1990s, binary Pt—Ru catalysts and oxidant mixing techniques were proposed various CO-tolerant materials such as zeolite, Pt—Mo, and sulfurized catalysts were investigated [58]. Another approach to reducing Pt loading and CL costs is to decrease the utilization of Pt. Studies have been carried out to increase the surface area of active catalysts. Another important issue is the durability of CLs. CLs are sensitive to material disruption during the process. The primary distortion mechanism occurs due to Pt aggregation or loss of active sites; Pt particles can dissolve in the ionomer and reassemble to form large particles [59]. Studies have shown that the durability of CLs is not limited to only corrosion resistance or H_2O_2 formation of the support material. However, uncleanliness in the reactants such as ammonia, carbon monoxide, and SO_2 can prevent the catalysts and reduce the active surface area [60].

Many studies have been conducted on alcohol-fed fuel cells and monometallic electrocatalysts. As shown in Fig. 3, while the anode part of the DAFCs is fed directly by a mixture of alcohol and water, the alcohol is broken down by forming an electrochemical reaction with the water formed in the cathode part. As a result of the decomposition of methanol, protons, electrons, and carbon dioxide are formed. The protons formed react on the oxygen

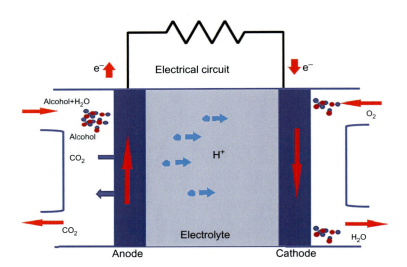

Fig. 3 DAFC operation [61].

supplied from the air to form water on the cathode. Thermodynamic potential as a result of these reactions causes a voltage to form in the external circuit, and electricity is generated.

The reactions that occur in the anode, cathode, and cell during the study of DAFCs are given here [61]:

$$Anode\,reaction : Alcohol + H_2O \rightarrow CO_2 + 6H^+ + 6e^-$$

$$Cathode\,reaction : 3/2O_2 + 6H^+ + 6e^- \rightarrow 3H_2O$$

$$Total\,reaction : Alcohol + 3/2O_2 \rightarrow CO_2 + 2H_2O$$

Although the alcohol used as a fuel in direct alcohol fuel cells (DAFCs) has many advantages, its electrochemical resistance is poor in practical applications. Therefore, transition metal nano-catalysts are used as anode catalysts to activate the alcohol in the fuel cell. In DAFCs, Pt and Pt group metals are used both as anode catalysts and as effective catalysts in the oxidation reaction of alcohol in an acidic medium [62–64]. It also has the disadvantage that the oxidized species that may form during the reaction can be poisoned very quickly due to adsorption to the surface. Because of these points, research on alternative catalysts has been initiated.

The cathode used in the performance tests of fuel cells is generally Pt/C. Li et al. emphasized that Ni and Ag metals can also be used as cathode electrocatalysts against corrosion in an alkaline environment [65]. The cathode electrocatalyst is expected to trigger an ORR. The most important problem encountered with cathode electrocatalysts is deactivation.

Investigating the effect of Pt loading amount on the cathode, Suda et al. showed that cell performance is not dependent on Pt loading and membrane thickness, as cell polarization curves and power densities [66]. This situation is also important in terms of the price of the fuel cell. However, because Pt is one of the most expensive materials that exists, replacing it with a cheaper active metal will increase the competitiveness of the fuel cells used in this study. Therefore, Suda et al. tested silver as an alternative cathode material [67]. The maximum power densities of the cell loaded with 1.6 mg/cm^{-2} Ag at the cathode and the cell loaded with 1.0 mg cm^{-2} Pt are close. This has shown that the boron hydride can be a suitable cathode electrocatalyst for the fuel cell [66]. It should be noted that the current density values obtained as a result of the electrochemical reactions also depend on the type of metal and method used in the electrode modification, the surface composition of the electrodes, the temperature, the pH of the medium, the concentration of the electroactive material, and the electrolyte.

4 Conclusion

With the technological and life standards developments in the world, energy demands are increasing rapidly. With the developing industry after the discovery of gasoline, the need for energy has increased more and more. More fossil fuels are consumed as the world population increases rapidly and more harmful gases are released into the atmosphere. Besides these harmful gases, the decrease in oil and natural gas reserves causes energy wars between countries. In developed countries, in order to reduce the dependence on fossil fuels for energy, the use of renewable energy sources and more efficient energy conversion technologies has been introduced into the agenda. The development of various types of fuel cells for use in a number of applications is becoming widespread.

Nanostructured materials have a comprehensive range of applications, such as optical, electronic, magnetic, biological, catalytic, and biomedical materials. Nanocarriers, sensors, nanomachines, and superior density data storage systems can be realized with unique functionalities. Monometallic nanoparticles are created when metal salt is reduced simultaneously in the presence of reducing reagent, stabilizer, and solvent. If a catalyst is able to bind reagents to its surface as strongly as the reaction can take place, it can be considered to be a good catalyst. Some metals such as silver catalysts cannot keep the reactive molecules on their surface strong enough. Another important criterion for a good catalyst is that the surface area/volume ratio is large, and the homogeneous nanoparticle distribution is excessive. As nanoparticles decrease in size, so does the ratio of their surfaces to their volumes. It has been found that the large surface area and large support porosity result in excellent catalytic efficiency, especially compared to the activity of state-of-the-art catalysts.

Catalysts are intended to be long-lasting, highly efficient, and available in the smallest amount per cell. Active centers are the most important elements in catalysts. The number and distribution of these active centers are extremely important for catalyst reactions. Studies have shown that the best catalysts for activeness and stability are Pt and Pt-containing catalysts. Electrocatalytic properties of metals can be determined by fermi dynamics. Fermi energy, the energy of the electrons moving in metals, is aimed to reduce the use of fossil fuels, and there has been an increasing number of studies about nanotechnology and fuel cells. In this way, energy production can be provided without causing any harm to nature.

References

[1] L. Gong, Q. Duan, J. Liu, M. Li, P. Li, K. Jin, J. Sun, Spontaneous ignition of high-pressure hydrogen during its sudden release into hydrogen/air mixtures, Int. J. Hydrog. Energy 43 (2018) 23558–23567, https://doi.org/10.1016/j.ijhydene.2018.10.226.

[2] P. Nikolaidis, A. Poullikkas, A comparative overview of hydrogen production processes, Renew. Sust. Energ. Rev. 67 (2017) 597–611, https://doi.org/10.1016/j.rser.2016.09.044.

[3] Y.H. Kwok, Y. Wang, M. Wu, F. Li, Y. Zhang, H. Zhang, D.Y.C. Leung, A dual fuel microfluidic fuel cell utilizing solar energy and methanol, J. Power Sources 409 (2019) 58–65, https://doi.org/10.1016/j.jpowsour.2018.10.095.

[4] Z. Daşdelen, Y. Yıldız, S. Eriş, F. Şen, Enhanced electrocatalytic activity and durability of Pt nanoparticles decorated on GO-PVP hybride material for methanol oxidation reaction, Appl. Catal. B Environ. 219 (2017) 511–516, https://doi.org/10.1016/j.apcatb.2017.08.014.

[5] L.J. Pettersson, R. Westerholm, State of the art of multi-fuel reformers for fuel cell vehicles: problem identification and research needs, Int. J. Hydrog. Energy 26 (2001) 243–264, https://doi.org/10.1016/S0360-3199(00)00073-2.

[6] M.N. Lakhoua, M. Harrabi, M. Lakhou, Application of system analysis for thermal power plant heat rate improvement, in: Thermal Power Plants – Advanced Applications, Intech, 2013.

[7] S. Eris, Z. Daşdelen, F. Sen, Enhanced electrocatalytic activity and stability of monodisperse Pt nanocomposites for direct methanol fuel cells, J. Colloid Interface Sci. 513 (2018) 767–773, https://doi.org/10.1016/j.jcis.2017.11.085.

[8] A.L. Dicks, D.A.J. Rand, Fuel Cell Systems Explained, Wiley, 2018, https://doi.org/10.1002/9781118706992.

[9] B. Şen, B. Demirkan, M. Levent, A. Şavk, F. Şen, Silica-based monodisperse PdCo nanohybrids as highly efficient and stable nanocatalyst for hydrogen evolution reaction, Int. J. Hydrog. Energy 43 (2018) 20234–20242, https://doi.org/10.1016/j.ijhydene.2018.07.080.

[10] M.V. Lototskyy, I. Tolj, L. Pickering, C. Sita, F. Barbir, V. Yartys, The use of metal hydrides in fuel cell applications, Prog. Nat. Sci. Mater. Int. 27 (2017) 3–20, https://doi.org/10.1016/j.pnsc.2017.01.008.

[11] R.H. Petrucci, W.S. Harwood, F.G. Herring, General Chemistry: Principles and Modern Applications, Prentice Hall, 2002.

[12] J.O. Bockris, A.K.N. Reddy, M. Gamboa-Aldeco, Modern Electrochemistry. Volume 2A, Fundamentals of Electrodics, Kluwer Academic, 2002.

[13] H. Liu, J. Zhang, Electrocatalysis of Direct Methanol Fuel Cells: From Fundamentals to Applications, John Wiley and Sons, 2009, https://doi.org/10.1002/9783527627707.

[14] F.C. Adams, C. Barbante, Nanoscience, nanotechnology and spectrometry, Spectrochim. Acta B At. Spectrosc. 86 (2013) 3–13, https://doi.org/10.1016/j.sab.2013.04.008.

[15] C.C. Koch, Nanostructured Materials: Processing, Properties and Potential Applications, Noyes Pub./William Andrew Pub, 2002.

[16] M. Faraday, The Bakerian Lecture: Experimental Relations of Gold (and Other Metals) to Light, 147, The Royal Society, 1857, pp. 145–181.

[17] M.-C. Daniel, D. Astruc, Gold Nanoparticles: Assembly, Supramolecular Chemistry, Quantum-Size-Related Properties, and Applications toward Biology, Catalysis, and Nanotechnology, American Chemical Society, 2003. https://doi.org/10.1021/cr030698+.

140 Chapter 8 Monometallic nanomaterials for direct alcohol fuel cells

[18] L. Razdolsky, Physical chemistry of nanoparticles, in: Razdolsky (Ed.), Phenomenological Creep Models of Composites and Nanomaterials, Taylor & Francis Group, 2019.

[19] S.K. Vashist, A.G. Venkatesh, K. Mitsakakis, G. Czilwik, G. Roth, F. von Stetten, Roland Zengerle, Nanotechnology-based biosensors and diagnostics: technology push versus industrial/healthcare requirements, BioNanoScience (2012) 115–126, https://doi.org/10.1007/s12668-012-0047-4.

[20] C.N.R. Rao, A.K. Cheetham, Science and technology of nanomaterials: current status and future prospects, J. Mater. Chem. 11 (2001) 2887–2894, https://doi.org/10.1039/b105058n.

[21] J. Alonso, J.M. Barandiarán, L. Fernández Barquín, A. García-Arribas, Magnetic nanoparticles, synthesis, properties, and applications, in: A.A. El-Gendy (Ed.), Magnetic Nanostructured Materials, Elsevier, 2018, pp. 1–40.

[22] A.A. El Gendy, J.M. Barandiaran, R.L. Hadimani, Magnetic Nanostructured Materials: From Lab to Fab, in:, A.A. El Gendy (Ed.), Magnetic Nanostructured Materials, Elsevier, 2018, p. xvii.

[23] K.K. Harish, V. Nagasamy, B. Himangshu, K. Anuttam, Metallic nanoparticle: a review, Biomed. J. Sci. Tech. Res. 4 (2018), https://doi.org/10.26717/BJSTR.2018.04.001011.

[24] V.V. Makarov, A.J. Love, O.V. Sinitsyna, S.S. Makarova, I.V. Yaminsky, M.E. Taliansky, N.O. Kalinina, "Green" nanotechnologies: synthesis of metal nanoparticles using plants, Acta Nat. 6 (2014) 35–44, https://doi.org/10.32607/20758251-2014-6-1-35-44.

[25] R. Grisel, K.-J. Weststrate, A. Gluhoi, B.E. Nieuwenhuys, Catalysis by Gold Nanoparticles, Springer, 2002.

[26] B. Çelik, Y. Yildiz, H. Sert, E. Erken, Y. Koşkun, F. Şen, Monodispersed palladium-cobalt alloy nanoparticles assembled on poly(N-vinyl-pyrrolidone) (PVP) as a highly effective catalyst for dimethylamine borane (DMAB) dehydrocoupling, RSC Adv. 6 (2016) 24097–24102, https://doi.org/10.1039/c6ra00536e.

[27] C. Gu, W.Q. Tao, M. Li, Y. Jiang, X.Q. Liu, P. Tan, L.B. Sun, Fabrication of multifunctional integrated catalysts by decorating confined Ag nanoparticles on magnetic nanostirring bars, J. Colloid Interface Sci. 555 (2019) 315–322, https://doi.org/10.1016/j.jcis.2019.07.098.

[28] E. Aneggi, C. de Leitenburg, M. Boaro, P. Fornasiero, A. Trovarelli, Catalytic applications of cerium dioxide, in: Cerium Oxide Synthesis, Properties and Applications, Elsevier, 2020, pp. 45–108, https://doi.org/10.1016/b978-0-12-815661-2.00003-7.

[29] K. Negi, A. Umar, M.S. Chauhan, M.S. Akhtar, Ag/CeO_2 nanostructured materials for enhanced photocatalytic and antibacterial applications, Ceram. Int. 45 (2019) 20509–20517, https://doi.org/10.1016/j.ceramint.2019.07.030.

[30] Ö. Karatepe, Y. Yildiz, H. Pamuk, S. Eris, Z. Dasdelen, F. Sen, Enhanced electrocatalytic activity and durability of highly monodisperse Pt@PPy-PANI nanocomposites as a novel catalyst for the electro-oxidation of methanol, RSC Adv. 6 (2016) 50851–50857, https://doi.org/10.1039/c6ra06210e.

[31] N. Nakashima, Carbon nanotube-based direct methanol fuel cell catalysts, in: N. Nakashima (Ed.), Nanocarbons for Energy Conversion: Supramolecular Approaches, Springer International Publishing, Cham, 2019, pp. 29–43.

[32] F.Y. Zhang, S.G. Advani, A.K. Prasad, Advanced high resolution characterization techniques for degradation studies in fuel cells, in: Polymer Electrolyte Fuel Cell Degradation, Elsevier Inc., 2012, pp. 365–421, https://doi.org/10.1016/B978-0-12-386936-4.10008-9.

[33] S. Eris, Z. Daşdelen, Y. Yıldız, F. Sen, Nanostructured polyaniline-rGO decorated platinum catalyst with enhanced activity and durability for methanol oxidation, Int. J. Hydrog. Energy 43 (2018) 1337–1343, https://doi.org/10.1016/j.ijhydene.2017.11.051.

[34] J. Andrews, A.K. Doddathimmaiah, Regenerative fuel cells, in: Materials for Fuel Cells, Elsevier Inc., 2008, pp. 344–385, https://doi.org/10.1533/9781845694838.344.

[35] N. Nakashima, T. Fujigaya, Carbon Nanotube-Based Fuel Cell Catalysts—Comparison with Carbon Black, Springer, Cham, 2019, pp. 1–28, https://doi.org/10.1007/978-3-319-92917-0_1.

[36] M. Nacef, M.L. Chelaghmia, A.M. Affoune, M. Pontié, Electrochemical investigation of glucose on a highly sensitive nickel-copper modified pencil graphite electrode, Electroanalysis 31 (2019) 113–120, https://doi.org/10.1002/elan.201800622.

[37] J. Yang, N. Nakashima, Carbon Nanotube-Based Non-Pt Fuel Cell Catalysts, Springer, Cham, 2019, pp. 277–293, https://doi.org/10.1007/978-3-319-92917-0_12.

[38] B. Çelik, G. Başkaya, H. Sert, Ö. Karatepe, E. Erken, F. Şen, Monodisperse Pt(0)/DPA@GO nanoparticles as highly active catalysts for alcohol oxidation and dehydrogenation of DMAB, Int. J. Hydrog. Energy 41 (2016) 5661–5669, https://doi.org/10.1016/j.ijhydene.2016.02.061.

[39] F. Şen, G. Gökagaç, Activity of carbon-supported platinum nanoparticles toward methanol oxidation reaction: role of metal precursor and a new surfactant, tert-octanethiol, J. Phys. Chem. C 111 (2007) 1467–1473, https://doi.org/10.1021/jp065809y.

[40] Y. Yildiz, E. Erken, H. Pamuk, H. Sert, F. Şen, Monodisperse Pt nanoparticles assembled on reduced graphene oxide: highly efficient and reusable catalyst for methanol oxidation and dehydrocoupling of dimethylamine-borane (DMAB), J. Nanosci. Nanotechnol. 16 (2016) 5951–5958, https://doi.org/10.1166/jnn.2016.11710.

[41] M.V.F. Delmonde, L.F. Sallum, N. Perini, E.R. Gonzalez, R. Schlögl, H. Varela, Electrocatalytic efficiency of the oxidation of small organic molecules under oscillatory regime, J. Phys. Chem. C 120 (2016) 22365–22374, https://doi.org/10.1021/acs.jpcc.6b06692.

[42] O.A. Petrii, G.A. Tsirlina, Electrocatalytic activity prediction for hydrogen electrode reaction: intuition, art, science, Electrochim. Acta 39 (1994) 1739–1747, https://doi.org/10.1016/0013-4686(94)85159-X.

[43] Y. Mao, J. Chen, H. Wang, P. Hu, Catalyst screening: refinement of the origin of the volcano curve and its implication in heterogeneous catalysis, Cuihua Xuebao/Chin. J. Catal. 36 (2015) 1596–1605, https://doi.org/10.1016/S1872-2067(15)60875-0.

[44] P. Yu, M. Pemberton, P. Plasse, PtCo/C cathode catalyst for improved durability in PEMFCs, J. Power Sources 144 (2005) 11–20, https://doi.org/10.1016/j.jpowsour.2004.11.067.

[45] Z.R. Ismagilov, M.A. Kerzhentsev, N.V. Shikina, A.S. Lisitsyn, L.B. Okhlopkova, C.N. Barnakov, M. Sakashita, T. Iijima, K. Tadokoro, Development of active catalysts for low Pt loading cathodes of PEMFC by surface tailoring of nanocarbon materials, Catal. Today (2005) 58–66, https://doi.org/10.1016/j.cattod.2005.02.007.

[46] J.L. Fernández, D.A. Walsh, A.J. Bard, Thermodynamic guidelines for the design of bimetallic catalysts for oxygen electroreduction and rapid screening by scanning electrochemical microscopy. M-Co (M: Pd, Ag, Au), J. Am. Chem. Soc. 127 (2005) 357–365, https://doi.org/10.1021/ja0449729.

142 Chapter 8 Monometallic nanomaterials for direct alcohol fuel cells

[47] R.G. González-Huerta, J.A. Chávez-Carvayar, O. Solorza-Feria, Electrocatalysis of oxygen reduction on carbon supported Ru-based catalysts in a polymer electrolyte fuel cell, J. Power Sources 153 (2006) 11–17, https://doi.org/10.1016/j.jpowsour.2005.03.188.

[48] C.R.K. Rao, D.C. Trivedi, Chemical and electrochemical depositions of platinum group metals and their applications, Coord. Chem. Rev. 249 (2005) 613–631, https://doi.org/10.1016/j.ccr.2004.08.015.

[49] P. Pharkya, A. Alfantazi, Z. Farhat, Fabrication using high-energy ball-milling technique and characterization of Pt-Co electrocatalysts for oxygen reduction in polymer electrolyte fuel cells, J. Fuel Cell Sci. Technol. 2 (2005) 171–178, https://doi.org/10.1115/1.1895985.

[50] J. Xie, D.L. Wood, D.M. Wayne, T.A. Zawodzinski, P. Atanassov, R.L. Borup, Durability of PEFCs at high humidity conditions, J. Electrochem. Soc. 152 (2005) A104, https://doi.org/10.1149/1.1830355.

[51] Y. Wang, K.S. Chen, J. Mishler, S.C. Cho, X.C. Adroher, A review of polymer electrolyte membrane fuel cells: technology, applications, and needs on fundamental research, Appl. Energy 88 (2011) 981–1007, https://doi.org/10.1016/j.apenergy.2010.09.030.

[52] Y. Yıldız, S. Kuzu, B. Sen, A. Savk, S. Akocak, F. Şen, Different ligand based monodispersed Pt nanoparticles decorated with rGO as highly active and reusable catalysts for the methanol oxidation, Int. J. Hydrog. Energy 42 (2017) 13061–13069, https://doi.org/10.1016/j.ijhydene.2017.03.230.

[53] N.A. Vante, H. Tributsch, Energy conversion catalysis using semiconducting transition metal cluster compounds, Nature 323 (1986) 431–432, https://doi.org/10.1038/323431a0.

[54] J.L. Fernández, V. Raghuveer, A. Manthiram, A.J. Bard, Pd-Ti and Pd-Co-Au electrocatalysts as a replacement for platinum for oxygen reduction in proton exchange membrane fuel cells, J. Am. Chem. Soc. 127 (2005) 13100–13101, https://doi.org/10.1021/ja0534710.

[55] B. Wang, Recent development of non-platinum catalysts for oxygen reduction reaction, J. Power Sources 152 (2005) 1–15, https://doi.org/10.1016/j.jpowsour.2005.05.098.

[56] Y. Liu, F. Yu, X.W. Wang, Z.B. Wen, Y.S. Zhu, Y.P. Wu, Nanostructured oxides as cathode materials for supercapacitors, Springer, 2016, pp. 205–269.

[57] S. Gottesfeld, A new approach to the problem of carbon monoxide poisoning in fuel cells operating at low temperatures, J. Electrochem. Soc. 135 (1988) 2651, https://doi.org/10.1149/1.2095401.

[58] S. Basu, Recent Trends in Fuel Cell Science and Technology, Springer, 2007.

[59] P.J. Ferreira, G.J. LaO', Y. Shao-Horn, D. Morgan, R. Makharia, S. Kocha, H.A. Gasteiger, Instability of Pt/C electrocatalysts in proton exchange membrane fuel cells, J. Electrochem. Soc. 152 (2005) A2256, https://doi.org/10.1149/1.2050347.

[60] S.J. Paddison, J.A. Elliott, Molecular modeling of the short-side-chain perfluorosulfonic acid membrane, J. Phys. Chem. A 109 (2005) 7583–7593, https://doi.org/10.1021/jp0524734.

[61] S. Dharmalingam, V. Kugarajah, M. Sugumar, Membranes for microbial fuel cells, in: Microbial Electrochemical Technology, Elsevier, 2019, pp. 143–194, https://doi.org/10.1016/b978-0-444-64052-9.00007-8.

[62] H. You, F. Zhang, Z. Liu, J. Fang, Free-standing Pt-Au hollow nanourchins with enhanced activity and stability for catalytic methanol oxidation, ACS Catal. (2014), https://doi.org/10.1021/cs500390s.

Chapter 8 Monometallic nanomaterials for direct alcohol fuel cells 143

[63] F. Şen, G. Gökağaç, Different sized platinum nanoparticles supported on carbon: an XPS study on these methanol oxidation catalysts, J. Phys. Chem. C 111 (2007) 5715–5720, https://doi.org/10.1021/jp068381b.

[64] S. Eris, Z. Daşdelen, F. Sen, Investigation of electrocatalytic activity and stability of Pt@f-VC catalyst prepared by in-situ synthesis for methanol electrooxidation, Int. J. Hydrog. Energy 43 (2018) 385–390, https://doi.org/10.1016/j.ijhydene.2017.11.063.

[65] Z.P. Li, B.H. Liu, K. Arai, K. Asaba, S. Suda, Evaluation of alkaline borohydride solutions as the fuel for fuel cell, J. Power Sources 126 (2004) 28–33, https://doi.org/10.1016/j.jpowsour.2003.08.017.

[66] Z.P. Li, B.H. Liu, K. Arai, S. Suda, Development of the direct borohydride fuel cell, J. Alloys Compd. 404–406 (2005) 648–652, https://doi.org/10.1016/j.jallcom.2005.01.130.

[67] Z.P. Li, B.H. Liu, K. Arai, S. Suda, A fuel cell development for using borohydrides as the fuel, J. Electrochem. Soc. 150 (2003) A868, https://doi.org/10.1149/1.1576767.

9

Bimetallic nanomaterials for direct alcohol fuel cells

Haydar Goksu[a], Muhammed Bekmezci[b,c], Vildan Erduran[b,c], and Fatih Şen[b]

[a]Kaynasli Vocational College, Duzce University, Duzce, Turkey. [b]Şen Research Group, Department of Biochemistry, Dumlupinar University, Kütahya, Turkey. [c]Department of Materials Science and Engineering, Faculty of Engineering, Dumlupinar University, Kütahya, Turkey

1 Introduction

The definition of nanomaterials is in the index of many international organizations [1]. This concept often includes specialized items to fulfill a specific function. Although it is generally understood that it includes particles smaller than 100 nm, it also encompasses all the work done to find or create this ideal substance [2]. Nanomaterials can be used in many different areas. Likewise, problems in these areas also have been addressed [3–7]. Some important points brought with it in its definitions include the following:

- The particles that have 1–100 nm particle size (a common point in most of these definitions) [1]
- Having an extremely large specific surface area
- Having fascinating and useful features
- Providing ideal results in structural and nonstructural applications
- Being stronger, softer materials

They are very chemically active materials [8]. In fact, an ideal nanoparticle can vary depending on the area that it needs to be used. These areas include material production, metallurgy, manufacturing sector, electrical and computer technologies, human and animal health sectors, the aviation sector, environment, environmental cleaning, environmental redesign, defense industry, new generation defense vehicles used in the defense industry, biotechnology, and the cosmetics sector. Even synthesis methods

Nanomaterials for Direct Alcohol Fuel Cells. https://doi.org/10.1016/B978-0-12-821713-9.00017-2
Copyright © 2021 Elsevier Inc. All rights reserved.

and steps, including metals used in synthesis, must be tailored to desired purpose. This process is highly multidisciplinary [9].

Generally, dimension is the common factor in these definitions [1]. Nanomaterials were the beginning of very important studies in the medical industry. A material system can be developed for a specific biomedical application. This approach has used electrospun nanofiber, a class of nanomaterials that has gained more attention in biomedical research, such as the extracellular matrix (ECM), the environment surrounding cells in animal tissue [10]. It is very difficult to find materials of such small size over and over again after application. Engineered nanomaterials are known to interact with biological entities. There is concern that they may cause serious toxic effects as a result of these interactions [11]. The same issue has manifested itself in many areas where nanotechnology is used [12–15].

Nanomaterial production is synthesized through a series of processes. Sometimes a single metal can function alone; at other times, up to two or three work together. Nanomaterials sometimes need support and sometimes can be successfully used in applications alone. Monometallic and bimetallic nanomaterial types are described next, and synthesis methods for bimetallic nanomaterials are detailed as well. In cases where technical features are sometimes insufficient, changes have been made in nanomaterials synthesis systems. Changes in the ductility, hardness, or ideal combination of materials will be made with various synthesis methods, and new nanomaterials will be obtained.

2 Monometallic nanomaterials

Monometallic nanoparticles consist of only one type of metal atoms with size ranging from 1 to 10 nm. They have different properties from those of the individual atoms and molecules of bulk metals [16] due to their small size, higher surface area, and quantum size. For example, silver is inactive in its bulk state; however, studies showed that its nanoparticles are a useful catalyzer in olefin hydrogenation. Due to their enhanced physical and chemical properties, they have been used in many applications in electronic [17], sensing, medical diagnosis, optical [18], and catalysis [19], as well as an antimicrobial agent [20, 21]. Nanoparticles of some noble metals, including Pt, Pd, Ag, and Au, exhibit good catalytic activity. Ferromagnetic transition metals, such as Fe, Co, and Ni, are widely used in various magnetic and spintronic devices [22]. Metal nanoparticles can be synthesized readily and modified chemically.

3 Bimetallic nanomaterials

The aim of obtaining bimetallic nanoparticles is to achieve ideal stability. Obtaining a suitable working environment is very valuable for these particles. For this purpose, many bimetallic studies have been started.

Two-alloy metals are easier to approximate to the desired properties than single-alloy metals. The alloying and mixing of metals and obtaining of nanomaterials have gained considerable value today [8]. There are different processes for the reduction of the noble metal cation to a less noble metal that are more than two centuries old. Bimetallic nanomaterials have complex structures and can take multiple forms [23–25]. Three key types of structures for bimetallic nanoparticles (i.e., core/shell, heterostructure or intermetallic, and alloy structures) can be categorized according to the mixing pattern of two metals [26].

Synthesizing bimetallic nanomaterials is a little more complicated and requires attention [27]. Bimetallic catalytic nanoparticles are theoretically capable of conducting extraordinary chemical reactions with monometallic catalysts. This is because the various components of the catalyst play a special role in the general reaction process [27]. Even when bimetallic nanoparticles are made of two separate metal atoms, atomic dispersion can have a significant impact on the final nanoparticles structures, which in turn can have a significant impact on their catalytic efficiency [25, 28].

In recent years, the synthesis and design of bimetallic nanomaterials are very important because of their catalytic activity, selectivity, and stability. Those nanomaterials can reach some conversions and/or transformations. Since they are not identical to the other prepared materials, they also have a different mechanism compared to the others. The plasmon absorption spectrum energy of the metallic mixture appears to be optimized, providing us with a multiuse method for biosensing. These characteristics vary from pure elementary particles and have specific optical, mechanical, thermal, and catalytic effects depending on the scale [29]. The ratio of the prepared bimetallic nanoparticles has a significant effect on the catalytic performance of the nanoparticle. The electronic effect characterizing the charge transfer is a critical point in bimetallic catalysts. The structural changes of the bimetallic nanoparticles can be provided by combining the various metals; thus, the extra degree of freedom can be achieved [29]. As mentioned earlier, the catalytic performance is further enhanced by exhibiting a synergistic effect beyond the effect of the present atoms [25, 30–33].

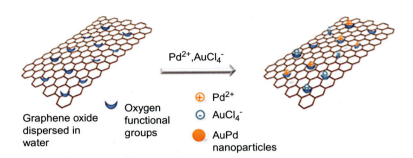

Fig. 1 Synthesis of AuPd nanoparticles [34].

The type, scale, and physicochemical properties of the resulting end nanomaterials are solely dependent upon the preparation techniques and conditions. Cubes, tetrahedrons, octahedrons, truncated octahedrons, icosahedrons, concentric cubicles, balls, spheres, and even stars could be formed as the nanoparticles of a monometal, as shown by Fig. 1.

3.1 Synthesis of bimetallic nanomaterials

A significant number of bimetallic catalysts are synthesized in diverse materials and properties, such as crown-jewel structure, hollow structure, heterostructure, core-shell structure, alloy structure, and porous structure. The form and scale of the nanoparticles are highly dependent on the process, which also influences the properties of the final material [35]. In the presence of appropriate stabilization techniques such as steric impediment and static electrical repulsive forces, bimetal nanoparticles are usually prepared by the simultaneous reduction of two metal ions.

a) Crown-jewel structure

On the surface of other metal atoms called the Crown-Jewel Structure, which is characterized as a single metal atom (a jewel). Here, jewel gold, which is a more precious and expensive element, affects the electronic state of the crown and offers higher catalytic activity than the crown cluster itself [36].

The main advantages of using the crown-jewel structure are the effective use of precious metal atoms and the improved catalytic performance of the catalyst. While the precious metal atoms are found on the surface of cheaper metals, they can actively perform the chemical reactions. Thus, the cost of operation is also reduced. For this reason, the synergistic effect of the preferred cheap metals is also significant. For example, nano-sized Pt catalysts are of significant interest in the chemical and petrochemical industries. In contrast to using monometallic Pt by itself, combining it with transition metals such as Fe, Co, Ni, and Pd improves catalytic activity and durability [31, 37–39].

In one study, PdCu nanomaterial was synthesized by chemical vapor deposition. Here, atomic Pd atoms are seen on the Cu (111) surface. Pd surface concentration can be adjusted by correct and repeatable metal flow control [40].

In another study, a nanomaterial was developed by combining Pd and Au atoms. The Au^{+3} concentration is critical in this reaction, carried out by heat treatment (Fig. 1) [34].

AgCu nanoparticles were obtained by heat treatment in the presence of silver nitrate and copper chloride. Economic AgCu nanoparticles are shown to be highly efficient via hydrogenation testing [41].

b) Hollow

The hollow nanostructures are increasingly attracting attention due to the unique structures that can be useful in the realization of multifunctional materials. The AgAu nanoparticle obtained by the reduction procedure using a surfactant such as polyvinylpyrrolidone (PVP) and cetyltrimethylammonium bromide (CTAB) can be used in many chemical reactions.

This example shows the controllable reaction processes, with small changes in the chemical environment and a synthetic pathway opened up for their preparation [42]. Hollow nanostructures distinguish themselves not only by their structural beauty, but, more important, by a number of interesting physicochemical properties, including low density, large surface area, high loading capacity, and reduced length scales for both mass and freight transport. These structures are very important for the formation of special ligand structures by forming structures such as special drug release systems [43]

c) Core-shell

Core-shell nanoparticles are formed by inserting various metal atoms onto a metal attached to a surface. Generally, reactions occur on the catalyst surface. That is, the molecules directly interact with the outermost metal. However, the synergistic effect of the metal in the nucleus contributes positively to the total activity of the catalyst. In the literature, the well-defined bimetallic AgPt@Ag core-shell nanoparticles were prepared using poly(ethylene imine; PEI) as the shape-directing and stabilizing agents. It is possible to observe scanning electron microscope/transmission electron microscopy (SEM/TEM) images and synthesis mechanism of the AgPt @ Ag nanoparticle [44].

In a study by Y. Mizukoshi et al., they determined that gold/palladium bimetallic nanoparticles in the determined ratio have high catalytic activity for hydrogenation of 4-pentenoic acid compared to monometallic nanoparticles. The behavior of bimetallic particles, when the gold-palladium ratio is 1:4, is about three times greater than that of monometallic palladium nanoparticles formed under similar conditions [45].

3.2 Alcohol oxidation catalyzed by bimetallic nanomaterials

Alcohol oxidation catalyzed by bimetallic nanomaterials are reactions with an industrial designation that trigger the oxidation of hydrocarbons by oxidation reactions. Molecular oxygen is often used in these reactions, which are particularly favored in the presence of ternary commercial molecules. However, a catalyst that activates molecular oxygen is much more important. For this purpose, bimetallic catalysts with higher activity and selectivity than monometallic catalysts are needed [46]. In recent years, bimetallic nanoparticles have been used to perform the alcohol oxidation reaction in alcohol fuel cells [47, 48]. Au nanoparticles, which are industrially prefabricated, are known to be an excellent catalyst for alcohol oxidation reaction (Scheme 1) [49].

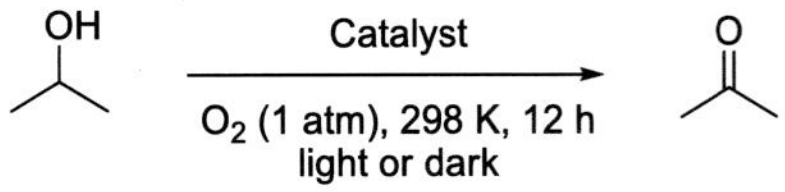

Scheme 1 The aerobic oxidation of 2-propanol to acetone [49]. Reprinted (adapted) with permission from T. Mallat, A. Baiker, Oxidation of alcohols with molecular oxygen on solid catalysts, Chem. Rev. 104 (2004) 3037–3058. https://doi.org/10.1021/cr0200116. Copyright (2020) American Chemical Society.

Tsunoyama et al. [50] studied PVP-stabilized Au nanoparticles, which showed that Au nanoparticles, had high catalytic activities and can be used to oxidize benzylic alcohols in water. Herein, PVP inhibited the agglomeration of Au nanoparticles and Au: PVP exhibited superior catalytic activities compared to larger particles. Several studies were done on bimetallic Au nanoparticles as catalysts. MAu/CB/P (M: Pt, Pd) nanoparticles obtained by the use of Au atoms, as well as Pt or Pd atoms, and attached to the carbon black/polymer surface are selective in alcohol oxidation reactions. In particular, the second metal used with Au completely changes the size of the reaction.

Dimitratos et al. [51] showed that using AuPd and AuPt bimetallic catalysts enhanced activity and selectivity for aerobic oxidations of alcohols. A positive synergistic effect was observed with Au/Pd catalysts, while Au/Pt catalysts showed a negative effect. The metal type and solvent system are essential parameters determining reactivity and selectivity. Of course, it is not possible to ignore the effect of surfactants in these reactions.

Hou et al. [52] studied polymer-stabilized bimetallic AuPd nanoparticles for aerobic alcohol oxidation of benzyl alcohol, 1-butanol, 2-butanol, 2-buten-1-ol, and 1,4-butanediol in aqueous solutions. They reported that PVP-stabilized, AuPd bimetallic nanoparticles showed higher activity than the pure monometallic nanoparticles.

The scientific community continues to study the effects on the reaction mechanism of bimetallic nanoparticles. Under the influence of MAu/CB/P nanoparticles, two different routes are proposed for the oxidation of primary alcohols. Via these routes, carboxylic acid and ester functional groups are formed. Solvent medium, metal type, and ratio directly affect these results, as shown in Scheme 2 [53].

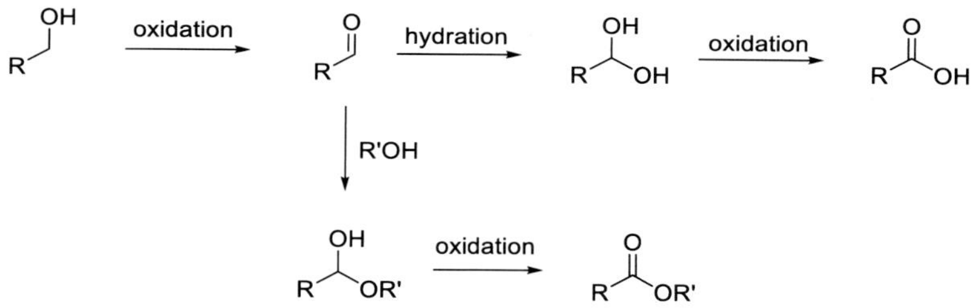

Scheme 2 Two possible reaction pathways for the oxidation of alcohols [53].

Bimetallic AuPd alloys performed selective aerobic oxidation of alcohols supported by MgO and MnO_2 that was formed by using $HAuCl_4.3H_2O$ as the Au source and $PdCl_2$ as the Pd source. Herein, Au and Pd were deposited on magnesium oxide (MgO), and manganese dioxide (MnO_2) supports using the sol-immobilization method. In this study, benzyl alcohol and 1-octanol were utilized for oxidation. The catalyst is reusable, stable, and efficient in aerobic oxidation of alcohols [54].

Various carbon materials such as carbon blacks, carbon nitride, carbon nanotubes, and carbon nanofibers are used as supports for metal catalysts. It was reported that graphene-supported transition metals improved the catalytic activity for oxidation reactions due to the strong interaction between graphene and metals [55, 56]. Wang et al. [57] studied the synthesis and electrocatalytic alcohol oxidation performance of magnetic Pd-Co bimetallic nanoparticles supported on reduced graphene oxide sheets. The results from this study support that the Co layer on Pd increased the activity of Pd and enhanced the alcohol oxidation kinetics. The catalytic performance is highly dependent on the interaction between Pd, Co, and graphene.

The carbon-supported Co_3O_4/Fe_3O_4 NPs were synthesized by heat treatment and under basic conditions. This corresponding catalyst was used for oxidation of alcohols, and promising results were obtained. Catalysts of hollow bimetallic oxide nanoparticles (HBNPs) synthesized as stabilized and effective catalysts for selective oxidation of alcohols with synthesized molecular oxygen are reported.

The catalyst exhibits an effective performance in the selective oxidation of alcohol. It is observed that no effective results are obtained from the reaction in oxygen and nitrogen atmosphere. Also, the dehydrogenation of alcohols by using oxygen atoms in the catalyst structure proved the formation of aldehydes [58].

Lee et al. synthesized the Pt-based nanosheets (S-PtPdPt$_{BN}$) for methanol oxidation reactions. The prepared materials were characterized by analytical techniques such as electron microscopy, X-ray diffraction, and photoelectron spectroscopy [59].

4 Conclusions

Nanomaterials are becoming important to our world. In this way, the researches in this area have been increased more and more rapidly. New developments and research studies will give more information about these production.

In short, the synergetic effects of two metal atoms in bimetallic nanoparticles are caused by many special properties. During many applications and reactions, including fuel cells, electrochemical sensors, biosensors, solar cells, medicine materials, and storage of hydrogen, their superior properties compare well against those of other ready-made materials. As they have special, superior properties to these other substances and are thus able to produce new materials with improved performance, operation, selectivity, longevity, reusability, and stability, they are expected to enhance their applications day to day in most applications in the near future.

References

[1] D.R. Boverhof, C.M. Bramante, J.H. Butala, S.F. Clancy, W.M. Lafranconi, J. West, S.C. Gordon, Comparative assessment of nanomaterial definitions and safety evaluation considerations, Regul. Toxicol. Pharmacol. 73 (2015) 137–150. https://doi.org/10.1016/j.yrtph.2015.06.001.

[2] W.G. Kreyling, M. Semmler-Behnke, Q. Chaudhry, A complementary definition of nanomaterial, Nano Today 5 (2010) 165–168. https://doi.org/10.1016/j.nantod.2010.03.004.

[3] H. Mudassir, et al., Bimetallic and trimetallic nanomaterials for hydrogen storage applications, in: F. Sen, A. Khan, A.M. Asiri (Eds.), Nanomaterials for Hydrogen Storage Applications, Elsevier, 2021. https://doi.org/10.1016/B978-0-12-819476-8.00004-9.

[4] M. Hassellöv, J.W. Readman, J.F. Ranville, K. Tiede, Nanoparticle analysis and characterization methodologies in environmental risk assessment of engineered nanoparticles, Ecotoxicology 17 (2008) 344–361. https://doi.org/10.1007/s10646-008-0225-x.

[5] A. López Rubio, M.J. Fabra Rovira, M. Martínez Sanz, L.G. Gómez-Mascaraque, Nanomaterials for Food Applications, Elsevier, 2018. https://doi.org/10.1016/C2017-0-01042-X.

[6] A. Jain, S. Ranjan, N. Dasgupta, C. Ramalingam, Nanomaterials in food and agriculture: an overview on their safety concerns and regulatory issues, Crit. Rev. Food Sci. Nutr. 58 (2018) 297–317. https://doi.org/10.1080/10408398.2016.1160363.

[7] J.C. Glenn, Nanotechnology: future military environmental health considerations, Technol. Forecast. Soc. Change. 73 (2006) 128–137. https://doi.org/10.1016/j.techfore.2005.06.010.

[8] M.G. Lines, Nanomaterials for practical functional uses, J. Alloys Compd. 449 (2008) 242–245. https://doi.org/10.1016/j.jallcom.2006.02.082.

[9] A.V. Herrera-Herrera, M.Á. González-Curbelo, J. Hernández-Borges, M.Á. Rodríguez-Delgado, Carbon nanotubes applications in separation science: a review, Anal. Chim. Acta 734 (2012) 1–30. https://doi.org/10.1016/j.aca.2012.04.035.

[10] Y. Xia, Nanomaterials at work in biomedical research, Nat. Mater. 7 (2008) 758–760. https://doi.org/10.1038/nmat2277.

[11] E. Valsami-Jones, I. Lynch, How safe are nanomaterials? Science 350 (2015) 388–389. https://doi.org/10.1126/science.aad0768.

[12] G. Pallas, W. Peijnenburg, J. Guinée, R. Heijungs, M. Vijver, Green and clean: reviewing the justification of claims for nanomaterials from a sustainability point of view, Sustainability. 10 (2018) 689. https://doi.org/10.3390/su10030689.

[13] M.T. Takeuchi, M. Kojima, M. Luetzow, State of the art on the initiatives and activities relevant to risk assessment and risk management of nanotechnologies in the food and agriculture sectors, Food Res. Int. 64 (2014) 976–981. https://doi.org/10.1016/j.foodres.2014.03.022.

[14] H. Rauscher, K. Rasmussen, B. Sokull-Klüttgen, Regulatory aspects of nanomaterials in the EU, Chemie Ing. Tech. 89 (2017) 224–231. https://doi.org/10.1002/cite.201600076.

[15] H. Bouwmeester, P. Brandhoff, H.J.P. Marvin, S. Weigel, R.J.B. Peters, State of the safety assessment and current use of nanomaterials in food and food production, Trends Food Sci. Technol. 40 (2014) 200–210. https://doi.org/10.1016/j.tifs.2014.08.009.

[16] R. Ferrando, J. Jellinek, R.L. Johnston, Nanoalloys: from theory to applications of alloy clusters and nanoparticles, Chem. Rev. 108 (2008) 845–910. https://doi.org/10.1021/cr040090g.

[17] S.H. Ko, I. Park, H. Pan, C.P. Grigoropoulos, A.P. Pisano, C.K. Luscombe, J.M.J. Fréchet, Direct nanoimprinting of metal nanoparticles for nanoscale electronics fabrication, Nano Lett. 7 (2007) 1869–1877. https://doi.org/10.1021/nl070333v.

[18] C.J. Murphy, T.K. Sau, A.M. Gole, C.J. Orendorff, J. Gao, L. Gou, S.E. Hunyadi, T. Li, Anisotropic metal nanoparticles: synthesis, assembly, and optical applications, J. Phys. Chem. B 109 (2005) 13857–13870. https://doi.org/10.1021/jp0516846.

[19] C.J. Jia, F. Schüth, Colloidal metal nanoparticles as a component of designed catalyst, Phys. Chem. Chem. Phys. 13 (2011) 2457–2487. https://doi.org/10.1039/c0cp02680h.

[20] S. Ahmed, M. Ahmad, B.L. Swami, S. Ikram, A review on plants extract mediated synthesis of silver nanoparticles for antimicrobial applications: a green expertise, J. Adv. Res. 7 (2016) 17–28. https://doi.org/10.1016/j.jare.2015.02.007.

[21] L.A. Jurado, R.M. Espinosa-Marzal, Insight into the electrical double layer of an ionic liquid on graphene, Sci. Rep. 7 (2017) 1–12. https://doi.org/10.1038/s41598-017-04576-x.

[22] S. Duan, R. Wang, Bimetallic nanostructures with magnetic and noble metals and their physicochemical applications, Prog. Nat. Sci. Mater. Int. 23 (2013) 113–126. https://doi.org/10.1016/j.pnsc.2013.02.001.

[23] J.P. Wilcoxon, B.L. Abrams, Synthesis, structure and properties of metal nanoclusters, Chem. Soc. Rev. 35 (2006) 1162–1194. https://doi.org/10.1039/b517312b.

[24] J.H. Bang, K.S. Suslick, Applications of ultrasound to the synthesis of nanostructured materials, Adv. Mater. 22 (2010) 1039–1059. https://doi.org/10.1002/adma.200904093.

[25] X. Liu, D. Wang, Y. Li, Synthesis and catalytic properties of bimetallic nanomaterials with various architectures, Nano Today 7 (2012) 448–466. https://doi.org/10.1016/j.nantod.2012.08.003.

[26] A. Dehghan Banadaki, A. Kajbafvala, Recent advances in facile synthesis of bimetallic nanostructures: an overview, J. Nanomater. 2014 (2014). https://doi.org/10.1155/2014/985948.

[27] U. Banin, Nanocrystals: tiny seeds make a big difference, Nat. Mater. 6 (2007) 625–626. https://doi.org/10.1038/nmat1993.

[28] S.E. Habas, H. Lee, V. Radmilovic, G.A. Somorjai, P. Yang, Shaping binary metal nanocrystals through epitaxial seeded growth, Nat. Mater. 6 (2007) 692–697. https://doi.org/10.1038/nmat1957.

[29] A. Nomura, S. Shin, O.O. Mehdi, J.M. Kauffmann, Preparation, characterization, and application of an enzyme-immobilized magnetic microreactor for flow injection analysis, Anal. Chem. 76 (2004) 5498–5502. https://doi.org/10.1021/ac049489v.

[30] J.W. Hong, D. Kim, Y.W. Lee, M. Kim, S.W. Kang, S.W. Han, Atomic-distribution-dependent electrocatalytic activity of au-Pd bimetallic nanocrystals, Angew. Chemie - Int. Ed. 50 (2011) 8876–8880. https://doi.org/10.1002/anie.201102578.

[31] F. Tao, M.E. Grass, Y. Zhang, D.R. Butcher, J.R. Renzas, Z. Liu, J.Y. Chung, B.S. Mun, M. Salmeron, G.A. Somorjai, Reaction-driven restructuring of Rh-Pd and Pt-Pd core-shell nanoparticles, Science 322 (2008) 932–934. https://doi.org/10.1126/science.1164170.

[32] J. Wu, A. Gross, H. Yang, Shape and composition-controlled platinum alloy nanocrystals using carbon monoxide as reducing agent, Nano Lett. 11 (2011) 798–802. https://doi.org/10.1021/nl104094p.

[33] Y. Yu, Q. Zhang, B. Liu, J.Y. Lee, Synthesis of nanocrystals with variable high-index Pd facets through the controlled heteroepitaxial growth of trisoctahedral au templates, J. Am. Chem. Soc. 132 (2010) 18258–18265. https://doi.org/10.1021/ja107405x.

[34] C. Wang, D. Astruc, Recent developments of metallic nanoparticle-graphene nanocatalysts, Prog. Mater. Sci. 94 (2018) 306–383. https://doi.org/10.1016/j.pmatsci.2018.01.003.

[35] A. Zaleska-Medynska, M. Marchelek, M. Diak, E. Grabowska, Noble metal-based bimetallic nanoparticles: the effect of the structure on the optical, catalytic and photocatalytic properties, Adv. Colloid Interf. Sci. 229 (2016) 80–107. https://doi.org/10.1016/j.cis.2015.12.008.

[36] H. Akbarzadeh, E. Mehrjouei, A. Masoumi, V. Sokhanvaran, Pt-Pd nanoalloys with crown-jewel structures: how size of the mother Pt cluster affects on thermal and structural properties of Pt-Pd nanoalloys? J. Mol. Liq. 249 (2018) 477–485. https://doi.org/10.1016/j.molliq.2017.11.040.

[37] A.T. Bell, The impact of nanoscience on heterogeneous catalysis, Science 299 (2003) 1688–1691. https://doi.org/10.1126/science.1083671.

[38] S.I. Sanchez, M.W. Small, J.M. Zuo, R.G. Nuzzo, Structural characterization of Pt-Pd and Pd-Pt core-shell nanoclusters at atomic resolution, J. Am. Chem. Soc. 131 (2009) 8683–8689. https://doi.org/10.1021/ja9020952.

[39] S.B.A. Hamid, R. Schlögl, The Impact of Nanoscience in Heterogeneous Catalysis, içinde: The Nano-Micro Interface, Wiley-VCH Verlag GmbH & Co. KGaA, Weinheim, Germany, 2015, pp. 405–430. https://doi.org/10.1002/9783527679195.ch20.

[40] G. Kyriakou, M.B. Boucher, A.D. Jewell, E.A. Lewis, T.J. Lawton, A.E. Baber, H.L. Tierney, M. Flytzani-Stephanopoulos, E.C.H. Sykes, Isolated metal atom geometries as a strategy for selective heterogeneous hydrogenations, Science 335 (2012) 1209–1212. https://doi.org/10.1126/science.1215864.

[41] M.N. Nadagouda, R.S. Varma, Green and controlled synthesis of gold and platinum nanomaterials using vitamin B2: density-assisted self-assembly of nanospheres, wires and rods, Green Chem. 8 (2006) 516–518. https://doi.org/10.1039/b601271j.

[42] E. González, J. Arbiol, V.F. Puntes, Carving at the nanoscale: sequential galvanic exchange and Kirkendall growth at room temperature, Science 334 (2011) 1377–1380. https://doi.org/10.1126/science.1212822.

[43] D. Wang, Hollow nanostructures, Adv. Mater. 31 (2019) 1904886. https://doi.org/10.1002/adma.201904886.

[44] X.Y. Zhu, A.J. Wang, S.S. Chen, X. Luo, J.J. Feng, Facile synthesis of AgPt@ag core-shell nanoparticles as highly active surface-enhanced Raman scattering substrates, Sensors Actuators B Chem. 260 (2018) 945–952. https://doi.org/10.1016/j.snb.2017.12.185.

[45] Y. Mizukoshi, T. Fujimoto, Y. Nagata, R. Oshima, Y. Maeda, Characterization and catalytic activity of core-shell structured gold/palladium bimetallic nanoparticles synthesized by the sonochemical method, J. Phys. Chem. B 104 (2000) 6028–6032. https://doi.org/10.1021/jp994255e.

[46] T. Punniyamurthy, S. Velusamy, J. Iqbal, Recent advances in transition metal catalyzed oxidation of organic substrates with molecular oxygen, Chem. Rev. 105 (2005) 2329–2363. https://doi.org/10.1021/cr050523v.

[47] Y. Mikami, A. Dhakshinamoorthy, M. Alvaro, H. García, Catalytic activity of unsupported gold nanoparticles, Catal. Sci. Technol. 3 (2013) 58–69. https://doi.org/10.1039/c2cy20068f.

[48] Y. Zhang, X. Cui, F. Shi, Y. Deng, Nano-gold catalysis in fine chemical synthesis, Chem. Rev. 112 (2012) 2467–2505. https://doi.org/10.1021/cr200260m.

[49] T. Mallat, A. Baiker, Oxidation of alcohols with molecular oxygen on solid catalysts, Chem. Rev. 104 (2004) 3037–3058. https://doi.org/10.1021/cr0200116.

[50] H. Tsunoyama, H. Sakurai, Y. Negishi, T. Tsukuda, Size-specific catalytic activity of polymer-stabilized gold nanoclusters for aerobic alcohol oxidation in water, J. Am. Chem. Soc. 127 (2005) 9374–9375. https://doi.org/10.1021/ja052161e.

[51] N. Dimitratos, A. Villa, D. Wang, F. Porta, D. Su, L. Prati, Pd and Pt catalysts modified by alloying with au in the selective oxidation of alcohols, J. Catal. 244 (2006) 113–121. https://doi.org/10.1016/j.jcat.2006.08.019.

[52] W. Hou, N.A. Dehm, R.W.J. Scott, Alcohol oxidations in aqueous solutions using au, Pd, and bimetallic AuPd nanoparticle catalysts, J. Catal. 253 (2008) 22–27. https://doi.org/10.1016/j.jcat.2007.10.025.

[53] K. Kaizuka, H. Miyamura, S. Kobayashi, Remarkable effect of bimetallic nanocluster catalysts for aerobic oxidation of alcohols: combining metals changes the activities and the reaction pathways to aldehydes/carboxylic acids or esters, J. Am. Chem. Soc. 132 (2010) 15096–15098. https://doi.org/10.1021/ja108256h.

[54] H. Alshammari, M. Alhumaimess, M.H. Alotaibi, A.S. Alshammari, Catalytic activity of bimetallic AuPd alloys supported MgO and MnO2 nanostructures and their role in selective aerobic oxidation of alcohols, J. King Saud Univ. - Sci. 29 (2017) 561–566. https://doi.org/10.1016/j.jksus.2017.03.003.

[55] Y.H. Lu, M. Zhou, C. Zhang, Y.P. Feng, Metal-embedded graphene: a possible catalyst with high activity, J. Phys. Chem. C 113 (2009) 20156–20160. https://doi.org/10.1021/jp908829m.

[56] G. Li, L. Jiang, B. Zhang, Q. Jiang, D.S. Su, G. Sun, A highly active porous Pt–PbOx/C catalyst toward alcohol electro-oxidation in alkaline electrolyte, Int. J. Hydrog. Energy 38 (2013) 12767–12773. https://doi.org/10.1016/J.IJHYDENE.2013.07.076.

[57] Y. Wang, Y. Zhao, J. Yin, M. Liu, Q. Dong, Y. Su, Synthesis and electrocatalytic alcohol oxidation performance of Pd-co bimetallic nanoparticles supported on graphene, Int. J. Hydrog. Energy 39 (2014) 1325–1335. https://doi.org/10.1016/j.ijhydene.2013.11.002.

[58] W. Gao, F. Li, H. Huo, Y. Yang, X. Wang, Y. Tang, P. Jiang, S. Li, R. Li, Investigation of hollow bimetal oxide nanomaterial and their catalytic activity for selective oxidation of alcohol, Mol. Catal. 448 (2018) 63–70. https://doi.org/10.1016/j.mcat.2018.01.028.

[59] Y.W. Lee, S.W. Han, K.Y. Lee, Site-selectively Pt-decorated PdPt bimetallic nanosheets characterized by electrocatalytic property for methanol oxidation, Mater. Chem. Phys. 214 (2018) 201–208. https://doi.org/10.1016/j.matchemphys.2018.04.072.

10

Ternary/quaternary nanomaterials for direct alcohol fuel cells

Elif Esra Altuner[a], Tugba Gur[b], and Fatih Şen[a]
[a]*Şen Research Group, Department of Biochemistry, Dumlupinar University, Kütahya, Turkey.* [b]*Vocational School of Health Services, Van Yuzuncu Yıl University, Van, Turkey*

1 Introduction

Generally, nanotechnology deals with very small and tiny materials. For this reason, nanotechnology is of great importance in revealing the biological and chemical properties of materials to the most important level of detail. Nanomaterials perform certain functions and superior properties in harmony with the target environment. To be able to explain and investigate these properties, the morphology and crystal structure of nanomaterials can be obtained by macroscopic and spectroscopic methods [1]. Further, different kinds of nanomaterials have been used to be able to increase the efficiency of the direct alcohol fuel cells (DAFCs). This chapter focuses on ternary/quaternary nanomaterials for DAFCs. Fuel cells are very popular in the field of science and technology due to both their environmentally friendly and significant energy savings [2].

Generally, fuel cells have various types: Proton exchange membrane fuel cells, direct methanol fuel cells, phosphoric acid fuel cells, alkaline fuel cells, molten carbonate fuel cells, and solid oxide fuel cells [3]. These types of fuel cells have some drawbacks. For instance, Direct methanol fuel cells have some problems such as methanol crossover, intermediate species poisoning, high polarization of the anode for the oxidation of methanol, and systems design [4]. Similar problems also can occur in other types of fuel cells. The use of the proton exchange membrane fuel cells also can have similar problems [5]. For instance, in molten carbon fuel cells, the cathode cannot show sufficient long-term stability

Nanomaterials for Direct Alcohol Fuel Cells. https://doi.org/10.1016/B978-0-12-821713-9.00001-9
Copyright © 2021 Elsevier Inc. All rights reserved.

during applications [6]. In solid oxide fuel cells, the gaps between different layers based on thermal expansion coefficients exceeding 200 nm cause problems in studies with these types of fuel cells [7]. Besides, DAFCs have become practical to use due to their cost and reaction kinetics in the environment [8]. The use of alcohol fuel cells is not problematic, and good for the environment as well [9].

2 Fuel cells

Today, due to the requirements of modern life and the demands of society, industrialization, energy consumption increases day by day. The one of the most important energy sources is fossil fuels, but they cause environmental pollution, threaten the environment, and negatively affect human life. For this reason, fuel cells are very important clean energy sources. They convert chemical energy to electrical energy [10]. In various studies, fuel cells are a potential energy source for renewable energy technologies. The relative energy unit of fuel cells is demonstrated in Fig. 1 [11]

Fuel cells are one of the most effective and useful energy source because they convert chemical energy directly into electrical energy [13]. The types of fuel cells are alkaline fuel cells (AFCs), molten carbonate fuel cells (MCFCs), phosphoric acid fuel cells (PAFCs), proton exhange membrane fuel cells (PEMFCs) and solid oxide fuel cells (SOFCs) [14]. Fuel cells use hydrogen, methane and some organic materials, carbon monoxide, and methanol

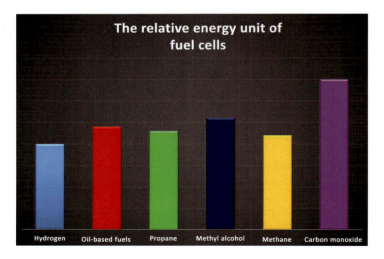

Fig. 1 Schematic representation of the ratio of types of fuel cells. [12]

(methyl alcohol) as fuel. To be able to activate oxidation of fuel in the fuel cells, many types of catalysts have been used in literature [15]. Platinum-, palladium-, and cobalt-based catalysts have significant nanocatalytic properties [16]. In order to decrease the cost of platinum-based catalysts, palladium- and cobalt-based materials are preferred in platinum alloys for DAFCs. Generally besides of the catalyst, temperature is also important parameter for fuel cells. Fuel cells with an operating temperature range of 120–150 °C are called "low-temperature fuel cells," those operating at 150–250 °C are "medium temperature fuel cells," and those operating at 650 °C and higher one are "high-temperature fuel cells" [17]. Therefore, alcohol fuel cells have gained great importance due to operation of the low temperature.

2.1 Direct alcohol fuel cells

DAFCs have become very important among alternative energy sources due to their high energy, quality, and efficiency, and they use many alcohols such as methanol, ethanol, propanol, ethanol, etc. [9,18,19]. The direct methanol fuel cell is the most prone to commercizalization within DAFCs. Further, C2–C4 alcohol-based fuel cells are also important but electrocatalytic research should be worked in order to break the C–C bond. This puts forward a new scientific research hypothesis in fuel cells other than methanol [20]. When various transition metals are used in the form of alloys as catalytic systems, they show good catalytic performance. There are so many catalysts such as monometallic, bimetallic, trimetallic, etc., including different transition metals [21].

Mostly, Pt-based catalytic systems have a very high catalytic effect. However, due to high cost of Pt-based system, they have been supported by carbon-based materials for alcohol oxidation reactions. Besides, to be able to decrease the cost of catalyst, bimetallic or trimetallic catalytic systems are used [22–25]. The anodic and cathodic reactions of direct methanol fuel cells are shown in Fig. 2. According to this schema, anode and cathode reactions are as follows:

Anode:

$$CH_3OH + 6\,OH^- \rightarrow CO_2 + 5\,H_2O + 6\,e^-$$

Cathode:

$$\frac{1}{2}O_2 + 3\,H_2O + 6\,e^- \rightarrow 6\,OH^-.$$

Direct ethanol fuel cells have also been in great demand after methanol fuel cells. Direct ethanol fuel cells are a group of alkaline

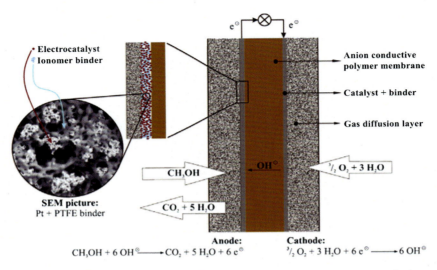

Fig. 2 Schematic of the catalyzation of anion exchange membrane direct methanol fuel cells [20].

fuel cells (Fig. 3). They use ethanol as a fuel instead of methanol which is more toxic. Ethanol-based fuels are important alternative to other alcohols due to the easier fuel to work with for widespread use by people [26].

Anode and cathode reactions of ethanol fuel cells are as follows:

Anode:

$$CH_2OH + 3 H_2O \rightarrow 2 CO_2 + 3 H_2O$$

Cathode:

$$O_2 + 12 H^+ + 12 e^- \rightarrow 6 H_2O$$

Total reaction:
$CH_2OH + 3 O_2 \rightarrow 2 CO_2 + 3H_2O$ [9].

Over the past several decades, researchers have also investigated the electrocatalyzation of ethylene glycol fuel cells. For this purpose, palladium, platinum, gold, and various other metals-based catalytic systems are worked in detail [27]. The anode-cathode catalytic systems of methanol, ethanol, and ethylene glycol fuel cells, as well as the details of DAFCs, are shown in Fig. 4. As shown in this figure, mostly Pt, Pt-M, Pd, Pd-M, Au, Au-M, etc.-based systems are used for anodic part of alcohol oxidation

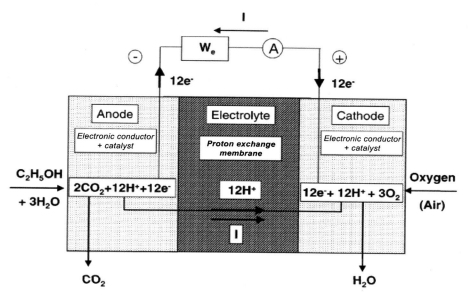

Fig. 3 The operation system of ethanol fuel cells [19].

Fuel	Anode catalysts	Cathode catalysts (alcohol tolerant)
Methanol	Pt, Pt-M (M = Ru,Pd,Au), Pt-M$_x$O$_y$(M =Ce,Ni,V), Pt-La$_{1-x}$Sr$_x$MO$_3$ (M = Co,Mn), Ni, Ni-M (M = Ru,Cu), Ni-complex, Pd, PdNi, AuLa$_{1-x}$Sr$_x$MO$_3$ (M = Co,Cu), La$_{2-x}$Sr$_x$NiO$_4$, Ni-La$_{2-x}$Sr$_x$NiO$_4$	Fe-TMPP, Ag, Ag-W$_2$C, Pd, Pd-Sn
Ethanol	Pd, Pd-M (M = Ru, Au, Sn, Cu),Pd-M$_x$O$_y$ (M = Ce, Zr, Mg, Co, Mn, Ni, In), Pd-(Ni-Zn), Pd-(Ni-Zn-P), Ru-Ni, Ru-Ni-Co.	Ag-W$_2$C, Pd, Pt-Ru
Ethylene glycol	Pt, Pt-M (M = Pb,Bi,Tl,Au), Au-M (M = Pb,Bi)	Pt-Pb, La$_{1-x}$Sr$_x$MnO$_3$

Fig. 4 The anode and cathode catalyzation processes of various alcohols in DAFCs [31].

reactions [28–30]. In ethylene glycol fuel cells, there are 10 electrons in the total oxidation reaction. Reactions follow consecutive and multiple sequences [31].

The total reaction is as follows:

$(CH_2OH)_2 + 2 H_2O \rightarrow 2CO_2 + 10H^+ + 10e^-$ [31].

The performances of platinum electrodes, which show the activation of reactions occurring directly in the alcohol fuel cell, have been emphasized in the literature [32].

2.2 Materials for direct alcohol fuel cells

Methanol and ethanol fuel cells are known as the best alcohol fuel cells due to their practicality, high density, and energy production. For this reason, researchers have turned to nanomaterials with good catalytic properties for these types of fuel cells [33]. Platinum (Pt) and palladium (Pd)-based alloys are the best known of these and are high-performance oxidation catalysts [34]. For example, Pt-Ru catalyst is one of the best-known catalyst for DAFCs [35].

Electron and proton transport in DAFCs results in an electrocatalytic reaction. In the search for materials that accelerate electrocatalytic reactions, researchers focused marginally on carbon supported materials to be able to increase the active surface area of the catalyst. For these reasons, quite a number of studies have been conducted on carbon. Researchers observed the good response of carbon in alkaline and acidic environments. In these trials, carbon was pretreated and modified. In addition to all these studies, commercial carbons were also tested, including Vulcan carbon, acetylene carbon black, and black pearl. Carbon nanofibers, carbon nanotubes (CNTs), microbeads, and mesocarbons are the best-performing carbon-based materials due to their catalytic activities for bimetallic and/or trimetallic alloys, etc. [36]. For instance, among the carbon-based catalysts, the Fe-Ni/C catalyst gave better results than the Pt/C catalyst for direct methanol fuel cells [37]. Multiwalled carbon nanotube (MWCNTs)-based catalysts have also been developed as another electrocatalyst system for DAFCs. This type of supporting material was developed for Pt – Ru/MWCNT and Pt – Sn/MWCNT [38].

2.2.1 Nanomaterials for direct alcohol fuel cells

Recently, there has been some interest in nanostructured materials rather than materials for catalysis reactions in DAFCs. Researchers have turned to experiments to explore this area. Nanostructured materials are examined in two classes:

carbon-based and noncarbon-based materials [39]. Carbon and carbon-based, functionalized nanocarbon materials are extremely effective catalysts for this purpose.

2.2.1.1 Nanostructured carbon

Structured carbons (the best-known and most effective Vulcan carbon) are very effective in platinum honey catalysis. They are preferred in most academic research and commercial applications due to their affordability, ease of use, and effective performance. However, it also has disadvantages, including that it is affected by organosulfur impurities and attacks nanoparticles, which can sometimes slow catalytic reactions [39]. These materials show high catalytic performance below 100 °C. At the same time, due to the porous structure of these materials, the problems of diffusion are greatly reduced. Above 100 °C, its catalytic activity is low [40].

2.2.1.2 Carbon nanotubes

CNTs have a hexagonal structure and a large surface area. At the same time, these materials have high electrical conductivity. Different trials have been done on the varieties. According to the research literature, the best performance was shown by platinum-ruthenium-based systems. Platinum-ruthenium-based CNTs show one of the best durability [41, 42]. In this system, attaching platinum to the inner and outer parts of a CNT-supported system will accelerate the electrocatalytic reaction, and thus a wide current density will be reached. At the same time, polyaniline and other support materials can be used as a hybrid system to CNTs [43–46].

2.2.1.3 Carbon nanofibers

Carbon nanofibers are materials that are very similar in structure to CNTs. Both support materials have a round structure and show very similar characteristics. They have sizes ranging from about 100 to 500 nm in characterization techniques. Carbon nanofibers have a surface area of 200 m^2/g, and the length range is between 0.1 and 1000 mm [47]. They serve as an important electrocatalyst supporting material due to their high electrical conductivity. They act as an electrocatalyst in thermal stability and electrochemical DAFCs [48–50]. These materials are formed as a result of the decomposition of hydrocarbons on metals [51].

2.2.1.4 Graphene

Graphene, another nanocarbon used in DAFCs, is composed of single, double, or even greater layers by its structure. It is in the form of a few layers. The structure of graphene is shown in Fig. 5 [52,53].

Rao et al. have revealed the structure of graphene with various characterization techniques [41]. Graphene oxide has been reported to oxidize methanol by reducing it to form a nanocomposite material with polyaniline [55]. The high chemical activity of graphene oxide has enabled them to research in various fields [56]. Graphene acts as a catalyst supported by various metals in both direct and indirect fuel cells, the main ones being platinum, palladium, cobalt, and copper [57, 58]. Studies have also been made with graphene with alloys of these metals.

In a number of studies, graphene was deposited on platinum and electrocatalytic oxidation in methanol fuel cells was investigated [59]. Graphene is supported by coating with independent cobalt denhydrides. Positive methanol oxidation against 0.07 V Ag/AgCl was demonstrated electronically. This oxidation activity has been observed to be much more intense than nickel [60]. Catalytic tests were carried out on fuel cells with palladium-cobalt metals attached to graphene. Graphene is a part of carbon materials and is a high-performance support nanomaterial for alcohol fuel cells directly from catalytic reactions [61]. Similarly, in direct ethanol fuel cell trials, a new, nanostructured material was developed by fixing gold-palladium metals on graphene. This nanomaterial has shown a highly efficient oxidation

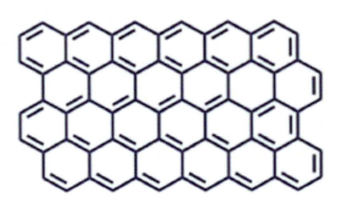

Fig. 5 The structure of graphene [54].

Chapter 10 Ternary/quaternary nanomaterials for direct alcohol fuel cells **165**

performance [62]. Such metals are attached to graphene, and trials continue.

2.2.1.5 Mesoporous nanocarbon

Mesoporous carbons are a class of nanocarbon called "porous carbons." Their sizes range from 2 to 50 nm. They are divided into two groups as sequential mesoporous carbons and irregular mesoporous carbons. This classification is due to their structure. They provide high-temperature catalytic oxidation in various metals and metal oxides in DAFCs. To produce mesoporous carbons, the raw-material carbon source is polymerized in an environment without oxygen. This polymerization is done under nitrogen or argon gas [63]. Until now, generally activated carbon or carbon black was used. Catalytic reaction studies with platinum-ruthenium-supported mesoporous nanocarbon in direct methanol fuel cells prepared with formic acid have been reported in the literature [64].

2.3 Ternary/quaternary nanomaterials for direct alcohol fuel cells

Fuel cells have an important place in today's technology for energy generation due to its many beneficial features, such as high energy, low cost, and environmental friendliness. It takes an even more important place, especially because alcohol is clean energy source and produces less pollution [65]. The use of platinum, which provides high catalytic activity, is widely observed in these fuel cells. However, it is a disadvantage that platinum and its precious metals are very expensive. In today's technological devices, multiple electrocatalysts are needed in fuel cells to increase energy, increase durability as much as possible, and reduce the cost of the catalysts. These catalysts can be worked as either double, triple, or quadruple, or even more.

It is extremely important to increase the efficiency of these nanomaterials to get the oxidation reaction with better efficiency and to observe the catalytic activity better [65]. Scientists have focused on binary, ternary, or even quarternary nanomaterials to increase the effectiveness of nanomaterials. Binary nanomaterials (palladium/platinum) and ternary nanomaterials (palladium/platinum/ruthenium) for DAFC studies has been reported [66]. Platinum-ruthenium couples can be cited as an example for such studies. The researchers examined the effect of platinum and ruthenium on nickel for ethanol oxidation in aqueous-alkaline media. Electrocatalytic results were supported by cyclic voltammetry, chronopotentiometry measurements, and impedance measurements. The percentage of elements in the binary system is determined by taking characterizations via techniques such as X-ray photoelectron spectroscopy (XPS) and scanning electron microscopy (SEM). In addition, in the images obtained from SEM, it was seen that the dual support materials were shaped like rods and disks [67]. Studies on the increase of electrocatalytic activity by creating a CeO_2/Pt binary system with CeO_2 nanoparticles to increase the activity of platinum for DAFC have been reported in the literature [68]. As mentioned before, there are studies on palladium/platinum and palladium/platinum/ruthenium for both binary and ternary systems.

2.3.1 Ternary nanomaterials for direct alcohol fuel cells

In DAFCs, there are some disadvantageous related to the catalysts such as high cost and poisoning of the catalytic systems. This causes inefficiency in performance of fuel cells. For this reason, To be able to eliminate these disadvantageous, binary, ternary, and quaternary nanomaterials have been used to further increase electrocatalytic activity.

In this section, ternary nanomaterials-based studies have been searched for DAFC systems. In these studies, carbon-supported, three metallic composites have been examined. As an example, platinum-palladium-cobalt triple system are shown as a ternary systems for direct methanol fuel cells [69].

When looking at the electrocatalytic activity for a long time, it was shown that PtPdCo/C catalysts act as catalysts at the anode. Because when the experiments are done between 0 and 1000 h in time, the power performance has decreased by only 14%. This shows that PtPdCo/C as a ternary catalyst give very good results [69]. A similar study was carried out to increase the catalytic performance of the anode electrode with the PSS/MnO_2/rGO for direct methanol fuel cells [70]. As mentioned earlier, there are studies on Pd/Pt and Pd/Pt/Ru in both binary and ternary systems [66].

Synthesis of Pt-Pt-Ni ternary metals as electrocatalysts was performed directly with ethanol fuel cells. This ternary system showed a great performance for ethanol oxidation and was found to be 206.93 mA/cm^2 and 1195.81 mA/mg. It appears to have a spiny star shape with the associated characterization techniques [71]. The use of Pt/Fe/Sn nanowires as electrocatalysts in DAFCs has brought a different perspective. These electrocatalysts provided one of the best oxidation reaction of ethanol and methanol and gave superior results as a catalytic performance [72].

Chapter 10 Ternary/quaternary nanomaterials for direct alcohol fuel cells **167**

In studies on the electrochemical oxidation of nickel and bismuth on palladium carbon (Pd/C), nickel formed more hydroxyl ions (OH) than bismuth, causing ethanol to be oxidized more than bismuth. However, the data obtained showed that it gives an electrocatalytic performance in this ternary system. It has also been reported that the catalytic reactions of an N-doped graphene-supported Pt/Ru/Fe system for methanol oxidation for direct methanol fuel cell show that it has good performance as an electrocatalyst. These studies concluded that the electrocatalytic performance for methanol oxidation increased by 30%–40%. Thus, its durability is increased by adding platinum, an expensive metal, to the system, especially in iron [73]. Likewise, it was observed that Pt/Pd/Cr) ternary system showed catalytic performance directly in methanol fuel cells [74]. By looking at methanol oxidation directly for methanol fuel cells, it has been determined that these alloys show catalytic reactions by using the alloy and supporting a reduced graphene oxide–supported function of Cu/Fe/Pt) [75].

2.3.2 Quarternary nanomaterials for direct alcohol fuel cells

Various metal ternary alloys have showed electrocatalytic reaction for DAFCs such as Pt/Ru/Co [76], Pt/Ru/Fe [77], Pt/Fe/Co [78], Pt/Pd/Co [69], and Pt-Ru-Fe [73]. To increase durability and further increase the quality of catalytic performance, academic studies on quaternary systems have been started and are being developed. Therefore, by synthesizing Pt/Ru/Fe/Co supported with N-doped graphene, quaternary electrocatalysts were formed in direct methanol fuel cells. The power signal is 778 mW/cm^2, and new quarternary catalysts have been developed to reduce costs [79]. Unlike these other substances, a new palladium electrocatalyst has been created with CNTs functionalized in an acidic medium. Palladium-phosphomolybdic acid-diallyldimethylammonium chloride-multiwalled CNTs showed significant catalytic performance [80]. Quarternary phosphonium polymers are suitable for catalytic reactions due to their energy potential and performance [81]. Two series of quaternary PtMnCuX/C (with X = Fe, Co, Ni, and Sn) and PtMnMoX/C (with X = Fe, Co, Ni, Cu, and Sn) were synthesized, and the electrocatalytic activities were investigated directly for ethanol fuel cells. These catalytic quaternary alloys have been subjected to various characterizations by determining elemental ratios with inductively coupled plasma (ICP) spectroscopy. Sn ions are oxidized in both quaternary systems. Electrochemical applications are supported by cyclic voltammetry and chronoamperometry measurements. Oxidation in CV cycles started from 0.059 V. Two series of quaternary PtMnCuX/C (with X = Fe, Co, Ni, and Sn) and PtMnMoX/C (with X = Fe, Co, Ni, Cu, and Sn) were synthesized, and the electrocatalytic activities were investigated directly for ethanol fuel cells. These catalytic quaternary alloys have been subjected to various characterizations by determining elemental ratios with ICP spectroscopy. Sn ions are oxidized in both quaternary systems. Electrochemical applications are supported by cyclic voltammetry and chronoamperometry measurements. Oxidation in CV cycles started from 0.059 V [82].

3 Conclusion

As a conclusion, in this chapter, ternary and quaternary nanomaterials are examined and it is seen that they have been prepared for increasing of electrocatalytic activity, decreasing of the cost in DAFCs. However, there are various obstacles, such as undesirable reactions in the anode and cathode of DAFC, and decreased their performance. For this reason, binary-, ternary-, and quaternary-supported electrocatalysts have been developed to solve these issues and ensure high performance. With these electrocatalysts, the performance of catalytic reactions in direct alcohol cells has become even more efficient. These types of materials also increase activation and efficiency of the alcohol oxidation reaction in DAFCs. In near future, it is thought that the importance of these types of catalytic system will be more and more and scientists need more studies about these types of materials.

References

[1] C. Bréchignac, P. Houdy, M. Lahmani, Nanomaterials and Nanochemistry, Springer Science & Business Media, 2008.

[2] K.I. Ozoemena, S. Chen, Nanomaterials for Fuel Cell Catalysis, Springer, 2016.

[3] A.M. Abdalla, et al., Nanomaterials for solid oxide fuel cells: a review, Renew. Sust. Energ. Rev. 82 (2018) 353–368.

[4] S. Surampudi, et al., Advances in direct oxidation methanol fuel cells, in: Across Conventional Lines, World Scientific, 2003, pp. 1226–1234. Selected Papers of George A Olah Volume 2.

[5] C.S. Kong, et al., Influence of pore-size distribution of diffusion layer on mass-transport problems of proton exchange membrane fuel cells, J. Power Sources 108 (1–2) (2002) 185–191.

[6] Q.M. Nguyen, Technological status of nickel oxide cathodes in molten carbonate fuel cells—a review, J. Power Sources 24 (1) (1988) 1–19.

[7] L. Blum, An analysis of contact problems in solid oxide fuel cell stacks arising from differences in thermal expansion coefficients, Electrochim. Acta 223 (2017) 100–108.

[8] E.H. Yu, U. Krewer, K. Scott, Principles and materials aspects of direct alkaline alcohol fuel cells, Energies 3 (8) (2010) 1499–1528.

[9] H.R. Corti, E.R. Gonzalez, Direct alcohol fuel cells. Materials, performance, durability and applications, in: H.R. Corti, E.R. Gonzalez (Eds.), Introduction to Direct Alcohol Fuel Cells, Springer, New York, 2014, p. 1.

[10] C.K. Dyer, Fuel cells for portable applications, J. Power Sources 106 (1–2) (2002) 31–34.

[11] X. Li, Principles of Fuel Cells, CRC Press, 2005.

[12] S. Mekhilef, R. Saidur, A. Safari, Comparative study of different fuel cell technologies, Renew. Sust. Energ. Rev. 16 (1) (2012) 981–989.

[13] U. Lucia, Overview on fuel cells, Renew. Sust. Energ. Rev. 30 (2014) 164–169.

[14] J. Benz, et al., Fuel cells in photovoltaic hybrid systems for stand-alone power supplies, in: 2nd European PV-Hybrid and Mini-Grid Conference, 2003.

[15] Ö. Karatepe, et al., Enhanced electrocatalytic activity and durability of highly monodisperse Pt@ PPy–PANI nanocomposites as a novel catalyst for the electro-oxidation of methanol, RSC Adv. 6 (56) (2016) 50851–50857.

[16] B. Şen, et al., Silica-based monodisperse PdCo nanohybrids as highly efficient and stable nanocatalyst for hydrogen evolution reaction, Int. J. Hydrog. Energy 43 (44) (2018) 20234–20242.

[17] V.S. Bagotsky, Fuel cells: problems and solutions, Vol. 56, John Wiley & Sons, 2012.

[18] F.A. Zakil, S. Kamarudin, S. Basri, Modified Nafion membranes for direct alcohol fuel cells: An overview, Renew. Sust. Energ. Rev. 65 (2016) 841–852.

[19] C. Lamy, et al., Recent advances in the development of direct alcohol fuel cells (DAFC), J. Power Sources 105 (2) (2002) 283–296.

[20] T. Jurzinsky, et al., Development of materials for anion-exchange membrane direct alcohol fuel cells, Int. J. Hydrog. Energy 40 (35) (2015) 11569–11576.

[21] B. Sen, et al., Monodisperse rutheniumcopper alloy nanoparticles decorated on reduced graphene oxide for dehydrogenation of DMAB, Int. J. Hydrog. Energy 44 (21) (2019) 10744–10751.

[22] Z. Daşdelen, et al., Enhanced electrocatalytic activity and durability of Pt nanoparticles decorated on GO-PVP hybride material for methanol oxidation reaction, Appl. Catal. B Environ. 219 (2017) 511–516.

[23] S. Şen, F. Şen, G. Gökağaç, Preparation and characterization of nano-sized Pt–Ru/C catalysts and their superior catalytic activities for methanol and ethanol oxidation, Phys. Chem. Chem. Phys. 13 (15) (2011) 6784–6792.

[24] F. Şen, G. Gökağaç, Activity of carbon-supported platinum nanoparticles toward methanol oxidation reaction: role of metal precursor and a new surfactant, tert-octanethiol, J. Phys. Chem. C 111 (3) (2007) 1467–1473.

[25] B. Çelik, et al., Monodisperse Pt (0)/DPA@ GO nanoparticles as highly active catalysts for alcohol oxidation and dehydrogenation of DMAB, Int. J. Hydrog. Energy 41 (13) (2016) 5661–5669.

[26] M. Kamarudin, et al., Direct ethanol fuel cells, Int. J. Hydrog. Energy 38 (22) (2013) 9438–9453.

[27] L. An, R. Chen, Recent progress in alkaline direct ethylene glycol fuel cells for sustainable energy production, J. Power Sources 329 (2016) 484–501.

[28] S. Eris, Z. Daşdelen, F. Sen, Enhanced electrocatalytic activity and stability of monodisperse Pt nanocomposites for direct methanol fuel cells, J. Colloid Interface Sci. 513 (2018) 767–773.

[29] F. Şen, G. Gökağaç, Different sized platinum nanoparticles supported on carbon: an XPS study on these methanol oxidation catalysts, J. Phys. Chem. C 111 (15) (2007) 5715–5720.

[30] A. Serov, C. Kwak, Recent achievements in direct ethylene glycol fuel cells (DEGFC), Appl. Catal. B Environ. 97 (1–2) (2010) 1–12.

[31] E. Antolini, E. Gonzalez, Alkaline direct alcohol fuel cells, J. Power Sources 195 (11) (2010) 3431–3450.

170 Chapter 10 Ternary/quaternary nanomaterials for direct alcohol fuel cells

[32] F. Vigier, et al., Electrocatalysis for the direct alcohol fuel cell, Top. Catal. 40 (1–4) (2006) 111–121.

[33] F. Şen, S. Şen, G. Gökağaç, Efficiency enhancement of methanol/ethanol oxidation reactions on Pt nanoparticles prepared using a new surfactant, 1, 1-dimethyl heptanethiol, Phys. Chem. Chem. Phys. 13 (4) (2011) 1676–1684.

[34] A. Sheikh, K.E.-A. Abd-Alftah, C. Malfatti, On reviewing the catalyst materials for direct alcohol fuel cells (DAFCs), Energy 1 (2014) 1–10.

[35] D.F. Gervasio, Fuel cells-direct alcohol fuel cells| new materials, in: Encyclopedia of Electrochemical Power Sources, Elsevier, 2009, pp. 420–427.

[36] T. Shuihua, et al., Review of new carbon materials as catalyst supports in direct alcohol fuel cells, Chin. J. Catal. 31 (1) (2010) 12–17.

[37] L. Osmieri, et al., Fe-N/C catalysts for oxygen reduction reaction supported on different carbonaceous materials. Performance in acidic and alkaline direct alcohol fuel cells, Appl. Catal. B Environ. 205 (2017) 637–653.

[38] A.L.M. Reddy, N. Rajalakshmi, S. Ramaprabhu, Cobalt-polypyrrole-multiwalled carbon nanotube catalysts for hydrogen and alcohol fuel cells, Carbon 46 (1) (2008) 2–11.

[39] Y. Wang, Nanomaterials for Direct Alcohol Fuel Cell, CRC Press, 2016.

[40] M.E. Gálvez, et al., Nanostructured carbon materials as supports in the preparation of direct methanol fuel cell electrocatalysts, Catalysts 3 (3) (2013) 671–682.

[41] C. Luo, et al., A review of the application and performance of carbon nanotubes in fuel cells, J. Nanomater. 2015 (2015).

[42] M. Borghei, et al., Enhanced performance of a silicon microfabricated direct methanol fuel cell with PtRu catalysts supported on few-walled carbon nanotubes, Energy 65 (2014) 612–620.

[43] T. Yoshitake, et al., Preparation of fine platinum catalyst supported on single-wall carbon nanohorns for fuel cell application, Phys. B Condens. Matter 323 (1–4) (2002) 124–126.

[44] W. Li, et al., Carbon nanotube film by filtration as cathode catalyst support for proton-exchange membrane fuel cell, Langmuir 21 (21) (2005) 9386–9389.

[45] J.E. Mink, M.M. Hussain, Sustainable design of high-performance microsized microbial fuel cell with carbon nanotube anode and air cathode, ACS Nano 7 (8) (2013) 6921–6927.

[46] B. Çelik, et al., Nearly monodisperse carbon nanotube furnished nanocatalysts as highly efficient and reusable catalyst for dehydrocoupling of DMAB and C1 to C3 alcohol oxidation, Int. J. Hydrog. Energy 41 (4) (2016) 3093–3101.

[47] K.P. De Jong, J.W. Geus, Carbon nanofibers: catalytic synthesis and applications, Catal. Rev. 42 (4) (2000) 481–510.

[48] L. Sikeyi, A. Adekunle, N. Maxakato, Electro-catalytic activity of carbon nanofibers supported palladium nanoparticles for direct alcohol fuel cells in alkaline medium, Electrocatalysis 10 (4) (2019) 420–428.

[49] M. Martínez-Huerta, M. Lázaro, Electrocatalysts for low temperature fuel cells, Catal. Today 285 (2017) 3–12.

[50] A. Suriani, et al., Synthesis of carbon nanofibres from waste chicken fat for field electron emission applications, Mater. Res. Bull. 70 (2015) 524–529.

[51] N. Rodriguez, A review of catalytically grown carbon nanofibers, J. Mater. Res. 8 (12) (1993) 3233–3250.

[52] Y. Yıldız, et al., Different ligand based monodispersed Pt nanoparticles decorated with rGO as highly active and reusable catalysts for the methanol oxidation, Int. J. Hydrog. Energy 42 (18) (2017) 13061–13069.

[53] Y. Yıldız, et al., Monodisperse Pt nanoparticles assembled on reduced graphene oxide: highly efficient and reusable catalyst for methanol oxidation

Chapter 10 Ternary/quaternary nanomaterials for direct alcohol fuel cells **171**

and dehydrocoupling of dimethylamine-borane (DMAB), J. Nanosci. Nanotechnol. 16 (6) (2016) 5951–5958.

[54] R.G. Bai, K. Muthoosamy, S. Manickam, A.H Alnaqbi, Graphene-based 3D scaffolds in tissue engineering: fabrication, applications, and future scope in liver tissue engineering, Researchgate 14 (2019), https://doi.org/10.2147/IJN.S192779.

[55] S. Eris, et al., Nanostructured polyaniline-rGO decorated platinum catalyst with enhanced activity and durability for methanol oxidation, Int. J. Hydrog. Energy 43 (3) (2018) 1337–1343.

[56] W. Gao, The Chemistry of Graphene Oxide, Springer, 2015, pp. 61–95.

[57] M. Bacon, S.J. Bradley, T. Nann, Graphene quantum dots, Part. Part. Syst. Charact. 31 (4) (2014) 415–428.

[58] A. Grizni, M. Aidin, Synthesis, determination of the structure and electrochemical performance of the palladium-cobalt alloy catalyst based on the graphene for usage in the direct alcohol fuel cells with ethanol fuel, Medbiotech Journal 3 (01) (2019) 22–25.

[59] N. Shang, et al., Platinum integrated graphene for methanol fuel cells, J. Phys. Chem. C 114 (37) (2010) 15837–15841.

[60] E.T. Sayed, et al., Facile and low-cost synthesis route for graphene deposition over cobalt dendrites for direct methanol fuel cell applications, J. Taiwan Inst. Chem. Eng. 115 (2020) 321–330.

[61] L.Y. Zhang, et al., Palladium-cobalt nanodots anchored on graphene: in-situ synthesis, and application as an anode catalyst for direct formic acid fuel cells, Appl. Surf. Sci. 469 (2019) 305–311.

[62] G.H. Jeong, et al., One-pot synthesis of au@ Pd/graphene nanostructures: electrocatalytic ethanol oxidation for direct alcohol fuel cells (DAFCs), RSC Adv. 3 (23) (2013) 8864–8870.

[63] H. Chang, S.H. Joo, C. Pak, Synthesis and characterization of mesoporous carbon for fuel cell applications, J. Mater. Chem. 17 (30) (2007) 3078–3088.

[64] J. Salgado, et al., Pt–Ru electrocatalysts supported on ordered mesoporous carbon for direct methanol fuel cell, J. Power Sources 195 (13) (2010) 4022–4029.

[65] H. Du, et al., Carbon nanomaterials in direct liquid fuel cells, Chem. Rec. 18 (9) (2018) 1365–1372.

[66] T. Arikan, A.M. Kannan, F. Kadirgan, Binary Pt–Pd and ternary Pt–Pd–Ru nanoelectrocatalysts for direct methanol fuel cells, Int. J. Hydrog. Energy 38 (6) (2013) 2900–2907.

[67] J. Bagchi, S.K. Bhattacharya, The effect of composition of Ni-supported Pt-Ru binary anode catalysts on ethanol oxidation for fuel cells, J. Power Sources 163 (2) (2007) 661–670.

[68] L. Yu, J. Xi, CeO2 nanoparticles improved Pt-based catalysts for direct alcohol fuel cells, Int. J. Hydrog. Energy 37 (21) (2012) 15938–15947.

[69] Y.-H. Cho, et al., PtPdCo ternary electrocatalyst for methanol tolerant oxygen reduction reaction in direct methanol fuel cell, Appl. Catal. B Environ. 154 (2014) 309–315.

[70] B. Baruah, A. Kumar, PEDOT: PSS/MnO2/rGO ternary nanocomposite based anode catalyst for enhanced electrocatalytic activity of methanol oxidation for direct methanol fuel cell, Synth. Met. 245 (2018) 74–86.

[71] G. Ren, et al., One-pot solvothermal preparation of ternary PdPtNi nanostructures with spiny surface and enhanced electrocatalytic performance during ethanol oxidation, J. Alloys Compd. 830 (2020) 154671.

[72] C. Wang, et al., High-density surface protuberances endow ternary PtFeSn nanowires with high catalytic performance for efficient alcohol electro-oxidation, Nanoscale 11 (39) (2019) 18176–18182.

[73] B. Cermenek, et al., Novel highly active carbon supported ternary PdNiBi nanoparticles as anode catalyst for the alkaline direct ethanol fuel cell, Nano Res. 12 (3) (2019) 683–693.

[74] K. Peng, et al., Carbon supported PtPdCr ternary alloy nanoparticles with enhanced electrocatalytic activity and durability for methanol oxidation reaction, Int. J. Hydrog. Energy 45 (43) (2020) 22752–22760.

[75] X. Zhang, et al., One-pot synthesis of ternary alloy CuFePt nanoparticles anchored on reduced graphene oxide and their enhanced electrocatalytic activity for both methanol and formic acid oxidation reactions, Electrochim. Acta 177 (2015) 93–99.

[76] J.S. Cooper, P.J. McGinn, Combinatorial screening of thin film electrocatalysts for a direct methanol fuel cell anode, J. Power Sources 163 (1) (2006) 330–338.

[77] M.K. Jeon, et al., Highly active PtRuFe/C catalyst for methanol electro-oxidation, Electrochem. Commun. 9 (9) (2007) 2163–2166.

[78] E. Antolini, Evaluation of the optimum composition of low-temperature fuel cell electrocatalysts for methanol oxidation by combinatorial screening, ACS Comb. Sci. 19 (2) (2017) 47–54.

[79] M. Rethinasabapathy, et al., Quaternary PtRuFeCo nanoparticles supported N-doped graphene as an efficient bifunctional electrocatalyst for low-temperature fuel cells, J. Ind. Eng. Chem. 69 (2019) 285–294.

[80] Z. Cui, et al., Pd nanoparticles supported on HPMo-PDDA-MWCNT and their activity for formic acid oxidation reaction of fuel cells, Int. J. Hydrog. Energy 36 (14) (2011) 8508–8517.

[81] S. Gu, et al., Quaternary phosphonium-based polymers as hydroxide exchange membranes, ChemSusChem: Chemistry & Sustainability Energy & Materials 3 (5) (2010) 555–558.

[82] M. Ammam, E.B. Easton, Quaternary PtMnCuX/C (X= Fe, co, Ni, and Sn) and PtMnMoX/C (X= Fe, co, Ni, cu and Sn) alloys catalysts: synthesis, characterization and activity towards ethanol electrooxidation, J. Power Sources 215 (2012) 188–198.

11

Catalysts for high-temperature fuel cells operated by alcohol fuels

Ali Cherif[a,b], Nimeti Doner[b], and Fatih Şen[a]
[a]Şen Research Group, Department of Biochemistry, Dumlupinar University, Kütahya, Turkey. [b]Department of Mechanical Engineering, Faculty of Engineering, Gazi University, Ankara, Turkey

1 Introduction

Nowadays, fuel cells are attracting tremendous interest, and institutions are researching their development in terms of fuel used, catalysts, processing type, configuration, and application domains.

Fuel cells can use a large range of fuels (namely, natural gas, alcohol, and other substances such as hydrazine and carbon in the form of graphite) [1–3]. Hydrogen is attracting increasing attention due to its many advantages, such as high specific power, and as a clean energy carrier, it is considered economically ideal due to its low cost-efficiency ratio [4, 5]. However, the lack of a safe technoeconomical storage process limits the effective implementation of hydrogen fuel cells. On the other hand, hydrocarbon fuel cells suffer from carbon deposition, which deactivates the catalyst and affects its stability [6]. This latter issue has been addressed by developing new catalysts via surface modification, doping, and alloying with coking-resistant metals [7–9]. In addition, novel fuel mixtures, as well as the use of precious metals and rare-earth materials, have been studied [10, 11]. Furthermore, indirect fuel cells are used to overcome the hydrocarbon drawbacks and the deactivation issue by using hydrogen generated from hydrocarbons via reforming, dehydrogination, or partial oxidation, or by electrolysis, which depends on renewable electricity generation; nevertheless, the compactness of the process can be strongly affected by applying indirect fueling [12–22].

Nanomaterials for Direct Alcohol Fuel Cells. https://doi.org/10.1016/B978-0-12-821713-9.00005-6
Copyright © 2021 Elsevier Inc. All rights reserved.

During the last several decades, research on fuel cells using alcohols (mainly ethanol and methanol), which are sustainable and abundant feedstocks, has been expanding, especially for electric vehicles due to their compactness and high efficiency [23, 24]. Moreover, the number of electrons that can be provided by ethanol is 12, while methanol can release 8 electrons in cases of complete oxidation [24, 25]. Water and carbon dioxide (CO_2), a greenhouse gas, are by-products of both alcohols; however, CO_2 can be avoided if partial oxidation takes place in ethanol fuel cells, producing acetaldehyde and two electrons or acetic acid and four electrons, depending on the operation conditions and the catalyst used [25]. This might affect the compactness of the process, rendering it ineffective for mobile applications. Fig. 1 depicts the ethanol fuel cell scheme, in which ethanol releases hydrogen ions and CO_2 on the anode side, while oxygen reacts with hydrogen, producing water on the cathode side [26].

Table 1 shows the property differences between the two main alcohols used: methanol and ethanol. The use of hydrogen as a liquid in not convenient in terms of energy efficiency. Moreover, the heat enthalpy and Gibbs energy of the two alcohols are much higher than those of hydrogen. Also, fuel cells fueled with methanol, for example, might achieve up to 90% of the fuel cell energy density supplied by hydrogen [27]. Hence, methanol, the simplest alcohol, is considered the most attractive for operating fuel cells,

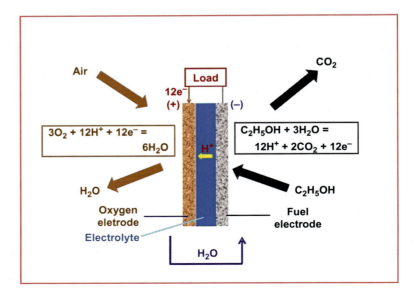

Fig. 1 Alcohol fuel cell (ethanol) [26].

Table 1 Properties of hydrogen, methanol, and ethanol [26].

Properties	H_2	CH_3OH	C_2H_5OH
$-\Delta G°$ (kJ/mol)	237	702	1325
$-\Delta H$ (kJ/mol)	286	726	1367
LHV (kWh/kg)	33	6.09	8.00
LHV (kWh/L)	2.96×10^{-3}	4.80	6.32
E^0_{cell} (V)	1.23	1.21	1.14
Energy stored (Ah/kg)	26,802	3350	2330
Energy stored (Ah/L)	2.40	2653	1841

with methanol cells providing as fast a conversion rate as hydrogen-fueled cells [28].

Many reviews have been conducted regarding alcohol fuel cells, focusing on process stability, application domain, system security and control, fuel category, and so on [29–34]. In this chapter, the catalysts used for alcohol fuel cells at high temperature are reviewed, revealing the issues presented by the catalysts employed and their suggested solutions, as well as the novelties of their design and synthesis.

2 High-temperature alcohol fuel cells

The main losses considered in high-temperature (HT) fuel cells are the Joule, leakage, activation, fuel crossover, internal current, and mass transport losses [35]. The only losses for a low-temperature alcohol fuel cell are the Joule and mass transfer losses; an HT alcohol fuel cell is affected by all these types of losses [36], which makes the process much more costly. However, a high temperature is indispensable to overcome the low heat conductivity of the electroite, and it is required to attain high efficiency. In alcohol solid oxide fuel cells (SOFCs), the operation should be 650–1000°C [37] and 800–1000°C for other hydrocarbons [38]. These temperature ranges allow the direct conversion of heavy hydrocarbons to electricity.

The HT operation of alcohol fuel cells affords the benefit of the advantages of a solid electrolyte, such as lower corrosion, minimal control of the electrolyte, and higher electron density. Moreover, the HT fuel cell is suitable for efficient CO_2 concentration and capture processes [39], and the released heat can be considered for

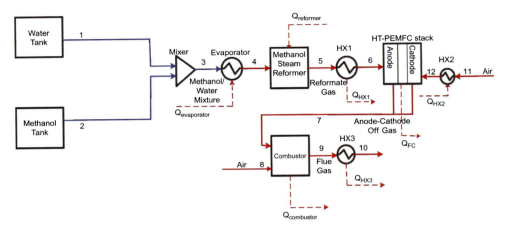

Fig. 2 Combining of fuel cells in an open cycle [42].

other uses and closed-cycle combinations, such as supplying heat for steam methane reforming for cheaper hydrogen production [40]. Also, the combination with cooling and heating systems is possible for fuel cells operating at high temperatures [41]. Fig. 2 shows the combination of fuel cells for potable water applications [42].

Another advantage of HT alcohol fuel cells is that precious or rare-earth metals are not necessary to reduce the activation energy; in contrast, advanced catalysts are needed as the operating temperature decreases. Also, high temperatures enforce the reaction kinetics and minimize catalyst poisoning. On the other hand, the high temperature in an alcohol SOFC often causes stability issues and decreases efficiency due to corrosion, transition, and expansion of the SOFC parts [43].

Not only is the SOFC considered less efficient at low temperatures, it is also unstable and prone to complex kinetics that are not yet understood [44, 45].

For the alcohol-fueled proton-exchange membrane fuel cell, the HT operation leads to better proton conductivity and chemical and thermal properties [26]. In addition, the contamination induced by the impurities of the inputs can be avoided when high temperatures are applied [46, 47]. Hence, this type of fuel cell is not suitable for mobile applications due to increased vibration. Nafion-based membranes might be considered as a solution to this issue; however, synthesis costs are high [48, 49].

In the case of alcohol-fueled polymer electrolyte membranes for fuel cells, operation at high temperatures is indispensable to attain acceptable performance. However, achieving temperatures higher than 90°C is a challenge, so many studies have been conducted to achieve this operating condition, while avoiding the membrane electrical conductivity losses and alcohol crossover [50, 51]. Many studies have been conducted to attain reasonable efficiency with slight temperature increases [52, 53].

Alkaline HT fuel cells operate at 90°C and above, with a higher membrane concentration of 0.85 wt% of KOH compared to 0.35 wt% concentration of low-temperature fuel cells [54, 55]. On the other hand, the alkaline anion-exchange membrane fuel cell is not been well known for HT operations; however, the operation of this type at high temperatures was introduced recently by Douglin et al. [56], who achieved an operating temperature of 110° C. This might open the door for more research into alcohol use in the future.

3 Operating catalyst

The catalyst is an intrinsic material for the alcohol fuel cell process to attain acceptable efficiency, high voltage, and lowering the activation energy. Many catalysts have been investigated and novel materials introduced. Moreover, the quality of the catalyst determines many features of the process, such as durability, stability, and process cost [27, 57–61]. Also, alcohol oxidation and electrooxidation over various catalysts have been studied [62–74]. Even though the high cost of fabrication of an alcohol HT-SOFC compared to other types of fuel cells is due to the high operating temperature, the materials are considered inexpensive [75]. This is ascribed to the low importance of the catalyst when it is operated at this temperature range. Overall, the most used materials for an HT-SOFC is an anode catalyst of Ni cermet, while the cathode is La, $SrMnO_3$-based perovskite, and Y_2O_3-ZrO_2 (YSZ) is the conventional electrolyte [76]. However, other materials have been studied, and other types of alcohol fuel cells have been operated at higher temperatures. The most significant issue of the use of Ni-based catalysts is the carbon deposition due to the high dehydrogenation of hydrocarbons when Ni particles are involved in the intermediate reaction. To overcome this issue, two pathways have been reported on in the literature. The first is the premixing of the fuel with water or another oxidizer, such as air [44], which affects the compactness, efficiency, and reaction kinetics of the process. The second pathway involves modifying conventional Ni-based materials [77–80], but this method affects the electronic and activity properties of the conventional catalyst.

178 Chapter 11 Catalysts for high-temperature fuel cells operated

Table 2 The reactions at the electrodes of direct alcohol fuel cells [82].

Direct ethanol fuel cells	Anode	$C_2H_5OH + 3H_2O \rightarrow 2CO_2 + 12H^+ + 12e^-$
	Cathode	$3O_2 + 12H^+ + 12e^- \rightarrow 6H_2O$
	Overall reaction	$C_2H_5OH + 3O_2 \rightarrow 2CO_2 + 3H_2O$
Direct methanol fuel cells	Anode	$CH_3OH + H_2O \rightarrow CO_2 + 6H^+ + 6e^-$
	Cathode	$1.5O_2 + 6H^+ + 6e^- \rightarrow 3H_2O$
	Overall reaction	$2CH_3OH + 3O_2 \rightarrow 4H_2O + 2CO_2$

Methanol is considered the preferred fuel among the prospective alcohols due to its many advantages over hydrogen, i.e., good solubility in electrolytes, abundance and low cost, and safe storage, transport, and control [81]. Among other alcohol fuels, only ethanol has been considered a candidate for advanced research [81].

Table 2 shows the reactions at the electrodes and the total reaction equation of direct alcohol fuel cells using ethanol and methanol [83, 84].

Not only is the efficiency of the catalyst important for a direct alcohol fuel cell, but also is its stability, which might be affected by several issues, such as coke deposition, catalyst poisoning, migration of catalyst particles, efficiency barriers, and hot and cold spots [45, 85–87]. Thus, many studies have been conducted to overcome these issues using a wide range of catalysts, support materials, and operating conditions.

Liu et al. studied the coke deposition problem over a traditional nickel/samarium-doped ceria (Ni/SDC) catalyst on a liquid methanol fuel cell anode at high and intermediate temperatures of between 400°C and 600°C [88], as well as for other fuels for comparision purposes (namely, hydrogen and ammonia). The highest operating temperature (650°C) showed the best stability and coking resistance, as well as the best power density, compared to the other tested temperatures of 550°C and 600°C, with a maximum voltage of 0.8 V [89].

Wang et al. [90] studied ethanol fuel cells over a Ni/Al_2O_3 catalyst using pyridine as an additive fuel, which occupies the anode's acidic zones, reducing the coke formation by 64% at a temperature at 600°C and pyridine concentration of 12.5%.

A conventional Ni catalyst is prone to affine carbon particles, which leads to greater coke deposition problems and makes its commercialization more inefficient, so that catalysts such as copper-based anodes and $La_{0.75}Sr_{0.25}Cr_{0.5}Mn_{0.5}O_3$ (LSCM) and Cu-Pd alloys are considered more suitable for commercialization

Chapter 11 Catalysts for high-temperature fuel cells operated **179**

and process durability [91]. Also, the protection of Ni particles using SDC or a wash coat of Ru-SDC can increase the resistance to coke deposition [89].

Hou et al. [92] studied a $Sm_{0.5}Ba_{0.5}MnO_{3-\delta}$ catalyst wash coat for operating an SOFC using H_2 and methanol as the feed at 850°C. Low efficiency was noted due to the low electrical conductivity of the catalyst, with about 150 mW/cm^2 power density. On the other hand, the power density was raised approximately threefold when methanol was used. This latter substance induced higher electrical conductivity due to slight carbon deposition. However, this might damage the catalyst for long-term usage.

Mejdoub et al. [93] found that 650°C was the optimum operating temperature for a methanol-fueled SOFC with approximatively complete methanol oxidation and full contribution of oxygen. In addition, coke formation was notably low when an SDC layer was integrated.

Li et al. [43] used a $La_2Ni_{0.9}Fe_{0.1}O_4$ anode for a methanol-fueled SOFC at a temperature of 800°C. They found that a degree of carbon disposition can be an asset for improving electrical conductivity. The results showed a high power density (166 mW/cm^2).

ZnO was incorporated into Ni crystals for the synthesis of $Ni_{1-x}Zn_xO/SDC$ to operate an HT-methanol SOFC at 700°C [94]. This catalyst configuration was also tested for hydrogen, Ni/SDC, and different concentrations of Ni ($x = 0.2$, 0.5, and 0.8) for comparison. A power density of 834 mW/cm^2 was achieved and coking resistance improved due to the presence of ZnO. The high efficiency was ascribed to the enhanced surface diffusion of active particles with ZnO.

Steil et al. [80] investigated a direct ethanol SOFC using a wash-coated layer of Ir catalyst supported on gadolinium-doped ceria mounted on an Ni/YSZ anode. They achieved a high stability for the fuel cell for 600 h at 850°C without the use of water, attaining similar performance when using hydrogen as fuel, with a current density of approximately 6.75 A/cm^2 in a stability test at 0.6 V.

Silva et al. [95] studied the impact of modifying an Ni-ceria anode with different dopants in an ethanol fuel cell at the high temperature of 900°C. They found that the performance of SOFCs depended strongly on the dopant used, as follows: Ni/CeO_2, Ni/CeGd, Ni/CeY, Ni/CePr, Ni/CeZr, and Ni/CeNb. The activity of the catalysts was limited by carbon deposition, with the lowest deactivation shown by Ni/CeNb, which exhibited high carbon deposition resistance due to the protection of the Ni particles by Nb_2O_6 in the $NiNb_2O_6$ phase. Sun et al. [96] also used carbon with an Ni-based porous media catalyst to facilitate the fabrication and dispersion of Pd particles for the operation of anion-exchange membrane direct liquid fuel cells.

The chromium-based catalyst $La_{0.75}Sr_{0.25}Cr_{0.50}Mn_{0.50}O_3$ (LSCM) was studied by Monteiro et al. for operating direct ethanol SOFCs at 800°C, and the impact of adding Ru particles was studied. This helped in coke removal, which improved stability without having a significant effect on the catalyst structure, as proved by X-ray diffusion.

An ethanol SOFC has shown low coke deposition when the temperature was controlled by the oxygen feed to achieve a final operation temperature of 800°C, exploiting conventional Ni-YSZ [97].

To overcome the operating temperature limit of a methanol-fueled proton-exchange membrane fuel cell, Lui et al. [98] analyzed the effect of polyvinyl alcohol for membrane support. The results showed low swelling by methanol and water depending on the composition of the polyvinyl alcohol film; moreover, this polymer showed high stability at temperatures of up to 250°C.

An external methanol reforming fuel cell is suitable for proton-exchange membrane fuel cells, especially for high heat exchange between the fuel cell and the reforming sections. For this situation, Herdem et al. [42] conducted a numerical study operating the fuel cell at 160°C, 170°C, and 180°C over a CuZn-based catalyst for steam methanol reforming, analyzing the effects of other parameters. The highest overall efficiency was attained with the highest applied operating temperature in the fuel cell section (240°C), with the lowest reforming section inlet temperature (180°C).

A polybenzimidazole-based membrane was used for HT membranes due to its proton conductivity properties and is convenient at high temperatures of up to 200°C, showing low methanol/ethanol permeability [99].

Precious-metal-based catalysts are generally used to enhance low-temperature fuel cells due to their low required activation energy, as well as their carbon deposition resistance, which is more harmful at low operating temperatures. Hence, their use in HT alcohol fuel cells is limited. For example, gold, silver, palladium, and platinum showed high efficiency for different fuels [9, 100–102].

4   Conclusion

In this chapter, the recent findings and results concerning the catalysts and catalytic anodes used for the activation of alcohol-fueled HT fuel cells have been summarized and discussed. HT fuel cell processes are beneficial for many reasons, including the possibility of wide integration in systems and plants. Even given the operating defects in this temperature range, such as coke deposition and swelling, the commercialization of all types of fuel cells is increasing, which leads to greater economic interest. The use of alcohol is also a focal point due to its properties, potential, and sustainability. However, its use is limited to only two alcohols: methanol and ethanol.

Further investigation and problem-solving of the use of other alcohols at high temperatures is indispensable to attain a greater variety of fuel usage and applications. Moreover, during the last two decades, the process has been less dependent on Ni-based catalysts, which show lower stability. Hence, the wider range of prospective materials for use as catalysts have led to more findings about process suitability and how to overcome the main impediments. Moreover, alcohol fuel cells have been limited to the conventional parallel electrode configuration; other configurations, such as spiral, leaflike configuration might be taken into account [103].

Many strategies have been adopted to recuperate the heat released from HT fuel cells. However, this has been very dependent on hydrogen production via reforming; consequently, other uses for this energy should be investigated.

References

[1] M. Bischoff, Molten carbonate fuel cells: a high temperature fuel cell on the edge to commercialization, J. Power Sources 160 (2) (2006) 842–845.

[2] X. Yan, et al., Direct N_2H_4/H_2O_2 fuel cells powered by nanoporous gold leaves, Sci. Rep. 2 (1) (2012) 941.

[3] G.-y. Liu, et al., Fuels for direct carbon fuel cells: present status and development prospects, Carbon 86 (2015) 371.

[4] D.L. Trimm, Z.I. Önsan, Onboard fuel conversion for hydrogen-fuel-cell-driven vehicles, Catal. Rev. 43 (1–2) (2001) 31–84.

[5] Y. Ligen, H. Vrubel, H. Girault, Energy efficient hydrogen drying and purification for fuel cell vehicles, Int. J. Hydrog. Energy 45 (18) (2020) 10639–10647.

[6] H. Mohammed, et al., Direct hydrocarbon fuel cells: a promising technology for improving energy efficiency, Energy 172 (2019) 207–219.

[7] R. da Paz Fiuza, M.A. da Silva, J.S. Boaventura, Development of Fe–Ni/YSZ–GDC electrocatalysts for application as SOFC anodes: XRD and TPR characterization and evaluation in the ethanol steam reforming reaction, Int. J. Hydrog. Energy 35 (20) (2010) 11216–11228.

[8] J. Qu, et al., One-pot synthesis of silver-modified sulfur-tolerant anode for SOFCs with an expanded operation temperature window, AICHE J. 63 (10) (2017) 4287–4295.

[9] D.Y. Jang, et al., Coke-free oxidation of methanol in solid oxide fuel cells with heterogeneous nickel–palladium catalysts prepared by atomic layer deposition, ACS Sustain. Chem. Eng. 8 (28) (2020) 10529–10535.

182 Chapter 11 Catalysts for high-temperature fuel cells operated

[10] J.S. Park, et al., Evidence of proton transport in atomic layer deposited Yttria-stabilized zirconia films, Chem. Mater. 22 (18) (2010) 5366–5370.

[11] H. Zhang, et al., Fuel additive injection system and methods for inhibiting coke formation, 2020. Google Patents.

[12] B. Şen, et al., Silica-based monodisperse PdCo nanohybrids as highly efficient and stable nanocatalyst for hydrogen evolution reaction, Int. J. Hydrog. Energy 43 (44) (2018) 20234–20242.

[13] R. Pitchai, K. Klier, Partial oxidation of methane, Catal. Rev. 28 (1) (1986) 13–88.

[14] A. Cherif, R. Nebbali, Numerical analysis on autothermal steam methane reforming: effects of catalysts arrangement and metal foam insertion, Int. J. Hydrog. Energy 44 (39) (2019) 22455–22466.

[15] F. Barbir, PEM electrolysis for production of hydrogen from renewable energy sources, Sol. Energy 78 (5) (2005) 661–669.

[16] B. Wang, J. Zhu, Z. Lin, A theoretical framework for multiphysics modeling of methane fueled solid oxide fuel cell and analysis of low steam methane reforming kinetics, Appl. Energy 176 (2016) 1–11.

[17] B. Şen, et al., High-performance graphite-supported ruthenium nanocatalyst for hydrogen evolution reaction, J. Mol. Liq. 268 (2018) 807–812.

[18] A. Cherif, R. Nebbali, L. Nasseri, CFD analysis of the metal foam insertion effects on SMR reaction over Ni/Al2O3 catalyst, in: Advances in Renewable Hydrogen and Other Sustainable Energy Carriers, Springer Singapore, Singapore, 2021.

[19] A. Cherif, R. Nebbali, L. Nasseri, CFD study of ATR reaction over dual Pt–Ni catalytic bed, in: Advances in Renewable Hydrogen and Other Sustainable Energy Carriers, Springer Singapore, Singapore, 2021.

[20] A. Cherif, R. Nebbali, L. Nasseri, Optimization of the Ni/Al_2O_3 and Pt/Al_2O_3 catalysts load in autothermal steam methane reforming, in: Advances in Renewable Hydrogen and Other Sustainable Energy Carriers, Springer Singapore, Singapore, 2021.

[21] B. Sen, et al., Monodisperse rutheniumcopper alloy nanoparticles decorated on reduced graphene oxide for dehydrogenation of DMAB, Int. J. Hydrog. Energy 44 (21) (2019) 10744–10751.

[22] B. Sen, et al., Monodisperse palladium nanocatalysts for dehydrocoupling of dimethylamineborane, Nano-Struct. Nano-Obj. 16 (2018) 209–214.

[23] K. Dircks, Recent advances in fuel cells for transportation applications, SAE Int. (1999).

[24] G.L. Soloveichik, Liquid fuel cells, Beilstein J. Nanotechnol. 5 (1) (2014) 1399–1418.

[25] F. Vigier, et al., Electrocatalysis for the direct alcohol fuel cell, Top. Catal. 40 (1–4) (2006) 111–121.

[26] S.P.S. Badwal, et al., Direct ethanol fuel cells for transport and stationary applications—a comprehensive review, Appl. Energy 145 (2015) 80–103.

[27] Y. Ru, et al., Durability of direct internal reforming of methanol as fuel for solid oxide fuel cell with double-sided cathodes, Int. J. Hydrog. Energy 45 (11) (2020) 7069–7076.

[28] E. Ventosa, et al., TiO_2 (B)/anatase composites synthesized by spray drying as high performance negative electrode material in li-ion batteries, ChemSusChem 6 (8) (2013) 1312–1315.

[29] W. Wang, et al., Recent advances in the development of anode materials for solid oxide fuel cells utilizing liquid oxygenated hydrocarbon fuels: a mini review, Energy Technol. 7 (1) (2019) 33–44.

[30] S.S. Munjewar, S.B. Thombre, R.K. Mallick, Approaches to overcome the barrier issues of passive direct methanol fuel cell—review, Renew. Sust. Energ. Rev. 67 (2017) 1087–1104.

Chapter 11 Catalysts for high-temperature fuel cells operated **183**

[31] T. Elmer, et al., Fuel cell technology for domestic built environment applications: state of-the-art review, Renew. Sust. Energ. Rev. 42 (2015) 913–931.

[32] V. Das, et al., Recent advances and challenges of fuel cell based power system architectures and control—a review, Renew. Sust. Energ. Rev. 73 (2017) 10–18.

[33] W.R.W. Daud, et al., PEM fuel cell system control: a review, Renew. Energy 113 (2017) 620–638.

[34] L. van Biert, et al., A review of fuel cell systems for maritime applications, J. Power Sources 327 (2016) 345–364.

[35] J. Larminie, A. Dicks, M.S. McDonald, Fuel Cell Systems Explained, vol. 2, J. Wiley, Chichester, UK, 2003.

[36] S. Basu, Fuel Cell Science and Technology, Springer, 2007.

[37] C. Song, Fuel processing for low-temperature and high-temperature fuel cells: challenges, and opportunities for sustainable development in the 21st century, Catal. Today 77 (1) (2002) 17–49.

[38] S. Chen, et al., An integrated system combining chemical looping hydrogen generation process and solid oxide fuel cell/gas turbine cycle for power production with CO_2 capture, J. Power Sources 215 (2012) 89–98.

[39] F. Wang, et al., A comprehensive review on high-temperature fuel cells with carbon capture, Appl. Energy 275 (2020) 115342.

[40] G. Diglio, et al., Modelling of sorption-enhanced steam methane reforming in a fixed bed reactor network integrated with fuel cell, Appl. Energy 210 (2018) 1–15.

[41] M. Mehrpooya, et al., Technical performance analysis of a combined cooling heating and power (CCHP) system based on solid oxide fuel cell (SOFC) technology—a building application, Energy Convers. Manag. 198 (2019).

[42] M.S. Herdem, S. Farhad, F. Hamdullahpur, Modeling and parametric study of a methanol reformate gas-fueled HT-PEMFC system for portable power generation applications, Energy Convers. Manag. 101 (2015) 19–29.

[43] P. Li, et al., Improve electrical conductivity of reduced La2Ni0.9Fe0.1O4+δ as the anode of a solid oxide fuel cell by carbon deposition, Int. J. Hydrog. Energy 40 (31) (2015) 9783–9789.

[44] H. Aslannejad, et al., Effect of air addition to methane on performance stability and coking over NiO–YSZ anodes of SOFC, Appl. Energy 177 (2016) 179–186.

[45] B.C. Yang, et al., Direct alcohol-fueled low-temperature solid oxide fuel cells: a review, Energy Technol. 7 (1) (2019) 5–19.

[46] X. Cheng, et al., A review of PEM hydrogen fuel cell contamination: impacts, mechanisms, and mitigation, J. Power Sources 165 (2) (2007) 739–756.

[47] Z. Zakaria, et al., New composite membrane poly(vinyl alcohol)/graphene oxide for direct ethanol-proton exchange membrane fuel cell, J. Appl. Polym. Sci. 136 (2) (2019).

[48] Y. Wang, et al., A review of polymer electrolyte membrane fuel cells: technology, applications, and needs on fundamental research, Appl. Energy 88 (4) (2011) 981–1007.

[49] B. Smitha, S. Sridhar, A.A. Khan, Solid polymer electrolyte membranes for fuel cell applications—a review, J. Membr. Sci. 259 (1–2) (2005) 10–26.

[50] X. Zhu, et al., Synthesis and properties of novel H-bonded composite membranes from sulfonated poly(phthalazinone ether)s for PEMFC, J. Membr. Sci. 312 (1–2) (2008) 59–65.

[51] M. Guo, et al., Preparation of sulfonated poly(ether ether ketone)s containing amino groups/epoxy resin composite membranes and their in situ crosslinking for application in fuel cells, J. Power Sources 195 (1) (2010) 11–20.

[52] J. Chen, et al., Crosslinking and grafting of polyetheretherketone film by radiation techniques for application in fuel cells, J. Membr. Sci. 362 (1–2) (2010) 488–494.

[53] C.-Y. Tseng, et al., Interpenetrating network-forming sulfonated poly(vinyl alcohol) proton exchange membranes for direct methanol fuel cell applications, Int. J. Hydrog. Energy 36 (18) (2011) 11936–11945.

[54] Y. Li, X. Sun, Y. Feng, Hydroxide self-feeding high-temperature alkaline direct formate fuel cells, ChemSusChem 10 (10) (2017) 2135–2139.

[55] S.A. Kalogirou, Industrial process heat, chemistry applications, and solar dryers, in: S.A. Kalogirou (Ed.), Solar Energy Engineering, second ed., Academic Press, Boston, 2014, pp. 397–429 (Chapter 7).

[56] J.C. Douglin, J.R. Varcoe, D.R. Dekel, A high-temperature anion-exchange membrane fuel cell, J. Power Sour. Adv. 5 (2020).

[57] C.-Y. Wang, Fundamental models for fuel cell engineering, Chem. Rev. 104 (10) (2004) 4727–4766.

[58] P. Li, et al., Improved activity and stability of $Ni-Ce0.8Sm0.2O1.9$ anode for solid oxide fuel cells fed with methanol through addition of molybdenum, J. Power Sources 320 (2016) 251–256.

[59] Ö. Karatepe, et al., Enhanced electrocatalytic activity and durability of highly monodisperse Pt@PPy–PANI nanocomposites as a novel catalyst for the electro-oxidation of methanol, RSC Adv. 6 (56) (2016) 50851–50857.

[60] S. Eris, et al., Nanostructured polyaniline-rGO decorated platinum catalyst with enhanced activity and durability for methanol oxidation, Int. J. Hydrog. Energy 43 (3) (2018) 1337–1343.

[61] S. Eris, Z. Daşdelen, F. Sen, Enhanced electrocatalytic activity and stability of monodisperse Pt nanocomposites for direct methanol fuel cells, J. Colloid Interface Sci. 513 (2018) 767–773.

[62] B. Sen, et al., Highly monodisperse RuCo nanoparticles decorated on functionalized multiwalled carbon nanotube with the highest observed catalytic activity in the dehydrogenation of dimethylamine-borane, Int. J. Hydrog. Energy 42 (36) (2017) 23292–23298.

[63] B. Çelik, et al., Nearly monodisperse carbon nanotube furnished nanocatalysts as highly efficient and reusable catalyst for dehydrocoupling of DMAB and C1 to C3 alcohol oxidation, Int. J. Hydrog. Energy 41 (4) (2016) 3093–3101.

[64] F. Şen, S. Şen, G. Gökağaç, Efficiency enhancement of methanol/ethanol oxidation reactions on Pt nanoparticles prepared using a new surfactant, 1,1-dimethyl heptanethiol, Phys. Chem. Chem. Phys. 13 (4) (2011) 1676–1684.

[65] Y. Yıldız, et al., Different ligand based monodispersed Pt nanoparticles decorated with rGO as highly active and reusable catalysts for the methanol oxidation, Int. J. Hydrog. Energy 42 (18) (2017) 13061–13069.

[66] Y. Yıldız, et al., Monodisperse Pt nanoparticles assembled on reduced graphene oxide: highly efficient and reusable catalyst for methanol oxidation and dehydrocoupling of dimethylamine-borane (DMAB), J. Nanosci. Nanotechnol. 16 (6) (2016) 5951–5958.

[67] S. Eris, Z. Daşdelen, F. Sen, Investigation of electrocatalytic activity and stability of Pt@f-VC catalyst prepared by in-situ synthesis for methanol electrooxidation, Int. J. Hydrog. Energy 43 (1) (2018) 385–390.

[68] F. Şen, G. Gökağaç, Different sized platinum nanoparticles supported on carbon: an XPS study on these methanol oxidation catalysts, J. Phys. Chem. C 111 (15) (2007) 5715–5720.

[69] Z. Daşdelen, et al., Enhanced electrocatalytic activity and durability of Pt nanoparticles decorated on GO-PVP hybride material for methanol oxidation reaction, Appl. Catal. B Environ. 219 (2017) 511–516.

Chapter 11 Catalysts for high-temperature fuel cells operated **185**

[70] S. Şen, F. Şen, G. Gökağaç, Preparation and characterization of nano-sized Pt–Ru/C catalysts and their superior catalytic activities for methanol and ethanol oxidation, Phys. Chem. Chem. Phys. 13 (15) (2011) 6784–6792.

[71] F. Sen, et al., Amylamine stabilized platinum(0) nanoparticles: active and reusable nanocatalyst in the room temperature dehydrogenation of dimethylamine-borane, RSC Adv. 4 (4) (2014) 1526–1531.

[72] B. Çelik, et al., Monodisperse Pt(0)/DPA@GO nanoparticles as highly active catalysts for alcohol oxidation and dehydrogenation of DMAB, Int. J. Hydrog. Energy 41 (13) (2016) 5661–5669.

[73] B. Çelik, et al., Monodispersed palladium–cobalt alloy nanoparticles assembled on poly(N-vinyl-pyrrolidone) (PVP) as a highly effective catalyst for dimethylamine borane (DMAB) dehydrocoupling, RSC Adv. 6 (29) (2016) 24097–24102.

[74] F. Şen, G. Gökağaç, Activity of carbon-supported platinum nanoparticles toward methanol oxidation reaction: role of metal precursor and a new surfactant, tert-octanethiol, J. Phys. Chem. C 111 (3) (2007) 1467–1473.

[75] S. Wang, S.P. Jiang, Prospects of fuel cell technologies, Natl. Sci. Rev. 4 (2) (2017) 163–166.

[76] B.S. Prakash, S.S. Kumar, S. Aruna, Properties and development of Ni/YSZ as an anode material in solid oxide fuel cell: a review, Renew. Sust. Energ. Rev. 36 (2014) 149–179.

[77] A. Atkinson, et al., Advanced anodes for high-temperature fuel cells, in: Materials for Sustainable Energy, 2010, pp. 213–223.

[78] J. Qu, et al., Stable direct-methane solid oxide fuel cells with calcium-oxide-modified nickel-based anodes operating at reduced temperatures, Appl. Energy 164 (2016) 563–571.

[79] S. McIntosh, R.J. Gorte, Direct hydrocarbon solid oxide fuel cells, Chem. Rev. 104 (10) (2004) 4845–4866.

[80] M.C. Steil, et al., Durable direct ethanol anode-supported solid oxide fuel cell, Appl. Energy 199 (2017) 180–186.

[81] C. Lamy, J.-M. Léger, S. Srinivasan, Direct methanol fuel cells: from a twentieth century electrochemist's dream to a twenty-first century emerging technology, in: Modern Aspects of Electrochemistry, Springer, 2002, pp. 53–118.

[82] K. Föger, Materials basics for fuel cells, in: M. Gasik (Ed.), Materials for Fuel Cells, Woodhead Publishing, 2008, pp. 6–63 (Chapter 2).

[83] S. Bose, et al., Polymer membranes for high temperature proton exchange membrane fuel cell: recent advances and challenges, Prog. Polym. Sci. 36 (6) (2011) 813–843.

[84] R.K. Shah, S.-P. Flow, Compact Heat Exchangers and Enhancement Technology for the Process Industries-2003, Begell House, Incorporated, 1905.

[85] K. Sasaki, The impact of fuels on solid oxide fuel cell anode lifetime, in: Solid Oxide Fuel Cell Lifetime and Reliability, 2017, pp. 37–50.

[86] H. Jeong, et al., Bimetallic nickel/ruthenium catalysts synthesized by atomic layer deposition for low-temperature direct methanol solid oxide fuel cells, ACS Appl. Mater. Interfaces 8 (44) (2016) 30090–30098.

[87] A. Pagidi, G. Arthanareeswaran, M.M. Seepana, Synthesis of highly stable PTFE-ZrP-PVA composite membrane for high-temperature direct methanol fuel cell, Int. J. Hydrog. Energy 45 (13) (2020) 7829–7837.

[88] B. Zhu, Advantages of intermediate temperature solid oxide fuel cells for tractionary applications, J. Power Sources 93 (1) (2001) 82–86.

[89] M. Liu, et al., Direct liquid methanol-fueled solid oxide fuel cell, J. Power Sources 185 (1) (2008) 188–192.

[90] W. Wang, et al., Coking suppression in solid oxide fuel cells operating on ethanol by applying pyridine as fuel additive, J. Power Sources 265 (2014) 20–29.

186 Chapter 11 Catalysts for high-temperature fuel cells operated

[91] S. Tao, J.T. Irvine, A redox-stable efficient anode for solid-oxide fuel cells, Nat. Mater. 2 (5) (2003) 320–323.

[92] N. Hou, et al., Sm0.5Ba0.5MnO3-δ anode for solid oxide fuel cells with hydrogen and methanol as fuels, Catal. Today 298 (2017) 33–39.

[93] F. Mejdoub, A. Elleuch, K. Halouani, Assessment of the intricate nickel-based anodic reactions mechanism within a methanol fed solid oxide fuel cell based on a co-ionic conducting composite electrolyte, J. Power Sources 414 (2019) 115–128.

[94] X. Zhi, et al., ZnO-promoted surface diffusion on NiO-Ce0.8Sm0.2O1.9 anode for solid oxide fuel cell, J. Power Sources 423 (2019) 290–296.

[95] A.A.A. da Silva, et al., Effect of the type of ceria dopant on the performance of Ni/CeO2 SOFC anode for ethanol internal reforming, Appl. Catal. B Environ. 206 (2017) 626–641.

[96] X. Sun, Y. Li, M.-J. Li, Highly dispersed palladium nanoparticles on carbon-decorated porous nickel electrode: an effective strategy to boost direct ethanol fuel cell up to 202 mW cm^{-2}, ACS Sustain. Chem. Eng. 7 (13) (2019) 11186–11193.

[97] J. Qu, et al., Ethylene glycol as a new sustainable fuel for solid oxide fuel cells with conventional nickel-based anodes, Appl. Energy 148 (2015) 1–9.

[98] C.-P. Liu, et al., Novel proton exchange membrane based on crosslinked poly(vinyl alcohol) for direct methanol fuel cells, J. Power Sources 249 (2014) 285–298.

[99] Q. Li, et al., High temperature proton exchange membranes based on polybenzimidazoles for fuel cells, Prog. Polym. Sci. 34 (5) (2009) 449–477.

[100] E. Antolini, Palladium in fuel cell catalysis, Energy Environ. Sci. 2 (9) (2009) 915–931.

[101] J. Qi, et al., PdAg/CNT catalyzed alcohol oxidation reaction for high-performance anion exchange membrane direct alcohol fuel cell (alcohol=methanol, ethanol, ethylene glycol and glycerol), Appl. Catal. B Environ. 199 (2016) 494–503.

[102] H.J. Jeong, et al., Platinum–ruthenium heterogeneous catalytic anodes prepared by atomic layer deposition for use in direct methanol solid oxide fuel cells, ACS Catal. 5 (3) (2015) 1914–1921.

[103] A. Coralli, et al., Fuel cells, in: Science and Engineering of Hydrogen-Based Energy Technologies, 2019, pp. 39–122.

12

Porous metal materials for polymer electrolyte membrane fuel cells

Fatma Aydın Ünal[a], Cisil Timuralp[b], Vildan Erduran[c,d], and Fatih Şen[c]

[a]*Metallurgical and Materials Engineering Department, Faculty of Engineering, Alanya Alaaddin Keykubat University, Alanya/Antalya, Turkey.* [b]*Mechanical Engineering Department, Eskisehir Osmangazi University, Eskişehir, Turkey.* [c]*Şen Research Group, Department of Biochemistry, Dumlupinar University, Kütahya, Turkey.* [d]*Department of Materials Science and Engineering, Faculty of Engineering, Dumlupinar University, Kütahya, Turkey*

1 Introduction

Fuels contain chemical energy, and this energy can be converted into electrical energy. In polymer fuel cell varieties, proton exchange membrane fuel cells are a valuable lot [1–4]. The catalyst for catalytic electrodes has significant characteristics, such as large surface, low resiliency, noisy power, and stability. Carbon nanotubes (CNTs) can be used as catalysts in fuel cells. Carbonnanotubes used as catalysts use elements such as iron, cobalt, and nickel, which accumulate on different substrates. Because of its relatively low cost, high resistance, and corrosion resistance in polymeric electrolyte membrane fuel cells (PEMFCs), rubber stainless steel is of significant importance. The PEMFC usually contains several main parts, such as cones, catalysts, catalyst supports, catalytic plates, gaseous diffusion layers, and current collectors [5, 6].

The knowledge in this chapter on the use of porous metal materials for polymer electrolyte membrane fuel cells is available.

Nanomaterials for Direct Alcohol Fuel Cells. https://doi.org/10.1016/B978-0-12-821713-9.00009-3
Copyright © 2021 Elsevier Inc. All rights reserved.

2 Polymer exchange membrane fuel cells (PEMFCs)

The fuel cell for proton-exchange membranes, also known as the "polymer electric membrane fuel cell (PEMFC)," usually operates at temperatures below 100°C with special polymeric electrical membranes [7, 8]. High energy transfer efficiency, low operating temperature, fast startup, pollutants with virtually zero by-products, portability, and user-friendliness are of great concern with fuel cells. PEMFCs are also very interesting.

It is of great importance that PEMFCs are interconnected with fuel cells; high transformation efficiency, low working temperature (approximately 100°C), quick inlet, contaminants with virtually zero by-product emissions of water, and portability and ease of use [7].

In the PEMFC, there is a membrane electrode assembly system (MEA). Fig. 1 displays a single MEA PEMFC. The cells for high voltages at MEA normally are compacted when stored in series with two mirrored flux surface plates. The MEA structure has catalyst walls, a gas diffusion layer (GDL), and an exchange membrane for protons. These MEA sections are separately developed and then integrated under high temperature and pressure [7,9].

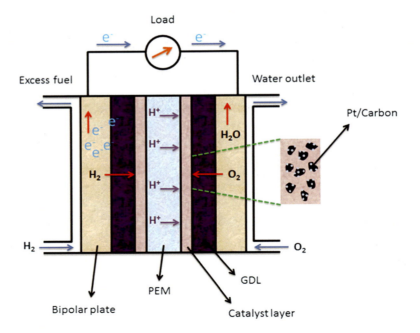

Fig. 1 Cross-sectional view of a PEMFC [10].

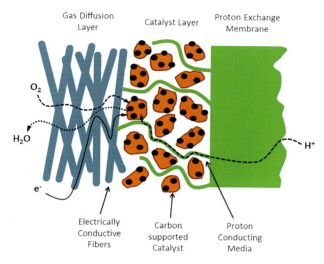

Fig. 2 Transport of gases, electrons, and protons in PEMs [11].

The diagram of the electrodes, which are considered as components extending from the gas channel to the surface of the membrane and up to the current collector, are shown in Fig. 2.

A good electrode is an electrode that correctly balances the transport processes necessary for a fuel cell in operation. The transportation of protons is carried out by three transport systems, responsible for the following:
- Moving to the catalyst from the membrane
- Presenting collector transportation to the catalyst
- The production of reactive and product gases of the GDLs, transferred by electrons and catalyst layers

In the catalyst layer, electrons, protons, and gases are generally in three phases. A part of the electrode optimization is planned to minimize the loss of propagation, such as the one that distributes the volume of the catalyst sheet appropriately for each step between the transport agents [12]. The intersection of the catalyst particles in such transport processes is of significant importance for the efficient activity of a PEM fuel pellet [1,3,10].

3 Fundamental components and materials of PEMFCs

The strict electrolyte ion-exchange membrane, porous and conductive gas diffusion plate, electrode (electrocatalyst), and the key components of PEMFC are cell plates on the interface between the support sheet and membrane [11, 13].

The perfluorinated composite membrane is based on the concept of reinforcing the perfluorinated membrane by incorporation of a fine, porous polytetrafluoroethylene (PTFE) film. It is critical that catalysts not containing platinum remain stable in the acidic environment of the PEMFC [14, 15]. It is necessary to ensure that the platinum-free catalysts retain stable in the PEMFC's acidic environment. Many metals, including cobalt (Co), and iron (Fe), have also been studied as substitute catalysts for platinum (Pt). The two metals demonstrated that metal ions can be used in conjunction with a macro-nitrogen macrocycle similar to the natural porphyrin mechanism as an efficient and sufficient catalyst in the PEMFC. It should be remembered that to maintain the proton and electron conductivity of the membrane active layer, the interaction between the catalyst, the reactive gas (oxygen or hydrogen), and the associated elements like polymer must stay at the highest degree. The catalyst particles are inserted into the porous conducting substance known as the "catalyst rear-blade" to ensure the best possible contact catalyst support layers (CSLs). To ensure that the proton conduction of the catalyst layer is provided by a Nafion with a porous structure, as seen in Fig. 3, CSLs, with its three-phase interface, has a big role to play in maintaining a successful balance between reagents and goods in the productive use of the catalyst. The most widely used catalysts for PEMFC include carbon black (Pt/C) catalyzed platinum nanoparticles with large surfaces. Moreover, carbon-based nanomaterials were explored as potential graphene superlattices (GSLs) with graphene, single-walled

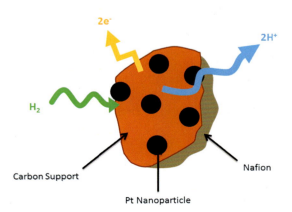

Fig. 3 Structure of the catalyst layer [11].

nanotubes (SWNTs), multiwalled carbon nanotubes (MWNTs), nanocorns, large-scale carbon nanofibers, and porous structures. Depending on the application form, the load number for Pt-based catalysts typically varies. Characteristic platinum loads in the electrode range from about 0.4 to about 0.8 mg of platinum per cm^2 [11, 16].

3.1 Fuel cell membranes

3.1.1 Preparation of filled porous polymeric membranes

There are some issues with the preparation of filled porous membranes with strong properties. These issues include the use of water-soluble proton conductive particles and the lack of dissolution of organic apolar solvents [5, 17].

3.1.1.1 Filled porous polymeric membranes

There are loud conductivity zirconium phosphate sulfophenylenphosphonates with similar function to Nafion [18, 19]. Some studies offered conclusions acquired than a porous Teflon membrane filled with $Zr(O_3P\text{-}OH)(O_3P\text{-}C_6H_4SO_3H)\cdot nH_2O$. The conductance of the filled porous membranes was seen to be the same as that shown by pure pellets of $Zr(O_3P\text{-}OH)(O_3P\text{-}C_6H_4SO_3H)\cdot nH_2O$.

The explanation for this is the improved conductivity of zirconium phosphonate produced in the pores due to a decrease in conductivity caused by the folded porous binding pathways. This demonstrates that the influence of the pores on the conductivity properties can be included in the desired parallel orientation of the cover lip particles generated by the pore surface (Fig. 4).

These consequences are promising for use at temperatures below 100°C. Because the properties of membranes can be compared with Nafion, but their use is cheaper, they can be prepared industrially. Because Teflon is more temperature tolerant than Nafion, the use of these full-membranes at temperatures between 130°C and 160°C is noteworthy [5].

3.1.2 Some general consideration of proton-conducting composite membranes

In studies involving hydrogen and indirect methanol cells at temperatures above 90–100°C, Alberti and Casciola emphasize that specific attention should be paid. Partial modification of the zirconium phosphate (ZrP) particles with the water contained in the pores can be beneficial for reducing the changes in membrane size under dehydration terms. In the absence of filler, it is expected that the collapse of the hydrophilic areas on dehydration

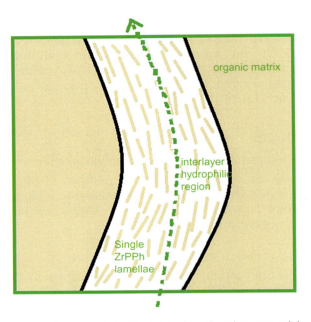

Fig. 4 A representative form of stratified zirconium phosphonate particles that are placed parallel to the pore surface in the membrane [20].

will harm the triple contact among the electronic conductor, the catalyst, and the proton conductor in the porous electrodes of the fuel cell. Using this information, the active layer of electrodes can be added to ZrP phosphonates or Nafion packed with silica to improve the productivity of Nafion membranes above 100°C in fuel cells [5]. For more effective membranes, porous polymer membranes with ZrP derivatives with a proton conductance greater than 10^{-2} S/cm can be used [5, 21–23].

3.1.3 Hybrid organic-inorganic membranes (nanocomposite membranes)

The method of preparation of composite membranes with inorganic and inorganic-organic proton conductive particles is very significant, as it will affect the microstructure of the membrane [24]. Properties of the particles, such as their specific size, specific surface area, type, surface acidity, shape, and their interaction with the polymer matrix are very important. These properties can cause large differences in membrane performance. Also, the enrichment of the present fillers (in particular nano-sized fillers) has been one of the most investigated subjects. The fillers may be separable into solid, nonporous fillers with zeolites such as SiO_2-TiO_2 nanoparticles, solid porous fillers such as porous metal oxides, metal-organic frameworks (MOFs) [25], and CNTs [26, 27]. The nanocomposite membranes based on different fillers will be described in the following sections [27, 28].

3.1.3.1 Nanocomposite PEMs with TiO_2

TiO_2 is a hygroscopic metal oxide that develops the cell performance, in conditions of high operating temperature, easier water management, and thermomechanical stability. The addition of TiO_2 had a very impressive and favorable effect on the proton conductivity of pure polystyrene porous membranes [27, 29].

3.1.3.2 Nanocomposite PEMs with MOFs

Pourzare et al. looked at novel Nafion-based composite membranes (PEM-1 and PEM-2) using two one-dimensional (1D) channel microporous MOFs as fillers; that is, CPO-27 (Mg) and MIL-53(Al) were investigated for PEMFC applications. The results showed an increase in water uptake and proton conductivity by 1.7 times and 2.1 times in size, respectively, as compared with the recurrent Nafion membrane. The CPO-27 (Mg)-Nafion composite membrane exhibited most power intensity values of $818 \, mW/cm^2$ and $591 \, mW/cm^2$, at $50°C$ and $80°C$, respectively [28].

3.1.3.3 Nanocomposite PEMs with other miscellaneous fillers

Mesoporous carbon-supported Pt catalysts showed perfect performance in PEMs [30]. First, nonprecious metal catalysts (NPMCs) have been obtained by precipitating the Fe/Co-Nx compound over nanoporous carbon black and ethylenediamine (EDA) as a nitrogen pioneer in an effort to minimize oxygen reaction. NPMCs have been generated. Two separate, nanoporous carbon catalysts assisted Black Ketjen EC300J (KJ300) and EC600JD (KJ600) as the nonprecious catalyst. The KJ600 has demonstrated advanced characteristics as semiwave capability and excellent selectivity compared with the KJ300 catalyst and is used as supporting material for FeCo/EDA-carbon. Moreover, hydrogen oxygen PEMFCs were surprisingly effective in this catalyst. A total of $0.37 \, A/cm^2$ at a power density of $0.44 \, W/cm^2$ was achieved at a cell voltage of $0.6 \, V$ in a fuel cell. A 0.41-V fixed voltage fuel cell lifetime test revealed strong expectations for up to $100 \, h$. The support materials are formed using carbon, high pores, and high surface area. The findings demonstrate that the material support used will contribute to more nitrogen content for oxygen reduction reaction (ORR). In PEMFCs, this will also allow carbon-backed materials [31]. Mesoporous Pt-catalysts, highly homogeneous catalytic metal delivery, and high electrical efficiency in fuel cell electrode PEMFC reactions have shown optimal performance thanks to the pore structures of the components. BET analyzes are helpful in providing structural knowledge about the catalyst, as well as support when the substance is porous [27, 32, 33].

3.2 Gas diffusion layer (GDL)

The PEMFC porous gas diffusion blade provides that the reactants are easily distributed to the catalyst substrate [32]. It forms a GDL from carbon paper or fabric with a thickness of 100–300 µm [33]. Characteristically, GDLs are covered by a PTFE blade to guarantee that gas diffusion sheet pores are not blocked by fluid water [10,34].

The effect of the addition of PTFE in water transport to various GDL materials was investigated by Benziger et al. Water flow did not occur because of the hydrophobic structure of PTFE until a pressure of 5–10 kPa was applied from GDL. By increasing the applied hydrostatic pressure, the water was allowed to flow through the smaller pores in the GDL. Water flow from less than 1% of the void volume in GDL is seen. The little pores in the GDL stay away from the water stream to allow gas to enter the catalyst sheet. The flow in the GDL can be classified as multiphase, especially on the cathode area where the water is formed. The GDL is subject to degradation as a result of long-term work. Metallic materials are used to prevent the deterioration of carbon-based GDLs. The little thickness and flat porosity of the metallic material used were able to develop water administration in PEMFCs [11, 35–37].

3.2.1 GDL improvement

There are many critical activities to the gas diffusion sheet. The electronic conductor is found mainly among the bipolar plates and catalyst sheets that are available. Gas diffusion layers should be made more porous, despite increasing electrical resistance, to boost public transport. As a general basis for the catalyst layer deposition, the porous gas diffusion blade is used [11, 38]. Carbon aerogels have recently been used to shape a pore substratum in layers of gas diffusion [39, 40]. PEM fuel cells have been used to apply carbon aerogels. On both sides of the gaseous diffusion layer, the 300-µm-thick sheets are fine-structured by a micron-thin sheet. The films are mixed to minimize touch resistance between the electrode and membrane and to reduce the current bipolar plate collector. In their analysis, the highest electrical conductance reached was 28 S/cm in a 80% porous specimen. In a few microns, the largest pores were obtained [11, 41].

3.2.2 Compound GDLs

The PTFE carbon fabric/paper is typically a composite GDL with a microporous Apolar substrate that is sandwiched in a carbon base [38] and a catalyst sheet [40]. The apolar lower layer is charged with designing transportation pathways and optimizing water conservation on the pore supports and catalyst surfaces. Qi and Kaufman believed that the microporosity of the substrate is the cause of the water management created. The scale and size of the macropore are normally determined by the carbon agglomeration. A carbon particle size of 30 nm in general indicated that stable water particles cannot form in these small, apolar pores [10,42]. Song et al. collected a lower micronic layer consisting of PTFE and carbon dust sandwiched between a wet-pore carbon paper and a thin sheet of film catalysts. When the microporous sublayer was being loaded optimally, the researchers found 3.5 mg/cm^2 and PTFE was 30% by weight. When using multiple carbon sheet stacks, the output gap decreased [10,43].

The PEMFC structure allows the electrodes to be porous in order to disperse the gas through the electric catalyst's efficient regions. Platinum are the most common forms of PEMFC catalysts. The electrode itself can have a high surface catalytic field to improve the reaction speed (flow intensity), as well as the polarization of electrodes [1, 44, 45]. The electrocatalyst is shown in Fig. 5 to allow the development of the porous gas diffusion electrode (GDE) structure. The electrode is mounted on a carbon foundation. It is critical that the GDE should be apolar in the cathodes (where the water is formed) so that the gas channels do not float with the electrode and thus contribute to the restrictions of mass transport. The relation of the carbon-based catalyst with PTFE was a hydrophobic first GDE [1].

Flow area designs are based on a variety of principles in PEM cells. The flows vary from zigzag to increasingly difficult designs from the simple parallel channels to the twisted flow, as seen in Fig. 6. In the twisted configuration, the flow snakes in a limited number of channels from one side of the cell to the other. This forms the cell's reactants into a long-flow pathway. Metal foam, sintered metal, metal fiber, or thin porous structures made of the mat in contact with the MEA are used in the open canals

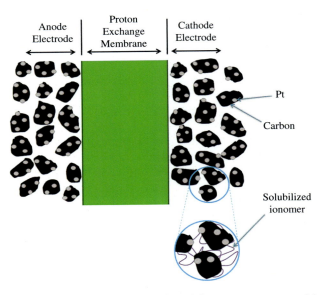

Fig. 5 Carbon-supported catalyst electrode and three separate parts of the active catalyst [1].

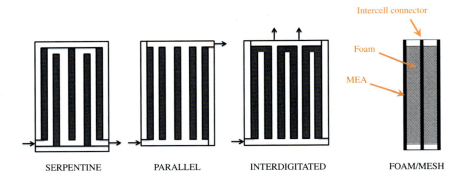

Fig. 6 Flow area designs used in fuel cells [1].

system. The dual-porous construction of this application is separated into a nonporous layer, slim conductive metal, or carbon. The alternative flow control is in the integrated flow region. This style has fingerlike channels with dead ends interwoven. The efficiency of interdigitated stream areas with a dead-end flow design in PEMFCs and direct methanol fuel cells has been announced. This evolution is related to the dead-end channels that shift the

Chapter 12 Porous metal materials for polymer electrolyte membrane fuel cells **197**

porous sheet transport process through forced convection instead of mostly diffusion [1].

Fly et al. have carried out ex-situ optical research using colored water in Reynolds air-conforming quantities in the distribution movement of porous metallic foam flow regions. The findings show that the moisture effectively deploys liquid across the flow spectrum, roughly matching the theoretical filling ratio to 70% flow field cover [42].

Fig. 7A displays the reconstructed scan of the metallic foam using X-ray computed tomography (CT) using a single pressurized flow channel, and the scanning electron microscopy (SEM) photo of the flow channel in Fig. 7B. Plastic deformation due to moisture compression in the area of the flow channel is illustrated in Fig. 7A and B. The resemblance between SEM and X-ray CT images shows segmentation validity. The cross-section reconstructed in the x-y plane and the corresponding deployment for porosity are shown in Fig. 7C and D. In the area below the flow channel, the porosity of the plastic deformation decreases from 84% to 15% when the flow channel is squeezed. The upper walls of the flow canal show a minimum reduction in porosity. This is an extremely porous direction that contributes to the flux of fluid into the flow channels through the foam mass. The minimum foam thickness is 0.3 mm, with a compression of 81.3% in the middle of the flow path. The findings of this study help report future integrated fuel cell designs using porous streams. Decreasing waiting times and parasites were accomplished and an advanced understanding of how various architectures influence the implementation of macroscopic flows [1, 46]. In contrast to other supporting materials, ORR obtained in the new carbon support materials for graphene showed excellent performance. Differences in preparatory techniques are the physical properties of the graphene. Graphene may have a mesoporous or microporous pore structure. Thus, the kinetics of electrochemical reactions in PEMFCs can be accelerated by it as an integral feature. Carbon support products are the primary supply of wide surface areas and strong conductance as a catalyst for catalytic ORR in PEMFCs. Carbon support materials, however, induce ORR kinetics, which as a result of surface oxidation undermines the efficiency of the catalyst [43, 44].

Absolute transformation in mesoporous carbon-supported catalysts such as CoN@MC (0.5) or cobalt acetate-incorporated catalysts such as CoAs@MC (0.05) and CoAs@MC (0.1) at 600°C was carried out by Mammadli. More than 80% of conversion values were obtained at the same temperature over cobalt chloride expressed with mesoporous carbon. Cobalt acetate integrated in porous-structured backed catalysts has the greatest

198 Chapter 12 Porous metal materials for polymer electrolyte membrane fuel cells

(a) 3D reconstruction of foam scan

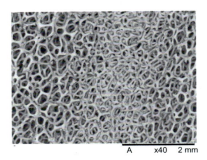

(b) SEM image of foam with pressed channel in centre

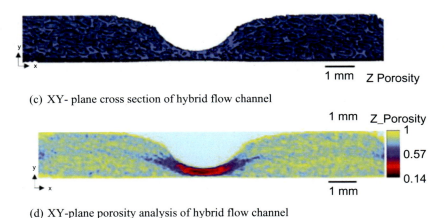

(c) XY- plane cross section of hybrid flow channel

(d) XY-plane porosity analysis of hybrid flow channel

Fig. 7 Three-dimensional (3D) reconstruction and SEM images of foam [42]. Reprinted (adapted) with permission from A. Fly, D. Butcher, Q. Meyer, M. Whiteley, A. Spencer, C. Kim, P.R. Shearing, D.J.L. Brett, R. Chen, Characterisation of the diffusion properties of metal foam hybrid flow-fields for fuel cells using optical flow visualisation and X-ray computed tomography, J. Power Sources, 395 (2018) 171–178, https://doi.org/10.1016/J.JPOWSOUR.2018.05.070. Creative Commons CC-BY license.

Chapter 12 Porous metal materials for polymer electrolyte membrane fuel cells **199**

impact on the extraction of hydrogen from ammonia under equivalent reaction conditions. It was seen that the activities of active carbon supported cobalt catalysts were lower than mesoporous carbon supported cobalt catalysts. This is because of variations in form [45].

3.2.3 Catalyst without carbon support

The support materials without carbon in the structure are mostly inorganic metal oxides. The enhanced interaction and stable morphology of inorganic metal oxides relative to other mineral oxides and metal nanoparticles show greater resistance to corrosion. At the same time, these materials have low electrical conductivity. Because of their limited porous shape, the surface region is less precise than porous supporting materials. The major inorganic additives included metal oxides, carbides, and layered silicate compounds, including tile minerals, used recently to support PEMFC supports [7, 47]. Khantimerov et al. studied the formation of CNTs from stainless steel powders in a pored granular structure and the deposition of metallic particles catalyzed by platinum and silver on carbohydrates of selected substrates [5–46, 48–54]. The benefit of this method is that this hybrid configuration shows the contents of the cell in which the three capabilities are integrated, after the growth of CNTs in pore pellets accompanied by CNTs decors by catalytic nanoparticles [5]. Fig. 8 shows a structure consisting of CNTs and metallic particles shaped in perforated rustproof steel pellets in a transmission electron microscopy (TEM) image.

The study of optical microscopy and electron microscopy revealed that CNTs form both the pellet surface and the surface pores of the subsurface layer (Fig. 9). CNTs have random alignment and have entangled on the surface of the pellet to form a thin "felt" mat.

A comparable effect of deposited nanoparticles measuring CNTs of about 50-nm dimensions was observed via TEM. As seen in Fig. 10B, nanometer holes are located on CNT walls instead of nanoparticles Fig. 10A indicates the original CNTs deposited Ni nanoparticles.

A hydrogen-oxygen fuel cell model was developed to track electrical characteristics of porous carbon-metal electrodes in place (Fig. 11). Two symmetrical sections were used in the laboratory model. Each section was fitted with a disklike anode and cathode. These disk electrodes are made of CNTs with porous stainless steel-based metal nanoparticles.

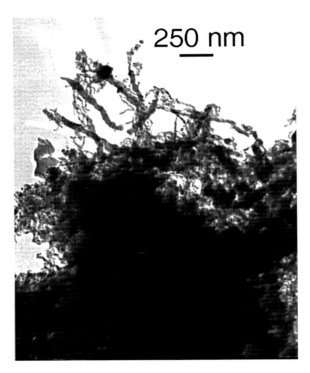

Fig. 8 TEM image of a CNT/metallic structure obtained in compacted steel structures. Black scales of platinum and silver metals have been identified [5]. Reprinted (adapted) with permission from S.M. Khantimerov, E.F. Kukovitsky, N.A. Sainov, N.M. Suleimanov, Fuel cell electrodes based on carbon nanotube/metallic nanoparticles hybrids formed on porous stainless steel pellets, Int. J. Chem. Eng. 2013 (2013) 1–4, https://doi.org/10.1155/2013/157098. Copyright © 2013 S.M. Khantimerov et al. This is an open access article distributed under the Creative Commons Attribution License.

The parameters of the electrodes used were as follows:
- Diameter: 10 mm
- Thickness: 2 mm
- Pt (for anode) and Ag (for cathode) loading: 0.75 ± 0.1 mg/cm^2

After being positioned between the electrodes, two halves of fuel cell along with bolster is connected to the PEM (Nafion 125, Aldrich). The porous anode and the cathode electrodes were supplied with hydrogen and oxygen gas. Hydrogen and oxygen have been shown to penetrate porous electrodes when excess pressure is applied (0.15 bar). The findings from the measurements of electrical features of the experimental fuel cell are seen in Fig. 12.

As a result, CNTs/metallic particular structures on compressed spaced pellets of stainless steel have been prepared by Khantimerov et al. CNTs were developed by chemical vapor deposition from a nickel catalyst impregnated with pellets of an extremely

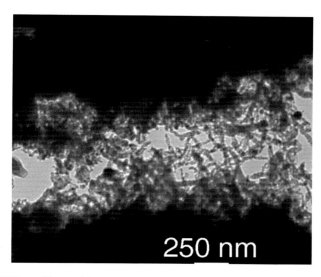

Fig. 9 A TEM image of CNT/metallic particles inside compressed stainless steel structures. For example, several "*black circles* with nanostructures" and platinum structures in the form of black scales are shown [5]. Reprinted (adapted) with permission from S.M. Khantimerov, E.F. Kukovitsky, N.A. Sainov, N.M. Suleimanov, Fuel cell electrodes based on carbon nanotube/metallic nanoparticles hybrids formed on porous stainless steel pellets, Int. J. Chem. Eng. 2013 (2013) 1–4, https://doi.org/10.1155/2013/157098. Copyright © 2013 S.M. Khantimerov et al. This is an open access article distributed under the Creative Commons Attribution License.

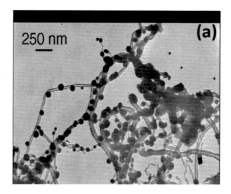 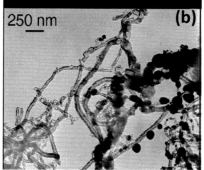

Fig. 10 A TEM image of unsupported CNTs with (A) deposited Ni nanoparticles and (B) through holes of nanometer scale on walls of CNTs [5]. Reprinted (adapted) with permission from S.M. Khantimerov, E.F. Kukovitsky, N.A. Sainov, N.M. Suleimanov, Fuel cell electrodes based on carbon nanotube/metallic nanoparticles hybrids formed on porous stainless steel pellets, Int. J. Chem. Eng. 2013 (2013) 1–4, https://doi.org/10.1155/2013/157098. Copyright © 2013 S.M. Khantimerov et al. This is an open access article distributed under the Creative Commons Attribution License.

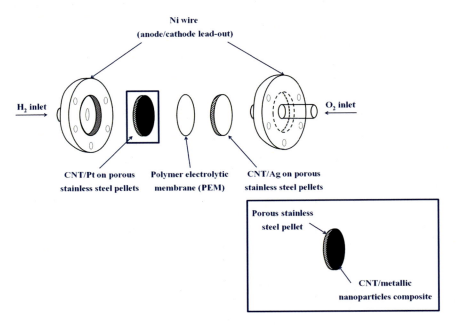

Fig. 11 Laboratory model of hydrogen-oxygen fuel cell [5]. Reprinted (adapted) with permission from S.M. Khantimerov, E.F. Kukovitsky, N.A. Sainov, N.M. Suleimanov, Fuel cell electrodes based on carbon nanotube/metallic nanoparticles hybrids formed on porous stainless steel pellets, Int. J. Chem. Eng. 2013 (2013) 1–4, https://doi.org/10.1155/2013/157098. Copyright © 2013 S.M. Khantimerov et al. This is an open access article distributed under the Creative Commons Attribution License.

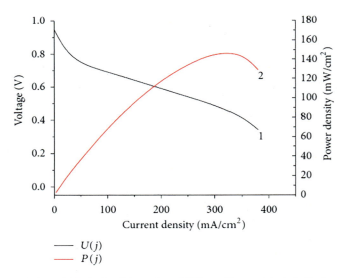

Fig. 12 Electrical characteristics of the fuel cell based on CNT/metallic particle hybrids formed on porous stainless steel pellets as electrodes. (1) $U(j)$; (2) $P(j)$ [5]. Reprinted (adapted) with permission from S.M. Khantimerov, E.F. Kukovitsky, N.A. Sainov, N.M. Suleimanov, Fuel cell electrodes based on carbon nanotube/metallic nanoparticles hybrids formed on porous stainless steel pellets, Int. J. Chem. Eng. 2013 (2013) 1–4, https://doi.org/10.1155/2013/157098. Copyright © 2013 S.M. Khantimerov et al. This is an open access article distributed under the Creative Commons Attribution License.

Chapter 12 Porous metal materials for polymer electrolyte membrane fuel cells **203**

disperse nickel acetate solution in ethanol accompanied by thermal processing. It was observed the effect of repository and holes in CNT walls on metallic agglomerations under the beam of the microscope electron. Integrating this application into the fuel cell system will make the design more applicable [5].

3.2.4 Carbon-supported, nonprecious catalysts

The most important reason in inhibiting the catalytic effect of NPMCs is the low catalytic effective field intensity [47, 55, 56]. Two basic approaches have been identified to eliminate this obstacle. These two fundamental approaches are synthesis of self-supporting catalysts with a high surface area and the production of carbon supports with pores of sufficiently high surface area [47].

Dodelet et al. stated that two different catalytic compounds, Me-N2-C (pyridinic type) and Me-N4-C (pyrrole type), are catalytically highly active, with carbon-supported catalysts and metal and nitrogen complexes. In addition, the metal ion in the macrocytes stated that the ORR plays an important role. In recent articles published by Dodelet et al., the use of microporous carbon in iron-based catalysts reported that Black Pearls 2000 caused a significant increase in ORR catalytic activity and increased site density. It has been proposed that two or four nitrogen functions coordinate Me-N-C catallytic regions are on the sides of opposing carbon cation walls [55, 56].

3.2.5 MEA preparation

The polymerization of NPMCs on two separate porous black ethyleneediamine supports of iron and cobalt obtained improved ORR operations [57]. A crucial parameter influencing ORR efficiency is that adequate time is needed to complete polymerization during the synthesis of the in situ synthesized NPMCs [58,59].

References

[1] K. Scott, A.K. Shukla, Polymer electrolyte membrane fuel cells: principles and advances, Rev. Environ. Sci. Biotechnol. 3 (2004) 273–280, https://doi.org/10.1007/s11157-004-6884-z.

[2] T. Taner, Alternative energy of the future: a technical note of PEM fuel cell water management, J. Fundam. Renew. Energy Appl. 05 (2015) 1–4, https://doi.org/10.4172/2090-4541.1000163.

[3] Z. Wan, H. Chang, S. Shu, Y. Wang, H. Tang, Z. Wan, H. Chang, S. Shu, Y. Wang, H. Tang, A review on cold start of proton exchange membrane fuel cells, Energies 7 (2014) 3179–3203, https://doi.org/10.3390/en7053179.

[4] R.A. Costa, J.R. Camacho, S.C. Guimarães, C.H. Salerno, The polymer electrolyte membrane fuel cell as electric energy source, steady state and dynamic behavior, Renew. Energy Power Qual. J. (2006), https://doi.org/10.24084/repqj04.223.

[5] S.M. Khantimerov, E.F. Kukovitsky, N.A. Sainov, N.M. Suleimanov, Fuel cell electrodes based on carbon nanotube/metallic nanoparticles hybrids formed on porous stainless steel pellets, Int. J. Chem. Eng. 2013 (2013) 1–4, https://doi.org/10.1155/2013/157098.

[6] W. Zhang, P. Sherrell, A.I. Minett, J.M. Razal, J. Chen, Carbon nanotube architectures as catalyst supports for proton exchange membrane fuel cells, Energy Environ. Sci. 3 (2010) 1286, https://doi.org/10.1039/c0ee00139b.

[7] N. Bhuvanendran, Advanced supporting materials for polymer electrolyte membrane fuel cells, in: Proton Exchange Membrane Fuel Cell, 2018, https://doi.org/10.5772/intechopen.71314.

[8] N. H. Jalani, Development of Nanocomposite Polymer Electrolyte Membranes for Higher Temperature PEM Fuel Cells. Digital WPI.

[9] N. Ahmadi, S. Rezazadeh, A. Dadvand, I. Mirzaee, Study of the effect of gas channels geometry on the performance of polymer electrolyte membrane fuel cell, Period. Polytech. Chem. Eng. 62 (2017) 97–105, https://doi.org/10.3311/PPch.9369.

[10] Y. Wang, K.S. Chen, J. Mishler, S.C. Cho, X.C. Adroher, A review of polymer electrolyte membrane fuel cells: technology, applications, and needs on fundamental research, Appl. Energy 88 (2011) 981–1007, https://doi.org/10.1016/j.apenergy.2010.09.030.

[11] S. Litster, G. McLean, PEM fuel cell electrodes, J. Power Sources 130 (2004) 61–76, https://doi.org/10.1016/j.jpowsour.2003.12.055.

[12] X. Sun, S.C. Simonsen, T. Norby, A. Chatzitakis, Composite membranes for high temperature PEM fuel cells and electrolysers: a critical review, Membranes 9 (2019), https://doi.org/10.3390/membranes9070083.

[13] C. Hartnig, L. Jörissen,, Polymer electrolyte membrane fuel cells, Mater. Fuel Cells (2008). *books.google.com.*

[14] I. Zamanillo Lopez, Hybrid Membranes for Fuel Cell, Materials, Université Grenoble Alpes, 2015.

[15] A. Baroutaji, J.G. Carton, M. Sajjia, A.G. Olabi, A. Baroutaji, Fuel cell power module view project crashworthiness and energy absorber design view project materials in PEM fuel cells, in: Reference Module in Materials Science and Materials Engineering, 2015, pp. 1–11, https://doi.org/10.1016/B978-0-12-803581-8.04006-6.

[16] O.T. Holton, J.W. Stevenson, The role of platinum in proton exchange membrane fuel cells, Platin. Met. Rev. 57 (2013) 259–271, https://doi.org/10.1595/147106713X671222.

[17] R. Gloukhovski, V. Freger, Y. Tsur, Understanding methods of preparation and characterization of pore-filling polymer composites for proton exchange membranes: a beginner's guide, Rev. Chem. Eng. 34 (2018) 455–479, https://doi.org/10.1515/revce-2016-0065.

[18] A. Kraytsberg, Y. Ein-Eli, Review of advanced materials for proton exchange membrane fuel cells, Energy Fuel 28 (2014) 7303–7330, https://doi.org/10.1021/ef501977k.

[19] M. Casciola, G. Alberti, A. Ciarletta, A. Cruccolini, P. Piaggio, M. Pica, Nanocomposite membranes made of zirconium phosphate sulfophenylenphosphonate dispersed in polyvinylidene fluoride: preparation and proton conductivity, Solid State Ionics 176 (2005) 2985–2989, https://doi.org/10.1016/j.ssi.2005.09.036.

Chapter 12 Porous metal materials for polymer electrolyte membrane fuel cells **205**

[20] G. Alberti, M. Casciola, Composite membranes for medium-temperature PEM fuel cells, Annu. Rev. Mater. Res. 33 (2003) 129–154, https://doi.org/10.1146/annurev.matsci.33.022702.154702.

[21] J. Wu, Investigation of Poly Composites and Their Potential as Proton Conductive Membranes, 2006.

[22] T. Mokrani, The Development of an Inorganic Direct Methanol Fuel Cell, 2004.

[23] M. Hattenberger, Composite Proton Exchange Membranes for Intermediate Temperature Fuel Cells, 2015.

[24] C. Laberty-Robert, K. Vallé, F. Pereira, C. Sanchez, Design and properties of functional hybrid organic–inorganic membranes for fuel cells, Chem. Soc. Rev. 40 (2011) 961, https://doi.org/10.1039/c0cs00144a.

[25] M. Songolzadeh, M.T. Ravanchi, M. Soleimani, Carbon dioxide capture and storage : a general review on adsorbents, Int. J. Chem. Mol. Nucl. Mater. Metall. Eng. 6 (2012) 900–907, https://doi.org/10.1016/S0002-9394(02)02072-X.

[26] K. Hunger, N. Schmeling, H.B.T. Jeazet, C. Janiak, C. Staudt, K. Kleinermanns, K. Hunger, N. Schmeling, H.B.T. Jeazet, C. Janiak, C. Staudt, K. Kleinermanns, Investigation of cross-linked and additive containing polymer materials for membranes with improved performance in pervaporation and gas separation, Membranes 2 (2012) 727–763, https://doi.org/10.3390/membranes2040727.

[27] B. Şen, B. Demirkan, M. Levent, A. Şavk, F. Şen, Silica-based monodisperse PdCo nanohybrids as highly efficient and stable nanocatalyst for hydrogen evolution reaction, Int. J. Hydrog. Energy 43 (2018) 20234–20242, https://doi.org/10.1016/j.ijhydene.2018.07.080.

[28] K. Pourzare, Y. Mansourpanah, S. Farhadi, Advanced nanocomposite membranes for fuel cell applications: a comprehensive review, Biofuel Research Journal 3 (2016) 496–513, https://doi.org/10.18331/BRJ2016.3.4.4.

[29] X.X. Wang, Z.H. Tan, M. Zeng, J.N. Wang, Carbon nanocages: a new support material for Pt catalyst with remarkably high durability, Sci. Rep. 4 (2015) 4437, https://doi.org/10.1038/srep04437.

[30] F.A. Viva, M.M. Bruno, E.A. Franceschini, Y.R.J. Thomas, G. Ramos Sanchez, O. Solorza-Feria, H.R. Corti, Mesoporous carbon as Pt support for PEM fuel cell, Int. J. Hydrog. Energy 39 (2014) 8821–8826, https://doi.org/10.1016/J.IJHYDENE.2013.12.027.

[31] S. Eris, Z. Daşdelen, F. Sen, Investigation of electrocatalytic activity and stability of Pt@f-VC catalyst prepared by in-situ synthesis for methanol electrooxidation, Int. J. Hydrog. Energy 43 (2018) 385–390, https://doi.org/10.1016/j.ijhydene.2017.11.063.

[32] M. Adnan, A. Dalod, M. Balci, J. Glaum, M.-A. Einarsrud, M.M. Adnan, A.R.M. Dalod, M.H. Balci, J. Glaum, M.-A. Einarsrud, In situ synthesis of hybrid inorganic–polymer nanocomposites, Polymers 10 (2018) 1129, https://doi.org/10.3390/polym10101129.

[33] G. Selvarani, A.K. Sahu, P. Sridhar, S. Pitchumani, A.K. Shukla, Effect of diffusion-layer porosity on the performance of polymer electrolyte fuel cells, J. Appl. Electrochem. 38 (2008) 357–362, https://doi.org/10.1007/s10800-007-9448-4.

[34] J. Benziger, J. Nehlsen, D. Blackwell, T. Brennan, J. Itescu, Water flow in the gas diffusion layer of PEM fuel cells, J. Membr. Sci. 261 (2005) 98–106, https://doi.org/10.1016/J.MEMSCI.2005.03.049.

[35] A.D. Santamaria, P.K. Das, J.C. MacDonald, A.Z. Weber, Liquid-water interactions with gas-diffusion-layer surfaces, J. Electrochem. Soc. 161 (2014) F1184–F1193, https://doi.org/10.1149/2.0321412jes.

[36] A. Tamayol, M. Bahrami, Water permeation through gas diffusion layers of proton exchange membrane fuel cells, in: Energy Syst. Anal. Thermodyn. Sustain. Nanoeng. Energy; Eng. to Address Clim. Chang. Parts A B, vol. 5, ASME, 2010, pp. 975–981, https://doi.org/10.1115/IMECE2010-40867.

[37] N.P. Brandon, D. Thompsett, Fuel cells compendium, Elsevier, 2005.

[38] M.A. Worsley, M. Stadermann, Y.M. Wang, J.H. Satcher Jr., T.F. Baumann, High surface area carbon aerogels as porous substrates for direct growth of carbon nanotubes, Chem. Commun. 46 (2010) 9253, https://doi.org/10.1039/c0cc03457f.

[39] F.J. Maldonado-Hódar, C. Moreno-Castilla, J. Rivera-Utrilla, Y. Hanzawa, Y. Yamada, Catalytic graphitization of carbon aerogels by transition metals, Langmuir (2000), https://doi.org/10.1021/LA991080R.

[40] Y.W. Chen-Yang, T.F. Hung, J. Huang, F.L. Yang, Novel single-layer gas diffusion layer based on PTFE/carbon black composite for proton exchange membrane fuel cell, J. Power Sources 173 (2007) 183–188, https://doi.org/10.1016/J.JPOWSOUR.2007.04.080.

[41] H.A. Kasat, Design and Development of Membrane Electrode Assembly for Proton Exchange Membrane Fuel Cell, 2016.

[42] A. Fly, D. Butcher, Q. Meyer, M. Whiteley, A. Spencer, C. Kim, P.R. Shearing, D.J.L. Brett, R. Chen, Characterisation of the diffusion properties of metal foam hybrid flow-fields for fuel cells using optical flow visualisation and X-ray computed tomography, J. Power Sources 395 (2018) 171–178, https://doi.org/10.1016/J.JPOWSOUR.2018.05.070.

[43] S. Eris, Z. Daşdelen, Y. Yıldız, F. Sen, Nanostructured polyaniline-rGO decorated platinum catalyst with enhanced activity and durability for methanol oxidation, Int. J. Hydrog. Energy 43 (2018) 1337–1343, https://doi.org/10.1016/j.ijhydene.2017.11.051.

[44] S. Eris, Z. Daşdelen, F. Sen, Enhanced electrocatalytic activity and stability of monodisperse Pt nanocomposites for direct methanol fuel cells, J. Colloid Interface Sci. 513 (2018) 767–773, https://doi.org/10.1016/j.jcis.2017.11.085.

[45] A. Mammadli, Kobalt Içerikli Karbon Nanomalzemeler Ile Amonyaktan Hidrojen Üretimi Çalişmalari, 2017.

[46] M. Kumar, Y. Ando, Chemical vapor deposition of carbon nanotubes: a review on growth mechanism and mass production, J. Nanosci. Nanotechnol. 10 (2010) 3739–3758, https://doi.org/10.1166/jnn.2010.2939.

[47] J.-Y. Choi, R.S. Hsu, Z. Chen, Highly active porous carbon-supported nonprecious metal–N electrocatalyst for oxygen reduction reaction in PEM fuel cells, J. Phys. Chem. C 114 (2010) 8048–8053, https://doi.org/10.1021/jp910138x.

[48] M.B. Kumbhare, V.S. Sapkal, R.S. Sapkal, International journal of basic and applied research PVAc-MgO Nanocomposite membranes for H2/CO2 Separation, UGC Appr. J. 8 (2018).

[49] G. Kumar, K.S. Nahm, Polymer nanocomposites-fuel cell applications, in: Advances in Nanocomposites—Synthesis, Characterization and Industrial Applications, 2011. *intechopen.com*.

[50] A. Jayakumar, M. Ramos, A. Al-Jumaily, A novel 3D printing technique to synthesise gas diffusion layer for PEM fuel cell application, in: ASME 2016 International Mechanical Engineering Congress and Exposition, vol. 6B, 2016, https://doi.org/10.1115/IMECE2016-65554.

[51] K.S. Dhathathreyan, N. Rajalakshmi, Polymer electrolyte membrane fuel cell, in: Recent Trends Fuel Cell Sci. Technol, Springer New York, New York, NY, 2007, pp. 40–115, https://doi.org/10.1007/978-0-387-68815-2_3.

[52] M. Wesselmark, Electrochemical Reactions in Polymer Electrolyte Fuel Cells, KTH, 2010.

[53] M.C. Williams, A. Suzuki, T. Hattori, R. Miura, H. Tsuboi, N. Hatakeyama, H. Takaba, M.C. Williams, A. Miyamoto, Porosity and Pt content in the catalyst layer of PEMFC: effects on diffusion and polarization characteristics, Int. J. Electrochem. Sci. (2010).

[54] B. Çelik, S. Kuzu, E. Erken, H. Sert, Y. Koşkun, F. Şen, Nearly monodisperse carbon nanotube furnished nanocatalysts as highly efficient and reusable catalyst for dehydrocoupling of DMAB and C1 to C3 alcohol oxidation, Int. J. Hydrog. Energy 41 (2016) 3093–3101, https://doi.org/10.1016/j.ijhydene.2015.12.138.

[55] J.-Y. Choi, R.S. Hsu, Z. Chen, Nanoporous carbon-supported Fe/Co-N electrocatalyst for oxygen reduction reaction in PEM fuel cells, ECS Trans. (2010) 101–112, https://doi.org/10.1149/1.3502342.

[56] J.-Y. Choi, L. Yang, T. Kishimoto, X. Fu, S. Ye, Z. Chen, D. Banham, Is the rapid initial performance loss of Fe/N/C non precious metal catalysts due to micropore flooding? Energy Environ. Sci. 10 (2017) 296–305, https://doi.org/10.1039/C6EE03005J.

[57] H. Peng, Z. Mo, S. Liao, H. Liang, L. Yang, F. Luo, H. Song, Y. Zhong, B. Zhang, High performance Fe- and N-doped carbon catalyst with graphene structure for oxygen reduction, Sci. Rep. 3 (2013) 1765, https://doi.org/10.1038/srep01765.

[58] H. Lee, M.J. Kim, T. Lim, Y.-E. Sung, H.-J. Kim, H.-N. Lee, O.J. Kwon, Y.-H. Cho, A facile synthetic strategy for iron, aniline-based non-precious metal catalysts for polymer electrolyte membrane fuel cells, Sci. Rep. 7 (2017) 5396, https://doi.org/10.1038/s41598-017-05830-y.

[59] J.-Y. Choi, Nanostructured Materials Supported Oxygen Reduction Catalysts in Polymer Electrolyte Membrane Fuel Cells, 2013.

13

Novel materials structures and compositions for alcohol oxidation reaction

Vildan Erduran[a,b], Muhammed Bekmezci[a,b], Merve Akin[a,b], Ramazan Bayat[a,b], Iskender Isik[b], and Fatih Şen[a]

[a]*Şen Research Group, Department of Biochemistry, Dumlupinar University, Kütahya, Turkey.* [b]*Department of Materials Science and Engineering, Faculty of Engineering, Dumlupinar University, Kütahya, Turkey*

1 Introduction

In recent years, energy demand is increasing in the industry due to the development of technology. Fossil fuels such as petroleum, coal, and natural gas are the main energy resources worldwide. These resources are very harmful to the environment and predominantly contribute to many global concerns, such as climate change, acid rain, depletion of the ozone layer, and health problems for human beings. All these factors have caused researchers to seek new, clean, and alternative energy conversion devices [1–3]. A fuel cell is a practical technology that converts the chemical energy of a fuel into electrical energy through electrochemical reactions. There are many types of fuel cells and one of the most important one is direct alcohol fuel cell (DAFC).

DAFCs use a proton exchange membrane (PEM), which makes them a subset of PEM fuel cell (PEMFC) technology as the electrolyte that separates the anode and cathode portions. Methanol, ethanol, ethylene glycol, and *n*-propanol are used in DAFCs as a fuel that has some advantages such as high power density, easy storage, transportation, and distribution of liquid fuel [4, 5].

The first one is direct methanol fuel cells (DMFCs), which have been investigated in detail. Thus, major advances have been made in the development of these fuel cells due to their high electrochemical kinetics [6]. In addition to DMFCs, studies have been carried out on direct ethanol fuel cells (DEFC's). Compared with DMFCs, DEFCs were found to have 50% less power density. This

Nanomaterials for Direct Alcohol Fuel Cells. https://doi.org/10.1016/B978-0-12-821713-9.00008-1
Copyright © 2021 Elsevier Inc. All rights reserved.

was not only true for ethanol; the same results were obtained from the *n*-propanol and ethylene glycol assays. However, to be able to obtain high efficiency in fuel cells, there is a need to develop electrocatalysts that may be most appropriate for the development of fuel cells and the achievement of better results. Unfortunately, a selective and active anode electrocatalyst for fuel cells has not been developed so far, and studies are ongoing. In the electrochemical process of the anode side of the DMFCs, which are the most popular among the DAFCs, methanol oxidation at the anode side generates carbon dioxide (CO_2) and produces six protons and electrons [7, 8]:

$$At the anode: CH_3OH + H_2O \rightarrow CO_2 + 6H^+ + 6e^- + heat \quad (1)$$

$$At the cathode: 3/2O_2 + 6H^+ + 6e^- \rightarrow 3H_2O + heat \quad (2)$$

$$The overall reaction: 3/2O_2 + CH_3OH \rightarrow CO_2 + 2H_2O + heat \quad (3)$$

Eqs. (1)–(3) demonstrate the anodic, cathodic, and overall reaction scheme of the oxidation reaction of methanol (MOR), not a detailed mechanism [9]. According to this mechanism, methanol is the fuel and CO_2 is the final product. Although methanol can reach CO_2 in many ways, the most preferred way involves stable compounds like formaldehyde (CH_2O), and formic acid (HCOOH). In this way, carbon monoxide (CO) may be generated as an intermediate product. Pt-based electrocatalysts are used as the most common anode electrocatalysts in DMFCs. Pt poisoned by adsorbed CO and reaction sites are occupied by this intermediate product, and accordingly, the methanol oxidation rate is reduced, which is an undesirable event. This event shows the importance of choosing an anode electrocatalyst. Various alloys of Pt with a combination of other metals are prepared and used as electrocatalysts for MOR. In addition, there are many efforts to develop new and cheaper anode electrocatalysts, as well as reducing the cost. MOR can be improved by selecting suitable electrocatalysts, which have good intrinsic catalytic activity and stability. In general, binary or ternary catalysts enhance the catalytic activity in comparison to corresponding individual metals, which is related to ligand effect (synergistic effect), modification of electronic properties, and geometric effect [10]. In this chapter, recent studies on the development of anode electrocatalysts in DAFCs are evaluated. Moreover, the effect of the electrocatalytic activity of the supporting materials such as carbon nanotubes (CNTs), carbon nanofibers (CNFs), graphene and oxide structures, and conducting polymers are reviewed.

2 Pt-based electrocatalysts

2.1 Alloy electrocatalysts

Pt is the best monometallic electrocatalyst in DAFCs [11, 12]. It is most frequently used as a model electrocatalyst in mechanistic and kinetic studies on alcohol oxidation reactions. Basically, methanol can either completely oxidize to CO_2 or oxidize incompletely to intermediate products such as CH_2O, HCOOH, and CO_{ads}. These products, especially CO_{ads} interact with Pt strongly, and reaction sites are occupied by this strong interaction, resulting in limited (or even eliminated) MOR kinetic activity. To overcome this difficulty, a promoter can be used due to effectively providing oxygen in some active form to achieve facile oxidation of the CO_{ads} on Pt. In the research literature, there are many approaches to Pt promotion. A first approach is to generate more Pt-O species on the Pt surface by incorporating certain metals with Pt to form alloys (e.g., $PtCr_3$ and Pt_3Sn). One another approach is to use adatoms obtained by underpotential deposition on the Pt surface. As a third approach, a base-metal oxide such as Nb, Zr, or Ta is combined with Pt. The usage of alloys of Pt with different metals such as Pt-Ru, Pt-Os, and Pt-Ir has been considered as a fourth approach. Among these alloys, Pt-Ru is one of the most studied alloys, but Pt-Ru-based ternary and quaternary alloys have been investigated due to their specific and effective catalytic activity. When compared with monometallic Pt, enhanced activity of these alloys has been attributed to both a bifunctional mechanism and a ligand effect (which can be either synergistic or electronic). The bifunctional mechanism can be explained by the fact that oxygen-containing species adsorb on other metals at lower potentials, thereby promoting the easy oxidation of CO to CO_2. In short, the catalytic activity of Pt-based electrocatalysts strongly depends on the alloy composition, structure, morphology, electroactive surface area, and surface porosity [13–15]. According to the ligand effect, the existence of the second metal changes the electronic level of M severing, decreasing M electron density and M-CO_{ads} strength [16].

Many studies focused on the development of the Pt-Ru catalyst, with optimization of the ratio of the Pt and Ru metals in the alloy. This modification not only reduces the cost of Pt, but can also annihilate CO poisoning by the DAFCs. For example, Kıvrak and Şahin [17] investigated the effect of Ru addition to Pt-Ru on MOR. They stated that the addition of Ru enhanced the MOR. The Pt-Ru (25:1) catalyst showed the best electrocatalytic activity, higher resistance to CO, and better long-term stability than Pt-Ru (3:1), Pt-Ru (1:1), and Pt. Tedsree. Thanatsiri [18] prepared the Pt-decorated Ru and Ru-decorated Pt core-shell nanocatalysts toward MOR. They indicated that maximum activities of Pt-Ru and Ru-Pt core shell are 3:1 and 1.2, respectively, and both samples contain Ru at about 33% and about 67% on the surface. In addition, Ru-Pt nanoparticles exhibited higher catalytic activity than that of obtained Pt-Ru nanoparticles due to the difference in electronic promotion at the interface. In another study by Bai [19], in the synthesis of platinum-ruthenium nanoparticles, PtRu/Ru was carried out by taking different amounts of Ru precursors to form heterostructures. The results of the methanol oxidation experiments showed that these heterostructures had high catalytic activity. It was also observed that PtRu alloys in a PtRu/Ru heterostructure played a role in increasing the adsorption of methanol. Thus, it is thought that PtRu nanocatalysts, which will be synthesized for this purpose, will give a new perspective. In addition to these studies, in the literature, other binary Pt-based electrocatalysts such as Pt-Sn and Pt-Pd have higher electrocatalytic activity for MOR compared to monometallic Pt. For example, Kadırgan et al. [20] studied the preparation of nanosized Pt-Pd/C catalysts and compared their electrocatalytic activity toward methanol and ethanol oxidation. Moreover, catalytic activity was compared with commercial Pt/C.

According to literature, Pt-Pd catalysts were more active for the oxidation of methanol than for ethanol. The significant improvement in the catalytic activities for methanol, and ethanol oxidation compared to that of the Pt/C can be associated with the higher electrochemical surface area and the synergistic effect between Pt and Pd. It should be noted that the preparation methods can affect the alcohol oxidation reaction. As an example, Santos et al. [21] investigated the catalytic activity of PtSn/C electrocatalysts by using two methods: The first is the polymeric precursor method (PPM), which obtained 92% alloy; and the second is a sol-gel method (SGM), which obtained 6% alloy.

Another results showed that materials have a similar chemical composition and similar mean crystalline sizes according to transmission electron microscopy (TEM) and X-ray diffraction (XRD) analysis, but they have different behaviors for ethanol oxidation reactions (EORs). The current density for EOR on the PPM material was nearly five times higher than that of the SGM material. Nonalloyed material led to CO_2 formation with slow kinetics. Rana et al. [22] reported that $Pd_{73}Pt_{27}$ nanowires (NWs) were prepared by a one-step aqueous method with a high aspect ratio and uniform diameter of ~26 nm, and this material showed 10 times more mass activity and 4.4 times more specific activity compared to commercial Pt/C toward MOR, as well as high electrochemical stability over commercial Pt/C. They attributed this high catalytic activity to synergy derived from alloying Pd-Pt. Ternary and even quaternary Pt-based alloys are other good examples, which could be successfully synthesized through a simple, one-step aqueous method [23]. The oxidation potential of small organic molecules is reduced by ternary Pt-based alloys, the C—C bond breaks even at low temperature, and then the reaction completely finishes. Therefore, the catalytic activity of ternary Pt-based alloys is higher than that of binary Pt-based alloys. The combination of tin (Sn) with Pt gives very good results on alcohol oxidation compared to the monometallic Pt. In fact, the oxidation of Sn on the catalyst surface can be useful to increase the level of alcohol electro-oxidation. As a result of the oxidation of Sn, adsorption of CH_3COO and CO in the adjacent Pt active sites occurs. In the oxidative removal of these adsorbed species, due to the presence of Sn, oxygen-containing species can be produced. Thus, the activity of alcohol electro-oxidation increases, even for low potentials [24]. Scott et al. [25] studied the methanol electro-oxidation on PtRu/Ti mesh and PtRuSn/Ti mesh. Surface areas of PtRu and PtRuSn were determined to be 1.7 and 3.0 m^2/g, respectively.

A significant increase of the surface area was obtained on PtRuSn with the addition of Sn. XRD analysis shows the incorporation of Sn into the PtRu crystal lattice. The lattice structure depends on the alloy formation. A lower lattice constant was observed with the addition of Ru to Pt and a higher lattice constant with the addition of Sn [26, 27]. Thermally prepared PtRuSn is composed of metal oxides, such as RuO_2 and SnO_2. Both RuO_2, and SnO_2 have a tetragonal rutile structure, and the difference of ionic radius of these metals is only 5%.

Thus, the bifunctional mechanism plays a significant role in promoting Pt catalytic activity with the addition of Sn into PtRu. Moreover, the interaction of Ru and Sn oxides may contribute to the performance of the catalyst [28]. Spinace et al. [29] compared the electrocatalytic activity of three catalysts (PtRu/C, PtSn/C, and PtRuSn/C) and measured the face-centered cubic (fcc) structure of Pt/C (0.3916 nm), PtRu/C (0.3886 nm), PtSn/C (0.3986 nm), and PtRuSn/C (0.3937 nm). The difference of lattice parameters of catalysts was attributed to the interactions of three metals. This feature was related to the alloying effect, which causes a synergistic effect of Ru as a water activator (a bifunctional mechanism) and Sn as an electronic modifier of Pt. They also stated that the performance of the prepared catalysts depended strongly on the preparation method. Antolini et al. [24] showed the enhancement of the electrocatalytic activity of the binary PtSn addition of Ru.

214 Chapter 13 Novel materials structures and compositions

RuO_xH_y plays an important role in practical DMFCs owing to the conduction of both protons and electrons and the innate expression of Ru-OH speciation [30].

Adsorption and decomposition of ethanol, and then the oxidation of ethanol in the ethanol oxidation mechanism during the production of the intermediates on the Pt surface, resulting in hydrated oxides being formed. The occurrence of this situation is most likely to be seen in the Ru or Sn regions as a result of the formation of oxygen species that cause these hydrated oxides. Fig. 1A–C shows the typical low- and high-resolution transmission microscopy (HRTEM) images, as well as the particle size distribution of the PtSnRu/C (1:1:0.3) catalyst that distributes uniformly over the surface. The particle size of PtSnRu/C (1:1:0.3) ranged between 1.5 and 8.0 nm, with a mean diameter of 3.2 nm, PtSnRu/C (1:1:1) ranged between 1 and 5 nm, with a mean diameter of 3.2 nm, and that of the PtSn catalyst ranged between 3 and 10 nm, with a mean diameter of 6.6 nm. Smaller particle sizes are obtained on the ternary catalyst than that of obtained binary PtSn/C, as confirmed by the HRTEM images in Fig. 1. The highest ethanol oxidation activity was obtained on the PtSnRu/C (1:1:0.3) catalyst, depending on the Ru content.

They ascribed the high catalytic activity of this catalyst to synergism that occurred between Sn and Ru oxides. This event can be explained by the fact that water dissociated at a lower potential concerning the PtSn and PtRu catalysts due to promoted hydroxyl species with the interaction of RuO_x-SnO_2. This interaction may weaken the bond between the catalyst surface and the hydroxyl. As a result of the interaction, the weakening of the bonds between the catalyst surface and hydroxyl and the increase in the number of these weak bonds indirectly affect the ethanol oxidation performance in a positive way. It has been observed that intermediate products such as acetaldehyde and CO, which are adsorbed with the increase of weak bonds, increase the electrooxidation in the electrocatalytic region at a low level. Thus, ethanol oxidation performance is increased. As mentioned earlier, smaller particle sizes result in an increase of the electro-catalytic active surface area, which could lead to enhancement of ethanol oxidation, especially at high current densities. Also, the presence of RuO_2 in the catalyst decreases the active surface area of Pt regions, which assumes that all the nonalloyed metals have an oxide structure.

Other studies regarding ternary Pt-based electrocatalysts have been carried out, such as Pt-Ir-Sn/C [31], PtSnM (M = Fe, Ni, Ru, and Pd) [32], Pt-Ni-Cr [33], Pt-Rh-Sn [34], Pt-Ru-Os [35], Pt-Cu-Fe [36], PtAuCu [37], Pt-Sn-Mo [38], Pt-Sn-Ce [39], Pt-Co-Cr [40], Pt-Ru-Ni [41], and Pt-Ru-Ta, Pt-Ru-Mo, and Pt-Ru-Rh [42]. Results

Chapter 13 Novel materials structures and compositions **215**

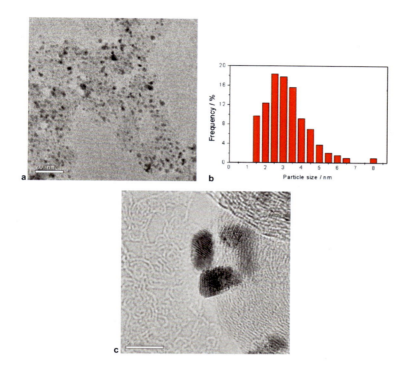

Fig. 1 HRTEM image at low (A) and high magnification (B) and histogram of particle size distribution (C) [24].

obtained from referred electrocatalysts showed high electrocatalytic activity toward alcohol oxidation. High catalytic activity is generally attributed to synergistic effects (i.e., electronic effect, cooperative effect); crystal structure; well dispersion of catalyst nanoparticles with limited agglomeration and higher mass activity; the formation of the porous structure, which enhance the specific surface area of the alloy; bifunctional mechanism; activation of water molecules at lower potentials; and the surface exophilic character. Alcohol oxidations were also investigated on quaternary electrocatalysts. For example, Fatih et al. [43] prepared the PtRuIrSn/C electrocatalyst with a known borohydride chemical reduction method and studied ethanol oxidation activity. They found higher ethanol oxidation activity than commercial PtRu/C in the potential region of 0.2–0.5 V. PtRuIrSn (1:1.5:0.5:2) showed the best performance concerning direct ethanol test cells, and found that its maximum specific power density was 29 mW/mg$_{Pt}$, followed by PtRuSn/C (1:0.8:2) with 19 mW/mg$_{Pt}$. Commercial PtRu/C showed the lowest power density (9 mW/mg$_{Pt}$). The presence of Sn in the quaternary electrocatalyst enhanced the overall fuel cell performance compared to PtRuIr/C and PtRu/C. It could improve the specific surface area. Moreover, the presence of Ir in

216 Chapter 13 Novel materials structures and compositions

the multisystem enhanced the maximum specific power density, which is a direct proportion of the activating and stabilizing effects of Ir. Further, Pt-Ru-Cu-Os-Ir [44], PtRuRhNi [45], and Pt-Ru-Os-Ir [46] showed excellent catalytic activity toward alcohol electro-oxidation.

2.2 Pt-oxide electrocatalysts

Another way to prepare Pt-based electrocatalyst to be able to decrease the CO_{ads} poisoning, obtain low price electrocatalysts and fast kinetic of alcohol oxidation reaction is the formation of metal oxide-based structure. These oxide structures allow the improvement of electrocatalytic activity of alcohol oxidation by leading to the adsorption of large quantities of OH species and having multiple valances that could reduce Pt poisoning and promote alcohol oxidation [47, 48]. One of the oxide structures is MnO_2, and it plays an important role in alcohol oxidation. For example, Li and Zhao [49] investigated the methanol oxidation activity of Pt nanoparticles composited on a MnO_2 NW arrayed electrode (PME). They found a higher ratio of the forward anodic peak current density (I_f) to the reverse anodic peak current density (I_b) ($I_f/I_b = 1.48$) than is obtained on the Pt NW arrayed electrode ($I_f/I_b = 1.08$), which indicates that more carbonaceous species are oxidized to CO_2 in the forward scan on PME than that on the Pt NW arrayed electrode. This may result indirectly in the presence of an active Mn^{+4}/Mn^{+3} transition. SnO_2, which has unique electrical and catalytic properties and stability, is used in many applications, such as gas sensors, lithium-ion batteries, and catalysis. High interaction with Pt leads to C—C bond breaking and OH generation; thus, it strongly oxidizes intermediates such as CO and shows better performance of $Pt/SnO_x/C$ catalysts than alloyed $PtSn_x/C$ [50]. Sedighi et al. [51] prepared the $Pt-CeO_2$ electrocatalyst by the electrodeposition technique and its catalytic activity was studied in ethanol containing basic media. They found that the onset potential of $Pt-CeO_2$ electrocatalysts (-0.59 V) is less than that of Pt (-0.48 V), indicating an improvement in the reaction kinetics and reduction of the ethanol oxidation overpotential. The peak mass activity of oxide electrocatalyst is 1413.7 mA/g_{Pt}, which is about three times higher than that of Pt (516.5 mA/g_{Pt}^{-1}). The superior catalytic activity of oxide electrocatalysts can be described as follows: The concentration of OH_{ads} species on the catalyst surface increases and can facilitate the EOR by the removal of the CH_3CO_{ads} and oxidation of the acetaldehyde into acetic acid. Many oxide-structured electrocatalysts such as $Pt-ZrO_2/C$ [52], $Pt-WO_3$ [53], $Pt-PbOx/C$ [54], and

Pt-NiO/C [55] have been studied, and this research found that alcohols are more easily adsorbed and oxidized on these electrocatalysts than commercial Pt by bifunctional mechanism, the multichannel structure of oxide, and synergistic effects, reducing the CO_{ads} poisoning tolerance.

3 Pt-free electrocatalysts

Pd offers an attractive alternative electrocatalyst to Pt in DAFCs because of its availability and low cost. Besides, the cost of Pd is about three times less than Pt, and it is more abundant on the Earth than Pt, and thus the catalyst cost can be significantly reduced. The crystal structure of Pd is very similar to Pt, and the atomic size is almost the same. Compared to Pt, both the catalytic activity and stability of Pd for the EOR in the alkaline electrolyte are higher [56, 57]. There are still many issues, such as completion of ethanol oxidation to CO_2 through 12 electrons and cleavage of the C—C bond, which is a very complex structure and requires high activation energy. Thus, much effort has been made to further improve the performance of Pd for alcohol oxidation. Some ways to promote the catalytic activity and stability of Pd are combining with metal oxides and alloying with other metals. Alloys formed by combining Pd and other metals play an important role in increasing the catalytic performance in the catalysts that are formed. However, it is known that anode catalysts see an increase in poison tolerance. Surface modifiers that can be combined with Pd to assist in the development of this catalytic activity are Ru, Sb, Pb, Bi, and As. High catalytic activity is attributed to the surface modification mechanism, leading to a decrease in adsorption of poisoning species on the surface, and a change to the electronic properties of the catalyst [58].

Surface modification depends strongly on the nanostructure of Pd-based electrocatalysts. These nanostructures can be dendritic NWs, a combination of core-shell, nanoflower-like, nanoplate arrays, thorn clusters, nanorods, and nanoflakes. As a result of the addition of various additives for the electrolyte to be prepared, it was determined that there was significant growth in Pd nanostructures [59, 60]. For example, Zhang et al. [61] prepared the Pd catalyst with the anodization process of Pd and investigated the electrocatalytic activity of this material toward methanol, ethanol, and formic acid oxidation. Results also were compared with flat Pd. According to scanning electron microscopy (SEM) images, Pd_{flat} has a very smooth surface, but Pd_{anod} has a highly rough surface with large quantities of nanoparticles. They found that the oxidation peak current density of Pd_{anod} (83.63 mA/cm^2) is 260 times higher than that of the oxidation peak current density of Pd_{flat} (0.32 mA/cm^2) in 1.0 M KOH + 0.5 M methanol. In addition, the electrocatalytic stability of prepared catalysts was performed by using cyclic voltammetry (CV).

They obtained that decay of current densities after 100 and 200 cycles is only 4.1% and 6.5% for Pd_{anod} and Pd_{flat}, respectively, indicating high catalytic stability. Both in ethanol and formic acid, higher catalytic activities were measured on Pd_{anod} than on Pd_{flat}. Lee et al. [62] demonstrated the synthesis of rodlike Pd nanostructures by using the electrochemical method and used these materials as electrocatalysts for the oxidation of methanol with the assistance of nicotinamide adenine dinucleotide (NAD$^+$). Fig. 2 illustrates the rodlike morphology, and the HRTEM image shows a fringe spacing of 0.23 nm corresponding to the (111) plane of the fcc Pd crystal. The ring pattern is also seen according to electron diffraction analysis of the selected area. Catalytic current densities are obtained on the Pd nanostructure and Pd nanorod catalysts as 6 and 19.72 mA/cm^2, respectively (Fig. 2C). The authors also found that the catalyst they prepared was successful catalysis of methanol with lower oxidation potential than methanol, triangular Pd rods, Pd nanoplate sequences, Pd porous flowers, Pd nanoflowers, and Pd urchins. In another study, Yi et al. [63] designed the novel porous Pd particles on a Ti sheet and examined its electrocatalytic activity in a 1.0 M NaOH + 1.0 Methanol solution. Porous network structures were obtained, and the sizes of the Pd nanoparticles changed between 80 and 220 nm. These changes are attributed to the presence of organic intermediates formed during the decomposition of the deduction agent, polyethylene glycol (PEG). Very high oxidation current densities (151 mA/cm^2) are exhibited on nano Pd-HCHO-EDTA. Also, electrochemical impedance spectroscopy (EIS) showed a significant decrease in the resistance on nano-Pd-HCHO-EDTA. They suggested that porous nano-Pd catalysts were promising alternatives to Pt electrodes applied in alkaline DEFCs.

Another effective route explored in the literature is to combine the second (binary Pd-based), third, or even fourth metal (ternary or quaternary-Pd based) with Pd. Most binary and ternary (quaternary) Pd-based electrocatalysts such as Pd-Bi/C [64], Ag-Pd/C [65], PdIr [66], Pd-Ni-Cu-P [67], Pd-Cu-Co [68], and Pd-Ni-Sn [69] showed high catalytic activity, poisoning tolerance to carbon intermediates, and stability toward the alcohol electrooxidation due to the synergistic effect, as well as the bifunctional mechanism. Much effort has been expended to reduce the cost of electrocatalysts that do not contain nonnoble metals. Among the

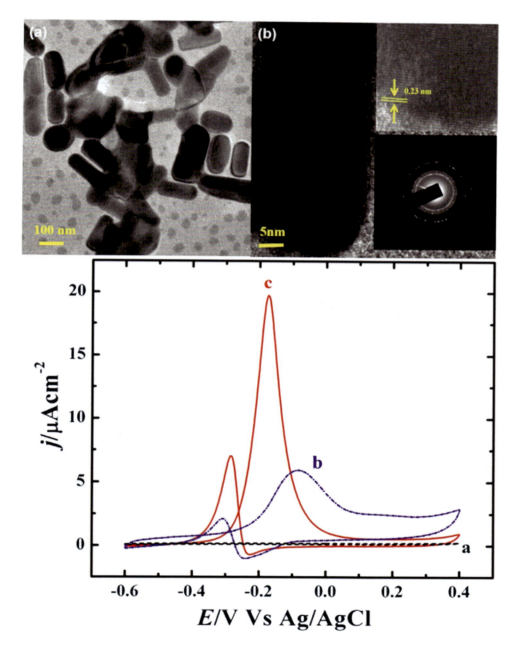

Fig. 2 (A) TEM and (B) HRTEM images of Pd nanorods. Inset in (B) is fringe spacing and an selected field electron diffraction pattern of the corresponding nanorod. (C) cyclic voltammograms obtained on (a) glassy carbon (GC), (b) Pd nanostructures, and (c) Pd nanorod based electrode toward 1.00 M methanol in 0.1 M NaOH at a scan rate of 10 mV/s [62].

nonnoble metals, Ni, Ir, and W compounds have become more attractive in recent years due to the exhibition of high catalytic activity for alcohol oxidation. For example, Maho et al. [70] reported the fabrication of nickel-copper/titanium nitride (Ni-Cu/TiN) and its methanol oxidation behavior. They stated that higher methanol oxidation was obtained on Ni-Cu/TiN than on Cu/TiN, Ni/TiN, Ni-Mn/TiN, and Ni-Co/TiN. Moreover, they found better stability on this electrocatalyst than on Ni-Cu/glassy carbon. They suggested that this electrocatalyst might be beneficial to develop high-performance of DMFCs at a low cost.

It should be noted that Ni has been commonly used for both anodic and cathodic reactions due to its surface electrochemistry. Generally, Ni-based materials exhibit high catalytic efficiency, low CO poisoning, and high catalytic stability. For example, Döner et al. [71, 72] fabricated the Ni/C, NiCd/C, and NiCdZn/C electrocatalysts by electrodeposition and investigated the methanol oxidation activities in 1.0 M methanol containing 1.0 M KOH solution. High catalytic activity and stability were found on the NiCdZn/C electrocatalyst, and its activity is related to the porous, large surface area. Alkaline leaching of Zn from the binary or ternary coating promotes the catalyst's surface. This chemical procedure is an effective way to get a greater surface area of catalysts [73–76].

Cesiulis et al. [77] developed the binary cobalt-tungsten/stainless steel catalyst, having different compositions and structures and tested as anodes for methanol oxidation. The prepared catalysts are either nanocrystalline or amorphouslike. Oxidation of methanol depended on the content of W in the alloy. They found that the crystalline Co-3 and Co-18 W alloys showed no methanol oxidation, and these materials dissolved in the test solution. Although the high content of W in the alloy prevented dissolution and it might be used as an anode catalyst for the oxidation of methanol. Sun et al. [78] studied the ethanol electro-oxidation on the IrSn catalyst and its catalytic activity compared to Pt/C and PtSn/C. According to CV results, oxidation peaks on the IrSn/C catalyst appeared at a more negative potential than those of Ir/C and PtSn/S, and high catalytic performance was obtained on this catalyst from chronoamperometry measurements. High activity depended on the promotion effect of added Sn.

4 Supporting materials

Supporting materials play a very important role in alcohol oxidation, as well as the fields of energy storage, environmental protection, and sustainable production of chemicals [79]. Many researchers devoted attention to the development of novel supporting materials to improve the dispersion, activity, and stability of electrocatalysts with a reduced price that is directly proportional to the reduced catalyst loading. Titanium oxide (TiO_2) nanotubes, conductive polymers, carbonaceous nanomaterials, metal oxides, carbides, and their hybrids, which are used as electrocatalysts in the field of sensors, photoelectrochemical cells, and electrolysis applications, also showed excellent catalytic activity for the electrooxidation of alcohols. Loading catalysts onto supporting materials has been proved to be an effective way to avoid the aggregation and maximize the surface area for both electron and mass transportation to and from electrocatalysts. To avoid aggregation, in other words, developing highly distributed catalyst nanoparticles with small size and narrow size distribution should be the main aim for high alcohol oxidation activity due to their large surface-to-volume ratio [80].

Carbonaceous materials are considered one of the most important groups of supporting materials for fuel cell applications, and different types of these materials have been widely used in this field. These supporting catalyst materials have some advantages for their possible applications: they can be in different shapes such as powders, spheres, or extrudates, they show good resistance against the acid or base environments, they have good porosity, and hydrophobic structures can be easily controlled by changing their surface chemistry until reaching certain limits, they possess high thermal and mechanical stabilities, they can easily combine with catalysts, they are very cheap, and they are very available [81]. In the past several decades, carbon blacks were almost extensively used as catalyst support [82, 83]. One of the most used carbon black is Vulcan XC72R, and it has some good properties, such as low cost, high availability, large surface area, good electronic conductivity, and good pore structure [84]. Nevertheless, this carbonaceous material, which should be activated before being used, has some disadvantages such as low corrosion resistance due to porous electrically conductive and aggregation of the catalyst [85]. Alternative carbonaceous materials to replace carbon black currently are being researched. With the development of nanotechnology, scientists have been interested in nanomaterials owing to their unique optical, electrical, thermal, and catalytic properties. These properties can be controlled easily due to the quantum size dimension of nanomaterials. Carbonaceous nanomaterials, including CNTs, CNFs, mesoporous carbon (MC), and graphene have been widely used as electrocatalyst supports in DAFCs, as well as sensors and solar cell applications. When these carbonaceous nanomaterials are used in DAFCs, they improve DAFCs due to their extraordinary properties, such as great superficial area, good electronic conductivity, perfect chemical inertness, low cost, mechanical strength, high stability, and strong interaction with catalysts. Thus, it is important to find the most suitable conditions for the selection of carbonaceous nanomaterials in developing DAFCs. These suitable conditions depend strongly on the preparation methods and synthesis of carbonaceous nanomaterials. Although there have been many efforts to understand the theoretical and experimental mechanisms, it is necessary to find out about novel carbonaceous nanomaterials as soon as possible.

4.1 Carbon-based supporting materials

4.1.1 Carbon blacks

Carbon black consists of ultrafine particles of 10–400 nm diameter, forms aggregates, and has a spherical shape. It contains carbon (90%–99%), oxygen (0.1%–10%), and hydrogen (0.2%–0.1%) atoms, and small amounts of sulfur and ash. In the crystalline structure, carbon atoms are joined by bonds that are covalent with each other, and the unsatisfied valences can be saturated by oxygen atoms. Therefore, the structure of carbon black contains oxygen-containing functional groups such as carboxyl, phenolic hydroxyl, and quinonic oxygen groups. These functional groups lead to some organic reactions such as methylation, esterification, and neutralization, as well as grafting sites for polymerization to the molecule. There are different types of carbon blacks, including channel black (BET surface area 906 m^2/g and particle size 29.4 nm), furnace black (BET surface area 79.6 m^2/g and particle size 15 nm), Acetylene black (BET surface area 65 m^2/g and particle size 40 nm), Vulcan XC72R (BET surface area 250 m^2/g, and particle size 40 nm) [86]. Carbon black is usually manufactured by partial combustion or thermal decomposition of natural gas, oil, or a mixture of the two at about 1000°C. The best-known form of carbon black is Vulcan XC72 and Vulcan XC72R (manufactured by Cabot Company). Carbon black is the first carbonaceous material used in fuel cell applications owing to its low cost, high availability, low ash content, high chemical, and electrochemical stability, high surface area, low density, and excellent electrical conductivity (2.77 S/cm).

Although carbon blacks are widely used in fuel cells for catalyst support, there is a decrease in the catalytic activity of the fuel cells due to the presence of organosulfur impurities and deep micropores. Moreover, it is unstable under the strong acid and base medium of a fuel cell, resulting in the corrosion of carbonaceous material and detachment of catalyst nanoparticles [87]. To reduce the corrosion of carbon black, some approaches can be used: (1) it is functionalized by sulfonates, carboxylates, tertiary amines, and steric polymer/oligomers [88, 89]; and (2) it can be doped by heteroatoms, and suitable doping may cause to enhance the performance of it [90]. Before doping with heteroatoms, the surface of the carbon black should generally be activated. The activation process is made in two ways: chemical and physical activation. Chemical activation is performed by using various oxidants, such as ozone, nitric acid, and hydrogen peroxide. This process creates a new pore structure, as well as high-reactive functional groups such as carboxylic acid/anhydride, lactone, and phenolic hydroxyl [91]. It changes the hydrophobic/hydrophilic structure and the polarity of the material [92, 93].

Oxygen-containing groups affecting the increase in performance of fuel cells also had an increasing effect on the distribution of catalyst nanoparticles [94]. For instance, Carmo et al. [95] studied how the chemical activation method altered methanol oxidation. Vulcan XC72R was used as catalyst support, and this material was treated with nitric acid to obtain a high-reactive oxygen-containing group. Methanol oxidation results showed that oxidation current increased on treated Vulcan XC72R-PtRu, and high activity was attributed to the better distribution of catalyst nanoparticles. The second activation is by a physical method involving the removal of the impurities on the carbon surface under an inert atmosphere at 800–1000°C or in air/steam at 400–500°C. When impurities are removed from the support surface, the electrochemical surface area (ECSA) of the electrocatalyst increases due to the presence of pores, etc., [96]. The formation of a carbon black with these two methods occurs because there is a strong interaction between catalyst and supporting material. The material used as a support in the catalyst; directly acts on the catalyst and enhances its activity. This strong interaction can depend on changing the electronic character of supporting material by catalyst nanoparticles and geometric effect [84].

Carbon blacks have been used as supporting material for catalysts in low-temperature fuel cells only since the 1990s. Low corrosion resistance and low utilization of catalysts limited the use of carbon blacks in fuel cells. As a result, other allotropic forms of carbonaceous nanomaterials must be explored for use in fuel cells.

4.1.2 Carbon nanotubes

One of the best-known carbonaceous nanomaterials that are used in fuel cell applications is CNTs. They were discovered in 1991, and since then, there are many interesting aspects of their

224 Chapter 13 Novel materials structures and compositions

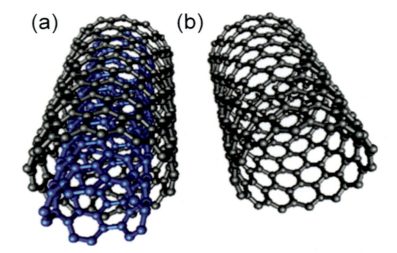

Fig. 3 Structures of (A) MWCNT and (B) SWCNT [97]. Reprinted (adapted) with permission from Y.-L. Zhao, J. F. Stoddart, Noncovalent functionalization of single-walled carbon nanotubes, Acc. Chem. Res. 42 (8) (2009) 1161–1171, https://doi.org/10.1021/ar900056z. Copyright (2020) American Chemical Society.

characterization and application in areas such as mechanics, optics, and electronics. There are two types of CNTs, which are formed two-dimensional (2D) nanostructures composed of single graphite sheets of hexagonally arranged carbon atoms rolled into a tubelike structure that looks like a latticework fence. A multiwalled carbon nanotube (MWCNT) is composed of multiple layers of graphite sheets, and a single-walled carbon nanotube (SWCNT) is composed of cylindrical single graphite sheets (Fig. 3) [97]. Simulation results show that the SWCNT structure is characterized by a chiral vector, which means its metallic or semiconducting properties.

The configuration of MWCNTs consists of diameters of a few tens of nanometers, with a spacing of 0.34 nm between the cylindrical single graphite sheets. Both of the CNTs have been extensively studied owing to several of their novel and unique characteristics, such as excellent flexibility, optical transparency, high electrical conductivity, extremely low weight, high surface area, high stability, and low cost for PEMFCs and DAFCs [98]. While MWCNTs are more conductive (10^3–10^5 S/cm) than SWCNT (10^2–10^6 S/cm), SWCNTs offer larger surface areas [99].

Among the various carbonaceous materials, nanostructured CNT supports have been used for enhancement in the electrocatalytic performance of DAFCs. It can be utilized from CNTs as a kind of alternative supporting material for low-temperature DAFCs owing to the abovementioned properties. The results showed that Pt/CNT electrodes have a much greater power density than those obtained on Pt/carbon black. This result is attributed to the higher retention of electrochemical area, smaller increment in interfacial charge transfer resistance, and slower degradation of the fuel cell performance. Namely, different power densities can be obtained when different carbonaceous supporting materials are used in fuel cells, resulting in different conversion efficiencies. Therefore, the kind of catalyst support used plays an important role in fuel cell applications [100]. According to other results, CNTs make the fuel cell more stable and durable, with higher corrosion resistance than the carbon black XC72 [101]. Before the use of CNTs as catalyst support, pristine CNTs were necessary in order to be functionalized like carbon black due to their chemical inertness and hydrophobic in nature. The functionalization of CNTs allows them to have the active binding site necessary for the nanoparticles to be fixed on them. As a result of their surface functionality, the CNTs allow metal nanoparticles or metal ions to exhibit uniform distribution with this binding ability [102]. It is understood that functionalized CNTs are needed because the later studies have very important functions. Therefore, in order to realize this process, numerous optimization studies and functionality have been made on CNTs. The most extensively used ways for functionalization are physical surface modification and chemical surface modification [103].

The chemical structure of the CNTs, the surface structure, and the physical properties of the surface are stimulated by physical surface modification. This is done by friction, breaking, milling, and mechanical processes such as ultrasonic. Chemical surface modification is divided into two methods: covalent and noncovalent functionalization. Noncovalent modification results from the combination of the delocalized π electrons of the CNT with conjugated π-π stacking interactions, Van der Waals compounds, electrostatic interactions, and hydrogen bonds. Interaction in this method is much weaker than covalent functionalization. It prevents change in the original structure of CNT. Covalent modification usually alters the structure of the tips and lateral walls of CNTs by harsh pretreatment of one or a mixture of strong acids such as concentrated HNO_3, H_2SO_4, HCl, or strong oxidizing reagents such as H_2O_2 and $KMnO_4$, and it generates functional groups in the sidewalls of CNTs [104–106]. Hybridization of sp^2 to sp^3 is shown in this method. Accordingly, this conversion induces disruption of the electronic structure of CNTs [107]. Such groups have been shown as suitable anchoring sites for metal nanoparticles on CNT support. These alteration methods provide a hydrophilic structure with more uniform metal nanoparticles and charge higher amounts of metal nanoparticles on CNTs compared to those prepared on pristine or as-received CNT soot under the same conditions [108]. MWCNTs are more suitable for chemical functionalization than SWCNTs due to modifying only the outer wall. Because CNTs tend to be concentrated, they are reduced by slowing the increase of the number of walls and nanocurvatures [109].

MWCNTs transform from a hydrophobic structure to hydrophilic by acid treatment, and a strong bond forms with metal nanoparticles and sidewalls of MWCNTs. Thus, electron transfer occurs from supporting material to metal nanoparticles, which can result in a decrease in the d-band vacancy. The synergistic effect between metal nanoparticles and supporting material causes more active binging sites for alcohol adsorption and enhances alcohol oxidation [110]. Fig. 4 shows the TEM images of Pt/MWCNT and PtRu/MWCNT electrocatalysts, which showed methanol electro-oxidation. All the samples were treated with

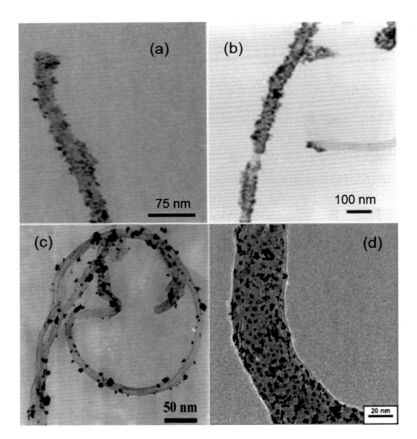

Fig. 4 TEM images of (A) Pt/MWCNT, (B) Pt-Ru/MWCNT [111], (C) Pt/MWCNT [112], and (D) PtRu/MWCNT [113].

concentrated acid. As shown in Fig. 4A and B, Pt and Ru nanoparticles were dispersed uniformly on treated MWCNTs due to strong interactions between metal nanoparticles and the graphene edges. The average particle size of Pt-Ru alloy clusters is 3–4 nm. Chemical functional groups such as hydroxyl and carboxyl generate anchoring sites due to chemical methods. The maximum power density of 29 mW/cm^2 is obtained on Pt-Ru/MWCNTs, which is higher than that of obtained Pt-Ru/Vulcan XC 72 (25 mW/cm^2) at 80°C.

In one study, high power density was obtained due to better conductivity and uniform dispersion of catalyst nanoparticles over MWCNTs [111]. In Fig. 4C, Pt and PtRu nanoparticles are well dispersed on the MWCNTs again, with a diameter of 5–10 or 1–6 nm [112, 113]. Xu et al. [113] prepared the PtRu/MWCNT electrocatalyst and investigated the methanol oxidation characteristic in a 1.0 M H_2SO_4 solution. This oxidation characteristic is compared with a commercial $Pt_{50}Ru_{50}$/C catalyst (E-TEK). They found the oxidation current densities as 114.3 and 54.0 mA/cm^2 for PtRu/MWCNT and E-TEK, respectively. They also examined the stabilities of prepared electrocatalysts under the same conditions between 0 and 1.0 V. In the initial stage, high current decay was seen in both electrocatalysts, and then the current densities were stable over 500-cycle scans. Peak currents remained at 85% for PtRu/MWCNT electrocatalysts, and 70% for E-TEK. In conclusion, a synthesized PtRu/MWCNT electrocatalyst exhibited excellent electrocatalytic activity and stability for methanol oxidation, which makes it a good candidate for application in DMFCs. MWCNTs treated with acid were also used as supporting materials for different mono, bi, ternary, and even quaternary electrocatalysts, which contain noble or nonnoble metallic materials. Examples include Pt, Pd, Ni, AuPt, PtFe, PtCo, Pd-Ag, NiPd, PdCu, PdCuSn, PtRuMo, Pb-PtCu, Pt/Sn, and Pt/Sn/PMo [114–124].

4.1.3 Carbon nanofibers

Several new catalyst supports were explored in the chemical industry that has been developed in the last several decades. Many of the new materials contain fibers made from various materials. The fiber structure as a supporting material could be beneficial due to the high surface-to-volume ratio, high void fractions, and high flexibility. Fiber supporting materials may have different geometries and have low cost. One of the important fibrous substrates is (CNFs), which are manufactured industrially. CNFs form from stacked graphene sheets, which induce high thermal and electrical conductivity. Due to possessing unique properties, CNFs are used in numerous applications such as hydrogen storage, catalyst support, drug delivery, and polymer reinforcement. CNFs are different from CNTs because CNFs do not have a hollow cavity structure, and they have a larger diameter and longer length (by 500 nm for diameter and by a few millimeters for lengths). Another different aspect is the exposure of active edge planes. Potential anchoring sites are only the edge planes in CNTs, but the predominant basal plane is exposed in the case of CNFs [98]. Both of them have good electrical and thermal conductivity (10^2–10^4 S/cm for platelet CNF). Fibers have very high tensile and compressive strength, high corrosion resistance, creep, and fatigue.

CNFs are two-dimensional and the higher aspect ratio substrates with high specific surface area for electrocatalysis. They also have greater purity and permeability to reactants [125]. These unique properties provide it for use in fuel cell applications as an attractive alternative supporting material in the automotive and aerospace industries. CNFs are used in fuel cells not only as a supporting material but also as a gas diffusion layer and bipolar plate [126].

Baker et al. investigated various supporting materials such as CNFs, active carbon, and γ-alumina in 1994, and Fe-Cu particles were doped on these supporting materials. The activity of this catalyst was more pronounced than that of the other two materials due to the intrinsic activity of this catalyst and the strong interaction between the catalyst and carbonaceous material [127]. There have been extensive efforts to investigate CNFs as supporting material in fuel cells since then. There are three kinds of CNFs, depending on the orientation of the nanofibers with respect to the growth axis: (1) ribbonlike CNF, (2) platelet CNF, and (3) herringbone (or stacked-cup). Herringbone CNFs are investigated more often than others due to having intermediate properties between parallel and platelet types. Herringbone CNFs have higher catalytic activity than the parallel and better durability than the platelet forms [128]. CNFs can be used as a supporting carbonaceous material, regardless of any pretreatment. Actually, platelets and herringbone forms contain potentially functional groups for metal anchoring. Even so, investigators functionalize CNFs to obtain more active binding sites for anchoring precursor metal ions with oxidative agents [129]. For example, Park et al. [130] reported the effect of chemical treatments on CNFs in various concentrations of nitric acid and coated CNFs with PtRu by using impregnation to investigate methanol oxidation. The study found that oxygen functional groups increased as the concentration of acid increased. The electrocatalytic performance of prepared catalyst with this treatment was also improved. Good results were obtained from the use of CNFs as supporting material for alcohol oxidation when compared with CNT and carbon black.

Kamarudin et al. [125] fabricated three types of carbonaceous materials: CB (Vulcan XC72R), CNF (Platelet CNF), and CB + CNF. They compared the effects of these supporting materials on the performance of DMFCs. A PtRu catalyst was loaded onto these materials. The results showed that the highest current density (119 mA/cm^2) was obtained on CNFs. The current densities were 46 mA/cm^2 for CB and 80 mA/cm^2 for CB + CNF. Accordingly, they suggested that CNF was a promising supporting material for anode layer fabrication for DMFCs. Bessel et al. [131] investigated the methanol oxidation behavior of CNFs supported by Pt for fuel cell applications at 40°C. In this study, they used platelet, ribbon, and Vulcan XC72 as supporting materials and found that less CO poisoning was measured on CNFs supported by Pt than traditional anode catalysts. This result was associated with specific crystallographic orientations of Pt particles on the highly tailored CNF structures.

4.1.4 Mesoporous carbon

The pore structure is seen only in solid substances, and this structure affects the behavior of the substances. The pore is proposed by Dubinin in 1960 [132] and then accepted by the International Union of Pure and Applied Chemistry (IUPAC) in 1985 [133]. There are three types of pores in terms of pore diameter. The first are micropores, which have a width smaller than 2 nm; the second is mesopores, whose size ranges from 2 to 50 nm; and the last is macropores, which are larger than 50 nm. Pores with a large variety of geometries, including slit-shaped, cylindrical, spherical conical, ink-bottle, or interstitial, can also be classified in terms of their shape. Porous carbon materials are irreplaceable for a variety of applications, such as pollutant adsorption, hydrogen storage, energy conversion, and electrochemical sensing owing to their high electrical and thermal conductivity, good chemical stability, low density, large surface area, and wide availability. Moreover, they are particularly used as supporting materials in batteries, fuel cells, and supercapacitors. Recently, numerous efforts have been made in carbon technology due to either exploring new porous carbon materials with mesostructures or developing and introducing new synthetic techniques. A new member of porous carbon materials is MC, which is interesting and attractive due to the large mass transfer limitation in microporous carbon materials. MC has an interesting surface morphological structure, with a large surface area

230 Chapter 13 Novel materials structures and compositions

by-products and these properties make it a very attractive carbon material for catalyst supports.

MC can be classified into two groups: ordered mesoporous carbon (OMC) and disordered mesoporous carbon (DOMC). OMC was explored in 1999 [134, 135], which is a versatile member of porous carbon materials with uniform pore channels and great surface area. Controllable pore size and large pore volume as well as during operation of the fuel cell of the reactants and improves mass transfer products [80, 136]. OMC was produced from several ordered mesoporous silicas (e.g., MCM-48, SBA-15, KIT-6, HMS, MSU-H), leading to cubic or hexagonal frameworks, narrow mesopore-size distributions, high-nitrogen, BET-specific surface areas (up to 1800 m^2/g), large pore volumes, and even mesoporous metal oxides by using a hard template (nanocasting) [84]. The production of OMC includes infiltration of the pores of the template with appropriate organic precursors such as furfuryl, alcohol, mesophase pitch, sucrose and acenaphthene, carbonization at high temperatures, and template removal in acid/base solutions. It could be used in the soft template method for OCM because the hard template method is very complex due to the used HF/NaOH for silica etching [137]. As a result, the structure of OMC depends on the template method, which leads to the synthesis of OMC, provided that there is a 3D-pore structure; otherwise, DOMC has occurred.

There are different types of OMCs, including ordered CMK-3 carbon that is the most tested in fuel cells as supporting material [84]. Yi et al. [138] investigated the methanol electro-oxidation on Pt/MC catalysts. In this study, the authors used MCM-48 and SBA-15 templating materials for the preparation of MCs with a hard template method. The MCs derived from MCM-48 and SBA-15 were named CMK-1 and CMK-3. Pt particles were loaded via the impregnation method, and methanol oxidation characteristics of prepared electrocatalysts were investigated using the CV method. The same experiments were done on conventional carbon support (Vulcan carbon) for comparison. The results demonstrated that the catalytic activity was seen in the following order: Pt/CMK-1 > Pt/CMK-3 > Pt/Vulcan. The high catalytic activity of Pt/CMK-1 depended on better dispersion of Pt particles on supporting materials than that of CMK-3. Shi et al. [139] synthesized the ordered MC (surface area 585 m^2/g) with Al-doped SBA-15. Then Pt and PtCo were loaded on OMC by chemical reduction. Prepared catalysts were labeled as Pt/GMC and PtCo/GMC. They found that mesostructured composites (Pt/GMC and PtCo/GMC) showed higher catalytic activity than commercial catalysts (Pt/C) toward methanol oxidation.

Chapter 13 Novel materials structures and compositions **231**

Similar to CNT and CNF, the MC surface can be activated with oxidative agents by chemical treatment. This treatment functionalizes the MC surface with oxygen groups, which improves the interaction of metal catalysts and supporting material surfaces, allowing uniform dispersion. It also reduces the ohmic and mass transfer resistance due to highly functionalized supporting materials [140, 141]. For instance, Pastor et al. [142] studied the methanol oxidation on Pt and Pt-Ru electrocatalysts supported on CMK-3, as well as Vulcan XC72 and E-TEK. SBA-15 was used as ordered mesoporous silica by using the hard template method. After this preparation, CMK-3 was treated with concentrated nitric acid at room temperature for 0.5 h to functionalize its surface morphology. The obtained electrocatalysts were Pt/CMK-3, Pt-Ru/CMK-3, Pt/Vulcan, and Pt-Ru/Vulcan. Pt-Ru/CMK-3 showed the best electrocatalytic activity toward methanol oxidation. Therefore, bimetallic catalysts supported on OMC make it a potential candidate for use in DMFC anodes.

4.1.5 Graphene

Graphene is known as the "mother of all carbon forms"; it is a 2D planer structure that designed a single layer of sp^2 hybridized graphite, having a thickness of 0.34 nm. This structure is a standard form of graphene. There are many methods of producing standard graphenes, such as exfoliation (physical or chemical), epitaxial growth via chemical vapor deposition (CVD), solvothermal synthesis with pyrolysis, chemical reduction, and electrochemical methods. It should be noted that the production of graphene is a complex process, and these methods result in a combination of optimum properties for all applications [143]. Graphene can be integrated into the implementation of electrochemical applications due to its high electrical ($\sim 10^6$ S/cm) and thermal conductivity (5000 W/m/K), a large specific surface area (2630 m^2/g), excellent heterogeneous electron transfer, and charge carrier rates, as well as low cost and wide availability [143–147]. Recently, graphene has been widely investigated as promising alternative catalyst support for low-temperature fuel cells. It possesses similar stable physical properties when compared with CNTs. But it has a larger surface area and lower cost than CNTs. Thus, it can replace CNTs and carbon black as attractive supporting materials for the metal electrocatalysts in fuel cells [84].

Both edge and basal planes enhance the surface area of graphene sheets, and thus these planes provide additional active sites for interaction with metal catalysts. Wang et al. [149] observed the smaller (2.83 nm) Pt nanoparticles loaded on a graphene sheet than those loaded on carbon black (4.21 nm) owing to the larger specific surface area of the graphene sheets. Li et al. [150] found the ECSAs of Pt/graphene and Pt/Vulcan as 44.6 and 30.1 m^2/g_{Pt}, respectively, and high ECSAs catalyzed the electrochemical methanol oxidation.

Graphene can easily be functionalized with heteroatom doping and surface functionalization. Heteroatoms such as nitrogen, boron, phosphorous, and sulfur are efficiently attached to graphene lattice, and functionalization modifies its electronic structure based on the size and hydrophilic structure of functional groups. Similar effects are also observed with the chemical functionalization method. Also, these modifications allow the anchoring of metal particles, enhancing the electron transfer between the metal catalyst and graphene support. Accordingly, modified graphene sheets either improve the catalytic activities of metal catalysts toward alcohol oxidation in DAFCs or prevent the aggregation of metal catalysts on the graphene support [151]. Consequently, both modification methods play a vital role in DAFC applications.

For example, Du et al. [152] synthesized boron-nitrogen doped graphene (BNG) by the two-step thermal annealing method, with Pt nanoparticles with a size of 2.3 nm anchored uniformly on the surface of BNG support. The authors observed that Pt/BNG exhibited excellent catalytic activity and stability toward MOR, which was related to the synergistic interaction between boron, nitrogen codoping, and graphene support. Codoping enhanced a number of oxygen-containing groups, which facilitated methanol oxidation by a bifunctional mechanism. This binary doping method could provide a new strategy for the preparation and fabrication of catalysts. Lu et al. [153] focused on the preparation of functionalized graphene with carboxylic sodium, which could provide more active centers, and examined the methanol oxidation characteristic of Pt-Pd/functionalized graphene. The obtained nanoelectrocatalyst showed higher catalytic activity and stability toward MOR compared to Pt or Pd blacks. The synergistic effect may affect the MOR, and carbon radical reaction may play an important role in the functionalization of graphene.

Graphene oxides (GOs), which are produced from graphene, constitute another important member of the graphene-graphite family. GO exhibits the graphitic layered structure. Graphene and GO can be easily converted to each other with the oxidation of graphite and reduction of GO by using various processes and agents. Groups such as oxygen-containing hydroxyl, carboxyl, and epoxy, which could make the substance amphiphilic, can

Chapter 13 Novel materials structures and compositions **233**

be attached at the edge planes of graphene oxide. This attachment could contribute to the high performance of electrocatalytic activities in fuel cell applications. There is a strong potential of GO to support metal or nonmetal catalysts, and further studies must be done to develop graphene and GO supports to obtain the best performance of DAFCs.

4.2 Conducting polymers

Polymers (neutral conjugated polymers) are macromolecules formed by many repeated units called "monomers." In principle, polymers are inherently insulators. Their conductivities are changing in the range of 10^{-10}–10^{-5} S/cm. This property provides for its use in the electronic industry. During the past decade, nanotechnology has emerged as an effective and active area of research owing to reducing the nanoscale of used materials. Recently, the utilization of polymeric materials with ionic and electronic conductivity in the industry has started using nanotechnology as the opposite of insulator polymeric materials. This introduction leads to the exploration of new smart materials for modern living standards. Polymers with electronic conductive are called "conducting polymers (CPs)" or "electroactive conjugated polymers," which are new classes of polymers. These materials as insulating polymers have an organic structure, are also known as "synthetic metals," and were first discovered in 2000 by Alan J. Heeger, Alan G. MacDiarmid, and Hideki Shirakawa. Scientists have made an incredible effort to research and develop nanostructured conducting polymers since their discovery. In particular, there is much interest in nanostructured conducting polymers in electrochemistry due to their key materials in electrochemical power sources to develop efficient, reliable, and environmentally friendly energy systems. In addition, conductive polymers have superior electrical, magnetic, and optical properties. It has derived by oxidizing or reducing the neutral polymers. Hereby, its electrical conductivity converts from the range of 10^{-10} to 10^{-5} S/cm to the range of 1 to 10^4 S/cm using chemical or electrochemical methods [154]. This electrical conductivity is due to the one-dimensional (1D) π-conjugated structure of double covalent bonds. For instance, undoped polyaniline (PANI) is an insulator polymer, and its conductivity is in the range of 10^{-10}–10^{-8} S/cm. The conductivity of doped PANI can increase to the range of 10^2–10^3 S/cm or higher. The doping form of PANI makes it a promising material for many industrial applications, including supercapacitors, fuel cells, sensors, and corrosion protector films [155]. Some conventional CPs are polyacetylene (PA), polypyrrole

234 Chapter 13 Novel materials structures and compositions

(PPY), PANI, and polythiophene (PTh), as well as their derivatives. These CPs are the most common ones used in the DAFC and PEMFC applications as supporting materials. They possess a high specific surface area, which provides an effective interaction between the surface of CP and the catalyst and more anchoring active sites, high conductivity, low cost, porous layer, and high stability. These unique properties of CPs promote their use in DAFCs by forming effective catalyst-support systems.

There are two main preparation methods for nanostructured CPs. The first is a template-free method that involves an easily formed nanotube or NW with low cost due to the existence of β-naphthalene sulfonic acid (β-NSA) as a dopant. Basically, the morphological structure of CPs depends on the monomer, dopant, oxidant, and condition of polymerization. The second is a template-based method that is classified into two categories. Shapes and sizes of nanostructured CPs can be controlled efficiently with this method [156]. In order to improve the supporting properties of CPs in DAFCs, their specific surface area, conductivity, stability, and strong interaction between catalyst nanoparticles with CP should be maximized with the abovementioned methods [157,158]. Hable and Wrighton [148,159] studied the methanol oxidation of Pt-Ru and Pt-Sn on synthesized PANI surface; a schematic illustration of incorporation of catalysts nanoparticles into polymer matrix was given in Fig. 5. They synthesized PANI on glassy carbon (GC) electrode and PANI

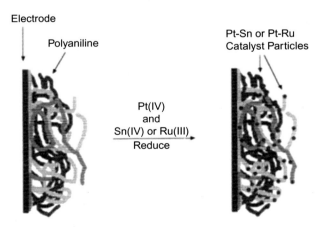

Fig. 5 Schematic representation of incorporation of Pt-Ru and Pt-Sn nanocatalysts into a polymer matrix [148]. Reprinted (adapted) with permission from C.T. Hable, M. S. Wrighton, Electrocatalytic oxidation of methanol and ethanol: a comparison of platinum-tin and platinum-ruthenium catalyst particles in a conducting polyaniline matrix, Langmuir 9 (11) (1993) 3284–3290, doi: 10.1021/la00035a085. Copyright (2020) American Chemical Society.

composed from fibrils. Fibril structures provided a porous and high surface area that may be exploited as catalyst support. Then Pt-Ru or Pt-Sn nanocatalyst particles were deposited into the polymer matrix, as illustrated in Fig. 5. This distribution of catalyst particles was quite uniform, and minimal aggregation of particles was seen in a polymer matrix. High catalytic activities were obtained on Pt-Ru or Pt-Sn modified PANI toward methanol oxidation.

Yoon et al. [160] synthesized PPY on graphitized carbon black (GCB) and Pt particles were supported in a PPY matrix by using the microwave-polyol process. Electrocatalytic performances of synthesized catalysts were investigated in 1.0 M H_2SO_4 containing 2 M methanol solution against methanol oxidation. Diameters of Pt particles were measured as 4.5 nm and 3.4 nm for Pt-GCB and Pt-Ppy/GCB, respectively. Pt particles were distributed homogeneously over the surface in the case of Ppy, and less agglomeration and a larger electrochemical active surface area were observed. These properties led to the increased catalytic performance of the Pt catalysts against methanol electro-oxidation.

In another study, Ojani et al. [161] demonstrated the electro-synthesis of 2-aminodiphenyllamine (2ADPA) on a carbon paste electrode (CPE) and incorporation of Cu particles into a polymer matrix via the electrodeposition method. The efficiency of this prepared catalyst system for the electro-catalytic oxidation of methanol in basic media was studied. A spherical structure with large-grain and porous polymer was observed to be a suitable dispersion of catalyst particles. The obtained results showed the fast electron transfer process in the modified polymer matrix. A modified polymer with Cu particles had excellent power with high methanol oxidation current density, and it may be unique.

4.3 Hybrid supporting materials

In recent years, different types of supporting materials have been combined to produce functional nanocomposite electrocatalysts with superior catalytic performance for DACF applications. This combination of different supporting materials is called "hybrid supports." CPs such as PANI, PPY, PTh, and poly(3,4-ethylenedioxythiophene) (PEDOT) were combined with carbonaceous materials, including SWCNTs, MWCNTs, CNFs, graphene, and GO. These combinations change the structure of supporting materials, and thus ideal and superior supporting materials are obtained for alcohol oxidation.

To date, many studies have been conducted on combining carbonaceous materials with CPs to create an ideal supporting material for metal nanoelectrocatalysts. For example, CNTs and graphene are innovative connective materials with CPs in the field of fuel cells. Zhao et al. [162] combined the graphene with Ppy and loaded Pd particles onto this binary supporting material. Graphene was obtained by reducing GO and PPy-graphene synthesized by in situ radical polymerization method on a GC electrode. For comparison, Pd/Vulcan XC72 and Pd/graphene catalysts were prepared under the same conditions. Catalytic activities of all the prepared electrocatalysts for methanol oxidation were investigated in 0.5 M NaOH containing 1 M methanol solution. Measured current densities were for forwarding peaks 205.3, 265.8, and 359.8 A/g_{Pd}, respectively. The highest current density was obtained on Pd/PPy-graphene. Also, stability tests of three types of electrocatalysts were investigated in 0.5 M NaOH containing 1 M methanol solution. Rapid current decay was seen on all prepared electrocatalysts in the initial stage, and after this period, a stable state was achieved. The authors stated that Pd/PPy-graphene was a more promising catalyst than the other catalysts in fuel cell applications. Sun et al. [163] reported the Pt nanoparticles combined MWCNT with polyindole (PIn) as a novel electrocatalyst system for use in the methanol oxidation reaction. First, MWCNT was activated by using the well-known acid oxidation method to obtain an oxygen functional group. PIn-functionalized MWCNT was synthesized by a chemical route, and the electrocatalytic activities and durability of Pt/PIn-MWCNT, Pt/MWCNT, and commercial Pt/C were evaluated for methanol oxidation. The maximum current density of 1013.5 mA/mg_{Pt} was measured on Pt/PIn-MWCNT, which was much higher than for others. Besides, higher mass transfer was observed on Pt/PIn-MWCNT than those of recently reported hybrid supports. This study demonstrated that Pt/PIn-MWCNT could be used in DMFC applications as an effective nanostructured electrocatalyst.

Another type of hybrid support is a combination of CPs with oxide materials. Conducting oxide materials that are emerging as a good candidate catalyst support for metal catalysts are important in heterogeneous catalysis systems. These supporting materials possess high thermal and electrochemical properties, as well as excellent corrosion resistance, in both alkaline and acidic media and stable in fuel cells. Conducting oxide materials are more stable and mechanically durable than carbonaceous materials. It is well known that metal oxide materials include RuO_2, SnO_2, TiO_2, and WO_3 [164], and these oxide structures can be combined with CPs through functionalized groups of CPs. Recently, Ma et al. [165] synthesized Pt-PANI/TiO_2-Vulcan XC72 and electrocatalytic activity and stability of this electrocatalyst toward methanol oxidation were investigated in 0.5 M H_2SO_4 containing 1.0 M methanol. It was proved the existence of Pt, TiO_2, and PANI by Fourier transform infrared spectra (FTIR) on the surface, energy-dispersive X-ray spectroscopy (EDS), and XRD. The electrocatalytic performance and stability of prepared electrocatalysts on the hybrid support were found to be much better than those of prepared Pt-TiO_2/Vulcan XC72. High catalytic activity and stability depended on (1) a decreased poisoning effect due to the improvement of catalyst dispersion, (2) decreasing agglomeration of Pt particles due to the PANI in the hybrid support, and (3) the acceleration effect of forming extra OH^- ions against the oxidation process. PANI also showed some synergistic effects for improving the catalytic activity and stability of electrocatalysts. In another study, Pang et al. [166] synthesized the PANI via a chemical route as the first supporting material. SnO_2, the second supporting material, was synthesized by the sol-gel method. Pt nanoparticles supporting PANI-SnO_2 electrocatalyst were prepared on a GC electrode to investigate the methanol electro-oxidation. For comparison, a Pt/SnO_2 electrode was also prepared. The authors deduced that Pt particles dispersed uniformly on the hybrid support, and Pt/PANI-SnO_2/GC electrocatalysts showed better electrocatalytic activity (i.e., larger ESA, better antipoisoning ability) in comparison to Pt/SnO_2/GC electrocatalyst.

Finally, this is an efficient way to combining oxide structures with carbonaceous materials for the practical application of fuel cells. Chu et al. [167] fabricated the Pd-In_2O_3/MWCNT by using a chemical reduction and hydrothermal process. The morphological structure was analyzed via TEM. Pd-In_2O_3 particles were distributed uniformly on the MWCNT surface with slight agglomeration, according to TEM images. Over potential (E_p) and the ethanol oxidation current density (J_p) of Pd-In_2O_3/MWCNT were 103 mV lower and 32 mA/cm^2 higher than that of the Pd/MWCNT catalyst, respectively. A significant increase in catalytic activity when it was added to In_2O_3 was observed. Also, higher stability was obtained on Pd-In_2O_3/MWCNT, indicating that In_2O_3 showed to be active sites for the formation of oxygen-containing groups, which accelerate the oxidation of intermediate species due to the bifunctional mechanism. The synergistic combination between Pd and In_2O_3 and the surface effect of In_2O_3 could also influence the catalytic properties of Pd-In_2O_3/MWCNT electrocatalysts.

5 Challenges and future perspectives

In recent years, depending on the developing economy and industry, energy demand has been increasing and energy resources are insufficient. For this reason, DAFCs constitute one

238 Chapter 13 Novel materials structures and compositions

of the most investigated energy conversion devices owing to their high power density, environmental friendliness, and longevity. The anode catalyst is a key factor in determining the performance of DAFCs. Over the past decade, many studies have focused on Pt and Pt-based materials, which are known to be the best catalysts for DAFCs. Unfortunately, some of their properties, such as high cost, high amount of Pt used in the electrode, and low abundance, have limited its use with the anode aspect of DAFCs. Moreover, Pt catalysts are more sensitive to CO poisoning than other kinds of catalysts. Thus, other non-Pt catalysts should be used in DAFC applications. Investigating novel Pt-free binary-ternary catalysts instead of Pt is strongly recommended.

Supporting materials for electrocatalysts play a vital role in ascertaining the electrocatalytic activity, stability, and price of DAFCs. Over the past decade, scientists have made many attempts to research nanostructured supporting materials such as carbonaceous materials, metal oxides, conducting polymers, and many hybrid supporting materials, as well as to improve the existing and also further develop such materials for use in DAFCs. On the other hand, surface functionalization of carbonaceous materials and polymer synthesis techniques are directly related to the activity and cost of DAFCs. The exploration of new structures and new supporting materials have recently picked up speed in the application of DAFCs. These new avenues offer promise for future developments.

6   Conclusions

Many materials have been reduced to nano size with the emergence of nanotechnology, and these nanomaterials have begun to be used in DAFCs. Thus, many nanoelectrocatalysts have been developed for DAFCs, and supporting materials with nano size also are used in DAFCs. In this chapter, many electrocatalysts used in DAFCs and their catalytic activities, stabilities, characterizations, and performances are reviewed in detail. Interactions between the nanocatalyst and nanostructured supporting materials are reviewed and the contribution of DAFCs are discussed. Recent studies have been conducted in this field to give supplementary information regarding anode electrocatalysts.

References

[1] S.K. Kamarudin, F. Achmad, W.R.W. Daud, Overview on the application of direct methanol fuel cell (DMFC) for portable electronic devices, Int. J.

Hydrog. Energy 34 (16) (2009) 6902–6916, https://doi.org/10.1016/j.ijhydene.2009.06.013.

[2] N. Radenahmad, A. Afif, P.I. Petra, S.M.H. Rahman, S.-G. Eriksson, A.K. Azad, Proton-conducting electrolytes for direct methanol and direct urea fuel cells—a state-of-the-art review, Renew. Sust. Energ. Rev. 57 (2016) 1347–1358, https://doi.org/10.1016/J.RSER.2015.12.103.

[3] L. Vasquez, Fuel Cell Research Trends, 2007.

[4] U.B. Demirci, Direct liquid-feed fuel cells: thermodynamic and environmental concerns, J. Power Sources 169 (2) (2007) 239–246, https://doi.org/10.1016/j.jpowsour.2007.03.050.

[5] C. Lamy, A. Lima, V. LeRhun, F. Delime, C. Coutanceau, J.-M. Léger, Recent advances in the development of direct alcohol fuel cells (DAFC), J. Power Sources 105 (2) (2002) 283–296, https://doi.org/10.1016/S0378-7753(01)00954-5.

[6] J. Garche, C.K. Dyer, Encyclopedia of Electrochemical Power Sources, Academic Press, 2009.

[7] L. Carrette, K.A. Friedrich, U. Stimming, Fuel cells—fundamentals and applications, Fuel Cells 1 (1) (2001) 5–39, https://doi.org/10.1002/1615-6854(200105)1:1<5::AID-FUCE5>3.0.CO;2-G.

[8] A. Hamnett, Mechanism and electrocatalysis in the direct methanol fuel cell, Catal. Today 38 (4) (1997) 445–457, https://doi.org/10.1016/S0920-5861(97)00054-0.

[9] V.S. Bagotzky, Y.B. Vassiliev, O.A. Khazova, Generalized scheme of chemisorption, electrooxidation and electroreduction of simple organic compounds on platinum group metals, J. Electroanal. Chem. 81 (2) (1977) 229–238, https://doi.org/10.1016/S0022-0728(77)80019-3.

[10] R. Solmaz, Electrochemical preparation and characterization of C/Ni–NiIr composite electrodes as novel cathode materials for alkaline water electrolysis, Int. J. Hydrog. Energy 38 (5) (2013) 2251–2256, https://doi.org/10.1016/J.IJHYDENE.2012.11.101.

[11] M.W. Breiter, Comparative voltammetric study of methanol oxidation and adsorption on noble metal electrodes in perchloric acid solutions, Electrochim. Acta 8 (12) (1963) 973–983, https://doi.org/10.1016/0013-4686(62)87051-0.

[12] J. Greeley, M. Mavrikakis, A first-principles study of methanol decomposition on Pt(111), J. Am. Chem. Soc. 124 (24) (2002) 7193–7201, https://doi.org/10.1021/ja017818k.

[13] M. Watanabe, S. Motoo, Electrocatalysis by ad-atoms. Part II. Enhancement of the oxidation of methanol on platinum by ruthenium ad-atoms, J. Electroanal. Chem. 60 (3) (1975) 267–273, https://doi.org/10.1016/S0022-0728(75)80261-0.

[14] T. Frelink, W. Visscher, J.A.R. van Veen, On the role of Ru and Sn as promotors of methanol electro-oxidation over Pt, Surf. Sci. 335 (1995) 353–360, https://doi.org/10.1016/0039-6028(95)00412-2.

[15] H. Liu, C. Song, L. Zhang, J. Zhang, H. Wang, D.P. Wilkinson, A review of anode catalysis in the direct methanol fuel cell, J. Power Sources 155 (2) (2006) 95–110, https://doi.org/10.1016/j.jpowsour.2006.01.030.

[16] T.-Y. Jeon, K.-S. Lee, S.J. Yoo, Y.-H. Cho, S.H. Kang, Y.-E. Sung, Effect of surface segregation on the methanol oxidation reaction in carbon-supported Pt-Ru alloy nanoparticles, Langmuir 26 (11) (2010) 9123–9129, https://doi.org/10.1021/la9049154.

[17] O. Sahin, H. Kivrak, A comparative study of electrochemical methods on Pt–Ru DMFC anode catalysts: the effect of Ru addition, Int. J. Hydrog. Energy 38 (2) (2013) 901–909, https://doi.org/10.1016/J.IJHYDENE.2012.10.066.

240 Chapter 13 Novel materials structures and compositions

[18] K. Tedsree, A. Thanatsiri, Comparative study on the catalytic activity between Pt-decorated Ru and Ru-decorated Pt core-shell nanocatalyst toward methanol electro-oxidation, Mater. Today Proc. 5 (5) (2018) 10954–10963, https://doi.org/10.1016/J.MATPR.2018.01.009.

[19] L. Bai, Synthesis of PtRu/Ru heterostructure for efficient methanol electro-oxidation: the role of extra Ru, Appl. Surf. Sci. 433 (2018) 279–284, https://doi.org/10.1016/j.apsusc.2017.10.026.

[20] F. Kadirgan, S. Beyhan, T. Atilan, Preparation and characterization of nanosized Pt–Pd/C catalysts and comparison of their electro-activity toward methanol and ethanol oxidation, Int. J. Hydrog. Energy 34 (10) (2009) 4312–4320, https://doi.org/10.1016/J.IJHYDENE.2009.03.024.

[21] J.C.M. Silva, et al., PtSn/C alloyed and non-alloyed materials: differences in the ethanol electro-oxidation reaction pathways, Appl. Catal. B Environ. 110 (2011) 141–147, https://doi.org/10.1016/j.apcatb.2011.08.036.

[22] M. Rana, P.K. Patil, M. Chhetri, K. Dileep, R. Datta, U.K. Gautam, Pd–Pt alloys nanowires as support-less electrocatalyst with high synergistic enhancement in efficiency for methanol oxidation in acidic medium, J. Colloid Interface Sci. 463 (2016) 99–106, https://doi.org/10.1016/j.jcis.2015.10.042.

[23] X. Zhao, J. Zhang, L. Wang, H.X. Li, Z. Liu, W. Chen, Ultrathin PtPdCu nanowires fused porous architecture with 3D molecular accessibility: an active and durable platform for methanol oxidation, ACS Appl. Mater. Interfaces 7 (47) (2015) 26333–26339, https://doi.org/10.1021/acsami.5b09357.

[24] E. Antolini, F. Colmati, E.R. Gonzalez, Effect of Ru addition on the structural characteristics and the electrochemical activity for ethanol oxidation of carbon supported Pt–Sn alloy catalysts, Electrochem. Commun. 9 (3) (2007) 398–404, https://doi.org/10.1016/J.ELECOM.2006.10.012.

[25] L.X. Yang, R.G. Allen, K. Scott, P. Christenson, S. Roy, A comparative study of ptru and ptrusn thermally formed on titanium mesh for methanol electro-oxidation, J. Power Sources 137 (2) (2004) 257–263, https://doi.org/10.1016/j.jpowsour.2004.06.028.

[26] L.X. Yang, C. Bock, B. MacDougall, J. Park, The role of the WO_x ad-component to Pt and PtRu catalysts in the electrochemical CH_3OH oxidation reaction, J. Appl. Electrochem. 34 (4) (2004) 427–438, https://doi.org/10.1023/B:JACH.0000016628.81571.f4.

[27] W. Zhou, et al., Pt based anode catalysts for direct ethanol fuel cells, Appl. Catal. B Environ. 46 (2) (2003) 273–285, https://doi.org/10.1016/S0926-3373(03)00218-2.

[28] J.C. Forti, P. Olivi, A.R. de Andrade, Electrochemical behavior of ethanol oxidation on a $Ti/Ru_{0.3}Ti_{(0.7-x)}Sn_xO_2$ electrode, J. Electrochem. Soc. 150 (4) (2003) E222, https://doi.org/10.1149/1.1556037.

[29] A.O. Neto, R.R. Dias, M.M. Tusi, M. Linardi, E.V. Spinacé, Electro-oxidation of methanol and ethanol using PtRu/C, PtSn/C and PtSnRu/C electrocatalysts prepared by an alcohol-reduction process, J. Power Sources 166 (1) (2007) 87–91, https://doi.org/10.1016/j.jpowsour.2006.12.088.

[30] J.W. Long, R.M. Stroud, K.E. Swider-Lyons, D.R. Rolison, How to make electrocatalysts more active for direct methanol oxidationavoid PtRu bimetallic alloys! J. Phys. Chem. B (2000), https://doi.org/10.1021/JP001954E.

[31] J. Tayal, B. Rawat, S. Basu, Bi-metallic and tri-metallic Pt–Sn/C, Pt–Ir/C, Pt–Ir–Sn/C catalysts for electro-oxidation of ethanol in direct ethanol fuel cell, Int. J. Hydrog. Energy 36 (22) (2011) 14884–14897, https://doi.org/10.1016/J.IJHYDENE.2011.03.035.

[32] T.S. Almeida, A.R. Van Wassen, R.B. VanDover, A.R. de Andrade, H.D. Abruña, Combinatorial PtSnM (M = Fe, Ni, Ru and Pd) nanoparticle catalyst library toward ethanol electrooxidation, J. Power Sources 284 (2015) 623–630, https://doi.org/10.1016/J.JPOWSOUR.2015.03.055.

[33] J.S. Cooper, M.K. Jeon, P.J. McGinn, Combinatorial screening of ternary Pt–Ni–Cr catalysts for methanol electro-oxidation, Electrochem. Commun. 10 (10) (2008) 1545–1547, https://doi.org/10.1016/J.ELECOM.2008.08.010.

[34] N. Erini, P. Krause, M. Gliech, R. Yang, Y. Huang, P. Strasser, Comparative assessment of synthetic strategies toward active platinum–rhodium–tin electrocatalysts for efficient ethanol electro-oxidation, J. Power Sources 294 (2015) 299–304, https://doi.org/10.1016/J.JPOWSOUR.2015.06.042.

[35] Y.M. Alyousef, M.K. Datta, S.C. Yao, P.N. Kumta, Complexed sol–gel synthesis of improved Pt–Ru–Os-based anode electro-catalysts for direct methanol fuel cells, J. Phys. Chem. Solids 70 (6) (2009) 1019–1023, https://doi.org/10.1016/J.JPCS.2009.05.014.

[36] Y.-X. Wang, C.-F. Liu, M.-L. Yang, X.-H. Zhao, Z.-X. Xue, Y.-Z. Xia, Concave Pt-Cu-Fe ternary nanocubes: one-pot synthesis and their electrocatalytic activity of methanol and formic acid oxidation, Chin. Chem. Lett. 28 (1) (2017) 60–64, https://doi.org/10.1016/J.CCLET.2016.05.025.

[37] X. Wang, L. Zhang, H. Gong, Y. Zhu, H. Zhao, Y. Fu, Dealloyed PtAuCu electrocatalyst to improve the activity and stability towards both oxygen reduction and methanol oxidation reactions, Electrochim. Acta 212 (2016) 277–285, https://doi.org/10.1016/j.electacta.2016.07.028.

[38] E. Lee, A. Murthy, A. Manthiram, Effect of Mo addition on the electrocatalytic activity of Pt-Sn-Mo/C for direct ethanol fuel cells, Electrochim. Acta 56 (3) (2011) 1611–1618, https://doi.org/10.1016/j.electacta.2010.10.086.

[39] J.M. Jacob, P.G. Corradini, E. Antolini, N.A. Santos, J. Perez, Electro-oxidation of ethanol on ternary Pt–Sn–Ce/C catalysts, Appl. Catal. B Environ. 165 (2015) 176–184, https://doi.org/10.1016/J.APCATB.2014.10.012.

[40] M.K. Jeon, J.S. Cooper, P.J. McGinn, Investigation of PtCoCr/C catalysts for methanol electro-oxidation identified by a thin film combinatorial method, J. Power Sources 192 (2) (2009) 391–395, https://doi.org/10.1016/J.JPOWSOUR.2009.02.087.

[41] X. Zhang, F. Zhang, R.-F. Guan, K.-Y. Chan, Preparation of Pt-Ru-Ni ternary nanoparticles by microemulsion and electrocatalytic activity for methanol oxidation, Mater. Res. Bull. 42 (2) (2007) 327–333, https://doi.org/10.1016/J.MATERRESBULL.2006.05.021.

[42] G.R. Salazar-Banda, K.I.B. Eguiluz, M.M.S. Pupo, H.B. Suffredini, M.L. Calegaro, L.A. Avaca, The influence of different co-catalysts in Pt-based ternary and quaternary electro-catalysts on the electro-oxidation of methanol and ethanol in acid media, J. Electroanal. Chem. 668 (2012) 13–25, https://doi.org/10.1016/j.jelechem.2012.01.006.

[43] K. Fatih, V. Neburchilov, V. Alzate, R. Neagu, H. Wang, Synthesis and characterization of quaternary PtRuIrSn/C electrocatalysts for direct ethanol fuel cells, J. Power Sources 195 (21) (2010) 7168–7175, https://doi.org/10.1016/J.JPOWSOUR.2010.05.038.

[44] X. Chen, et al., Multi-component nanoporous platinum–ruthenium–copper–osmium–iridium alloy with enhanced electrocatalytic activity towards methanol oxidation and oxygen reduction, J. Power Sources 273 (2015) 324–332, https://doi.org/10.1016/J.JPOWSOUR.2014.09.076.

[45] K.-W. Park, J.-H. Choi, S.-A. Lee, C. Pak, H. Chang, Y.-E. Sung, PtRuRhNi nanoparticle electrocatalyst for methanol electrooxidation in direct methanol fuel cell, J. Catal. 224 (2) (2004) 236–242, https://doi.org/10.1016/J.JCAT.2004.02.010.

[46] Y.M. Alyousef, M.K. Datta, K. Kadakia, S.C. Yao, P.N. Kumta, Sol–gel synthesis of Pt-Ru-Os-Ir based anode electro-catalysts for direct methanol fuel cells, J. Alloys Compd. 506 (2) (2010) 698–702, https://doi.org/10.1016/J.JALLCOM.2010.07.046.

242 Chapter 13 Novel materials structures and compositions

[47] L.D. Burke, O.J. Murphy, The electrooxidation of methanol and related compounds at ruthenium dioxide-coated electrodes, J. Electroanal. Chem. 101 (3) (1979) 351–361, https://doi.org/10.1016/S0022-0728(79)80046-7.

[48] M.B. de Oliveira, L.P.R. Profeti, P. Olivi, Electrooxidation of methanol on PtMyOx (M = Sn, Mo, Os or W) electrodes, Electrochem. Commun. 7 (7) (2005) 703–709, https://doi.org/10.1016/J.ELECOM.2005.04.024.

[49] G.-Y. Zhao, H.-L. Li, Electrochemical oxidation of methanol on Pt nanoparticles composited MnO2 nanowire arrayed electrode, Appl. Surf. Sci. 254 (10) (2008) 3232–3235, https://doi.org/10.1016/J.APSUSC.2007.10.086.

[50] L. Jiang, L. Colmenares, Z. Jusys, G.Q. Sun, R.J. Behm, Ethanol electrooxidation on novel carbon supported Pt/SnOx/C catalysts with varied Pt:Sn ratio, Electrochim. Acta 53 (2) (2007) 377–389, https://doi.org/10.1016/J.ELECTACTA.2007.01.047.

[51] M. Sedighi, A.A. Rostami, E. Alizadeh, Enhanced electro-oxidation of ethanol using $Pt–CeO_2$ electrocatalyst prepared by electrodeposition technique, Int. J. Hydrog. Energy 42 (8) (2017) 4998–5005, https://doi.org/10.1016/j.ijhydene.2016.12.014.

[52] Y. Bai, et al., Electrochemical oxidation of ethanol on Pt–ZrO2/C catalyst, Electrochem. Commun. 7 (11) (2005) 1087–1090, https://doi.org/10.1016/J.ELECOM.2005.08.002.

[53] M.-Q. Shi, G.-H. Song, P.-P. Yang, Y.-Q. Chu, C.-A. Ma, A new biomass template to prepare multi-channel structure of WO3 and its application for methanol electro-oxidation, Mater. Lett. 153 (2015) 124–127, https://doi.org/10.1016/J.MATLET.2015.04.014.

[54] G. Li, L. Jiang, B. Zhang, Q. Jiang, D.S. Su, G. Sun, A highly active porous Pt–PbOx/C catalyst toward alcohol electro-oxidation in alkaline electrolyte, Int. J. Hydrog. Energy 38 (29) (2013) 12767–12773, https://doi.org/10.1016/J.IJHYDENE.2013.07.076.

[55] R.S. Amin, R.M.A. Hameed, K.M. El-Khatib, M.E. Youssef, A.A. Elzatahry, Pt–NiO/C anode electrocatalysts for direct methanol fuel cells, Electrochim. Acta 59 (2012) 499–508, https://doi.org/10.1016/J.ELECTACTA.2011.11.013.

[56] E. Antolini, E.R. Gonzalez, Alkaline direct alcohol fuel cells, J. Power Sources 195 (11) (2010) 3431–3450, https://doi.org/10.1016/j.jpowsour.2009.11.145.

[57] F. Ren, et al., Clean method for the synthesis of reduced graphene oxide-supported PtPd alloys with high electrocatalytic activity for ethanol oxidation in alkaline medium, ACS Appl. Mater. Interfaces 6 (5) (2014) 3607–3614, https://doi.org/10.1021/am405846h.

[58] I.G. Casella, M. Contursi, Characterization of bismuth adatom-modified palladium electrodes. The electrocatalytic oxidation of aliphatic aldehydes in alkaline solutions, Electrochim. Acta 52 (2) (2006) 649–657, https://doi.org/10.1016/j.electacta.2006.05.048.

[59] Y.-F. Li, J.-J. Lv, M. Zhang, J.-J. Feng, F.-F. Li, A.-J. Wang, A simple and controlled electrochemical deposition route to urchin-like Pd nanoparticles with enhanced electrocatalytic properties, J. Electroanal. Chem. 738 (2015) 1–7, https://doi.org/10.1016/J.JELECHEM.2014.11.017.

[60] D. Xu, X. Yan, P. Diao, P. Yin, Electrodeposition of vertically aligned palladium nanoneedles and their application as active substrates for surface-enhanced Raman scattering, J. Phys. Chem. C 118 (18) (2014) 9758–9768, https://doi.org/10.1021/jp500667f.

[61] J. Sun, Y. Wang, C. Zhang, T. Kou, Z. Zhang, Anodization driven enhancement of catalytic activity of Pd towards electro-oxidation of methanol, ethanol and formic acid, Electrochem. Commun. 21 (1) (2012) 42–45, https://doi.org/10.1016/j.elecom.2012.04.023.

Chapter 13 Novel materials structures and compositions **243**

[62] A.K. Das, N.H. Kim, D. Pradhan, D. Hui, J.H. Lee, Electrochemical synthesis of palladium (Pd) nanorods: an efficient electrocatalyst for methanol and hydrazine electro-oxidation, Compos. Part B 144 (2018) 11–18, https://doi.org/10.1016/J.COMPOSITESB.2018.02.017.

[63] Q. Yi, F. Niu, L. Sun, Fabrication of novel porous Pd particles and their electroactivity towards ethanol oxidation in alkaline media, Fuel 90 (8) (2011) 2617–2623, https://doi.org/10.1016/J.FUEL.2011.03.038.

[64] J. Cai, Y. Huang, Y. Guo, Bi-modified Pd/C catalyst via irreversible adsorption and its catalytic activity for ethanol oxidation in alkaline medium, Electrochim. Acta 99 (2013) 22–29, https://doi.org/10.1016/J.ELECTACTA.2013.03.059.

[65] S.T. Nguyen, et al., Enhancement effect of ag for Pd/C towards the ethanol electro-oxidation in alkaline media, Appl. Catal. B Environ. 91 (1–2) (2009) 507–515, https://doi.org/10.1016/J.APCATB.2009.06.021.

[66] X. Xu, X. Wang, S. Huo, Z. Chen, H. Zhao, J. Xu, Facile synthesis of PdIr nanoporous aggregates as highly active electrocatalyst towards methanol and ethylene glycol oxidation, Catal. Today 318 (2018) 157–166, https://doi.org/10.1016/j.cattod.2017.09.054.

[67] R.C. Sekol, et al., Pd-Ni-cu-P metallic glass nanowires for methanol and ethanol oxidation in alkaline media, Int. J. Hydrog. Energy 38 (26) (2013) 11248–11255, https://doi.org/10.1016/j.ijhydene.2013.06.017.

[68] F. Yang, et al., Reduced graphene oxide supported Pd-cu-co trimetallic catalyst: synthesis, characterization and methanol electrooxidation properties, J. Energy Chem. (2019) 72–78, https://doi.org/10.1016/j.jechem.2018.02.007.

[69] S. Jongsomjit, P. Prapainainar, K. Sombatmankhong, Synthesis and characterisation of Pd–Ni–Sn electrocatalyst for use in direct ethanol fuel cells, Solid State Ionics 288 (2016) 147–153, https://doi.org/10.1016/J.SSI.2015.12.009.

[70] Y.-H. Mao, C.-Y. Chen, J.-X. Fu, T.-Y. Lai, F.-H. Lu, Y.-C. Tsai, Electrodeposition of nickel-copper on titanium nitride for methanol electrooxidation, Surf. Coat. Technol. 350 (2018) 949–953, https://doi.org/10.1016/J.SURFCOAT.2018.03.048.

[71] A. Döner, E. Telli, G. Kardaş, Electrocatalysis of Ni-promoted cd coated graphite toward methanol oxidation in alkaline medium, J. Power Sources 205 (2012) 71–79, https://doi.org/10.1016/j.jpowsour.2012.01.020.

[72] A. Döner, G. Kardaş, A novel, effective and low cost electrocatalyst for direct methanol fuel cells applications, Int. J. Hydrog. Energy 40 (14) (2015) 4840–4849, https://doi.org/10.1016/j.ijhydene.2015.02.039.

[73] A. Döner, R. Solmaz, G. Kardaş, Enhancement of hydrogen evolution at cobalt-zinc deposited graphite electrode in alkaline solution, Int. J. Hydrog. Energy 36 (13) (2011) 7391–7397, https://doi.org/10.1016/j.ijhydene.2011.03.083.

[74] E. Telli, A. Döner, G. Kardaş, Electrocatalytic oxidation of methanol on Ru deposited NiZn catalyst at graphite in alkaline medium, Electrochim. Acta 107 (2013) 216–224, https://doi.org/10.1016/j.electacta.2013.05.113.

[75] R. Solmaz, G. Kardaş, Fabrication and characterization of NiCoZn–M (M: ag, Pd and Pt) electrocatalysts as cathode materials for electrochemical hydrogen production, Int. J. Hydrog. Energy 36 (19) (2011) 12079–12087, https://doi.org/10.1016/J.IJHYDENE.2011.06.101.

[76] A.O. Yüce, A. Döner, G. Kardaş, NiMn composite electrodes as cathode material for hydrogen evolution reaction in alkaline solution, Int. J. Hydrog. Energy 38 (11) (2013) 4466–4473, https://doi.org/10.1016/J.IJHYDENE.2013.01.160.

244 Chapter 13 Novel materials structures and compositions

[77] E. Vernickaite, N. Tsyntsaru, H. Cesiulis, Electrodeposited Co-W alloys and their prospects as effective anode for methanol oxidation in acidic media, Surf. Coat. Technol. 307 (2016) 1322–1328, https://doi.org/10.1016/J.SURFCOAT.2016.07.049.

[78] L. Cao, G. Sun, H. Li, Q. Xin, Carbon-supported IrSn catalysts for direct ethanol fuel cell, Fuel Cells Bull. 2007 (11) (2007) 12–16, https://doi.org/10.1016/S1464-2859(08)70142-1.

[79] P. Wu, Y. Huang, L. Zhou, Y. Wang, Y. Bu, J. Yao, Nitrogen-doped graphene supported highly dispersed palladium-lead nanoparticles for synergetic enhancement of ethanol electrooxidation in alkaline medium, Electrochim. Acta 152 (2015) 68–74, https://doi.org/10.1016/j.electacta.2014.11.110.

[80] S.H. Joo, et al., Ordered nanoporous arrays of carbon supporting high dispersions of platinum nanoparticles, Nature 412 (6843) (2001) 169–172, https://doi.org/10.1038/35084046.

[81] L.M. Esteves, H.A. Oliveira, F.B. Passos, Carbon nanotubes as catalyst support in chemical vapor deposition reaction: a review, J. Ind. Eng. Chem. 65 (2018) 1–12, https://doi.org/10.1016/J.JIEC.2018.04.012.

[82] U.A. Paulus, U. Endruschat, G.J. Feldmeyer, T.J. Schmidt, H. Bönnemann, R.J. Behm, New PtRu alloy colloids as precursors for fuel cell catalysts, J. Catal. 195 (2) (2000) 383–393, https://doi.org/10.1006/jcat.2000.2998.

[83] O. Antoine, Y. Bultel, R. Durand, Oxygen reduction reaction kinetics and mechanism on platinum nanoparticles inside Nafion®, J. Electroanal. Chem. 499 (1) (2001) 85–94, https://doi.org/10.1016/S0022-0728(00)00492-7.

[84] E. Antolini, Carbon supports for low-temperature fuel cell catalysts, Appl. Catal. B Environ. 88 (1–2) (2009) 1–24, https://doi.org/10.1016/j.apcatb.2008.09.030.

[85] X. Yuan, X.-L. Ding, C.-Y. Wang, Z.-F. Ma, Use of polypyrrole in catalysts for low temperature fuel cells, Energy Environ. Sci. 6 (4) (2013) 1105, https://doi.org/10.1039/c3ee23520c.

[86] N. Tsubokawa, Functionalization of carbon black by surface grafting of polymers, Prog. Polym. Sci. 17 (3) (1992) 417–470, https://doi.org/10.1016/0079-6700(92)90021-P.

[87] Y. Shao-Horn, W.C. Sheng, S. Chen, P.J. Ferreira, E.F. Holby, D. Morgan, Instability of supported platinum nanoparticles in low-temperature fuel cells, Top. Catal. 46 (3–4) (2007) 285–305, https://doi.org/10.1007/s11244-007-9000-0.

[88] M. Toupin, D. Bélanger, Thermal stability study of aryl modified carbon black by in situ generated diazonium salt, J. Phys. Chem. C (2007), https://doi.org/10.1021/JP066868E.

[89] D. Shao, Z. Jiang, X. Wang, SDBS modified XC-72 carbon for the removal of Pb(II) from aqueous solutions, Plasma Process. Polym. 7 (7) (2010) 552–560, https://doi.org/10.1002/ppap.201000005.

[90] J.P. Paraknowitsch, A. Thomas, Doping carbons beyond nitrogen: an overview of advanced heteroatom doped carbons with boron, sulphur and phosphorus for energy applications, Energy Environ. Sci. 6 (10) (2013) 2839–2855, https://doi.org/10.1039/c3ee41444b.

[91] M.M. Antxustegi, A.R. Pierna, N. Ruiz, Chemical activation of Vulcan® XC72R to be used as support for NiNbPtRu catalysts in PEM fuel cells, Int. J. Hydrog. Energy 39 (8) (2014) 3978–3983, https://doi.org/10.1016/j.ijhydene.2013.04.061.

[92] F. Rodríguez-Reinoso, I. Rodríguez-Ramos, C. Moreno-Castilla, A. Guerrero-Ruiz, J.D. López-González, Platinum catalysts supported on activated carbons. I. Preparation and characterization, J. Catal. 99 (1) (1986) 171–183, https://doi.org/10.1016/0021-9517(86)90210-1.

[93] Z. Tang, Q. Li, G. Lu, The effect of plasma pre-treatment of carbon used as a Pt catalyst support for methanol electrooxidation, Carbon 45 (1) (2007) 41–46, https://doi.org/10.1016/j.carbon.2006.08.010.

[94] N.P. Subramanian, et al., Studies on co-based catalysts supported on modified carbon substrates for PEMFC cathodes, J. Power Sources 157 (1) (2006) 56–63, https://doi.org/10.1016/j.jpowsour.2005.07.031.

[95] M. Carmo, M. Linardi, J. Guilherme, R. Poco, Characterization of nitric acid functionalized carbon black and its evaluation as electrocatalyst support for direct methanol fuel cell applications, Appl. Catal. A Gen. (2009), https://doi.org/10.1016/j.apcata.2008.12.010.

[96] S. Samad, et al., Carbon and non-carbon support materials for platinum-based catalysts in fuel cells, Int. J. Hydrog. Energy 43 (16) (2018) 7823–7854, https://doi.org/10.1016/j.ijhydene.2018.02.154.

[97] Y.-L. Zhao, J.F. Stoddart, Noncovalent functionalization of single-walled carbon nanotubes, Acc. Chem. Res. 42 (8) (2009) 1161–1171, https://doi.org/10.1021/ar900056z.

[98] S. Sharma, B.G. Pollet, Support materials for PEMFC and DMFC electrocatalysts-a review, J. Power Sources 208 (2012) 96–119, https://doi.org/10.1016/j.jpowsour.2012.02.011.

[99] S. Mallakpour, E. Khadem, Carbon nanotube–metal oxide nanocomposites: fabrication, properties and applications, Chem. Eng. J. 302 (2016) 344–367, https://doi.org/10.1016/J.CEJ.2016.05.038.

[100] T. Matsumoto, et al., Reduction of Pt usage in fuel cell electrocatalysts with carbon nanotube electrodes, Chem. Commun. (7) (2004) 840, https://doi.org/10.1039/b400607k.

[101] X. Wang, W. Li, Z. Chen, M. Waje, Y. Yan, Durability investigation of carbon nanotube as catalyst support for proton exchange membrane fuel cell, J. Power Sources 158 (1) (2006) 154–159, https://doi.org/10.1016/J.JPOWSOUR.2005.09.039.

[102] K. Lee, J. Zhang, H. Wang, D.P. Wilkinson, Progress in the synthesis of carbon nanotube- and nanofiber-supported Pt electrocatalysts for PEM fuel cell catalysis, J. Appl. Electrochem. 36 (5) (2006) 507–522, https://doi.org/10.1007/s10800-006-9120-4.

[103] N. Yang, X. Chen, T. Ren, P. Zhang, D. Yang, Carbon nanotube based biosensors, Sensors Actuators B Chem. 207 (2015) 690–715, https://doi.org/10.1016/J.SNB.2014.10.040.

[104] A.V. Herrera-Herrera, M.Á. González-Curbelo, J. Hernández-Borges, M.Á. Rodríguez-Delgado, Carbon nanotubes applications in separation science: a review, Anal. Chim. Acta 734 (2012) 1–30, https://doi.org/10.1016/j.aca.2012.04.035.

[105] G.V. Dubacheva, C.-K. Liang, D.M. Bassani, Functional monolayers from carbon nanostructures—fullerenes, carbon nanotubes, and graphene—as novel materials for solar energy conversion, Coord. Chem. Rev. 256 (21–22) (2012) 2628–2639, https://doi.org/10.1016/J.CCR.2012.04.007.

[106] M.R. Axet, O. Dechy-Cabaret, J. Durand, M. Gouygou, P. Serp, Coordination chemistry on carbon surfaces, Coord. Chem. Rev. 308 (2016) 236–345, https://doi.org/10.1016/J.CCR.2015.06.005.

[107] I. Capek, Dispersions based on carbon nanotubes—biomolecules conjugates, in: Carbon Nanotubes - Growth and Applications, InTech, 2011.

[108] W. Zhang, P. Sherrell, A.I. Minett, J.M. Razal, J. Chen, Carbon nanotube architectures as catalyst supports for proton exchange membrane fuel cells, Energy Environ. Sci. 3 (9) (2010) 1286, https://doi.org/10.1039/c0ee00139b.

[109] X. Liu, M. Wang, S. Zhang, B. Pan, Application potential of carbon nanotubes in water treatment: a review, J. Environ. Sci. 25 (7) (2013) 1263–1280, https://doi.org/10.1016/S1001-0742(12)60161-2.

[110] S. Shahgaldi, J. Hamelin, Improved carbon nanostructures as a novel catalyst support in the cathode side of PEMFC: a critical review, Carbon 94 (2015) 705–728, https://doi.org/10.1016/J.CARBON.2015.07.055.

[111] N. Jha, A. Leelamohanareddy, M. Shaijumon, N. Rajalakshmi, S. Ramaprabhu, Pt–Ru/multi-walled carbon nanotubes as electrocatalysts for direct methanol fuel cell, Int. J. Hydrog. Energy 33 (1) (2008) 427–433, https://doi.org/10.1016/j.ijhydene.2007.07.064.

[112] G. Wu, Y.-S. Chen, B.-Q. Xu, Remarkable support effect of SWNTs in Pt catalyst for methanol electrooxidation, Electrochem. Commun. 7 (12) (2005) 1237–1243, https://doi.org/10.1016/J.ELECOM.2005.07.015.

[113] M.-W. Xu, Z. Su, Z.-W. Weng, Z.-C. Wang, B. Dong, An approach for synthesizing PtRu/MWCNT nanocomposite for methanol electro-oxidation, Mater. Chem. Phys. 124 (1) (2010) 785–790, https://doi.org/10.1016/J.MATCHEMPHYS.2010.07.061.

[114] J.R. Rodriguez, et al., Synthesis of Pt and Pt-Fe nanoparticles supported on MWCNTs used as electrocatalysts in the methanol oxidation reaction, J. Energy Chem. 23 (4) (2014) 483–490, https://doi.org/10.1016/S2095-4956(14)60175-3.

[115] H. An, et al., Electrocatalytic performance of Pd nanoparticles supported on TiO_2-MWCNTs for methanol, ethanol, and isopropanol in alkaline media, J. Electroanal. Chem. 741 (2015) 56–63, https://doi.org/10.1016/J.JELECHEM.2015.01.015.

[116] D.M. Han, Z.P. Guo, R. Zeng, C.J. Kim, Y.Z. Meng, H.K. Liu, Multiwalled carbon nanotube-supported Pt/Sn and Pt/Sn/PMo12 electrocatalysts for methanol electro-oxidation, Int. J. Hydrog. Energy 34 (5) (2009) 2426–2434, https://doi.org/10.1016/J.IJHYDENE.2008.12.073.

[117] G.-P. Jin, Y.-F. Ding, P.-P. Zheng, Electrodeposition of nickel nanoparticles on functional MWCNT surfaces for ethanol oxidation, J. Power Sources 166 (1) (2007) 80–86, https://doi.org/10.1016/J.JPOWSOUR.2006.12.087.

[118] X. Guo, D.-J. Guo, X.-P. Qiu, L.-Q. Chen, W.-T. Zhu, A simple one-step preparation of high utilization AuPt nanoparticles supported on MWCNTs for methanol oxidation in alkaline medium, Electrochem. Commun. 10 (11) (2008) 1748–1751, https://doi.org/10.1016/J.ELECOM.2008.09.005.

[119] D.-J. Guo, S.-K. Cui, Hollow PtCo nanospheres supported on multi-walled carbon nanotubes for methanol electrooxidation, J. Colloid Interface Sci. 340 (1) (2009) 53–57, https://doi.org/10.1016/J.JCIS.2009.08.030.

[120] Y. Zhang, Q. Yi, H. Chu, H. Nie, Catalytic activity of Pd-Ag nanoparticles supported on carbon nanotubes for the electro-oxidation of ethanol and propanol, J. Fuel Chem. Technol. 45 (2017) 475–483, https://doi.org/10.1016/s1872-5813(17)30026-9.

[121] M.G. Hosseini, R. Mahmoodi, D.-E. Vahid, M.G. Hosseini, R. Mahmoodi, D.-E. Vahid, Energy 161 (C) (2018).

[122] F. Zhu, et al., High activity of carbon nanotubes supported binary and ternary Pd-based catalysts for methanol, ethanol and formic acid electro-oxidation, J. Power Sources 242 (2013) 610–620, https://doi.org/10.1016/J.JPOWSOUR.2013.05.145.

[123] S. Chen, F. Ye, W. Lin, Effect of operating conditions on the performance of a direct methanol fuel cell with PtRuMo/CNTs as anode catalyst, Int. J. Hydrog. Energy 35 (15) (2010) 8225–8233, https://doi.org/10.1016/J.IJHYDENE.2009.12.085.

[124] Y. Huang, J. Cai, S. Zheng, Y. Guo, Fabrication of a high-performance Pb–PtCu/CNT catalyst for methanol electro-oxidation, J. Power Sources 210 (2012) 81–85, https://doi.org/10.1016/J.JPOWSOUR.2012.03.002.

[125] A.M. Zainoodin, S.K. Kamarudin, M.S. Masdar, W.R.W. Daud, A.B. Mohamad, J. Sahari, High power direct methanol fuel cell with a porous carbon nanofiber anode layer, Appl. Energy 113 (2014) 946–954, https://doi.org/10.1016/J.APENERGY.2013.07.066.

[126] A.L. Dicks, The role of carbon in fuel cells, J. Power Sources 156 (2) (2006) 128–141, https://doi.org/10.1016/J.JPOWSOUR.2006.02.054.

[127] N.M. Rodriguez, M.-S. Kim, R.T.K. Baker, Carbon nanofibers: a unique catalyst support medium, J. Phys. Chem. 98 (50) (1994) 13108–13111, https://doi.org/10.1021/j100101a003.

[128] J.-S. Zheng, X.-S. Zhang, P. Li, J. Zhu, X.-G. Zhou, W.-K. Yuan, Effect of carbon nanofiber microstructure on oxygen reduction activity of supported palladium electrocatalyst, Electrochem. Commun. 9 (5) (2007) 895–900, https://doi.org/10.1016/J.ELECOM.2006.12.006.

[129] G.-H. An, H.-J. Ahn, Pt electrocatalyst-loaded carbon nanofibre–Ru core–shell supports for improved methanol electrooxidation, J. Electroanal. Chem. 707 (2013) 74–77, https://doi.org/10.1016/J.JELECHEM.2013.08.024.

[130] S.-Y. Lee, J.-M. Park, S.-J. Park, Roles of nitric acid treatment on PtRu catalyst supported on graphite nanofibers and their methanol electro-oxidation behaviors, Int. J. Hydrog. Energy 39 (29) (2014) 16468–16473, https://doi.org/10.1016/J.IJHYDENE.2014.04.181.

[131] C.A. Bessel, K. Laubernds, N.M. Rodriguez, R.T.K. Baker, Graphite nanofibers as an electrode for fuel cell applications, J. Phys. Chem. B 105 (6) (2001) 1121–1122, https://doi.org/10.1021/jp003280d.

[132] M.M. Dubinin, The potential theory of adsorption of gases and vapors for adsorbents with energetically nonuniform surfaces, Chem. Rev. 60 (2) (1960) 235–241, https://doi.org/10.1021/cr60204a006.

[133] K.S.W. Sing, Reporting physisorption data for gas/solid systems with special reference to the determination of surface area and porosity (recommendations 1984), Pure Appl. Chem. 57 (4) (1985) 603–619, https://doi.org/10.1351/pac198557040603.

[134] R. Ryoo, S.H. Joo, S. Jun, Synthesis of highly ordered carbon molecular sieves via template-mediated structural transformation, J. Phys. Chem. B 103 (1999) 7743–7746, https://doi.org/10.1021/Jp991673a.

[135] J. Lee, S. Yoon, T. Hyeon, S.M. Oh, K.B. Kim, Synthesis of a new mesoporous carbon and its application to electrochemical double-layer capacitors, Chem. Commun. (21) (1999) 2177–2178, https://doi.org/10.1039/a906872d.

[136] S. Fabing, J. Zeng, X. Bao, Y. Yu, J.Y. Lee, X.S. Zhao, Preparation and characterization of highly ordered graphitic mesoporous carbon as a Pt catalyst support for direct methanol fuel cells, Chem. Mater. (2005), https://doi.org/10.1021/CM0502222.

[137] P. Zhang, J. Zhang, S. Dai, Mesoporous carbon materials with functional compositions, Chem. Eur. J. 23 (9) (2017) 1986–1998, https://doi.org/10.1002/chem.201602199.

[138] H. Kim, P. Kim, K. Choi, J. Yi, Preparation of Pt/mesoporous carbon catalysts and their application to the methanol electro-oxidation, Stud. Surf. Sci. Catal. 159 (2006) 609–612, https://doi.org/10.1016/S0167-2991(06)81670-8.

[139] X. Cui, J. Shi, L. Zhang, M. Ruan, J. Gao, PtCo supported on ordered mesoporous carbon as an electrode catalyst for methanol oxidation, Carbon 47 (1) (2009) 186–194, https://doi.org/10.1016/J.CARBON.2008.09.054.

[140] B. Liu, S. Creager, Silica–sol-templated mesoporous carbon as catalyst support for polymer electrolyte membrane fuel cell applications, Electrochim. Acta 55 (8) (2010) 2721–2726, https://doi.org/10.1016/J.ELECTACTA.2009.12.044.

[141] L. Calvillo, M. Gangeri, S. Perathoner, G. Centi, R. Moliner, M.J. Lázaro, Synthesis and performance of platinum supported on ordered mesoporous carbons as catalyst for PEM fuel cells: effect of the surface chemistry of the support, Int. J. Hydrog. Energy 36 (16) (2011) 9805–9814, https://doi.org/10.1016/j.ijhydene.2011.03.023.

248 Chapter 13 Novel materials structures and compositions

[142] J.R.C. Salgado, F. Alcaide, G. Álvarez, L. Calvillo, M.J. Lázaro, E. Pastor, Pt–Ru electrocatalysts supported on ordered mesoporous carbon for direct methanol fuel cell, J. Power Sources 195 (13) (2010) 4022–4029, https://doi.org/10.1016/J.JPOWSOUR.2010.01.001.

[143] M. Pumera, Electrochemistry of graphene: new horizons for sensing and energy storage, Chem. Rec. 9 (4) (2009) 211–223, https://doi.org/10.1002/tcr.200900008.

[144] A.K. Geim, K.S. Novoselov, The rise of graphene, Nat. Mater. 6 (3) (2007) 183–191, https://doi.org/10.1038/nmat1849.

[145] D. Chen, L. Tang, J. Li, Graphene-based materials in electrochemistry, Chem. Soc. Rev. 39 (8) (2010) 3157, https://doi.org/10.1039/b923596e.

[146] D.A.C. Brownson, C.E. Banks, Graphene electrochemistry: an overview of potential applications, Analyst 135 (11) (2010) 2768, https://doi.org/10.1039/c0an00590h.

[147] M. Liang, L. Zhi, Graphene-based electrode materials for rechargeable lithium batteries, J. Mater. Chem. 19 (33) (2009) 5871, https://doi.org/10.1039/b901551e.

[148] C.T. Hable, M.S. Wrighton, Electrocatalytic oxidation of methanol and ethanol: a comparison of platinum-tin and platinum-ruthenium catalyst particles in a conducting polyaniline matrix, Langmuir 9 (11) (1993) 3284–3290, https://doi.org/10.1021/la00035a085.

[149] F. Han, X. Wang, J. Lian, Y. Wang, The effect of Sn content on the electrocatalytic properties of Pt–Sn nanoparticles dispersed on graphene nanosheets for the methanol oxidation reaction, Carbon 50 (15) (2012) 5498–5504, https://doi.org/10.1016/J.CARBON.2012.07.039.

[150] Y. Li, L. Tang, J. Li, Preparation and electrochemical performance for methanol oxidation of pt/graphene nanocomposites, Electrochem. Commun. 11 (4) (2009) 846–849, https://doi.org/10.1016/J.ELECOM.2009.02.009.

[151] D.K. Perivoliotis, N. Tagmatarchis, Recent advancements in metal-based hybrid electrocatalysts supported on graphene and related 2D materials for the oxygen reduction reaction, Carbon 118 (2017) 493–510, https://doi.org/10.1016/J.CARBON.2017.03.073.

[152] Y. Sun, et al., Boron, nitrogen co-doped graphene: a superior electrocatalyst support and enhancing mechanism for methanol electrooxidation, Electrochim. Acta 212 (2016) 313–321, https://doi.org/10.1016/J.ELECTACTA.2016.06.168.

[153] X. Yan, T. Liu, J. Jin, S. Devaramani, D. Qin, X. Lu, Well dispersed Pt–Pd bimetallic nanoparticles on functionalized graphene as excellent electro-catalyst towards electro-oxidation of methanol, J. Electroanal. Chem. 770 (2016) 33–38, https://doi.org/10.1016/J.JELECHEM.2016.03.033.

[154] A.G. MacDiarmid, 'Synthetic metals': a novel role for organic polymers (Nobel lecture), Angew. Chem. Int. Ed. 40 (14) (2001) 2581–2590, https://doi.org/10.1002/1521-3773(20010716)40:14<2581::AID-ANIE2581>3.0.CO;2-2.

[155] X. Lu, W. Zhang, C. Wang, T.-C. Wen, Y. Wei, One-dimensional conducting polymer nanocomposites: synthesis, properties and applications, Prog. Polym. Sci. 36 (5) (2011) 671–712, https://doi.org/10.1016/J.PROGPOLYMSCI.2010.07.010.

[156] R. Megha, et al., Conducting polymer nanocomposite based temperature sensors: a review, Inorg. Chem. Commun. 98 (2018) 11–28, https://doi.org/10.1016/J.INOCHE.2018.09.040.

[157] R. Trevisan, et al., PEDOT nanotube arrays as high performing counter electrodes for dye sensitized solar cells. Study of the interactions among

Chapter 13 Novel materials structures and compositions **249**

electrolytes and counter electrodes, Adv. Energy Mater. 1 (5) (2011) 781–784, https://doi.org/10.1002/aenm.201100324.

[158] T.H. Lee, et al., High-performance dye-sensitized solar cells based on PEDOT nanofibers as an efficient catalytic counter electrode, J. Mater. Chem. 22 (40) (2012) 21624, https://doi.org/10.1039/c2jm34807a.

[159] C.T. Hable, M.S. Wrighton, Electrocatalytic oxidation of methanol by assemblies of platinum/tin catalyst particles in a conducting polyaniline matrix, Langmuir 7 (7) (1991) 1305–1309, https://doi.org/10.1021/la00055a001.

[160] S.S. Jeon, W.B. Han, H.H. An, S.S. Im, C.S. Yoon, Polypyrrole-modified graphitized carbon black as a catalyst support for methanol oxidation, Appl. Catal. A Gen. 409–410 (2011) 156–161, https://doi.org/10.1016/J.APCATA.2011.09.044.

[161] R. Ojani, J.-B. Raoof, Y. Ahmady-Khanghah, Copper-poly(2-aminodiphenylamine) as a novel and low cost electrocatalyst for electrocatalytic oxidation of methanol in alkaline solution, Electrochim. Acta 56 (9) (2011) 3380–3386, https://doi.org/10.1016/J.ELECTACTA.2010.12.082.

[162] Y. Zhao, L. Zhan, J. Tian, S. Nie, Z. Ning, Enhanced electrocatalytic oxidation of methanol on Pd/polypyrrole–graphene in alkaline medium, Electrochim. Acta 56 (5) (2011) 1967–1972, https://doi.org/10.1016/J.ELECTACTA.2010.12.005.

[163] R.-X. Wang, Y.-J. Fan, L. Wang, L.-N. Wu, S.-N. Sun, S.-G. Sun, Pt nanocatalysts on a polyindole-functionalized carbon nanotube composite with high performance for methanol electrooxidation, J. Power Sources 287 (2015) 341–348, https://doi.org/10.1016/J.JPOWSOUR.2015.03.181.

[164] E. Antolini, Composite materials: an emerging class of fuel cell catalyst supports, Appl. Catal. B Environ. 100 (3–4) (2010) 413–426, https://doi.org/10.1016/J.APCATB.2010.08.025.

[165] F. Yang, et al., Polyaniline-functionalized TiO2–C supported Pt catalyst for methanol electro-oxidation, Synth. Met. 205 (2015) 23–31, https://doi.org/10.1016/J.SYNTHMET.2015.03.017.

[166] H. Pang, C. Huang, J. Chen, B. Liu, Y. Kuang, X. Zhang, Preparation of polyaniline–tin dioxide composites and their application in methanol electro-oxidation, J. Solid State Electrochem. 14 (2) (2010) 169–174, https://doi.org/10.1007/s10008-009-0892-4.

[167] D. Chu, et al., High activity of Pd–In2O3/CNTs electrocatalyst for electrooxidation of ethanol, Catal. Commun. 10 (6) (2009) 955–958, https://doi.org/10.1016/J.CATCOM.2008.12.041.

14

Synthesis and characterization of nanocomposite membranes for high-temperature polymer electrolyte membranes (PEM) methanol fuel cells

Fatma Aydın Ünal[a], Vildan Erduran[b,c], Ramazan Bayat[b,c], Sadin Ozdemir[d], and Fatih Şen[b]

[a]Metallurgical and Materials Engineering Department, Faculty of Engineering, Alanya Alaaddin Keykubat University, Alanya/Antalya, Turkey. [b]Şen Research Group, Department of Biochemistry, Dumlupinar University, Kütahya, Turkey. [c]Department of Materials Science and Engineering, Faculty of Engineering, Dumlupinar University, Kütahya, Turkey. [d]Mersin University, Food Processing Programme, Technical Science Vocational School, Mersin, Turkey

1 Introduction

Energy is very important today, and most of it is derived from fossil fuels; however, it is predicted that fossil fuels will become depleted soon [1, 2]. Efforts are being made to produce greener energy resources because of the decrease in fossil fuels, the increase in the amount of carbon dioxide (CO_2) in the air, and the increase in other environmental pollution [3–5]. Polymer electrolyte membrane fuel cells (PEMFCs) are of interest due to the damage that depleted fossil fuels cause in nature. These fuel cells can be preferred because they reduce the damage to nature and offer fairly good efficiency [6–8].

The working principle of PEMFCs is based on perfluorosulfonic acid (PFSA) ionomeric membranes operating at 80°C. In recent years, studies have been underway on fuel cells operating above 100°C [7].

PEMFCs have the following processes, which go with their use at high temperatures:

(1) Electrochemical mobility occurring in Ant and cathode

(2) Simplification of the use of water used in the system

(3) Making the cooling system useful

(4) Waste heat recovery systems

(5) Lower-quality reformed hydrogen [9]

In these membranes, temperatures of 100°C and above poses problems. Due to high-temperature water evaporation and changes in mechanical properties, proton conductivity decreases. Therefore, proton-conducting membranes operating at high temperatures have been developed. Li et al. state that polymer membranes that can still be used at high temperatures are divided into three categories: "modified PFSI membranes, alternative sulfonated polymers and inorganic composite membranes, and acid-based complex membranes." In studies with perfluorosulfonated ionomer (PFSI) membranes, DMFCs and H_2/O_2 (air) cells performed very well at 120°C/atmospheric pressure and 150°C/3–5 atmospheric pressure. Thermal stability and conductivity at temperatures above 100°C is positively influenced by sulfonated hydrocarbons and their inorganic composites. For DMFCs and H_2/O_2 (air), PEMFCs at temperatures up to 200°C under atmospheric pressure, acid-based polymer membranes, particularly H_3PO_4-doped polybenzimidazole (PBI), are shown [10, 11].

Anhydrous proton-conducting polymers can increase the working temperature of the fuel cell. It has been documented that phosphoric acid-doped PBIs are promising high-temperature fuel cell membranes. Low gas permeability, strong proton conductivity, and superior thermal stability were shown by composite membranes at high temperatures [12]. Due to the outstanding high-temperature efficiency, PBI membranes with PA electrolytes have been used. High-temperature guarantees that the correct electrode kinetics are tolerant for fuel impurities such as CO [13, 14].

2 Why high-temperature polymer electrolyte membranes?

Developing alternative membranes that can be operated above 100°C can be a solution for deficiencies related to the low-temperature PEMFC technology based on PFSA membranes, as follows:

(1) Electrical kinetics must be improved for both electrode reactions.

(2) The operation of PEMFCs can be simplified, as it contains only a single water phase above 100°C.

(3) The necessary cooling system is easy to install, and it is also practical as a result of the growing temperature ramp between the coolant and the fuel cell stack.

(4) The overall system efficiency increases significantly due to the heat recovery system.

(5) At 80°C, CO is 10–20 ppm, and CO can rise significantly from 1000 ppm at 130°C, while at 200°C, CO is increased to 30,000 ppm. This high CO resistance enables a hydrogen cell to be directly used by a basic reformator.

(6) For reformation of methanol and hydrogen desorption of new storage material, which has an operating temperature similar to fuel cells, the necessary temperature is approximately 200°C. The fuel cells can either be integrated with a methanol reformer or a hydrogen storage tank with a high volume. PEMFC technology is predicted to be very stable, as well as requiring less maintenance and showing stronger reactions [11].

3 Opportunities and challenges for high-temperature PEMFCs

Due to their low or zero emissions, high energy density, and high energy efficiency, PEMFCs, including direct methanol fuel cells (DMFCs), are considered to be one of the most promising energy conversion systems. Furthermore, there are significant technical challenges in terms of marketing. In particular, expense, inadequate longevity, and versatility are among those challenges. There have been significant attempts to overcome these challenges over the past couple of decades. One approach is to operate PEMFCs above 90°C. The next-generation fuel cell technology for reducing reliability costs in terms of evolved reaction kinetics, catalyst toleration, heat rejection, and water control is the high-temperature proton exchange membrane fuel cell (HT-PEMFCs) controlled at >90°C [15].

4 Properties of composite membranes

Membrane components for fuel cells are critical [16]. While this proton change depends on the efficiency of the fuel cells, it is considered the cost-determining factor [17–19]. One of the most commonly used membranes is PFSA membranes. These membranes, however, have a number of detrimental implications, such as minute humidity and/or maximum saturation, poor conductivity, and max [20, 21]. To overcome these problems, there has been an increasing interest in developing different membranes [22–24]. Five groups are formed using these membranes, depending on the primary chain compounds and functional groups: nonfluorinated acid ionomer membranes [25], PFSA membranes [26], partially fluorinated acid ionomer membranes [27], alkaline ionomer membranes [28, 29], and PBI/H_3PO_4 membranes [30]. The main purpose of the structural modification and the mixing of the present membranes is to increase their ionic conductivity under dry conditions, and especially at high temperatures, to enhance their chemical-mechanical stability [31, 32]. Information on nanocomposite membranes that synthesize and characterize high-temperature PEMFCs are given next.

5 Synthesis and characterization

5.1 Synthesis and characterization of PBI membranes

A novel, high-temperature electrolyte membrane, which was utilized in the fragmented block copolymer of PBIs, was prepared by Benicewicz et al. [33]. The membrane showed very high acid loading and excellent conductivity, as well as the best mechanical properties and fuel cell performance [13, 34]. Compared to using pure hydrogen as a fuel, a resin containing 40% H_2 showed a very modest loss. This is especially notable when considering the resin containing 2000 ppm CO. An O_2 increase of 70 mV is measured at 0.2 A/cm^2 when H_2 is used as a fuel [14].

A similar effect and a similar oxygen gain are noteworthy with the minimal losses associated with reform. Operating with hydrogen and air at 15 psi, the same performance was shown with hydrogen and oxygen at atmospheric pressure.

In one study, a para-oriented PBIs was synthesized in poly(2,-2′-(p-phenylene)-5,5′-bibenzimidazole), polyphosphoric acid in high molecular weights. The membrane produced films with high acid levels using the new sol-gel method, without redissolution or isolation of the polymer. The utilized membrane had good thermal properties, high acid content, good mechanical properties, and high conductivity (0.23 S/cm at 160°C). These acid-doped membrane fuel cells demonstrated excellent performance and durability over a period of more than 2500 h (160°C) [35, 36].

The relationship between the temperature and conductivity of the phosphoric acid-supported p-PBI membrane. For reliable and reasonable results, the second conductivity measurement was performed immediately after the first operation or preheating. The conductivity tested in the second heating flow is about

0.23 S/cm at 160°C, which is much higher than the published data. Cell temperatures ranged between 80°C and 160°C, and the cell was tested on H_2/air. As expected, power densities and performance increased with increasing temperatures. Long-term measurements at ambient pressure using H_2 and air at 160°C revealed a decrease from 0.648 to 0.595 V and 53 mV and a current density of 0.2 A/cm^2 after operating at 160°C for 2500 h [10, 14]. Although companies perform research to find their own polymer electrolyte membranes, the most popular and widely used one is Nafion. Nafion has become the industry standard owing to its perfect proton conductivity, resistance to chemical reactions, good mechanical properties, acidity, and relatively low production costs. Nafion is formed by the perfluronation of polyethylene, which produces the hydrophobic compound polytetrafluoroethylene (PTFE, or Teflon). To produce electrolytic properties, sulfonate side chains are added to the PTFE backbone. The SO_3^- ends of the side chains form hydrophilic regions in hydrophobic material, which allows the membrane to absorb sufficient water to separate the acid groups and make the protons more mobile as they move in the liquid phase. To transport the protons efficiently, the membrane should be very thin, and 50 μm is set to be the lowest limit [14, 37].

At high temperatures, sulfonated poly(arylene ether sulfone)s have adequate conductivity, medium water absorption, and perfect thermal stability. Therefore, it has good potential for use in fuel cells. By adding inorganic structures to these membranes, it can provide high-temperature performance; however, it can increase important properties such as the water absorption capacity of the membrane. Smallparticle size can be obtained by in situ formations of inorganics. In this way, the original membrane structure is preserved. Hill et al. [38] presented a synthesis method for in situ composite membranes consisting of disulfated poly(arylene ether sulfone)/zirconium phenyl phosphonate designed as PEM.

Disulfonated poly(arylene ether sulfonate) and zirconium phenyl phosphonate, as composite membranes, have been synthesized, and the incorporation of an inorganic additive has been shown to reduce water uptake (WU) despite the decrease in conductivity. Thermal stability, measured by thermogravimetric analysis (TGA), has also increased after the addition of ZrPP. The ZrPP amount in the membrane can be controlled by the immersion time and zirconium chloride solution concentration. The composite membrane, containing 3% ZrPP, performed better on pure BPSH-35 and Nafion 1135 [14, 38].

Mishra et al. suggested that "the cost, electrical conductivity and water retention properties of silicate-based nanoparticles compared to other nanoparticles" were studied [39]. Hybrid membrane systems are considered to be a favorable solution to the disadvantage of water-swellable membranes, such as fluorosulfonic acid copolymers. Hybrid membranes, as the organic phase for the preparation of the cation exchange, include chitosan, PBI, sulfonated poly(2,6-dimethyl-1,4-phenylene oxide) (SPPO), and polyvinyl alcohol (PVA); there are many studies using these polymer types. Among these membranes, PVA is a highly preferred polymer due to its good flexibility, low cost, film-forming properties, and water solubility. Furthermore, in the polymer chain, there are C-OH groups that can react with the $-Si(OR)_3$ groups through the sol-gel; that makes PVA the perfect starter product for the production of inorganic-organic hybrid membranes. In this study, the membrane exhibited good mechanical properties, chemical stability, and high proton conductivity. Due to hydroxyl (-OH) groups of PVA-supported membranes, they are not durable in water. Formaldehyde was added to the solution to increase the mechanical properties of the membranes and to increasing the number of crosslinking. Characterization tests were carried out by applying appropriate techniques to these membranes [22].

In the research literature, the most commonly used technique for membrane production is the modified sol-gel technique, which is replaced by the classical sol-gel technique. The technique based on the dissolution of the basic PVA was applied directly to the solution of the basic polymer solution in a solution containing the active groups for the more homogeneous binding of the active groups to the structure. TiO_2 was added to membranes in different ratios (5%, 10%, 15%, 20%, 25%, 30%, and 40%). After the degree of phosphorylation of the base polymer was adjusted to 25%, an appropriate amount of acid and a 2:1 (acid-formaldehyde) ratio of formaldehyde was added to the solution. The most significant issue with this method is the pouring of the membrane before the salt formed by the reaction between phosphoric acid and formaldehyde. Therefore, the duration and temperature of the mixing process after the phosphorylation process are important.

5.1.1 Water uptake capacities

The membranes should have a high-water-uptake capacity for high proton conductivity and low resistance. In this study, water absorption capacity tests were performed at 60°C, 70°C, and 80°C at room temperature. The produced membrane structures were dried at 100°C and weighted after reaching a constant value. They were kept in hot water for 1 day at above-mentioned temperatures. One day later, the membranes were removed from the hot water and separated from the excess water bubble. The water retention capacities of the synthesized membranes were calculated as shown in Eq. (1) [22].

$$\%\text{Water uptake} = \frac{m_{\text{wet}} - m_{\text{dry}}}{m_{\text{dry}}} \times 100 \tag{1}$$

Due to the hydrophobic properties of TiO_2, the water retention ability of membranes is not expected to change with temperature. In the study, the water intake of the membranes did not change much as the temperature increased. The capacity of the membranes reached a stable value with increased additives. The water absorption ability of the TiO_2-containing membrane with hydrophobic properties did not change with increasing temperatures. Thus, the characteristics of the membranes were dominant and the membranes exhibited a constant structure. This condition shows high temperature stability [22].

Water appeared on the cathode side of the membranes. If the membranes are swollen due to water absorption, the protons need to move longer and show higher diffusional resistance. According to the results, the proton conductivity of the membranes is decreased. Membranes must have a high level of WU without swelling [40].

For the swelling properties of the membranes, the WU capacities at 60°C, 70°C, and 80°C were measured. The swelling capacities of the membranes were determined in terms of increased width and size ratio. For swelling tests, membrane samples were prepared by cutting a 2 cm × 2 cm membrane and Sheen Brand Minitest 3100 was employed to measure the membrane thickness. Membranes were remeasured after storage in water. The diameters and surface areas of the membranes were examined in detail, and the results showed that changes in these properties were reduced by increasing temperature.

As the temperature of the prepared membranes increased, the change in diameter and surface areas was observed to decrease. Because the surface area and diameter of the membranes are too small to vary at temperatures of more than 60°C, membranes do not swell at this temperature level [22].

5.1.2 Ion exchange capacity

The ion exchange capacity (IEC) is known as the measurement of conductivity of proton [41, 42]. The membranes produced were immersed in 30 mL or 0.1 N NaOH solution at isothermal temperatures for a period of time. Afterward, membranes were discharged from the NaOH solution. The titration process was performed with 0.1 N HCl solution. Eq. (2) was used to find ion exchange capacities as follows:

$$IEC = \frac{(N_{NaOH} \times V_{NaOH}) - (N_{HCl} \times V_{HCl})}{m_{membrane}} \tag{2}$$

5.1.3 Electrochemical impedance spectroscopy

The most selective and best property of the membranes is the presence of proton conductivity [43]. Solartron 1260, 1287 new test cells were used to calculate the membrane proton conductivity. The impedance spectroscopy was conducted using two probe methods. The frequency was set as 1 MHz to 100 Hz, humidity was 100% and at room conditions. After this characterization, the resistances of the membranes were obtained. Eq. (3) indicates that resistivity of membranes were used to calculate the proton conductivity:

$$\alpha = \frac{t_{membrane}}{RA} \tag{3}$$

All the proton conductivity results acquired at different temperatures are compatible with the Arrhenius equation. Eq. (5) was acquired through the linearization of Eq. (4), and was used to calculate the activation energies of membranes:

$$\alpha = \alpha_0 e^{-Ea/RT} \tag{4}$$

$$\ln \alpha = \ln \alpha_o - Ea/RT \tag{5}$$

The proton conductivity of the TiO_2-doped membranes changed with the temperature. According to the proton conductivity results, the mobility of protons in TiO_2 doped membranes greatly increased.

Wu et al. [44] conducted a study on different additive rates, up to 15%. The proton conductivity of the membranes was investigated by using greater additive ratios. It was observed that TiO_2 acts as a support substance and the active substance hypophosphoric acid increases its properties by 20%. According to the results obtained from Fenton measurements, when the contribution rate is higher than 20%, it causes negative effects on the membrane. Yang et al. [45] indicated that with the addition of TiO_2 up to 10% PVA, and using polystyrene sulfonic acid as an active group, a constant reduction in proton conductance occurred. The value of maximum proton conductivity was found as 0.00354 S/cm at 70°C. We compared the data shown here with the data obtained in that study and found that it was significantly less. The proton conductivity of the TiO_2 membrane increased in direct proportion to the temperature. The membranes, which doped at 15% TiO_2, showed higher proton conductivity than others [22].

5.1.4 Fourier transform infrared analyses

The structural properties of membranes were analyzed using Jasco FTIR 480+ [46]. The highest peak observed between 3100 and 3600 cm^{-1} in the Fourier transform infrared (FTIR) spectrum of TiO_2-supported membranes belongs to OH. The C=O expansion peak was observed at 1700 cm^{-1}, which is caused by the ester functional groups. According to a number of studies, the peak belonging to Ti nanoparticles is expected to be seen at 474 cm^{-1}. However, this peak was observed at a wavelength of 500 cm^{-1}. The increasing amount of additives caused an increase in the density of the peaks. As can be seen from the absorbance data of the peaks, it is lower than the peak in the organic structure; the spectra obtained from the peaks are not significantly visible.

By increasing the amount of support materials, the mechanical stability values showed a decrease in the tensile strength of the membranes. The tensile strength of the Nafion membrane was measured by Na et al. [47] and found to be 37 and 180 MPa to the Young's modulus. Commercially obtained Nafion membrane values were measured at 36.82 and 176 MPa. Li et al. [11] synthesized PBI-based membranes and determined that the maximum Young's modulus value was 550 MPa. Tang et al. [48] found that Young's modulus value of the membrane rising by WU capacity was reduced. The results indicate a directly proportional relationship between Young's modulus values and the WU capacities [22].

5.1.5 Fenton tests

To define the oxidative stability of the membranes, Fenton experiments were carried out. The test weight loss assessment was administered at intervals of 20 h. The oxidative stability of the additive samples increased in direct proportion to the increasing amount of additive, up to 15% of the amount of support material. The membrane encoded as 15T showed a similar durability to the Nafion membrane. However, for additional rates of more than 15%, the membrane underwent rapid dissolution in the oxidative environment. Ghassemzadeh et al. [49] conducted a Fenton test on Nafion 117 membrane and observed no reduction in the density of peaks in nuclear magnetic resonance spectrums [22].

In conclusion, PVA-based membranes with different ratios (5%, 10%, 15%, 20%, 25%, 30%, and 40%) of hypophosphoric acid and TiO_2 were synthesized. The ion exchange capacity values remained unchanged, which is a constant beyond a certain contribution rate. There was a decrease of 20% in proton conductivity values. This is because the detection of Fenton test results is the predominant contributor. According to the results of the FTIR spectrum, the support material could be added to the membrane in a positively way. Mechanical stability measurements showed that the membranes had the desired level of mechanical stability. Characterization tests showed that the 15% TiO_2-doped membrane had better qualities than other membranes. For instance, this membrane has a proton conductivity of 0.023 S/cm at room temperature, an ion exchange capacity of 1.04 meq/g, and a water absorption capacity of 45%. The structural properties of the membrane show better proton conductivity than other membranes and are 0.030 S/cm under better fuel cell conditions. It is concluded that the 15T coded membrane is promising for the utilization of electrolytes in fuel cells [22]. Şahin and Ar obtained an organic/inorganic membrane, which enhanced the quality compared to the Nafion membrane. Therefore, phosphonate PVA support membranes were synthesized. For these membranes, various characterization tests, such as determination of ion exchange power and WU power, tensile capacity, Fenton measurements, electrochemical impedance spectroscopy FTIR, and mechanical tests, were performed. The results showed that 15% TiO_2-doped membrane has better properties, such as highest proton conductivity (0.03 S/cm), ion exchange capacity (1.04 meq/g), and water intake (45%). It has been concluded that any membrane having these properties is appropriate for fuel cell applications [22].

In recent years, higher-temperature PEMFCs have been extensively studied for the purpose of developing newer proton conductive membranes [50]. Nafion, a leading, conventional electrolyte proton polymer membrane, is cost effective, mechanically brittle, and conductible only when immersed in water at temperatures greater than 100°C. The higher-temperature operation of fuel cells has advantages in many systems, such as reduction in the size of heat exchangers and ease of incorporation with reformers. Therefore, the development of membranes that are stable in physical and chemical properties at higher temperatures (above 100°C) and for the production of low-cost fuel cells is a promising research area. It includes the modification of polymers, the assembly of various inorganic hygroscopic nanoparticles, and the design of next-generation polymer systems [6]. Malhotra and Datta [51] introduced the addition of inorganic solid acids to

Nafion-like standard polymeric ion exchange membranes, as well as assisting the binary functionalities of water-holding capacity and providing acidic areas. Thus, they are able to supplement Nafion membranes, which contain heteropolyacids that exhibit high fuel cell performance at high temperatures (120°C) and low relative humidity (RH). Ramani et al. [22, 52] recently demonstrated that heat treatment can be used in the stabilization of Nafion-phosphotungstic acid (PTA) membranes. Thus, PTA deformation can be eliminated [13]. Nafion can be replaced by the addition of hygroscopic oxides like TiO_2 and SiO_2 or inorganic powder acids such as ZrO_2/SO_4, to increase WU capacity. Based on this analysis, which was carried out on higher-temperature membranes, an expression strategy can be inferred. Either a polymer solution mixture and an inorganic powder or colloidal mass in an in-pocket state or the creation of inorganic sections in which the membrane is used as a pattern is extremely promising for nanocomposite membranes. The advantage of the in situ technique is that the particle size can be controlled by the concentration of the precursors. In situ techniques are generally based on solar-gel reactions in membrane pores between organometallic and water. This study demonstrates the potential for synthesizing nanocomposite membranes using a sol-gel structure with increased hydration, along with high hydration and conductivity at low RH conditions. In this study, Nafion/TiO_2, Nafion/ZrO_2, and Nafion/SiO_2 nanocomposite membranes were synthesized by the in situ sol-gel technique. In different RH conditions, WU is compared to Nafion, which remains constant for proton conductivity. These nanocomposite membranes have also been characterized using TGA and dynamic mechanical analysis (DMA) to detect deformation and glass transition temperatures (T_g) [53].

6 Synthesis and characterization of nanocomposite membranes via the sol-gel method

ZrO_2, SiO_2, and TiO_2 were used for the preparation of composite by in situ sol-gel synthesis methods in PEMs. In this procedure, the nanocomposite that was obtained was indicated to be PEM Nafion-ZrO_2 sol-gel. Similarly, Ti (IV) tertbutoxide and tetraethyl orthosilicate (TEOS) were used as precursors in the synthesis of Nafion-TiO_2 sol-gel and Nafion-SiO_2 sol-gel membranes, respectively. The membrane prepared through this technique is homogeneous and completely transparent compared to the membranes

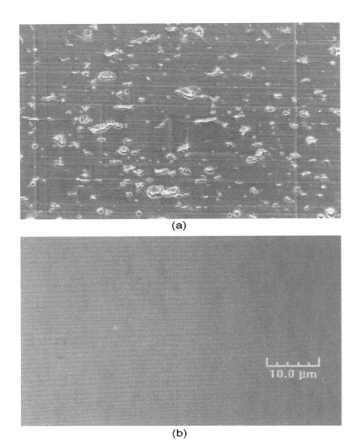

Fig. 1 SEM images of membranes synthesized using the doping method and in situ. (A) Nafion membranes that are modified with zirconium dioxide (ZrO$_2$). (B) The phase separation of the polymer electrolyte membrane (PEM) used for zirconium dioxide (ZrO$_2$)-modified Nafion sol-gel, which is transparent and homogeneous [53].

synthesized by the previous casting techniques due to the presence of large particles. Transmission electron microscopy (TEM) images for the membranes synthesized by two techniques are shown in Fig. 1. The membrane synthesized by the casting technique varies in the range of 5–15 μm. Sol-gel membranes did not exhibit X-ray scattering, and they also had larger Zr particles. The structure of the sol-gel membranes did not exhibit any confirmatory oxidation of Zr in response to the presence of the Nafion membrane in the pores. These membranes are thought to have nanosized zirconia particles in the membrane pores [53, 54].

6.1 Water uptake measurements

The tapered element oscillating microequilibrium was used to test the WU of compound PEMs compared to the unmodified Nafion membrane. The frequency difference of the tapered

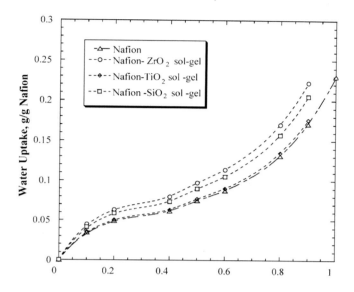

Fig. 2 WU versus the activity of water vapor for nanocomposite Nafion-MO$_2$ and Nafion membranes at 90°C [53].

element oscillating bed was taken and the change on sample mass of tapered element oscillating microbalance (TEOM) was tested. The RH was checked by mixing the measured flows with stream of dry helium. Calibration was performed with an RH meter. The membrane was divided into thin strips (measuring 1.5 mm × 1.5 mm). Quartz wool was put in the oscillatory TEOM test bed to vibrate the membranes. WU was measured at RH values from 0% to 90% at 90°C and 0% to 40% RH at 120°C for all the samples. After loading each sample, helium gas with suitable RH was exposed, and it was observed that the real-time mass change was determined when the equilibrium amount of water on the membrane was adsorbed [6, 55]. Figs. 2 and 3 show the WU results of the nanocomposite membranes at 90°C and 120°C, respectively. All the Nafion-MO$_2$ nanocomposites showed better WU than the unmodified Nafion membrane at the given RH and temperatures. The Nafion-ZrO$_2$ nanocomposite exhibited higher WU than the Nafion membrane. The WU was measured at about 33% and 45%, at 90°C and 120°C, respectively [53].

6.2 Ion-exchange capacity measurements

A sample of 0.2 g of PEM was immersed into 1 M CH$_3$CO$_2$NH$_4$ for a day, and NH$_4^+$ in NH$_4$Cl for 1 h more. The replacement of NH$_4^+$ with Na$^+$ was achieved with the addition of 2 mL of 5 M NaOH solution. Using a calibrated ammonia electrode, it is considered that the amount of NH$_4^+$ released is exactly measurable, and thus, it gives a correct measure of the ion exchange capacity [13, 53].

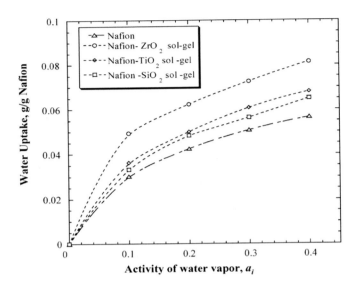

Fig. 3 WU versus the activity of water vapor (a_i) for Nafion membranes and nanocomposite Nafion-MO$_2$ at 120°C [53].

6.3 Ex situ conductivity testing

Solectron SI 1260 FRA was used for conductivity measurements. The tests were performed under 10 mV voltage and at a frequency range of 0.01–106 Hz. The z-axis provided true intersection, membrane strength, and conductivity. A composite membrane sample was placed in a humidity-controlled chamber to measure the conductivity between two electrodes on both sides. The humidity of the chamber was monitored using a dew-point/temperature probe. An air stream was saturated with water by bubbling it through a humidifier [56]. In room conditions, for reaching the targeted partial water pressure, the temperature was set to 90°C and 120°C, respectively. The conductivity of PEM was tested at 10%–90% RH at 90°C, while the RH range at 120°C was similar to dry conditions between 10% and 40% and was the same as that used in WU tests [53, 57].

The selected type was used as a one-sided ELAT gas diffusion electrode. Using software, the feed gas mass flow rate was programmed to the stoichiometry-dependent flow rates. The gas pressure of the reagents was monitored by manometers [53, 58]. Fig. 4 shows fuel cell performance for four membranes in humidified environments at 80°C, and Fig. 5 shows the same in dry conditions (Tcell: 110°C; humidifiers: 80°C). Fig. 6 shows the fuel cell activity of the Nafion-ZrO$_2$ nanocomposite membrane at Tcell of 135°C, and Thumidifiers of 80°C and 90°C. The Nafion-ZrO$_2$ nanocomposite membrane exhibited much better proton conductivity

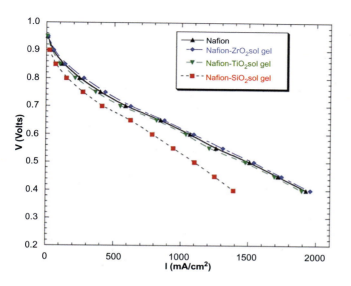

Fig. 4 The performance results for Nafion 112 MEA versus Nafion-MO$_2$ sol-gel, composite MEA. Tcell = 80°C. THumidifier = 80°C [53].

than Nafion under the same conditions—the result of better water intake capacity, stronger acid regions, and higher surface water content. Therefore, compared to other membranes, a slight improvement on the fuel cell performance at 80°C was observed for the Nafion-ZrO$_2$ sol-gel nanocomposite. However, Nafion-SiO$_2$, in contrast to the results reported in the literature, subsequently showed a lower performance than Nafion. At 110°

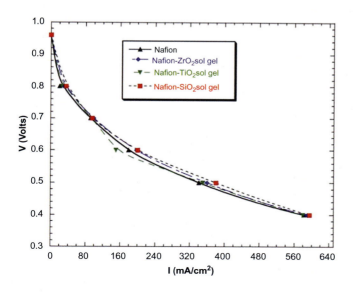

Fig. 5 The performance results for Nafion 112 MEA versus Nafion-MO$_2$ sol-gel. Composite MEA. Tcell = 110°C, THumidifier = 80°C [53].

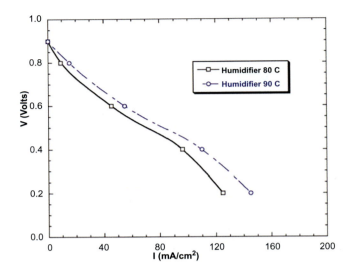

Fig. 6 The performance results for Nafion-ZrO$_2$ sol-gel composite MEA. Tcell = 135°C (*circle*: THumidifier = 90°C; *square*: THumidifier = 80°C) [53].

C, TiO$_2$ and ZrO$_2$ membranes have produced a flow of about 30–40 mA/cm^2 at 0.5 V compared to Nafion membranes. SiO$_2$-based membranes showed a similar performance to Nafion, as seen in the graph. At 135°C, current densities for ZrO$_2$ membranes were obtained under dry and hot conditions.

6.4 Conductivity measurements

Figs. 7 and 8 indicate conductivity measurements at 90°C and 120°C for nanocomposite membranes, respectively. Nafion-ZrO$_2$ sol-gel-nanocomposite demonstrated better conductivity than Nafion in two trials during water operation. Both Ti and Si nanoparticles showed slightly less conductivity than Nafion at 90°C, while Zr membranes showed a significant increase in water activity. In contrast with Nafion, Nafion-SiO$_2$ sol gel nanopartite demonstrated poorer conductivity. However, Zr membranes indicated approximately 8%–10% higher conductivity at 120°C. The other nanocomposite showed higher WU performance. However, acidity was similar and conductivity was less with Nafion. Therefore, conductivity, water intake, and ion exchange capacity results showed that higher water intake leads to higher conductivity [53].

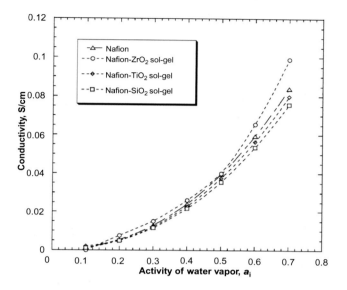

Fig. 7 Water vapor and conductivity activity of Nafion and Nafion at MO_2 membranes in a nanocomposite structure at 90°C [53].

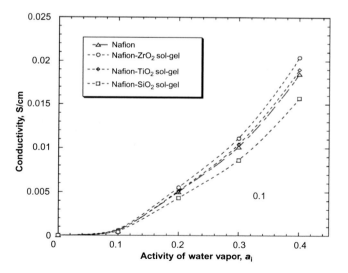

Fig. 8 Water vapor and conductivity activity of Nafion and Nafion at MO_2 membranes in a nanocomposite structure at 120°C [53].

6.5 Thermomechanical characterization

TEM was employed to determine the morphology of the prepared nanocomposite membranes. TG measurement of nanocomposite membranes was carried out using Thermal Analysis Instruments 2050. The temperature interval was from 25°C to 700°C, and the heating rate was 20°C/min. A DMA device

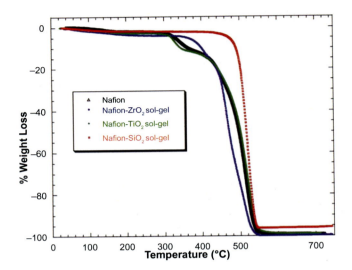

Fig. 9 Thermogravimetric analysis graph of Nafion at MO_2 and Nafion nanocomposite membranes [53].

(Thermal Analysis 2980) was used to determine the glass transition temperature of nanocomposite membranes; the temperature range was from 25°C to 175°C [53]. TG thermograms of nanocomposite are shown in Fig. 9 relative to the Nafion membrane. Fig. 9 also shows how to sustain a temperature of around 310°C in about 90% of the weight of all membranes. Around 310°C, all the membranes decoupled rapidly and started to drop in weight. The temperature of the dissociation varied depending on the structure of the inorganic support material in the pores of the Nafion membrane. The TiO_2 membranes do not show much improvement compared to Nafion in terms of the thermal decomposition temperature. The DMA thermogram for an unmodified Nafion membrane nanocomposite membrane is shown in Fig. 10; the TGs of all nanocomposite membranes appear to shift to higher temperatures.

Furthermore, T_g appears to be increasing for the nanocomposite. Here, it shows the thermomechanically excellent stability from an unmodified Nafion membrane. The data from TGA and DSC for fuel cell high-temperature studies show an improved potential performance of nanocomposite membranes. However, all nanocomposite membranes should be verified in the PEMFC system for better thermal-mechanical properties and stability of nanocomposite membranes in the Nafion membrane [59].

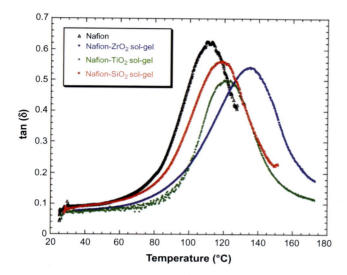

Fig. 10 DMA graph of Nafion at MO₂ and Nafion nanocomposite membranes [53].

7 Characterization and synthesis of the polybenzimidazole-base membranes

One of the most substantial polymers used in the production of membranes for HT-PEMFCs is PBI. The preparation methods include polymerization from PPA [60], methane sulfonic acid [61], and microwave-assisted organic synthesis [62, 63].

The most commonly used PBI membrane as an electrolyte for HT-PEMFCs is poly [2,20-(*m*-phenylene)-5,5-(benzimidazole)], which is shown in Fig. 11 and labeled as PBI_4N. A product obtained by reacting 3,30-diaminobenzidine and dicarboxylic acid contain two basic nitrogen units.

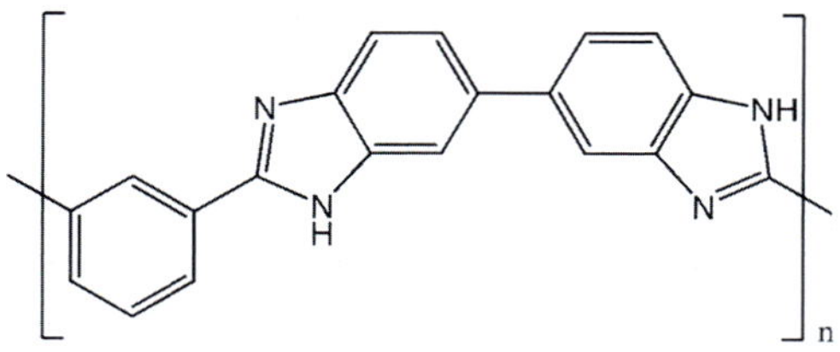

Fig. 11 Chemical structure of poly[2,20-(*m*-phenylene)-5,50-(bibenzimidazole)] [63]. Reprinted (adapted) with permission from E. Quartarone, S. Angioni, P. Mustarelli, Polymer and composite membranes for proton-conducting, high-temperature fuel cells: a critical review, Materials (Basel) 10 (2017) 687. doi:10.3390/ma10070687. Copyright (2020) American Chemical Society.

It is possible to modify the physical and chemical properties of the final membrane to change the chemical structure of the starting monomers, their mechanical properties, thermo-oxidative stability, polymer backbone, polymer solubility, and properties of methanol permeability [16, 25, 28]. PBIs can be crosslinked to enhance the mechanical strength of doped membranes. Various types of membrane crosslinking processes were subjected to covalent, ionic covalent, and mixed ionic covalent bonding. However, it is observed that if the membrane toughness increases, the polymer reduces its solubility in standard cast solvents [15, 38, 63].

In the investigation of new nanocomposite proton conductive membranes, SnO_2 ceramic powders with surface functionality were synthesized as an additive in Nafion-based polymer systems. Various synthetic routes were investigated to obtain suitable nanometer-sized sulfated tin oxide particles. N_2 adsorption and TGA was analyzed by using Fourier transform infrared (FTIR), X-ray diffraction (XRD), scanning electron microscopy (SEM), and Raman spectroscopy. Atomic force microscopy (AFM), dynamic mechanical analysis (DMA), water uptake (WU) analyses, thermal tests, and ionic exchange capacity (IEC) experiments were used for nanocomposite membranes. The structure of the tin oxide precursor and the preparation technique play an important role in the determination of the ceramic solid structure and the particle size distribution [63–65].

7.1 X-ray diffraction analysis

Both bare and sulfated XRD models of synthesized ceramic samples are given in Fig. 12. The XRD models of all materials are consistent with the tetragonal, rutile structure of tin oxide in bulk. For all the synthesized samples, there were large ranges of peaks indicating that the particles are in the nano size range. Calculation of average crystallite sizes ($<d>$) is based on the standard expansion profile. According to the obtained data, the values were 1.7 nm for the F110 sample, 2.8 nm for the F110S sample, 5.0 nm for the E400 sample, and 6.1 nm for the E400S sample, respectively. It can be seen that the sizes of the crystallized oxides in the E400S and F110S cases are larger than those of the naked, untouched oxide (E400 and F110, respectively) [65].

7.2 Thermal gravimetric analysis (TGA)

The sulfate functional groups amount, which is chemically bound to the surfaces, is detected with thermal gravimetric (TG) tests, as shown in Fig. 13.

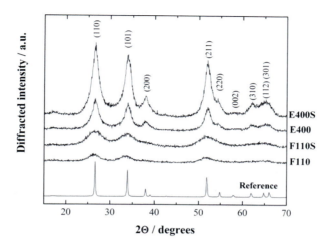

Fig. 12 XRD graph of sulfate-modified and nonsulfated samples for compounds with SnO$_2$ [66]. Reprinted (adapted) with permission from R. Scipioni, D. Gazzoli, F. Teocoli, O. Palumbo, A. Paolone, N. Ibris, S. Brutti, M.A. Navarra, Preparation and characterization of nanocomposite polymer membranes containing functionalized SnO2 additives, Membranes (Basel) 4 (2014) 123–42. doi:10.3390/membranes4010123. Copyright (2020) American Chemical Society.

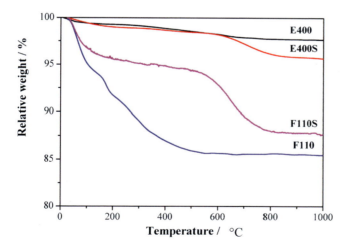

Fig. 13 TGA results for a synthesized four-oxide compound [66]. Reprinted (adapted) with permission from R. Scipioni, D. Gazzoli, F. Teocoli, O. Palumbo, A. Paolone, N. Ibris, S. Brutti, M.A. Navarra, Preparation and characterization of nanocomposite polymer membranes containing functionalized SnO2 additives, Membranes (Basel) 4 (2014) 123–42. doi:10.3390/membranes4010123. Copyright (2020) American Chemical Society.

The first mass loss was seen in all the samples before 250°C at different thermal ranges. The largest weight loss was observed for the F110 sample. A second weight loss was observed at a temperature range of 400–500°C for the E400 sample, which was thought to be the result of residual organic traces belonging to the organometallic precursor. Another loss of mass was observed for only two sulfated samples, E400S and F110S, above 600°C, which was thought to be due to the removal of sulfate groups attached to the surface of tin oxide particles. The loss of mass for the E400S sample was about 3%, while a loss of 7% was observed for F110S. Compared to E400S, the concentration of the larger, sulfate groups in F110S was parallel to the smaller, crystalline size of the intact F110 naked oxide before sulfating. According to the findings, the functional sulfate group is chemically dissolved on the surface of SnO_2 and increased by nanometricization of the oxide particles [65].

7.3 Scanning electron microscopy images

The scanning electron microscopy (SEM) images at 20k magnifications for the four oxides are shown in Fig. 14.

Nanometric crystallites aggregation causes the formation of large particles, which can be measured by X-ray powder diffraction data. This image is confirmed by the surface areas, the measurements are $100 \pm 1\,m^2/g$ (F110S), $160 \pm 5\,m^2/g$ (F110), $52 \pm 3\,m^2/g$ (E400S), $65 \pm 2\,m^2/g$ (E400). The findings are parallel to the diameters of spherical particles measured for the F110S, F110, E400S, and E400 samples in the range of 9, 5, 17, and 13 nm, respectively. It has been observed that the sulfated E400S sample has about half the surface area of the 110S sample. This discrepancy constitutes roughly two layers of the F110S sulfate group compared with the E400S sample [66].

7.4 Fourier transform infrared analysis

FTIR spectra are given in Fig. 15, in which three key features are monitored. The samples indicate a large composite chain, which is due to the water molecules of the surfaces, and the bending modus OH groups, according to the vibration mode of the Sn-O-SAn oxide grid, which is about $1620\,cm^{-1}$ and less than $800\,cm^{-1}$ [66].

7.5 Raman spectroscopy

Fig. 16 shows the results of Raman spectroscopy of the F110S and F110 samples obtained from the surface vibration modes. For comparison, the spectrum of bulk SnO_2 has also been illustrated.

The F110 Raman range basically consists of a wide, solid band centered about $570\,cm^{-1}$ and a low-density spectrum at 770, 640, 435, and $350\,cm^{-1}$. The measurements for the four first-order

Chapter 14 Synthesis and characterization of nanocomposite membranes **273**

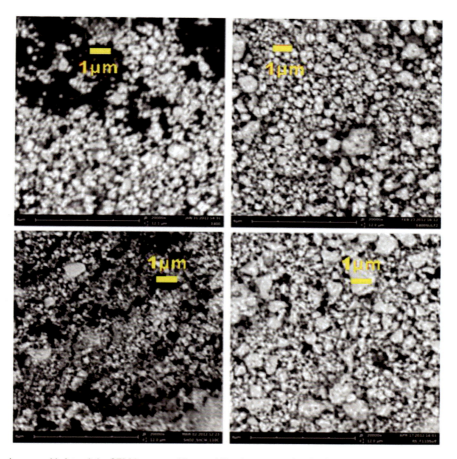

Fig. 14 *Top-down and left to right:* SEM images of four oxidized compounds obtained from F110, F110S, E400, and E400S samples, respectively [66]. Reprinted (adapted) with permission from R. Scipioni, D. Gazzoli, F. Teocoli, O. Palumbo, A. Paolone, N. Ibris, S. Brutti, M.A. Navarra, Preparation and characterization of nanocomposite polymer membranes containing functionalized SnO2 additives, Membranes (Basel) 4 (2014) 123–42. doi:10.3390/membranes4010123. Copyright (2020) American Chemical Society.

Raman active modes (Eg, B1g, B2g, A1g) are 479 cm^{-1} (Eg), 100 cm^{-1} (B1g), 779 cm^{-1} (B2g), and 638 cm^{-1} (A1g), for crystalline SnO$_2$ (with a rectangular, rutile structure) estimated according to factor group analysis. Bands between 570 and 350 cm^{-1} are based in compliance with the findings of XRD and morphological analyses on the interface or surface finance modes of nerostructured tin-oxide material.

274 Chapter 14 Synthesis and characterization of nanocomposite membranes

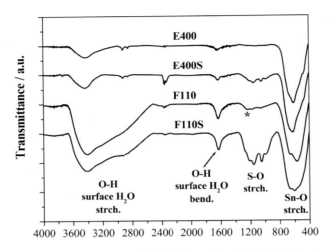

Fig. 15 FTIR spectra of the sulfated and bare oxides. [*] Bending mode of Sn-OH [66]. Reprinted (adapted) with permission from R. Scipioni, D. Gazzoli, F. Teocoli, O. Palumbo, A. Paolone, N. Ibris, S. Brutti, M.A. Navarra, Preparation and characterization of nanocomposite polymer membranes containing functionalized SnO2 additives, Membranes (Basel). 4 (2014) 123–42. doi:10.3390/membranes4010123. Copyright (2020) American Chemical Society.

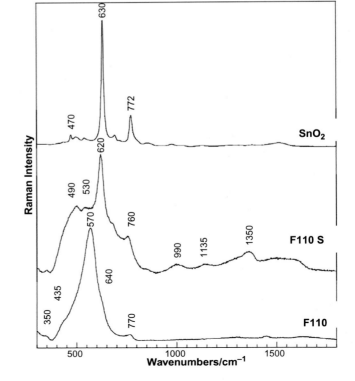

Fig. 16 Raman spectra of F110S and F110 samples given for the SnO$_2$ compound [66]. Reprinted (adapted) with permission from R. Scipioni, D. Gazzoli, F. Teocoli, O. Palumbo, A. Paolone, N. Ibris, S. Brutti, M.A. Navarra, Preparation and characterization of nanocomposite polymer membranes containing functionalized SnO2 additives, Membranes (Basel). 4 (2014) 123–42. doi:10.3390/membranes4010123. Copyright (2020) American Chemical Society.

While waiting for a transition to lower fluctuations by reducing the particle size, the powerful band at 570 cm^{-1} may be related to the A1g mode of the tin-oxide structure, while about 435 cm^{-1} can consist of detectable bands of about 770 cm^{-1}, and 640 cm^{-1} is based on the normal internal phonon modes of the bulk tin-oxide structure. For the F110 sample, the sulfation procedure provided a robust change in the Raman spectra. Furthermore, the changes in the structure of tin oxide and the presence of sulfate species in the range of 930–1600 cm^{-1} in the F110S sample, which appeared in the 300–800 cm^{-1} dimensional region, were determined. The surface is under the entire region, representative of phonon modes. The various bands, with a range of 900–1500 cm^{-1}, approximately 1360, 1150, and 1000 cm^{-1}, reveal the existence of various surface sulfate types [66].

7.6 Thermal analysis

As expected from the TGA test results, two major weight reductions were observed between 200°C and 300°C. In the first step, water was removed from the polymer matrix. The composite membranes, which have sulfated oxide fillers, N-F110S, and N-E400S, showed slightly less weight loss than membranes with nude oxides, N-E400, and N-F110, respectively, as shown in Fig. 17 [66].

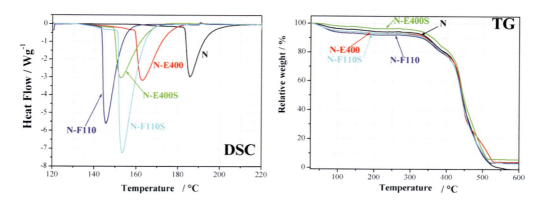

Fig. 17 Results obtained from DSC and TGA of various membrane samples [66]. Reprinted (adapted) with permission from R. Scipioni, D. Gazzoli, F. Teocoli, O. Palumbo, A. Paolone, N. Ibris, S. Brutti, M.A. Navarra, Preparation and characterization of nanocomposite polymer membranes containing functionalized SnO2 additives, Membranes (Basel) 4 (2014) 123–42. doi:10.3390/membranes4010123. Copyright (2020) American Chemical Society.

The DSC experiments indicate the expected wide endothermic peaks for all samples between 140°C and 190°C. The subsequent thermal effects are due to ionic disruptions in the polymer membrane of Nafion. With the addition of inorganic support materials, all the membranes showed lower transition temperatures and greater enthalpy changes than the membrane of the Nafion criteria. Therefore, the addition of tin oxide-based inorganic support substances causes a water increase in the composite membrane structure. The N-F110 sample has more water content than the others, which is in conformity with TGA data and WU results [66].

7.7 Atomic force microscopy analysis

AFM topograms of the five membranes in dry conditions are shown in Fig. 18. Every specimen is compared to the sizes of the various round nanoparticles obtained from the images of the five AFM micrographs of different sizes [66].

7.8 Energy distributions, mechanics, and storage modules of nanocomposite membranes

Next, the heating was tested starting from the cooling "wet membrane" state. The storage modulus (E) of the N-F110, N-F110S, N-E400, and N-E400S membranes and their energy distributions (tan δ) are shown in Fig. 19 [66].

The heating operation initiated from −160°C, and alpha relaxation was around 110°C in F110 and E400; however, it changed slightly at the higher temperature of 120°C for the E400 sample. The peak added to the alpha relaxation is tested on all membranes (as shown on the right side of the figure) at approximately 80°C, and it shows thermal hysteresis [66].

For the heating in the range of −90°C to 30°C, all nanocomposite membranes represent the upper surface, which is wide with the rise of the module. These criteria were observed in a wet, unfolded Nafion membrane. The fluid attached to the walls of the hydrophilic areas is created by the friction between the water in the channels and the solid ice [66].

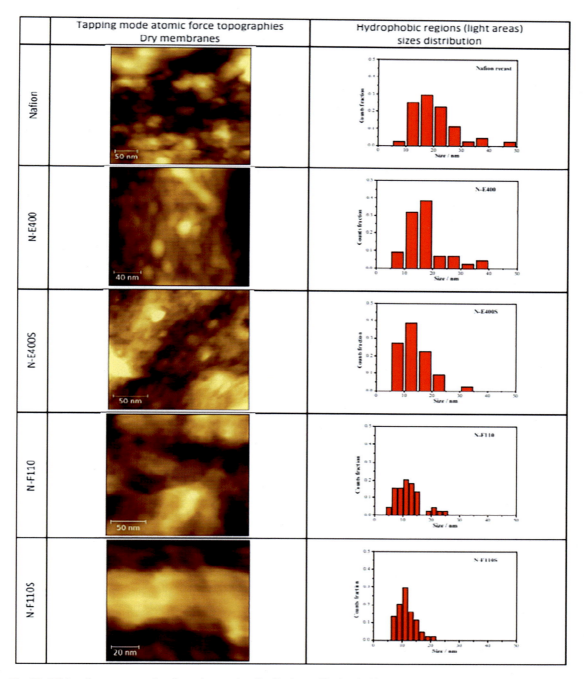

Fig. 18 AFM surface topography of membrane size distributions of hydrophobic areas obtained from images taken from five dried membrane touch mode images [66]. Reprinted (adapted) with permission from R. Scipioni, D. Gazzoli, F. Teocoli, O. Palumbo, A. Paolone, N. Ibris, S. Brutti, M.A. Navarra, Preparation and characterization of nanocomposite polymer membranes containing functionalized SnO2 additives, Membranes (Basel). 4 (2014) 123–42. doi:10.3390/membranes4010123. Copyright (2020)

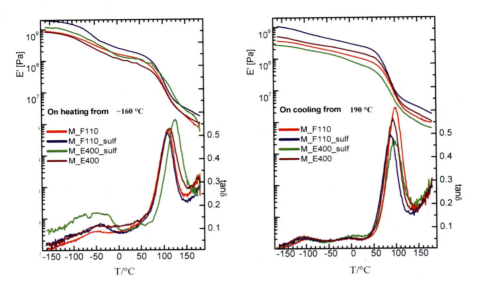

Fig. 19 The nanocomposite tested in the storage model with $f = 1$ Hz is cooling at $-160°C$ and elastic energy distribution graphs are generated during heating at $190°C$ [66]. Reprinted (adapted) with permission from R. Scipioni, D. Gazzoli, F. Teocoli, O. Palumbo, A. Paolone, N. Ibris, S. Brutti, M.A. Navarra, Preparation and characterization of nanocomposite polymer membranes containing functionalized SnO2 additives, Membranes (Basel) 4 (2014) 123–42. doi:10.3390/membranes4010123. Copyright (2020) American Chemical Society.

8 Conclusion

In this chapter, nanocomposite ceramic-doped polymer membranes were explored and characterized. The synthesis of sulfated surfaces and nonsulfated surfaces and highly hydrophilic, nanosized tin oxide fillers was performed. Two types of nanometric tin oxide in crystallite size and surface area were produced with different precursors. Both substances were sulfated with a special technique to obtain superacidic tin oxide nanoparticles. All the inorganic materials were obtained in the form of the Nafion matrix to create the required nanocomposite, hybrid membrane structures. The particle size of the support material and the sulfate ratio were used to determine the degree of hydration that determines the final properties of membranes. The proton exchange is extremely promising for applications of the proposed systems as sophisticated electrolytes in membrane fuel cells [34].

References

[1] S. Eris, Z. Daşdelen, Y. Yıldız, F. Sen, Nanostructured polyaniline-rGO decorated platinum catalyst with enhanced activity and durability for methanol oxidation, Int. J. Hydrog. Energy 43 (2018) 1337–1343, https://doi.org/10.1016/j.ijhydene.2017.11.051.

[2] N. Kızıltas, Y. Karatas, M. Gulcan, S.D. Mustafov, F. Sen, Hydrogen generation by hydrolysis of NaBH4 using nanocomposites, in: Nanomater. Hydrog. Storage Appl, Elsevier, 2021, pp. 231–248, https://doi.org/10.1016/b978-0-12-819476-8.00020-7.

[3] B. Çelik, G. Başkaya, H. Sert, Ö. Karatepe, E. Erken, F. Şen, Monodisperse Pt(0)/DPA@GO nanoparticles as highly active catalysts for alcohol oxidation and dehydrogenation of DMAB, Int. J. Hydrog. Energy 41 (2016) 5661–5669, https://doi.org/10.1016/j.ijhydene.2016.02.061.

[4] E. Kuyuldar, S.S. Polat, H. Burhan, S.D. Mustafov, A. Iyidogan, F. Sen, Monodisperse thiourea functionalized graphene oxide-based PtRu nanocatalysts for alcohol oxidation, Sci. Rep. 10 (2020) 7811, https://doi.org/10.1038/s41598-020-64885-6.

[5] O. Alptekin, B. Sen, A. Savk, U. Ercetin, S.D. Mustafov, M.F. Fellah, F. Sen, Use of silica-based homogeneously distributed gold nickel nanohybrid as a stable nanocatalyst for the hydrogen production from the dimethylamine borane, Sci. Rep. 10 (2020) 1–12, https://doi.org/10.1038/s41598-020-64221-y.

[6] N.H. Jalani, Development of Nanocomposite Polymer Electrolyte Membranes for Higher Temperature PEM Fuel Cells, Worcester Polytechnic Institute, 2006.

[7] H. Wang, H. Li, X.-Z. Yuan, PEM Fuel Cell Failure Mode Analysis, 2011, https://doi.org/10.1201/b11112.

[8] S. Swier, M.T. Shaw, R.A. Weiss, Design of polymer blends for proton-exchange membranes in fuel cells, Preprint Paper Am. Chem. Soc. Div. Fuel Chem. 49 (2) (2004) 532–533.

[9] A. Ersöz, A dynamic simulation study of a small-scale hydrogen production system for a high temperature proton exchange fuel cell, in: Prog. Exergy, Energy, Environ, Springer International Publishing, Cham, 2014, pp. 913–926, https://doi.org/10.1007/978-3-319-04681-5_87.

[10] J. Escorihuela, Ó. Sahuquillo, A. García-Bernabé, E. Giménez, V. Compañ, Phosphoric acid doped polybenzimidazole (PBI)/zeolitic imidazolate framework composite membranes with significantly enhanced proton Conductivity under low humidity conditions, Nanomaterials (Basel) 8 (2018), https://doi.org/10.3390/nano8100775.

[11] Q. Li, R. He, J.O. Jensen, N.J. Bjerrum, Approaches and recent development of polymer electrolyte membranes for fuel cells operating above 100°C, Chem. Mater. 15 (2003) 4896–4915, https://doi.org/10.1021/cm0310519.

[12] H. Zhang, R. Chen, L.S. Ramanathan, E. Scanlon, L. Xiao, E.W. Choe, B.C. Benicewicz, Synthesis, characterization and fuel cell performance of poly(2,2′-(P-Phenylene)-5,5′-Bibenzimidazole) as a high temperature fuel cell membrane, Fuel Chem. 49 (2004) 588–589.

[13] E. Scanlon, Polybenzimidazole Based Segmented Block Copolymers for High Temperature Fuel Cell Applications, 2005.

[14] J.A. Schaeffer, L.-D. Chen, J.P. Seaba, A Simulink model for calculation of fuel cell stack performance, ACS Div. Fuel Chem. 49 (2004) 787–788.

[15] C. Song, S. (Rob) Hui, J. Zhang, High-temperature PEM fuel cell catalysts and catalyst layers, in: PEM Fuel Cell Electrocatal. Catal. Layers, Springer London, London, 2008, pp. 861–888, https://doi.org/10.1007/978-1-84800-936-3_18.

[16] N.Y. Yusuf, M.S. Masdar, D. Nordin, T. Husaini, Challenges in biohydrogen technologies for fuel cell application, Am. J. Chem. 5 (2015) 40–47, https://doi.org/10.5923/c.chemistry.201501.06.

[17] K. Feng, Z. Li, H. Sun, L. Yu, Y. Cai, Y. Wu, P.K. Chu, C/CrN multilayer coating for polymer electrolyte membrane fuel cell metallic bipolar plates, J. Power Sources 222 (2013) 351–358, https://doi.org/10.1016/J.JPOWSOUR.2012.08.087.

[18] C. He, S. Desai, G. Brown, S. Bollepalli, PEM fuel cell catalysts: cost, performance, and durability, Electrochem. Soc. Interface (2005) 41–44.

[19] H.M. Yu, C. Ziegler, M. Oszcipok, M. Zobel, C. Hebling, Hydrophilicity and hydrophobicity study of catalyst layers in proton exchange membrane fuel cells, Electrochim. Acta 51 (2006) 1199–1207, https://doi.org/10.1016/J.ELECTACTA.2005.06.036.

[20] M.-Y. Lim, K. Kim, M.-Y. Lim, K. Kim, Sulfonated poly(arylene ether sulfone) and perfluorosulfonic acid composite membranes containing perfluoropolyether grafted graphene oxide for polymer electrolyte membrane fuel cell applications, Polymers (Basel) 10 (2018) 569, https://doi.org/10.3390/polym10060569.

[21] A. Aricò, D. Sebastian, M. Schuster, B. Bauer, C. D'Urso, F. Lufrano, V. Baglio, Selectivity of direct methanol fuel cell membranes, Membranes (Basel) 5 (2015) 793–809, https://doi.org/10.3390/membranes5040793.

[22] A. Şahin, İ. Ar, Synthesis and characterization of polyvinyl alcohol based and titaniumdioxide doped nanocomposite membrane, J. Therm. Sci. Technol. 34 (2014) 153–162.

[23] H. Beydaghi, M. Javanbakht, A. Badiei, Cross-linked poly(vinyl alcohol)/sulfonated nanoporous silica hybrid membranes for proton exchange membrane fuel cell, J. Nanostruct. Chem. 4 (2014) 97, https://doi.org/10.1007/s40097-014-0097-y.

[24] A.R. Hakim, A. Purbasari, T.D. Kusworo, E. Listiani, Composite sPEEK with nanoparticles for fuel cell's applications: review, in: Proceeding Int. Conf. Chem. Mater. Eng. 2012, 2012, pp. 1–11.

[25] V. Neburchilov, J. Martin, H. Wang, J. Zhang, A review of polymer electrolyte membranes for direct methanol fuel cells, J. Power Sources 169 (2007) 221–238, https://doi.org/10.1016/j.jpowsour.2007.03.044.

[26] Y. Chang, Y.-B. Lee, C. Bae, Partially fluorinated sulfonated poly(ether amide) fuel cell membranes: influence of chemical structure on membrane properties, Polymers (Basel) 3 (2011) 222–235, https://doi.org/10.3390/polym3010222.

[27] Q. Li, D. Aili, H.A. Hjuler, J.O. Jensen, (Eds.), High Temperature Polymer Electrolyte Membrane Fuel Cells Approaches, Status, and Perspectives, Springer International Publishing, 2016.

[28] J.R. Varcoe, M. Beillard, D.M. Halepoto, J.P. Kizewski, S. Poynton, R.C.T. Slade, Membrane and electrode materials for alkaline membrane fuel cells, ECS Trans. 16 (2008) 1819–1834, https://doi.org/10.1149/1.2982023.

[29] R.C.T. Slade, J.P. Kizewski, S.D. Poynton, R. Zeng, J.R. Varcoe, Alkaline membrane fuel cells, in: Fuel Cells, Springer New York, New York, NY, 2013, pp. 9–29, https://doi.org/10.1007/978-1-4614-5785-5_2.

[30] T.-L. Leon Yu, H.-L. Lin, Preparation of PBI/H 3 PO 4-PTFE Composite Membranes for High Temperature Fuel Cells, 2010.

[31] B. Muriithi, D.A. Loy, Processing, morphology, and water uptake of Nafion/ex situ Stöber silica nanocomposite membranes as a function of particle size, ACS Appl. Mater. Interfaces 4 (2012) 6766–6773, https://doi.org/10.1021/am301931e.

[32] K.T. Adjemian, S. Srinivasan, J. Benziger, A.B. Bocarsly, Investigation of PEMFC operation above 100 C employing perfluorosulfonic acid silicon oxide composite membranes, J. Power Sources 109 (2002) 356–364.

[33] L. Xiao, H. Zhang, T. Jana, E. Scanlon, R. Chen, E. Choe, L.S. Ramanathan, S. Yu, B.C. Benicewicz, Synthesis and characterization of pyridine-based polybenzimidazoles for high temperature polymer electrolyte membrane fuel cell applications, Fuel Cells 5 (2005) 287–295.

[34] J.A. Mader, B.C. Benicewicz, Synthesis and properties of segmented block copolymers of functionalised polybenzimidazoles for high-temperature PEM fuel cells, Fuel Cells 11 (2011) 222–237.

Chapter 14 Synthesis and characterization of nanocomposite membranes **281**

[35] K. Fishel, Hydroxy Polybenzimidazoles for High Temperature Fuel Cells and the Solution Polymerization of Polybenzimidazole, 2015.

[36] S. Yu, H. Zhang, L. Xiao, E. Choe, B.C. Benicewicz, Synthesis of poly (2, 2′-(1,4-phenylene) 5,5′-bibenzimidazole)(Para-PBI) and phosphoric acid doped membrane for fuel cells, Fuel Cells 9 (2009) 318–324.

[37] E. Choi, J.A. Moss, Observation of a polymer electrolyte membrane fuel cell degradation under dynamic load cycling, Worcester Polytech. Inst. (2009).

[38] M.L. Hill, B.R. Einsla, Y.S. Kim, J.E. McGrath, Synthesis and characterization of sulfonated poly(arylene ether sulfone)/zirconium phenylphosphonate composite membranes for proton exchange membrane fuel cell applications, in: ACS Div. Fuel Chem. Prepr, 2004, pp. 584–585.

[39] A.K. Mishra, S. Bose, T. Kuila, N.H. Kim, J.H. Lee, Silicate-based polymer-nanocomposite membranes for polymer electrolyte membrane fuel cells, Prog. Polym. Sci. (2012), https://doi.org/10.1016/j.progpolymsci.2011.11.002.

[40] Q. Duan, S. Ge, C.Y. Wang, Water uptake, ionic conductivity and swelling properties of anion-exchange membrane, J. Power Sources (2013), https://doi.org/10.1016/j.jpowsour.2013.06.095.

[41] F. Müller, C.A. Ferreira, D.S. Azambuja, C. Alemán, E. Armelin, Measuring the proton conductivity of ion-exchange membranes using electrochemical impedance spectroscopy and through-plane cell, J. Phys. Chem. B (2014), https://doi.org/10.1021/jp409675z.

[42] H. Prifti, A. Parasuraman, S. Winardi, T.M. Lim, M. Skyllas-Kazacos, Membranes for redox flow battery applications, Membranes (Basel) 2 (2012) 275–306, https://doi.org/10.3390/membranes2020275.

[43] S.L. Sahlin, S.S. Araya, S.J. Andreasen, S.K. Kær, Electrochemical impedance spectroscopy (EIS) characterization of reformate-operated high temperature PEM fuel cell stack, Int. J. Power Energy Res. 1 (2017) 20–40, https://doi.org/10.22606/ijper.2017.11003.

[44] Z. Wu, G. Sun, W. Jin, H. Hou, S. Wang, Q. Xin, Nafion® and nano-size TiO_2–SO_{42}– solid superacid composite membrane for direct methanol fuel cell, J. Membr. Sci. 313 (2008) 336–343, https://doi.org/10.1016/J.MEMSCI.2008.01.027.

[45] C.-C. Yang, W.-C. Chien, Y.J. Li, Direct methanol fuel cell based on poly(vinyl alcohol)/titanium oxide nanotubes/poly(styrene sulfonic acid) (PVA/nt-TiO_2/PSSA) composite polymer membrane, J. Power Sources 195 (2010) 3407–3415, https://doi.org/10.1016/J.JPOWSOUR.2009.12.024.

[46] Y. Hosakun, K. Halász, M. Horváth, L. Csóka, V. Djoković, ATR-FTIR study of the interaction of CO_2 with bacterial cellulose-based membranes, Chem. Eng. J. 324 (2017) 83–92, https://doi.org/10.1016/J.CEJ.2017.05.029.

[47] T. Na, K. Shao, J. Zhu, H. Sun, D. Xu, Z. Zhang, C.M. Lew, G. Zhang, Composite membranes based on fully sulfonated poly(aryl ether ketone)/epoxy resin/different curing agents for direct methanol fuel cells, J. Power Sources 230 (2013) 290–297, https://doi.org/10.1016/J.JPOWSOUR.2012.12.082.

[48] Y. Tang, A. Kusoglu, A.M. Karlsson, M.H. Santare, S. Cleghorn, W.B. Johnson, Mechanical properties of a reinforced composite polymer electrolyte membrane and its simulated performance in PEM fuel cells, J. Power Sources 175 (2008) 817–825, https://doi.org/10.1016/J.JPOWSOUR.2007.09.093.

[49] L. Ghassemzadeh, K.D. Kreuer, J. Maier, K. Müller, Evaluating chemical degradation of proton conducting perfluorosulfonic acid ionomers in a Fenton test by solid-state 19F NMR spectroscopy, J. Power Sources 196 (2011) 2490–2497, https://doi.org/10.1016/j.jpowsour.2010.11.053.

[50] X. Zhu, Y. Liu, L. Zhu, Polymer composites for high-temperature proton-exchange membrane fuel cells, in: Polym. Membr. Fuel Cells, Springer US, Boston, MA, 2008, pp. 1–26, https://doi.org/10.1007/978-0-387-73532-0_7.

[51] S. Malhotra, R. Datta, Membrane-supported nonvolatile acidic electrolytes allow higher temperature operation of proton-exchange membrane fuel cells, J. Electrochem. Soc. 144 (1997) L23, https://doi.org/10.1149/1.1837420.

[52] V. Ramani, H. Kunz, J. Fenton, Investigation of Nafion®/HPA composite membranes for high temperature/low relative humidity PEMFC operation, J. Membr. Sci. 232 (2004) 31–44, https://doi.org/10.1016/J.MEMSCI.2003.11.016.

[53] N.H. Jalani, K. Dunn, R. Datta, Synthesis and characterization of Nafion®-MO2 (M = Zr, Si, Ti) nanocomposite membranes for higher temperature PEM fuel cells, Electrochim. Acta 51 (2005) 553–560, https://doi.org/10.1016/J.ELECTACTA.2005.05.016.

[54] R. Sigwadi, Zirconia Based/Nafion Coposite Membranes for Fuel Cell Applications, 2013.

[55] N.H. Jalani, R. Datta, The effect of equivalent weight, temperature, cationic forms, sorbates, and nanoinorganic additives on the sorption behavior of Nafion®, J. Membr. Sci. 264 (2005) 167–175, https://doi.org/10.1016/J.MEMSCI.2005.04.047.

[56] N. GÜR, Synthesis of Zeolite Beta for Composite Membranes, Master's Thesis, ODTÜ, Ankara, 2006. 85 p.

[57] T.M. Thampan, N.H. Jalani, P. Choi, R. Datta, Systematic approach to design higher temperature composite PEMs, J. Electrochem. Soc. 152 (2005) A316, https://doi.org/10.1149/1.1843771.

[58] W. Lee, C.-H. Ho, J.W. Van Zee, M. Murthy, The effects of compression and gas diffusion layers on the performance of a PEM fuel cell, J. Power Sources 84 (1999) 45–51, https://doi.org/10.1016/S0378-7753(99)00298-0.

[59] M.J. Parnian, S. Rowshanzamir, J. Alipour Moghaddam, Investigation of physicochemical and electrochemical properties of recast Nafion nanocomposite membranes using different loading of zirconia nanoparticles for proton exchange membrane fuel cell applications, Mater. Sci. Energy Technol. 1 (2018) 146–154, https://doi.org/10.1016/J.MSET.2018.06.008.

[60] J.-F. Masson, Brief review of the chemistry of polyphosphoric acid (PPA) and bitumen, Energy Fuel 22 (2008) 3560, https://doi.org/10.1021/ef8005433.

[61] S. Angioni, D.C. Villa, P. Mustarelli, E. Quartarone, Polybenzimidazoles with enhanced basicity: a chemical approach for durable membranes, in: High Temp. Polym. Electrolyte Membr. Fuel Cells, Springer International Publishing, Cham, 2016, pp. 239–250, https://doi.org/10.1007/978-3-319-17082-4_11.

[62] A. Chawla, R. Kaur, A. Goyal, Importance of microwave reactions in the synthesis of novel benzimidazole derivatives: a review, J. Chem. Pharm. Res. 3 (2011).

[63] E. Quartarone, S. Angioni, P. Mustarelli, Polymer and composite membranes for proton-conducting, high-temperature fuel cells: a critical review, Materials (Basel) 10 (2017) 687–10.3390/ma10070687.

[64] H.B. Park, Y.M. Lee, Polymeric membrane materials and potential use in gas separation, in: Adv. Membr. Technol. Appl, John Wiley & Sons, Inc., Hoboken, NJ, 2008, pp. 633–669, https://doi.org/10.1002/9780470276280.ch24.

[65] I. Nicotera, C. Simari, L. Coppola, P. Zygouri, D. Gournis, S. Brutti, F.D. Minuto, A.S. Aricò, D. Sebastian, V. Baglio, Sulfonated graphene oxide platelets in Nafion nanocomposite membrane: advantages for application in direct methanol fuel cells, J. Phys. Chem. C 118 (2014) 24357–24368, https://doi.org/10.1021/jp5080779.

[66] R. Scipioni, D. Gazzoli, F. Teocoli, O. Palumbo, A. Paolone, N. Ibris, S. Brutti, M.-A. Navarra, Preparation and characterization of nanocomposite polymer membranes containing functionalized SnO2 additives, Membranes (Basel) 4 (2014) 123–142, https://doi.org/10.3390/membranes4010123.

15

Fabrication and properties of polymer electrolyte membranes (PEM) for direct methanol fuel cell application

Fatma Aydın Ünal[a], Vildan Erduran[b,c], Cisil Timuralp[d], and Fatih Şen[b]

[a]Metallurgical and Materials Engineering Department, Faculty of Engineering, Alanya Alaaddin Keykubat University, Alanya/Antalya, Turkey. [b]Şen Research Group, Department of Biochemistry, Dumlupinar University, Kütahya, Turkey. [c]Department of Materials Science and Engineering, Faculty of Engineering, Dumlupinar University, Kütahya, Turkey. [d]Mechanical Engineering Department, Eskisehir Osmangazi University, Eskişehir, Turkey

1 Introduction

In the past several years, fuel cell technology has attracted increasing attention [1, 2]. Fuel cells potentially have higher yield. Many fuel cells that have been investigated and developed for many years are now ready to be used for commercial purposes. Hydrogen is one of the most popular fuel cells for obtaining electrical energy from chemical energy. However, it has several disadvantages concerning distribution, production, and storage [3]. Hydrogen to be used in fuel cells need to be stored, and there is no effective and practical method to accomplish it. Although liquefaction is a possibility, its very low gravimetric density leads to a decrease in energy density (Table 1).

As indicated in Table 1, in alternative fuel cells, methanol (CH_3OH) is acceptable, although it is less reactive than hydrogen. Because of its high energy density, it is quite possible for it to be stored and distributed. MeOH (methanol) is liquid under atmospheric conditions [2–4]. Significant financial investments have been made in the development and transformation of direct methanol fuel cells (DMFCs) by such well-known companies as Sony, Motorola, Toshiba, Samsung, Siemens, and Nokia [5].

Nanomaterials for Direct Alcohol Fuel Cells. https://doi.org/10.1016/B978-0-12-821713-9.00015-9
Copyright © 2021 Elsevier Inc. All rights reserved.

284 Chapter 15 Fabrication and properties of polymer electrolyte membranes

Table 1 Energy densities of fuels used in direct polymer electrolyte fuel cells [3].

Fuel	Fuel cell reaction	Energy density (Wh/cm^3)
Hydrogen (liquid at -273°C)	$H_2 + 0.5O_2 \rightarrow H_2O$	2.7
Hydrogen (gas at 20 MPa)	$H_2 + 0.5O_2 \rightarrow H_2O$	0.5
Methanol (liquid)	$CH_3OH + 1.5O_2 \rightarrow CO_2 + 2H_2O$	4.8
Ethanol (liquid)	$C_2H_5OH + 3O_2 \rightarrow 2CO_2 + 3H_2O$	6.3
Ethylene glycol (liquid)	$C_2O_2H_6 + 2.5O_2 \rightarrow 2CO_2 + 3H_2O$	5.9

Reprinted (adapted) with permission from V.S. Silva, A.M. Mendes, L.M. Madeira, S.P. Nunes, Membranes for Direct Methanol Fuel Cell Applications: Analysis Based on Characterization, Experimentation and Modeling, (2005). Copyright (2005) Research Signpost, Advances in Fuel Cells, 2005: ISBN: 81-308-0026-8 Editor: Xiang-Wu Zhang.

Compared to conventional, lithium-ion, polymer-based rechargeable batteries, DMFCs contain about 10 times more energy density. Due to this kind of performance, it is aimed to increase the talk time for mobile phones, increase the life of the fuel cartridges of laptops, and increase the power required for such devices to meet consumer demand [2–5]. As a result of their high power density and conversion efficiency, low emission, and low levels of contamination, there has been increased interest in DMFCs [6–8]. The polymer electrolyte membrane (PEM) is the essential part of the DMFC [8, 9]. Most of these membranes, which are still commercially available for polymer electrolyte membrane fuel cells (PEMFCs), are perfluorosulfonic acid (PFSA)-based polymer membranes (e.g., Nafion, Flemion, and Acipex). The properties of these PEMs, such as good physical and chemical stability and high proton conductivity, provide many advantages over a wide relative humidity (RH) range at moderate operating temperatures. In addition, several disadvantages limit the use of Nafion, including environmental inconsistency, high cost, and high permeability to methanol [10–12]. Methanol can be fully electro-oxidized to CO_2 at temperatures below 100°C, and it also has better energy density compared to other fuels [13, 14]. Researchers have planned to produce closed systems using a dilute methanol solution to reduce problems such as flammability and toxicity in the development of fuel cells. Since the necessary corrections were made, methanol has become an indirect source of hydrogen in both DMFCs and in PEMFCs, as well as in other direct fuel cells [2–15].

PEM is one of the important components in many electrochemical devices [2, 4–16]. The polymer electrolyte permits the passage of protons from the anode to the cathode, eliminating obstacles in the direct contact of the anode/cathode electrolytes in PEMFCs [17, 18]. PEMFCs are the most used perfluorinated Nafion membrane developed by DuPont. However, this membrane has a number of disadvantages such as high cost, low ionic conductivity at low humidity, and susceptibility to deterioration at high temperatures [2, 4–22]. Various polymers have been investigated in addition to Nafion in order to produce PEM [23]. Among these polymer membranes, eco-friendly and low-cost polymers can be promising.

In view of this, the natural polymer, chitosan, and the synthetic polymer, polyvinyl alcohol (PVA), are noteworthy. Chitosan is the second biopolymer obtained by alkaline deacetylation of chitin, and it is abundant in nature. It is a biomaterial membrane used for ultrafiltration, pervaporation, and reverse osmosis. It is both abundant in nature and environmentally friendly and its production is low cost. PVA is a remarkable material for the development of cross-membrane electrolytes. Because it is an inexpensive polymer with high-density functional -OH groups with chemical cross-linking potential. These polymers allow cross-linking of the polymer chain by grafting the hydrophilic group, chemical resistance, good film-forming properties, and high chain cross-linking [2].

In DMFCs, oxygen in the air is reduced in the cathode, but instead of hydrogen, the liquid methanol is directly oxidized in the anode. Therefore, the DMFC system can be an attractive alternative to direct hydrogen or reforming systems because it does not require a hydrogen storage tank or reformer [24, 25]. Fuel cells running directly in methanol or in various liquid fuels without converting fuel to hydrogen gas are DMFCs. They contain two electrodes that are divided by a proton exchange membrane and connected via an external circuit that allows free energy to be converted directly into electrical energy from the chemical reaction of methanol with air or oxygen. PEMFC, known as proton exchange membrane, is used in a DMFC, fed with an aqueous methanol solution. Methoxide oxidation takes place in the anode part and oxygen reduction in the cathode part. These catalytic electrodes are separated from each other by a membrane that transmits protons from the anode to the cathode. This membrane prevents diffusion of other compounds. The membrane combination with the electrode is called the "membrane electrode assembly (MEA)," and it was made with Pt-Ru black in the anode part, Nafion membrane in the middle, and Pt black in the cathode section. The structure of the cell is illustrated in Fig. 1 [26].

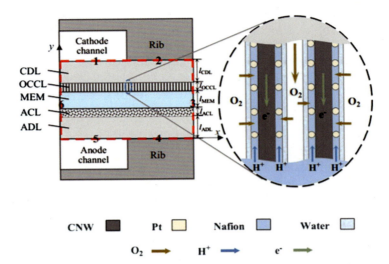

Fig. 1 The structure of a direct methanol-oxygen fuel cell (DMFC) [27].

Reactions

Anodic reaction (oxidation): $CH_3OH + 7H_2O \rightarrow CO_2 + 6H_3O^+ + 6e^- \rightarrow$ $E^0 = 0.02V$
Cathodic reaction (reduction): $3/2O_2 + 6H_3O^+ + 6e^- \rightarrow 9H_2O \rightarrow$ $E^0 = 1.23V$ [27]
Overall reaction: $CH_3OH + 3/2O_2 \rightarrow CO_2 + 2H_2O \rightarrow$ $E^0 = 1.21V$

A further advantage of DMFCs is the direct conversion of the chemical energy of the reaction from CH_3OH to electrical energy (relative to conventional combustor). In the running DMFCs, PEM performs three basic functions:

(1) An ion conductor between the cathode and anode,
(2) A separator for the fuel (methanol) and oxidant (oxygen or air)
(3) An insulator between the anode and cathode

PEM has many desirable characteristics, such as chemical and electrochemical stability, adequate mechanical power, low reactive liquid/gas permeability, and high proton conductivity. These traits guarantee that it can be used in the operation of DMFCs [28–30].

2 Basics of the DMFC

The DMFC consists mainly of a solid electrolyte between the two electrodes, the cathode and the anode [31]. The catalysts frequently used as the DMFC anode and cathode catalysts are Pt/Ru

$(\sim 2 \text{ mg/cm}^2)$ and Pt $(\sim 0.1 \text{ mg/cm}^2)$. The proton exchange membrane is used directly in the electrolyte function in methanol fuel cells. The electrooxidation reaction of MeOH is a slow process. The corresponding reaction is given here:

$$\text{Anode reaction}: \text{CH}_3\text{OH} + \text{H}_2\text{O} \xrightarrow{\text{Pt/Ru}} \text{CO}_2 + 6\text{H}^+ + 6\text{e}^-$$

$$\text{Cathode reaction}: \frac{3}{2}\text{O}_2 + 6\text{H}^+ + 6\text{e}^- \xrightarrow{\text{Pt}} 3\text{H}_2\text{O}$$

$$\text{Overall reaction}: \text{CH}_3\text{OH} + \frac{3}{2}\text{O}_2 \rightarrow \text{CO}_2 + 2\text{H}_2\text{O}$$

Fig. 1 shows the basic operating principles of DMFC: The anode is provided with methanol and water. CO_2 is converted to protons and electrons in the anode. The electrons generating the anode reaction is transmitted through the external circuit system. Protons are transferred to the cathode zone through the proton exchange membrane. The transported protons and electrons form water in the cathode by combining with oxygen (air) [3].

3 Requirements for DMFC membranes

Here is a list of common properties for both PEM and DMFC:
- High operating temperature
- The low methanol diffusion coefficient of the membrane of $<5.6 \times 10^{-6} \text{ cm}^2/\text{s}$ at 25°C or lower methanol passage of $<10^{-6} \text{ mol/min/cm}$
- High ionic conductivity ($>80 \text{ mS/cm}$)
- Mechanical stability and increased chemical content for increased CO tolerance at temperatures above 80°C
- Low ruthenium (Ru) diagonal transition formed by the anode catalyst containing ruthenium (Ru)
- Low-cost proton exchange compared to PEMFCs

4 Materials and properties

The polymer electrolyte or proton exchange membrane in PEMFCs is typically a polymer film formed using an inert polymer containing an acidic group [32]. The membrane is positioned between two porous gas diffusion layers (GDLs) and the electrodes [33]. All these structures form an MEA, which performs serial alignment with bipolar plates with gas flow areas. The membranes used must be homogeneous, not defective; have high mechanical properties; and have low permeability to oxygen, which acts as a barrier against reacting gases. Also, the membranes used must have oxidation and chemical stability to water shortage or reduction in use.

First, a polymer or sulfonated crosslinked polystyrene electrode containing sulfonated phenolic resin is used for PEMFCs. In addition, nowadays perfluorinated sulfonic acid-containing ionomers (PFSAs) are used. The polymers are produced by copolymerization of a perfluorinated vinyl ether monomer, which contains a sulfonyl fluoride in the side chain of polytetrafluoroethylene (PTFE) (C_2F_4). It is then necessary to convert the sulfonyl fluoride to the sulfonic acid group. For the chlorine alkali process, these PFSAs have found a home as an electrolyte for PEMFCs. The membranes used in chlorine-alkaline cells are thick, while membranes less than 50 μm thick are used in fuel cells and have high conductivity and good handling properties. Chlorine alkali membranes are often used for the sulfonyl fluoride form.

The next treatment is acidification, hydrolysis, and washing of the solution. The mixture of water and alcohol is discharged by dispersion of PEMFC membranes in acid form to produce fine, high-quality films. Although it is not a requirement for the proper selection of solvents, the membranes must have very good mechanical properties, such as drying, temperature and casting conditions. The high purity required for casting membranes must be maintained for sufficient mechanical properties. It is well known that "superacidic" perfluorosulfonic acid groups disrupt organic solvents in cast dispersions upon heating. Membranes that contain nonconductive material, such as those in porous PTFE films, need to be strengthened. Thinner membranes have good mechanical properties. The hydration of the membranes has a big impact on the performance of the fuel cell. Because the incoming gas flow needs to be moistened, water is required to sustain high conductivity of these membranes. Membranes used in operating fuel cell systems can absorb half (50%) and all (100%) of the water. Hydrated Nafion absorbs 21 water molecules per sulfonic acid group.

Despite proton conductivity above 0.1 S/cm, membrane resistance is 75% greater compared to the total resistance of the cell. When a membrane is dried, the conductivity drops and the cell resistance may rise. The need for additional money for parasitic power losses and humidifiers is one of the disadvantages of gas humidification. The humidifying process reduces the overall performance of the system, diluting reactive gases. In addition, high-gas humidification may contribute to flooding at high current densities.

Another limitation regarding the humidity requirements of the membranes is related to the range of operating temperatures. At temperatures below 1008°C, the PFSA membranes used as the standard are sufficiently moistened. It is the CO found in renewed hydrocarbon membranes, which has a poisonous effect against Pt (platinum) catalysts [33].

4.1 Category of the materials used in PEMs

Improving redox (reduction-oxidation) flow cell membranes is important for fuel cell performance. In fuel cells, the membrane materials used can be separated into five groups:

(i) Perfluorinated ionomers
(ii) Partially fluorinated polymers
(iii) Nonfluorinated hydrocarbons
(iv) Nonfluorinated membranes with aromatic backbone
(v) Acid-base blends [34]

Several evaluations of the chemicals used in PEMs are available [35].

5 Membrane and its properties

The membrane has a thin layer of electrolyte that transmits protons from anode to cathode. The desired properties of membrane materials are high ionic conductivity, prohibiting electron transportation, and crossover of oxygen reactant from the cathode and hydrogen fuel from the anode. The membrane material should be chemically stable. Perfluorosulfonic acid-based membranes are commonly used, especially Nafion, developed by DuPont in the 1960s [36]. Nafion has charge sites for proton transport due to functional sulfonic acid groups. Other perfluorinated polymer materials used in PEM fuel cell applications include Gore-Select (WL Gore and Associates), Neosepta-F (Tokuyama), Asiplex (Asahi Chemical Industry), and Flemion (Asahi Glass Company). Furthermore, membrane materials capable of operating at high temperatures (100–200°C) are used in high-temperature PEMFCs, which has the benefits of better catalyst tolerance to CO for the fuel cell and a better cooling strategy. Nafion-based membranes are expensive due to the complexity of their production processes [16].

Electrolyte Membrane: The membranes used in DMFC studies are known as "proton exchange membranes" due to their H^+ ion

content. For these membranes to be functional, they must have the following properties:
(1) High proton activity
(2) Mechanical durability
(3) Electrochemical reliability during processing
(4) High electrical performance and low fuel usage
(5) A fairly small amount of water diffusion transfer
(6) Easy fabricability

Nafion membranes are very widely used in DMFC production, as they meet the required high-level properties. The adverse aspects of these membranes are that as a result of rapid water loss and F^- ion loss caused by the presence of OH^- ions, low conductivity occurs at high temperatures, which in turn increases the production costs [37, 38].

5.1 DuPont Nafion membranes

As shown in Fig. 2, the polymeric structure is preferred in most PEFC and DMFC applications [39]. Nafion consists of a PTFE backbone endowed with perfluoroether side chains terminated by a sulfonic acid group. The PEMFC lasts longer than 60,000 h, demonstrating that the Nafion membrane is highly durable. In addition, it has high stability and ionic conductivity [40]. Nafion membranes show thermal and chemical stability imparted by the hydrophobic property of PTFE: the PTFE structure in its main skeleton. For protons to form channels, they must have side chains modified with perfluoride-containing hydrophilic sulfonic acid in the termination groups ($-SO_3H$). Protons move from anode to cathode at 80°C with 90–120 m/S high conductivity and 34%–100% RH [41, 42].

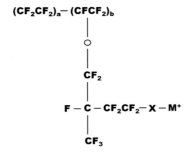

Fig. 2 The structure of Nafion [39].

In DMFC applications, Nafion-based membranes have a number of drawbacks:

(1) High MeOH and Ru transition in platinum-ruthenium anode regions
(2) Be expensive [41–44]
(3) Loss of proton conductivity at temperatures above 80°C [7]
(4) High humidification requirements [10, 24–45]

Although the most widely used membrane in PEMFCs was Nafion 112, Nafion 117 was chosen for use in DMFCs [10, 22–48]. Cyclic voltammetry studies were performed using Nafion 112, Nafion 1035, Nafion 1135, and Nafion 117 [42–49]. DuPont's alternative proton-conducting membrane technologies have led to the expansion of Nafion products to enhance competition. DuPont introduced Gen IV MEA, the latest technology for DMFC applications. The new technology offered shows a twofold increase in durability, a lower catalyst-loading requirement than previous MEAs, and a 20% increase in power density [42].

5.2 Membranes in contact with fluorine without using Nafion

5.2.1 Chemical DOW-XUS membranes

The Dow product membranes, one of the fluorinated membranes, are prepared by polymerization of tetrafluoroethylene with a vinylene monomer. The Dow membranes are prepared by polymerization of tetrafluoroethylene (TFE) with vinylene monomers. The specific conductivities of the membranes at 850 and 800 equivalent weights (EW) of moles per mole of ion exchange regions at EW are 0.12 and 0.2 $mol^{-1} cm^{-1}$, respectively [42–50]. Compared to Nafion 117, the MCO values are higher [42]. Dow membranes have greater methanol permeability ($4 \times 10^{-10} A cm^{-2}$) than does Nafion 117 ($2.7 \times 10^{-10} A cm^{-2}$) [50].

5.2.2 3P energy membranes

The PFSA membrane developed by the German 3P energy company shows 20 times less methanol permeability compared to Nafion membranes. The PFSA membrane has both high power density and high methanol concentration in DMFCs. In addition, other features such as mechanical strength and long-life use 3P membranes have not been used in any previous study [42–50].

5.3 Composite membranes used with fluoride

The creation of composite membranes is one of the approaches that can improve the performance of perfluorinated membranes [42]. Penner and Martin prepared the first composite membranes by impregnating Gore-Tex with Nafion. The composite fluorinated membranes can be prepared with the addition of: (1) Hydrophilic inorganic substances such as clay, zeolite, and SiO_2, to prevent high-water content and to avoid Nafion dehydration at high temperatures; (2) solid acids, such as zirconium phosphate (ZrP) and sulfonated ZrO_2, which increase the concentration in the acid regions to trigger proton migration in the membrane; (3) heteropolyacids, such as phosphotungstic acid ($H_3PW_{12}O_{40}$), and phosphoric acid. Nafion membranes show thermal and chemical stable properties caused by the hydrophobic property of PTFE. These structures also can be prepared with the addition of polymer, such as PTFE, polyvinylidene fluoride (PVDF), and polybenzimidazole (PBI) [50].

5.3.1 Composite membranes with inorganic organic structure

5.3.1.1 Nafion-based membranes modified with zirconium (IV) hydrogen phosphate (Zr (HPO$_4$)$_2$) (Nafion@Zr (HPO$_4$)$_2$)

Nafion-Zr membranes are prepared using commercial films similar to Nafion 115. The purchased film is modified with the help of ZrP with a replacement reaction containing Zr ions. Soaking of the membrane in the H_3PO_4 solution allows ZrP to precipitate. As a result, insoluble ZrP is retained in pores. Nafion-Zr membranes are kept an isothermal temperature of 150°C using dry oxidants. The membrane resistance of $0.08 \, \Omega \, cm^2$ was obtained. The maximum power densities required for DMFC were 380 and 260 mW/cm^2. The ZrP additive increased the maximum operating temperature with water retention properties. At the same time, it increased its dry weight and membrane thickness by 23% and 30%, respectively [42]. DMFC open-circuit voltage (OCV) were measured in the range of 0.86–0.87 V between 120° C and 150°C with oxygen and air during the procedure [10]. To understand the conductivity of modified ZrP membranes, surface and crystallinity morphologies should be known. ZrP particles are uniformly distributed on the membrane, with a size of 1161 nm. The pore size of these membranes is larger than in the Nafion membrane during hydration. Nafion @ zirconium membranes have 100% relative humidity (RH) and proton conductivity at room temperature [42].

Chapter 15 Fabrication and properties of polymer electrolyte membranes **293**

5.3.1.2 Nafion, silica, and molybophosphoric acid modification

Changes in membrane fuel cell applications by the addition of Nafion membrane silica increased membrane performance. For Nafion-silica membranes, solutions are prepared with tetraethylortosilicate (TEOS), silica powder, phosphotungstic acid (PWA), dithenylsilate (DPS), and silica oxide, and then these solutions are mixed with the Nafion solution [42].

5.3.1.3 Nafion nanocomposite membranes modified with polyfurfuryl alcohol (PFA)

Nafion membranes are produced at the industrial level as a result of polymerization of furfuryl alcohol [51].

5.3.2 Composite membranes with acid-base structure

Acid-base composite membranes can maintain high proton conductivity at elevated temperatures without suffering from dehydration effects [50].

5.3.2.1 Nafion@Polypyrrole membranes

These membranes have obtained from the in situ polymerization of pyrrole onto Nafion in the presence of Fe(III), an oxidizing agent in hydrogen peroxide. Fe (III) oxidation-modified membranes with low MCO exhibit high resistance and low performance compared to Nafion-based DMFCs.

5.3.2.2 Pall IonClad membranes

Tetra-fluoroethyl/perflouropropylene-based Pall IonClad membranes are developed by Pall Gelman Sciences. Ionclad R-1010 and IonClad R-410 can block the transition of methanol about three times better than Nafion 117.

5.4 Unmodified composite membranes with fluorine

5.4.1 Composite membranes with inorganic and organic structure

Proton exchange membranes consisting of organic or inorganic composites have been developed to eliminate the disadvantages of the state-of-the-art membranes produced.

5.4.1.1 Membranes with SiO_2/SiO_2 gel, acid, and polyvinylidene fluoride

Membranes with nanoscale proton conductivity are modified PVDF membranes. These membranes have high surface areas and two to four times more ionic conductivity than Nafion. As a result of the smaller pore size, the methanol transition of PVDF membranes is two to four times less than that of Nafion. These membranes are operated in a wider temperature range (0–90°C). The MEAs using the Pt-Ru anode loaded in 1 M methanol at 80°C and 4–6 mg/cm^2 have reached a power density of 85 mW/cm^2. The cost of modified PVDF membranes is about $4 per square meter, which is cheaper than Nafion. Modified PVDF membranes can be bent without breaking at 180°C; in addition, they have high mechanical stability, high thermal stability (0–100°C), stable membrane size as a result of water absorption and 50%–90% by volume that can be produced with acid solutions.

5.4.1.2 SPEEK modified with silane or silica structures

Sulfonated polyether ether ketone (SPEEK)-based membranes, heteropolyacids, and amino-modified silanes have been developed as a result of the hydrolysis of the oxide phase produced by the dispersion of the surface-modified silica.

5.4.1.3 Polybenzimidazole-based membranes

PBI is a basic polymer (pKa = 5.5). Single-base polymer electrolytes with oxidative and thermal stability are treated with acid, and mechanical flexibility is achieved at high temperatures (<200°C). At temperatures above 100°C, PBI membranes have several advantages over Nafion
- Proton conductivity is high.
- When water content is compared, Nafion is 0.6, while PBI is about 0. This shows that the effects on proton transport and electroosmotic friction are low.
- The thermal operational range is higher.
- Nafion has a thickness of 210 μm, whereas PBI has a thickness of 80 μm. This means that it has a low methane passage.
- They are cheap compared to Nafion.

5.4.2 Acid-base membranes formed by direct modification of structures having polymer skeletons

5.4.2.1 Acidic-based composite membranes formed by direct modification of structures having polymeric skeletons

5.4.2.1.1 Polyphosphazene membranes with sulfonate Modification of sulfonated polyphosphate (sPPZ) is accomplished by sulfonated poly [bis (3-methylphenoxy) phosphazene]. It provides higher conduction of protons. In addition, the coefficient of methanol diffusion is 1.6×10^{-8} cm^2/s, while it is 6.5×10^{-6} cm^2/s in Nafion. As these values indicate, it has a lower methanol diffusion coefficient in 30°C and 1 M methanol compared to sPPZ Nafion. The advantages of these membranes with a thickness of 200 m are as follows:

- The cost is low.
- The membranes have a proton conductivity of less than 30% compared to Nafion 117 in high sulfonation.
- Low water and methanol differences in cross-linked polyphosphate (1.2×10^{-7} cm^2/s) are available.

5.4.2.2 Acid-based membranes as a result of direct modification of the polymer skeleton

5.4.2.2.1 ETFE-SA membranes Polyethylene-sub-tetrafluoroethylene (ETFE) polymer is produced by Nowofol GmbH. These membranes are prepared by sulfonation without irradiation and other polymer vaccines. Membranes typically have a semicrystalline structure. Crossover effects of crystallites decreased the osmotic pressure of sulfonated membranes.

ETFE-SA has more chemical crosslinks than Nafion during irradiation and sulfonation. In room conditions (at 25°C), the conductivity of ETFE-SA and Nafion 115 is 10 and 40 mS/cm, respectively. MCO for 1 and 2 M methanol solutions at 25°C, 0.3 and 0.6 mol/min/cm $\times 10^8$ for EFTE, 6.7 and 12 mol/min/cm for Nafion 115, respectively. The ETFE-SA membranes are 35 μm thick and Nafion 115 is 127 μm thick; the thinner membrane reduces the cost.

5.4.2.2.2 SPEEK membranes SPEEK is one of the most attractive membrane materials for use in PEMFCs due to its high thermal stability, high proton conductivity, and low cost. One of the most widely used processing methods for preparing SPEEK PEMs is solution casting employing organic solvents, including dimethylacetamide. The use of low-boiling-point compounds reduces the risk of solvent residue in the casting membranes and ensures easy membrane production [52].

5.4.3 Membrane and acid-based membranes used in direct polymerization and monomer structures

Fluorine-modified membranes are formed by the direct use of sulfonate groups or sulfonic acids present in the functionalized monomer building blocks of the novel polymer structures. The cost of membranes produced by sulfonation is less than the cost of modified membranes.

5.4.3.1 Polyarylene ether sulfone-based membranes

Polyarylene ether sulfone-based membranes include biphenol-based polyarylene ether sulfone (PES), 6FCN-35 membrane, a polyarylene ether benzonitrile, BPSH-40 membrane (40, molar fraction of sulfonic acid), and hexafluoroisopropylidene.

5.4.3.2 Composite membranes such as SPEEK/sulfonated polysulfone and polybutyl/PBI

Composite membranes as the main chain polymer are prepared by mixing the sulfonated PSU Udel (sPSU) or basic polymers such as sulfonated PEEK Victex. These prepared composite membranes exhibit comparable performance with poly (4-vinyl pyridine) (P4VP) or PBI membranes with Nafion 105 at $T = 110°C$, $j = 0.5$ cm^{-2} at 500 mV. In addition, composite membranes have higher methanol transition compared to Nafion 117, at 150 mA/cm^2.

5.4.3.3 Asymmetric acrylic membranes

An acrylic asymmetric membrane can be prepared by using three polymers, which are poly (butyl methacrylate) (PBMA), 4-vinyl phenol-*co*-methyl methacrylate (P (4-VP-MMA)), and Paraloid B-82 acrylic copolymer resin. The main features of the membrane design are as follows:

- The acrylic polymer in methanol has low solubility.
- As a methanol barrier, it uses Paraloid B-82 acrylic copolymer resin.
- Poly (butyl methacrylate) and Paraloid B-82 resin serve as walls due to their structural stability and flexibility.
- The 4VP layers in the P (4-VP-MMA) membrane show a hydrophilic character and serve as a channel in the transmission of protons in the TCP.

5.4.3.4 Polyethylene and styrene or PVDF-based membranes

PVDF-based membranes and 125-thick, low, density polyethylene (LDPE) copolymers were developed at Grandfield University. Its resistance factor is higher than those of Nafion 10, PVDF, and LDPE membranes. It has low cost, but it has low methanol diffusion coefficients at 2 M methanol and $T = 20°C$ (compared to 0.11×106 for Nafion $\ll 0.05 \times 106$ cm^2/s). The lifetime use is not possible, and it does not have mechanical durability.

5.4.3.5 Sulfonated polyarylether ketone membranes

Polyarylether ketones consist of sequences of ether and carbonyl linkage between phenyl rings, which can be either ether-rich or keton-rich. The most common materials are polyetherketone (PEK), and polyetherketone (PEEK).

5.4.3.6 Polyfuel polycarbon membranes

PolyFuel industrialized new membranes in 2005, with thicknesses of 62 and 45 m and power densities of 60 and 80 mW/cm^2, respectively. These membranes are polycarbonate membranes used in passive DMFCs. Polycarbonate membranes have catalyst loading at 40°C, 0.28 V, and 8 mg/cm^2. The fixed working time of these membranes is 5000 h, and the working time for commercially available fuel cells is 2000–3000 h. The prepared membrane has longer usage times than commercial fuel cells. The poly-fuel membranes do not cause any changes in water flow. In automatic control panels, they provide approximately 57 mA/cm^2 methanol transition.

5.5 Membranes suitable for use in DMFC applications

The membranes mentioned next were developed for proton exchange membrane fuel cells. However, these membranes can be used for DMFCs. These membranes include resin-modified polystyrene sulfonate, Ballard membranes, and fluorocarbon membranes from Hoku Scientific [42].

6 Fabrication methods of nanocomposite membranes

Among the various approaches for introducing inorganic fillers into an ionomer matrix, the blending, infiltration, and sol-gel approaches have attracted a good deal of attention due to their wide availability. These approaches are described in the following sections and illustrated in Fig. 3 [31].

6.1 Blending method

The simplest method to produce polymer/inorganic nanocomposites is direct mixing of the nanoparticles into the polymer matrix. Mixing is done by melt blending or solution blending (Fig. 3A).

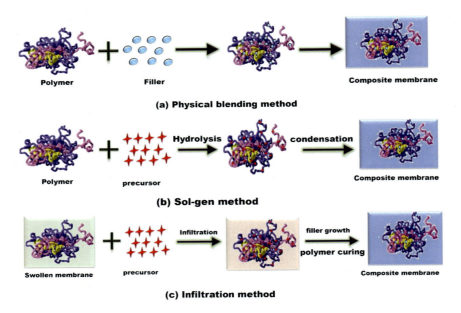

Fig. 3 Three typical methods for fabricating composite membranes: (A) physical blending; (B) sol-gel; and (C) infiltration [31]. Reprinted (adapted) with permission from K. Pourzare, Y. Mansourpanah, S. Farhadi, Advanced nanocomposite membranes for fuel cell applications: a comprehensive review. Biofuel Res. J. 3 (2016) 496–513. https://doi.org/10.18331/BRJ2016.3.4.4. Copyright (2016).

6.2 Sol-gel method

The sol-gel method is a way to synthesize at low temperatures. It is also widely used to synthesize organic-inorganic nanocomposites (Fig. 3B). This process is generally done by hydrolysis and condensation reactions of metal alkoxides, M (OR)n (M = Ti, Si, VO, Zr, W, Mo, Al, Zn, Ce, Sn; and R = Me, Et,...) inside a polymer dissolved in nonaqueous or aqueous solutions. These reactions are shown in Eqs. (1), (2):

$$M(OR)_4 + 4H_2O \rightarrow M(OH)_4 + 4ROH \tag{1}$$

$$mM(OH)_4 \rightarrow (MO_2)_m + 2mH_2O \tag{2}$$

Because silicon alkoxides are not sensitive to hydrolysis, gelation is likely to occur within a few days when water is added. Therefore, hydrolysis and condensation proceed independent of catalysts for nonsilicate metal alkoxides, while for silicon alkoxides, the presence of acid or base catalysts is necessary. Four factors affecting the kinetic reactions, and consequently the final structure and properties of the product, include molar ratio of water/silane, tape catalyst, temperature, and the nature of the solvent [31].

6.3 Infiltration method

In situ or infiltration techniques were used for the preparation of organic-inorganic nanocomposites. The nanocomposite membranes were then acquired by the growth of filler, removal of impurities, and polymer curing. The isolation impact caused by the polymer network may prevent the unwanted agglomeration of nanoparticles, but it also may lead to controlled particle size and uniform distribution (Fig. 3C).

7 Conclusions

In this chapter, DMFC systems with features such as easy operation and low temperature sensitivity were discussed. These systems are very important to the development of energy converters. The factors that prevent the commercialization of DMFC sites include the sensitivity of the polymer to osmotic swelling of the electrolyte membrane, transition to methanol, high cost, and limited operating temperature. The production and properties of PEMs used in DMFC applications, consisting mainly of a solid electrolyte between two electrodes (cathode and anode), are detailed.

References

[1] X. Huang, Z. Zhang, J. Jiang, Fuel cell technology for distributed generation: an overview, in: 2006 IEEE Int. Symp. Ind. Electron, IEEE, 2006, pp. 1613–1618, https://doi.org/10.1109/ISIE.2006.295713.

[2] P.N. de Oliveira, A.M.M. Mendes, P.N. de Oliveira, A.M.M. Mendes, Preparation and characterization of an eco-friendly polymer electrolyte membrane (PEM) based in a blend of sulphonated poly(vinyl alcohol)/chitosan mechanically stabilised by nylon 6,6, Mater. Res. 19 (2016) 954–962, https://doi.org/10.1590/1980-5373-MR-2016-0387.

[3] V.S. Silva, A.M. Mendes, L.M. Madeira, S.P. Nunes, Membranes for Direct Methanol Fuel Cell Applications: Analysis Based on Characterization, Experimentation and Modeling, 2005.

[4] S. Eris, Z. Daşdelen, F. Sen, Investigation of electrocatalytic activity and stability of Pt@f-VC catalyst prepared by in-situ synthesis for methanol electrooxidation, Int. J. Hydrog. Energy 43 (2018) 385–390, https://doi.org/10.1016/j.ijhydene.2017.11.063.

[5] V.S. Correia, F. Silva, A. Miguel, M. Mendes, L. Miguel, P. Madeira, Direct Methanol Fuel Cell: Analysis Based on Experimentation and Modeling, 2005.

[6] M.P. Hogarth, G.A. Hards, Direct methanol fuel cells: technological advances and further requirements, Platin. Met. Rev. 40 (1996) 150–159, https://doi.org/10.1149/1.1837166.

[7] P. Umsarika, S. Changkhamchom, N. Paradee, A. Sirivat, P. Supaphol, P. Hormnirun, Proton exchange membrane based on sulfonated poly (aromatic imide-

co-aliphatic imide) for direct methanol fuel cell, Mater. Res. 21 (2017) 1–8, https://doi.org/10.1590/1980-5373-mr-2017-0823.

[8] Ö. Karatepe, Y. Yildiz, H. Pamuk, S. Eris, Z. Dasdelen, F. Sen, Enhanced electrocatalytic activity and durability of highly monodisperse Pt@PPy-PANI nanocomposites as a novel catalyst for the electro-oxidation of methanol, RSC Adv. 6 (2016) 50851–50857, https://doi.org/10.1039/c6ra06210e.

[9] X. Yang, Y. Liu, S. Li, X. Wei, L. Wang, Y. Chen, A direct borohydride fuel cell with a polymer fiber membrane and non-noble metal catalysts, Sci. Rep. 2 (2012) 567, https://doi.org/10.1038/srep00567.

[10] Y.-S. Ye, J. Rick, B.-J. Hwang, Y.-S. Ye, J. Rick, B.-J. Hwang, Water soluble polymers as proton exchange membranes for fuel cells, Polymers 4 (2012) 913–963, https://doi.org/10.3390/polym4020913.

[11] S. Eris, Z. Daşdelen, Y. Yıldız, F. Sen, Nanostructured polyaniline-rGO decorated platinum catalyst with enhanced activity and durability for methanol oxidation, Int. J. Hydrog. Energy 43 (2018) 1337–1343, https://doi.org/10.1016/j.ijhydene.2017.11.051.

[12] F. Şen, S. Şen, G. Gökağaç, Efficiency enhancement of methanol/ethanol oxidation reactions on Pt nanoparticles prepared using a new surfactant, 1,1-dimethyl heptanethiol, Phys. Chem. Chem. Phys. 13 (2011) 1676–1684, https://doi.org/10.1039/c0cp01212b.

[13] J. Tayal, B. Rawat, S. Basu, Bi-metallic and tri-metallic Pt-Sn/C, Pt-Ir/C, Pt-Ir-Sn/C catalysts for electro-oxidation of ethanol in direct ethanol fuel cell, Int. J. Hydrog. Energy 36 (2011) 14884–14897, https://doi.org/10.1016/j.ijhydene.2011.03.035.

[14] M. Sgroi, F. Zedde, O. Barbera, A. Stassi, D. Sebastián, F. Lufrano, V. Baglio, A. Aricò, J. Bonde, M. Schuster, M.F. Sgroi, F. Zedde, O. Barbera, A. Stassi, D. Sebastián, F. Lufrano, V. Baglio, A.S. Aricò, J.L. Bonde, M. Schuster, Cost analysis of direct methanol fuel cell stacks for mass production, Energies 9 (2016) 1008, https://doi.org/10.3390/en9121008.

[15] Y. Yıldız, S. Kuzu, B. Sen, A. Savk, S. Akocak, F. Şen, Different ligand based monodispersed Pt nanoparticles decorated with rGO as highly active and reusable catalysts for the methanol oxidation, Int. J. Hydrog. Energy 42 (2017) 13061–13069, https://doi.org/10.1016/j.ijhydene.2017.03.230.

[16] Y. Wang, K.S. Chen, J. Mishler, S.C. Cho, X.C. Adroher, A review of polymer electrolyte membrane fuel cells: technology, applications, and needs on fundamental research, Appl. Energy 88 (2011) 981–1007, https://doi.org/10.1016/j.apenergy.2010.09.030.

[17] M. Ji, Z. Wei, M. Ji, Z. Wei, A review of water management in polymer electrolyte membrane fuel cells, Energies 2 (2009) 1057–1106, https://doi.org/10.3390/en20401057.

[18] S. Litster, G. Mclean, PEM fuel cell electrodes, J. Power Sources 130 (2004) 61–76, https://doi.org/10.1016/j.jpowsour.2003.12.055.

[19] C. Houchins, G. Kleen, J. Spendelow, J. Kopasz, D. Peterson, N. Garland, D. Ho, J. Marcinkoski, K. Martin, R. Tyler, D. Papageorgopoulos, U.S. DOE progress towards developing low-cost, high performance, durable polymer electrolyte membranes for fuel cell applications, Membranes 2 (2012) 855–878, https://doi.org/10.3390/membranes2040855.

[20] D. Dissertations, A. Dissertations, T.M. Thampan, Digital WPI Design and Development of Higher Temperature Membranes for PEM Fuel Cells Repository Citation, 2003.

[21] N. Ahmad Nazir, Modification and Characterization of Nafion Perfluorinated Ionomer Membrane for Polymer Electrolyte Fuel Cells, 2011.

Chapter 15 Fabrication and properties of polymer electrolyte membranes **301**

[22] R. Sigwadi, Zirconia Based/Nafion Composite Membranes For Fuel Cell Applications, 2013.

[23] M.A. Hickner, H. Ghassemi, Y.S. Kim, B.R. Einsla, J.E. McGrath, Alternative polymer systems for proton exchange membranes (PEMs), Chem. Rev. 104 (2004) 4587–4612, https://doi.org/10.1021/cr020711a.

[24] M.H.D. Othman, A.F. Ismail, A. Mustafa, Recent development of polymer electrolyte membranes for direct methanol fuel cell application—a review, Malays. Polym. J. 5 (2010) 1–36.

[25] P. Sennequier, Signal Conditioning for Electrochemical Sensors, 2017, pp. 1–27.

[26] N. Ramkrishna Joshi, Development in Direct Methanol-Oxygen Fuel Cell (DMFC), 2014.

[27] J. Jiang, Y. Li, J. Liang, W. Yang, X. Li, Modeling of high-efficient direct methanol fuel cells with order-structured catalyst layer, Appl. Energy 252 (2019) 113431, https://doi.org/10.1016/j.apenergy.2019.113431.

[28] Y. Na, F. Zenith, U. Krewer, Increasing fuel efficiency of direct methanol fuel cell systems with feedforward control of the operating concentration, Energies 8 (2015) 10409–10429, https://doi.org/10.3390/en80910409.

[29] S. Kang, S.J. Lee, H. Chang, Mass balance in a direct methanol fuel cell, J. Electrochem. Soc. 154 (2007) B1179, https://doi.org/10.1149/1.2777109.

[30] M. Sajgure, B. Kachare, P. Gawhale, S. Waghmare, & G. Jagadale, Int. J. Curr. Eng. Technol. Direct methanol fuel cell: a review.

[31] K. Pourzare, Y. Mansourpanah, S. Farhadi, Advanced nanocomposite membranes for fuel cell applications: a comprehensive review, Biofuel Res. J. 3 (2016) 496–513, https://doi.org/10.18331/BRJ2016.3.4.4.

[32] A. El-kharouf, A. Chandan, M. Hattenberger, B.G. Pollet, Proton exchange membrane fuel cell degradation and testing: review, J. Energy Inst. 85 (2012) 188–200, https://doi.org/10.1179/1743967112Z.00000000036.

[33] S.J. Hamrock, M.A. Yandrasits, Proton exchange membranes for fuel cell applications, J. Macromol. Sci. Polym. Rev. 46 (2006) 219–244, https://doi.org/10.1080/15583720600796474.

[34] H. Prifti, A. Parasuraman, S. Winardi, T.M. Lim, M. Skyllas-Kazacos, Membranes for redox flow battery applications, Membranes 2 (2012) 275–306, https://doi.org/10.3390/membranes2020275.

[35] N. Awang, A.F. Ismail, J. Jaafar, T. Matsuura, H. Junoh, M.H.D. Othman, M.A. Rahman, Functionalization of polymeric materials as a high performance membrane for direct methanol fuel cell: a review, React. Funct. Polym. 86 (2015) 248–258, https://doi.org/10.1016/J.REACTFUNCTPOLYM.2014.09.019.

[36] R. Vecci, Fluid Machines and Energy Systems Engineering, Mechanics and Engineering Advanced Sciences - Project n° 2, 2013, https://doi.org/10.6092/unibo/amsdottorato/5814.

[37] H. Junoh, J. Jaafar, M.N.A. Mohd Norddin, A.F. Ismail, M.H.D. Othman, M.A. Rahman, N. Yusof, W.N. Wan Salleh, H. Ilbeygi, A review on the fabrication of electrospun polymer electrolyte membrane for direct methanol fuel cell, J. Nanomater. 2015 (2015) 1–16, https://doi.org/10.1155/2015/690965.

[38] L. Zhang, S.-R. Chae, Z. Hendren, J.-S. Park, M.R. Wiesner, Recent advances in proton exchange membranes for fuel cell applications, Chem. Eng. J. 204–206 (2012) 87–97, https://doi.org/10.1016/J.CEJ.2012.07.103.

[39] S.J. Peighambardoust, S. Rowshanzamir, M. Amjadi, Review of the Proton Exchange Membranes for Fuel Cell Applications, Elsevier Ltd, 2010, https://doi.org/10.1016/j.ijhydene.2010.05.017.

[40] M. Piga, New Hybrid Inorganic-Organic Proton Conducting Membranes for PEMFC: Synthesis, Properties and Conduction Mechanisms, 2012.

[41] N.H. Jalani, Digital WPI Development of Nanocomposite Polymer Electrolyte Membranes for Higher Temperature PEM Fuel Cells, 2006.

[42] V. Neburchilov, J. Martin, H. Wang, J. Zhang, A review of polymer electrolyte membranes for direct methanol fuel cells, J. Power Sources 169 (2007) 221–238, https://doi.org/10.1016/j.jpowsour.2007.03.044.

[43] N. Shaari, S.K. Kamarudin, S. Basri, L.K. Shyuan, M.S. Masdar, D. Nordin, Enhanced proton conductivity and methanol permeability reduction via sodium alginate electrolyte-sulfonated graphene oxide bio-membrane, Nano-scale Res. Lett. 13 (2018) 82, https://doi.org/10.1186/s11671-018-2493-6.

[44] A. Aricò, D. Sebastian, M. Schuster, B. Bauer, C. D'Urso, F. Lufrano, V. Baglio, Selectivity of direct methanol fuel cell membranes, Membranes 5 (2015) 793–809, https://doi.org/10.3390/membranes5040793.

[45] Q. Li, R. He, J.O. Jensen, N.J. Bjerrum, Approaches and recent development of polymer electrolyte membranes for fuel cells operating above 100°C, Chem. Mater. 15 (2003) 4896–4915, https://doi.org/10.1021/cm0310519.

[46] M.H.D. Othman, A.F. Ismail, A. Mustafa, Physico-Chemical Study of Sulfonated Poly(Ether Ether Ketone) Membranes for Direct Methanol Fuel Cell Application, 2007.

[47] L. Carrette, K.A. Friedrich, U. Stimming, Fuel cells—fundamentals and applications, Fuel Cells 1 (2001) 5–39, https://doi.org/10.1002/1615-6854(200105)1:1<5::AID-FUCE5>3.0.CO;2-G.

[48] S. Slade, S.A. Campbell, T.R. Ralph, F.C. Walsh, Ionic conductivity of an extruded Nafion 1100 EW series of membranes, J. Electrochem. Soc. 149 (2002) A1556, https://doi.org/10.1149/1.1517281.

[49] J. Ling, O. Savadogo, Comparison of methanol crossover among four types of Nafion membranes, J. Electrochem. Soc. 151 (2004) A1604, https://doi.org/10.1149/1.1789394.

[50] H. Luo, Proton Conducting Polymer Composite Membrane Development for Direct Methanol Fuel Cell Applications, 2008.

[51] S.P. Jiang, Nanostructured and Advanced Materials for Fuel Cells, 2013.

[52] S. He, Y. Lin, H. Ma, H. Jia, X. Liu, J. Lin, Preparation of sulfonated poly(ether ether ketone) (SPEEK) membrane using ethanol/water mixed solvent, Mater. Lett. 169 (2016) 69–72, https://doi.org/10.1016/j.matlet.2016.01.099.

16

Carbonaceous nanomaterials (carbon nanotubes, fullerenes, and nanofibers) for alcohol fuel cells

Vildan Erduran[a,b], Muhammed Bekmezci[a,b], Ramazan Bayat[a,b], Iskender Isik[b], and Fatih Şen[a]

[a]Şen Research Group, Department of Biochemistry, Dumlupinar University, Kütahya, Turkey. [b]Department of Materials Science and Engineering, Faculty of Engineering, Dumlupinar University, Kütahya, Turkey

1 Introduction

The need for energy increases 4%–5% every year due to the growing population, growth in the industry, and developments in living standards throughout the world. By 2050, it is expected that the energy need will increase by 80%. Today, most of the world's energy needs are met by using fossil fuels. However, these fuels are decreasing gradually, and their use harms nature. The need for alternative, renewable energy that will meet energy needs without harming nature is increasing day by day [1, 2]. Fuel cells such as hydrogen, methanol, natural gas, and petroleum, which convert chemical energy into direct heat and electrical energy with high efficiency via electrochemical methods, are called fuel cells.

Today, energy production is provided by turbine generators and fossil fuels. However, these systems have problems such as cost, resource availability, harm to nature, and continuity of work. Fuel cells stand out in alternative energy production due to their positive features [3]. They produce electricity as a result of an electrochemical reaction, as in batteries and accumulators; the fuel contained in these devices is converted to electrical energy [4]. The electrical efficiency of fuel cells varies between 45% and 60%. In addition, as a result of the evaluation of the waste heat resulting from the electrochemical reaction, electrical efficiency can be up to 70%–80% [5]. Different types of fuels can be converted to electrical energy by reduction (cathodic) and oxidation

Nanomaterials for Direct Alcohol Fuel Cells. https://doi.org/10.1016/B978-0-12-821713-9.00021-4
Copyright © 2021 Elsevier Inc. All rights reserved.

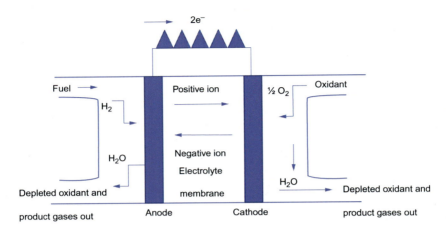

Fig. 1 The general scheme of a fuel cell [6].

(anodic) reactions in fuel cells (Fig. 1). The standard operation of a fuel cell consists of an oxidation reaction on the anode and an oxygen elimination reaction on the cathode [6].

Proton exchange membrane fuel cells (PEMFCs) and phosphoric acid fuel cells (PAFCs) use H_2 as fuel, whereas alcohols such as ethanol and methanol are used as fuel for direct alcohol fuel cells (DAFCs). Such fuel cells are often called "low-temperature fuel cells" because the working temperatures are below 200°C. Fuel cells operating at high temperatures (i.e., higher than 600°C) are known as solid oxide fuel cells (SOFCs) and molten carbonate fuel cells (MCFCs); they use hydrogen, natural gas, hydrocarbon, or coal as fuel. Fuel cells operated at low temperatures are of particular interest in the consumer market due to their light operating conditions and high safety factors [7]. In addition, these fuels cannot be oxidized without catalysts because of their low operating temperatures. The cathode also needs a catalyst to reduce O_2. It has been traditionally used as a catalyst in fuel cells, including noble metals such as Pt and Pd or their Pt/Ru and Pt/Pd alloys. Pt and Pd or their alloys, Pt/Pd, Pt/Ru, containing precious metals are used as catalysts in fuel cells [8–10].

In addition, just as their high costs prevent the commercialization of fuel cells, they narrow the application areas required for researchers to make the most of catalysts. For this reason, studies have been conducted on the nanoscaling of catalyst particles to create effective areas per unit of mass.

However, metallic catalyst particles carried to nanoscale must be loaded onto support materials that prevent them from being collected and/or sintered during synthesis and fuel cell operation. Fuel cell and hydrogen storage studies with different nanoparticles have also been performed [11–17]; improved electrocatalytic activities of nanocomposites were examined [18–23].

The ideal support material is expected to have properties such as (1) high electrical conductance; (2) interaction between catalysts and support materials; (3) high surface area; (4) a mesoporosity shape that will not be blocked by the ionomer, allowing the reactants to reach the catalyst nanoparticles easily; (5) acceptable abrasion resistance; and (6) simple recovery of the catalyst [24].

Because the high interaction between the catalyst and the support material can increase the activity of the catalyst, this reduces catalyst loss and manages the charge transport. The support material also increases catalyst performance by removal catalyst poisoning (CO, S). The support material, in some cases, also affects the particle size of the catalyst. Therefore, determining the support material to use is important and has a great impact on the catalyst's cost-effectiveness, life, performance, and behavior. For high catalyst effectiveness, we can subclassify the studies as follows: (1) research of Pt-based bimetallic and trimetallic catalysis processes and other nonprecious metals to reduce Pt dependence, and (2) developing catalyst support.

Many electrocatalyst support materials have been investigated. The fact that they have a wide surface area, high electronic conductivity, good chemical robustness, and high mesopores that allow good reactant flow and high metal distribution has made regular mesoporous carbons and nanostructured carbon materials such as carbon nanotubes (CNTs) and carbon nanofibers (CNFs) the center of attention [25].

2 Carbon nanomaterials

Carbon (C) is the most flexible element in the periodic table [26]. Therefore, the capacity of carbon orbitals to hybridize in sp., sp^2, and sp^3 configurations paves the way to a variety of allotropes [27]. Carbon has allotropes of different sizes, such as one-dimensional (1D) carbon nanotubes, two-dimensional (2D) graphene. Different nano sizes give these carbons excellent features compared to conventional macroscopic C; these materials were used as modern fuel cell catalysts and are discussed next.

2.1 Carbon nanotubes

CNTs are cylinder-shaped macromolecules with a scale as small as a few nanometers and as large as 20 cm wide. The walls of these tubes form hexagonal carbon atom lattices that are identical to the atomic planes of graphite. At their ends, they are bound with half of a fullerene-like compound. In the most common case, a CNT consists of a compact, multicylinder arrangement. Such multiwalled nanotubes (MWCNTs) may have diameters of up to 100 nm. A special case of such an MWCNT is the double-walled CNT, composed of only two condensed cylinders. Such nanotubes can achieve diameters of up to 100 nm [28,29]. The double-walled CNT, composed of only two concentric cylinders, is a rare example of such an MWCNT. CNTs are 2D nanostructures that exist when a single sheet of hexagonal C atoms is coiled (Fig. 2) [30].

They may be single-walled carbon nanotubes (SWCNTs) or MWCNTs. While the SWCNT has a higher surface area, the MWCNT is more conductive [31–33]. For catalyst deposition onto CNTs, techniques such as impregnation [34], ultrasonic assisting [35], polyol technique [36], spraying [37] colloidal [38], ion exchange [39], and electrochemical deposition [40] have been used.

Pure CNTs are chemically inert, and metal nanoparticles can't hold on to the surface. Therefore, research has focused on studies that enhance the hydrophilic characteristics of CNTs and the catalyst support relationship, making their surfaces functional with oxygen groups [32]. Ultrasonic treatment is another technique used to modify the surface of CNTs and achieve the formation

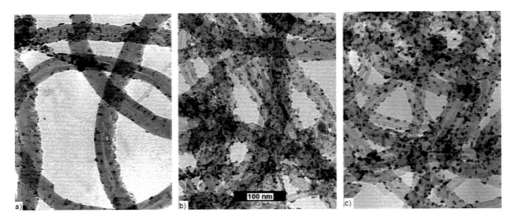

Fig. 2 CNT-supported catalysts (10% (A), 20% (B), and 30% (C) by weight) loaded with different amounts of Pt [30]
Reprinted (adapted) with permission from Y. Xing, Synthesis and electrochemical characterization of uniformly-dispersed high loading Pt nanoparticles on sonochemically-treated carbon nanotubes, J. Phys. Chem. B 108 (2004) 19255–19259. https://doi.org/10.1021/jp046697i. Copyright (2004) American Chemical Society.

of smaller and uniform nanoparticles [35]. These materials were not ideal candidates for support only in heterogeneous catalysts [41].

Lately, carbon nanotubes have been proposed to replace conventional carbon powders in fuel cells and have been directly used as electrode materials for both oxidation and reduction reactions in alcohol fuel cells [42–44]. SWNTs often create bundles that reduce the existing surface areas to support Pt particles. MWNTs are more often used for catalysts in fuel cells than SWNTs because of both their low production costs and conductivity, as mentioned previously.

Rajesh et al. examined the oxidation of methanol catalyzed by Pt, PtRu, and Pt-WO$_3$ nanoparticles supported in CNTs and Vulcan XC-72R. Pt-WO$_3$/CNTs have shown better catalytic efficiency and durability than Pt-Ru/Vulcan XC-72R and Pt-Ru/Vulcan XC-72R [20]. Possible reasons for this include the ability of CNT-based catalysts to create more efficient ternary-phase boundaries and specific metal support interactions; also, CNT-based catalysts had good conductivity and more organic cleanliness than conventional C black-based ones. According to Britto et al., a wide number of CNT defects, such as hexagons at the nanotube end and pairs of the hexagon-heptagon defects, were found to be useful in the oxygen reduction reaction [45].

Recently, Lin et al., [46], used the supercritical fluid (SCfs) method to set the Pt/CNT catalyst for DAFCs. It has been declared that SCfs technology can result in a cleaner, cheaper, and higher-quality products (and processes).

In producing high-performance working electrodes, the use of CNTs as catalyst support for DAFCs creates some problems. When the electrode is produced with a traditional ink procedure, it is estimated that approximately 20%–30% of the Pt catalyst is used due to the difficulty of reagents entering the internal electrocatalytic regions [47].

Carmo et al. investigated powder metal catalysts, MWNTs, and SWNTs, and high-surface-area C powders (Vulcan XC-72) supported catalysts for proton exchange membrane fuel cells (PEMFCs) and direct methanol fuel cells (DMFCs) for low-temperature fuel cells [48–50]. In CNTs, Ni and Fe materials have reduced CO poisoning. PtRu catalysts with CNT support have performed well in $H_2 + 100$ rpm oxidation (100 mV and 1 A/cm^2). This result is very close to the value obtained with Vulcan XC-72R.

In DMFCs, especially for PtRu/MWNT, power densities exceeding 100 mW/cm^2 were obtained at 90°C and 0.3 MPa. All the results showed that the materials produced by the installation of PtRu/MWCNT catalysts on MWCNTs produced the best results, and the performance was better than the classic Vulcan XC-72R. It is estimated that this type of carbon-supported material will be cheaper with the advancement of CNT technology.

In single-celled DMFC tests, Sun et al. [51–53] achieved very good results by using a CNT-assisted, Pt-Ru anode catalyst, a Nafion 117 membrane, graphite as an electrode collector, and an aqueous solution of CH_3OH as fuel. They indicated that they achieved successful test results in accordance with the high conductivity of the CNT, good physical morphology properties of the catalyst, and a compatible Mo-Ru/CNT composition. The anode and cathode were coated with 4 mg/cm^2 Pt-Ru/CNT electrocatalyst, resulting in a power density of 60°C and 0.21 V with more than 60 mW/cm^2. These values were found to be much better than those achieved by Pt-Ru/C carbon catalysts used in the market.

Recently, some new techniques [50–53] have been developed for 3D nanostructures that can benefit from the structural and electronic advantages of CNTs. With this method, Pt or Pt alloys are assumed to accumulate directly to CNT-based catalyst supports.

In addition to CNTs, new carbon forms such as fullerenes and nanofibers, which have become more common lately for the purpose of developing the efficiency of fuel cells, have attracted attention in recent years [54,55].

2.2 Carbon nanofibers

There are some distinctions between traditional carbon fiber (CFs) and CNFs. The most evident difference is their size. While the diameter of CFs is several micrometers, the diameter of the CNFs varies between 50 and 200 nm. Another difference is their structure and production methods. Typical conventional CFs are obtained from high-power polyacrylonitrile (PAN) or mesophase pitch. Preparation of various conditions, such as the oxidation atmosphere, selected raw materials, and heat treatment temperatures, form CFs with different properties. However, CNFs are mainly prepared in two ways: catalytic vapor deposition (CVD) growth and electrospinning.

Two types of CNF are produced by the CVD growth method: conical and platelet (Fig. 3). Conical CNFs exhibit most of the open edges on the outer surface, and the inner hollow core provides high chemical reactivity [56].

CNFs are thinner than CNTs and have no gaps. Their diameters are larger than CNTs, and they are divided into three groups, depending on the orientation of the nanofibers according to their growth axis. Platelet, ribbon, and fishbone structures are shown in Fig. 4 [58].

Chapter 16 Carbonaceous nanomaterials for alcohol fuel cells **309**

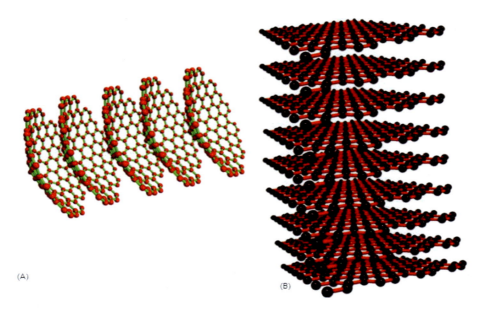

Fig. 3 (A) Conical CNF structure, (B) platelet CNF structure [57]. Reprinted (adapted) with permission from L. Feng, N. Xie, J. Zhong, Carbon nanofibers and their composites: a review of synthesizing, properties and applications, Materials (Basel) 7 (2014) 3919–3945. https://doi.org/10.3390/ma7053919. Copyright (2014) Materials (Basel).

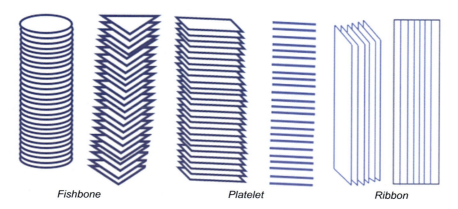

Fig. 4 Types of CNFs [46]. Reprinted (adapted) with permission from D.A. Bulushev, Noble metal nanoparticles on carbon fibers special issue, in: Advanced Catalysis in Hydrogen Production from Formic Acid and Methanol, View project, (2009) 3195–3205. https://doi.org/10.1081/E-ENN2-120034032. Copyright (2009) In *Dekker Encyclopedia of Nanoscience and Nanotechnology,* Second Edition. Taylor and Francis: New York.

These forms of nanofibers were tested as support materials for fuel cells in catalysts. Particles loaded with 5% Pt by weight were accumulated on the platelet and ribbon types of CNF and showed similar activities with 25% Pt/Vulcan XC-72R by weight [59]. In addition, CNF-supported electrocatalysts are less susceptible to CO poisoning than C black-supported Pt.

The cause of this high activity in CNFs is the crystallographic orientation of CNFs, which interact well with catalyst particles. On the surface of herringbone-structured CNFs [54,60], another Pt-Ru bimetallic catalyst was synthesized. Pt-Ru/CNFs demonstrated the most activity at 0.23 A/cm^2 DMFC at 0.4 V when compared with the same catalyst loaded onto MWNTs and SWNTs.

Guo et al. [61], used oxidized and reduced carbon nanofibers (OCNFs and RCNFs) as support for Pt-Ru catalysts in DAFCs, and examined in detail the structure, surface, and electrocatalytic properties of PtRu/OCNFs and PtRu/RCNFs. While PtRu/OCNFs performed better in DMFCs at low and medium current densities than PtRu/RCNF, the opposite was the case for high current densities. For all current densities, PtRu/OCNF anode DMFCs were found to perform better than DMFCs using catalysts currently on the market.

2.3 Fullerenes

Carbon is one of the most common elements in the world. Pure carbon exists in two main crystalline forms: graphite and diamond. Fulgens are hollow structures made of carbon. They can be found in spheres, ellipsoids, or tubes. Fullerenes are made of graphene and have a graphitelike structure connected by hexagonal, pentagonal, or heptagonal rings [62, 63]. Fullerenes, on the other hand, are known for their molecular forms [64]. The molecular structure of C_{60} is included in Fig. 5 [65].

Buckminsterfullerene, or C_{60}, is the smallest form of fullerene and the most abundant in nature. It is a spherical molecule

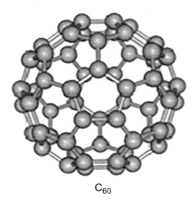

Fig. 5 The structure of fullerene C_{60} [66].

consisting of 60 carbon atoms. The sphere is hollow and has pentagons and hexagons on its surface. C_{60} was first obtained in 1985 by Smalley, Curl, and Kroto by evaporating and condensing graphite in an inert gas environment; this discovery earned them the 1996 Nobel Prize in Chemistry.

Among other fullerenes, C_{60} and C_{70} have a more stable structure, so the research literature contains information about their properties, especially in solution and solid state. The heat of producing of C_{60} and C_{70} was calorimetrically examined and found to be 10.16 kcal/mol per carbon atom for C_{60} and 9.65 kcal/mol per carbon atom for C_{70} [67]. For graphite, this temperature is 0 kcal/mol, while for diamonds, it is 0.4 kcal/mol [68]. Therefore, fullerenes are less steady than graphite and diamond. Each fullerene consists of 12 pentagons and 20 hexagons with a C atom number of 2 (10 + M). This structure is a result of Euler's theorem [69]. Of all the fullerenes, C_{60} is the most stable [70]. The color of C_{60} in toluene and benzene solution is dark violet.

To be able to modify chemically fullerenes, they must be present in solution. C_{60} and C_{70} are soluble in many organic solvents. With the increase in diameter, the solubility of fullerenes decreases. Fullerenes are virtually insoluble in acetone, alcohol, and ether, and their solubility is expressed in mg/mL. C_{70} has a 50% lower resolution in these solvents than C_{60}. Fullerene derivatives, on the other hand, have more solubility than pure C_{60} and C_{70}. To chemically modify fullerenes, they need to be soluble in solution. C_{60} and C_{70} are soluble in many organic solvents [71]. But acetone, alcohol, and ether are also very little soluble [72]. Because of its energy, electron transfer process, and sensitivity to chemical reactions, C_{60} is a remarkable substance in electrochemistry [73]. The high surface area of these fullerene films mainly makes them well suited for fuel cell applications.

Vinodgopal et al. conducted research examining the ability of C_{60} clusters to function as a new generation of C support for fuel cell applications. It is aimed to develop the efficiency of optically transparent electrodes (OTE)/C_{60}/Pt electrodes by modifying electrochemical deposition methods [55].

In other experiments, C products have been tested as substrate-providing catalysts for fuel cells. The usage of large-surface-area C nanomaterials (CNT, CNF, and fullerenes) has developed modern C-based materials. These new supports outperformed commercial carbon blacks in many studies, through unique structures. Because C nanomaterials are prepared by the bottom-up synthesis method, they are encouraging as catalyst support for fuel cells. Thus, theoretically, it may have a high surface area.

3 Conclusion

The need for energy is increasing rapidly every day. The inadequate traditional methods, the depletion of fossil fuels, pollution of nature caused by fossil fuels, and the existing systems are insufficient to meet these needs. Fuel cells have the potential to solve problems in existing systems. Fuel cells produce electricity as a result of an electrochemical reaction. The electrical efficiency of fuel cells varies between 45% and 60%. Different types of fuels can be converted to electrical energy by reduction (cathodic) and oxidation (anodic) reactions in fuel cells.

Carbon is the most flexible element in the periodic table. CNTs of varying types have been used as supporting agents in fuel cells. Developing and using new materials originating from carbon as a catalyst will potentially solve the energy problem by eliminating the existing problems with using fuel cells.

References

[1] P. Nikolaidis, A. Poullikkas, A comparative overview of hydrogen production processes, Renew. Sust. Energ. Rev. 67 (2017) 597–611, https://doi.org/10.1016/j.rser.2016.09.044.

[2] L. Gong, Q. Duan, J. Liu, M. Li, P. Li, K. Jin, J. Sun, Spontaneous ignition of high-pressure hydrogen during its sudden release into hydrogen/air mixtures, Int. J. Hydrog. Energy 43 (2018) 23558–23567, https://doi.org/10.1016/j.ijhydene.2018.10.226.

[3] Y.H. Kwok, Y. Wang, M. Wu, F. Li, Y. Zhang, H. Zhang, D.Y.C. Leung, A dual fuel microfluidic fuel cell utilizing solar energy and methanol, J. Power Sources 409 (2019) 58–65, https://doi.org/10.1016/j.jpowsour.2018.10.095.

[4] L.J. Pettersson, R. Westerholm, State of the art of multi-fuel reformers for fuel cell vehicles: problem identification and research needs, Int. J. Hydrog. Energy 26 (2001) 243–264, https://doi.org/10.1016/S0360-3199(00)00073-2.

[5] M.N. Lakhoua, M. Harrabi, M. Lakhou, Application of system analysis for thermal power plant heat rate improvement, Therm. Power Plants Adv. Appl. (2013), https://doi.org/10.5772/55498.

[6] A. Kirubakaran, S. Jain, R.K. Nema, A review on fuel cell technologies and power electronic interface, Renew. Sust. Energ. Rev. 13 (2009) 2430–2440, https://doi.org/10.1016/j.rser.2009.04.004.

[7] P. Costamagna, S. Srinivasan, Quantum jumps in the PEMFC science and technology from the 1960s to the year 2000, J. Power Sources 102 (2001) 242–252, https://doi.org/10.1016/S0378-7753(01)00807-2.

[8] M. Neergat, D. Leveratto, U. Stimming, Catalysts for direct methanol fuel cells, Fuel Cells 2 (2002) 25–30, https://doi.org/10.1002/1615-6854(20020815)2:1<25::AID-FUCE25>3.0.CO;2-4.

[9] X. Ren, High performance direct methanol polymer electrolyte fuel cells, J. Electrochem. Soc. 143 (1996) L12, https://doi.org/10.1149/1.1836375.

[10] N.V. Long, T. Duy Hien, T. Asaka, M. Ohtaki, M. Nogami, Synthesis and characterization of Pt–Pd alloy and core-shell bimetallic nanoparticles for direct methanol fuel cells (DMFCs): enhanced electrocatalytic properties of well-

Chapter 16 Carbonaceous nanomaterials for alcohol fuel cells **313**

shaped core-shell morphologies and nanostructures, Int. J. Hydrog. Energy 36 (2011) 8478–8491, https://doi.org/10.1016/j.ijhydene.2011.03.140.

[11] S. Şen, F. Şen, G. Gökağaç, Preparation and characterization of nano-sized Pt-Ru/C catalysts and their superior catalytic activities for methanol and ethanol oxidation, Phys. Chem. Chem. Phys. 13 (2011) 6784–6792, https://doi.org/10.1039/c1cp20064j.

[12] Z. Daşdelen, Y. Yıldız, S. Eriş, F. Şen, Enhanced electrocatalytic activity and durability of Pt nanoparticles decorated on GO-PVP hybride material for methanol oxidation reaction, Appl. Catal. B Environ. 219 (2017) 511–516, https://doi.org/10.1016/j.apcatb.2017.08.014.

[13] B. Sen, B. Demirkan, B. Şimşek, A. Savk, F. Sen, Monodisperse palladium nanocatalysts for dehydrocoupling of dimethylamineborane, Nano-Struct. Nano-Obj. 16 (2018) 209–214, https://doi.org/10.1016/j.nanoso.2018.07.008.

[14] B. Sen, E. Kuyuldar, A. Şavk, H. Calimli, S. Duman, F. Sen, Monodisperse ruthenium–copper alloy nanoparticles decorated on reduced graphene oxide for dehydrogenation of DMAB, Int. J. Hydrog. Energy 44 (2019) 10744–10751, https://doi.org/10.1016/j.ijhydene.2019.02.176.

[15] B. Şen, B. Demirkan, M. Levent, A. Şavk, F. Şen, Silica-based monodisperse PdCo nanohybrids as highly efficient and stable nanocatalyst for hydrogen evolution reaction, Int. J. Hydrog. Energy 43 (2018) 20234–20242, https://doi.org/10.1016/j.ijhydene.2018.07.080.

[16] B. Şen, B. Demirkan, A. Savk, R. Kartop, M.S. Nas, M.H. Alma, S. Sürdem, F. Şen, High-performance graphite-supported ruthenium nanocatalyst for hydrogen evolution reaction, J. Mol. Liq. 268 (2018) 807–812, https://doi.org/10.1016/j.molliq.2018.07.117.

[17] S. Eris, Z. Daşdelen, F. Sen, Investigation of electrocatalytic activity and stability of Pt@f-VC catalyst prepared by in-situ synthesis for methanol electrooxidation, Int. J. Hydrog. Energy 43 (2018) 385–390, https://doi.org/10.1016/j.ijhydene.2017.11.063.

[18] Ö. Karatepe, Y. Yildiz, H. Pamuk, S. Eris, Z. Dasdelen, F. Sen, Enhanced electrocatalytic activity and durability of highly monodisperse Pt@PPy-PANI nanocomposites as a novel catalyst for the electro-oxidation of methanol, RSC Adv. 6 (2016) 50851–50857, https://doi.org/10.1039/c6ra06210e.

[19] Y. Yildiz, E. Erken, H. Pamuk, H. Sert, F. Şen, Monodisperse Pt nanoparticles assembled on reduced graphene oxide: highly efficient and reusable catalyst for methanol oxidation and dehydrocoupling of dimethylamine-borane (DMAB), J. Nanosci. Nanotechnol. 16 (2016) 5951–5958, https://doi.org/10.1166/jnn.2016.11710.

[20] Y. Yıldız, S. Kuzu, B. Sen, A. Savk, S. Akocak, F. Şen, Different ligand based monodispersed Pt nanoparticles decorated with rGO as highly active and reusable catalysts for the methanol oxidation, Int. J. Hydrog. Energy 42 (2017) 13061–13069, https://doi.org/10.1016/j.ijhydene.2017.03.230.

[21] F. Şen, S. Şen, G. Gökağaç, Efficiency enhancement of methanol/ethanol oxidation reactions on Pt nanoparticles prepared using a new surfactant, 1,1-dimethyl heptanethiol, Phys. Chem. Chem. Phys. 13 (2011) 1676–1684, https://doi.org/10.1039/c0cp01212b.

[22] S. Eris, Z. Daşdelen, Y. Yıldız, F. Sen, Nanostructured polyaniline-rGO decorated platinum catalyst with enhanced activity and durability for methanol oxidation, Int. J. Hydrog. Energy 43 (2018) 1337–1343, https://doi.org/10.1016/j.ijhydene.2017.11.051.

[23] B. Çelik, S. Kuzu, E. Erken, H. Sert, Y. Koşkun, F. Şen, Nearly monodisperse carbon nanotube furnished nanocatalysts as highly efficient and reusable catalyst for dehydrocoupling of DMAB and C1 to C3 alcohol oxidation, Int. J.

Hydrog. Energy 41 (2016) 3093–3101, https://doi.org/10.1016/j.ijhydene.2015.12.138.

[24] N. Jha, A. Leelamohanareddy, M. Shaijumon, N. Rajalakshmi, S. Ramaprabhu, Pt–Ru/multi-walled carbon nanotubes as electrocatalysts for direct methanol fuel cell, Int. J. Hydrog. Energy 33 (2008) 427–433, https://doi.org/10.1016/j.ijhydene.2007.07.064.

[25] E. Antolini, Composite materials: an emerging class of fuel cell catalyst supports, Appl. Catal. B Environ. 100 (2010) 413–426, https://doi.org/10.1016/J.APCATB.2010.08.025.

[26] G.A. Silva, et al., BMC Neurosci. 4 (2008) 1–4, https://doi.org/10.1186/1471-2202-9-S3-S4.

[27] D. Pantarotto, J.P. Briand, M. Prato, A. Bianco, Translocation of bioactive peptides across cell membranes by carbon nanotubes, Chem. Commun. 4 (2004) 16–17, https://doi.org/10.1039/b311254c.

[28] S. Iijima, T. Ichihashi, Single-shell carbon nanotubes of 1-nm diameter, Nature 363 (1993) 603–605, https://doi.org/10.1038/363603a0.

[29] K. Balasubramanian, M. Burghard, Chemically functionalized carbon nanotubes, Small 1 (2005) 180–192, https://doi.org/10.1002/smll.200400118.

[30] J.M. Tascón, Carbon-based catalyst support in fuel cell applications, in: Nov. Carbon Adsorbents, Elsevier Ltd, 2012, p. 564, https://doi.org/10.1016/B978-0-08-097744-7.00018-1.

[31] G. Che, B.B. Lakshmi, E.R. Fisher, C.R. Martin, Carbon nanotubule membranes for electrochemical energy storage and production, Nature 393 (1998) 346–349, https://doi.org/10.1038/30694.

[32] S. Maass, F. Finsterwalder, G. Frank, R. Hartmann, C. Merten, Carbon support oxidation in PEM fuel cell cathodes, J. Power Sources 176 (2008) 444–451, https://doi.org/10.1016/j.jpowsour.2007.08.053.

[33] D.B. Mawhinney, V. Naumenko, A. Kuznetsova, J.T. Yates, J. Liu, R.E. Smalley, Infrared spectral evidence for the etching of carbon nanotubes: ozone oxidation at 298 K [7], J. Am. Chem. Soc. 122 (2000) 2383–2384, https://doi.org/10.1021/ja994094s.

[34] T. Matsumoto, T. Komatsu, K. Arai, T. Yamazaki, M. Kijima, H. Shimizu, Y. Takasawa, J. Nakamura, Reduction of Pt usage in fuel cell electrocatalysts with carbon nanotube electrodes, Chem. Commun. (2004) 840, https://doi.org/10.1039/b400607k.

[35] C. Yang, X. Hu, D. Wang, C. Dai, L. Zhang, H. Jin, S. Agathopoulos, Ultrasonically treated multi-walled carbon nanotubes (MWCNTs) as PtRu catalyst supports for methanol electrooxidation, J. Power Sources 160 (2006) 187–193, https://doi.org/10.1016/j.jpowsour.2006.05.015.

[36] S.L. Knupp, W. Li, O. Paschos, T.M. Murray, J. Snyder, P. Haldar, The effect of experimental parameters on the synthesis of carbon nanotube/nanofiber supported platinum by polyol processing techniques, Carbon (2008), https://doi.org/10.1016/j.carbon.2008.05.007.

[37] J. Lee, J. Kim, T. Hyeon, Recent progress in the synthesis of porous carbon materials, Adv. Mater. 18 (2006) 2073–2094, https://doi.org/10.1002/adma.200501576.

[38] L. Guo, L. Zhang, J. Zhang, J. Zhou, Q. He, S. Zeng, X. Cui, J. Shi, Hollow mesoporous carbon spheres—an excellent bilirubin adsorbent, Chem. Commun. (2009) 6071–6073, https://doi.org/10.1039/b911083f.

[39] K. Yasuda, Y. Nishimura, The deposition of ultrafine platinum particles on carbon black by surface ion exchange—increase in loading amount, Mater. Chem. Phys. 82 (2003) 921–928, https://doi.org/10.1016/j.matchemphys.2003.08.020.

Chapter 16 Carbonaceous nanomaterials for alcohol fuel cells **315**

[40] T.M. Day, P.R. Unwin, N.R. Wilson, J.V. Macpherson, Electrochemical templating of metal nanoparticles and nanowires on single-walled carbon nanotube networks, J. Am. Chem. Soc. 127 (2005) 10639–10647, https://doi.org/10.1021/ja051320r.

[41] J.Z. Luo, L.Z. Gao, Y.L. Leung, C.T. Au, The decomposition of NO on CNTs and 1 wt% Rh/CNTs, Catal. Lett. 66 (2000) 91–97, https://doi.org/10.1023/A:1019035220233.

[42] W. Li, C. Liang, W. Zhou, J. Qiu, Z. Zhou, G. Sun, Q. Xin, Preparation and characterization of multiwalled carbon nanotube-supported platinum for cathode catalysts of direct methanol fuel cells, J. Phys. Chem. B 107 (2003) 6292–6299, https://doi.org/10.1021/jp022505c.

[43] B. Rajesh, K. Ravindranathan Thampi, J.M. Bonard, N. Xanthopoulos, H.J. Mathieu, B. Viswanathan, Carbon nanotubes generated from template carbonization of polyphenyl acetylene as the support for electrooxidation of methanol, J. Phys. Chem. B 107 (2003) 2701–2708, https://doi.org/10.1021/jp0219350.

[44] Z. Liu, J.Y. Lee, W. Chen, M. Han, L.M. Gan, Physical and electrochemical characterizations of microwave-assisted polyol preparation of carbon-supported PtRu nanoparticles, Langmuir 20 (2004) 181–187, https://doi.org/10.1021/la035204i.

[45] P.J. Britto, K.S.V. Santhanam, A. Rubio, J.A. Alonso, P.M. Ajayan, Improved charge transfer at carbon nanotube electrodes, Adv. Mater. 11 (1999) 154–157, https://doi.org/10.1002/(SICI)1521-4095(199902)11:2<154::AID-ADMA154>3.0.CO;2-B.

[46] Y. Lin, X. Cui, C. Yen, C.M. Wai, Platinum/carbon nanotube nanocomposite synthesized in supercritical fluid as electrocatalysts for low-temperature fuel cells, J. Phys. Chem. B 109 (2005) 14410–14415, https://doi.org/10.1021/jp0514675.

[47] G. Girishkumar, K. Vinodgopal, P.V. Kamat, Carbon nanostructures in portable fuel cells: single-walled carbon nanotube electrodes for methanol oxidation and oxygen reduction, J. Phys. Chem. B 108 (2004) 19960–19966, https://doi.org/10.1021/jp046872v.

[48] M. Carmo, V.A. Paganin, J.M. Rosolen, E.R. Gonzalez, Alternative supports for the preparation of catalysts for low-temperature fuel cells: the use of carbon nanotubes, J. Power Sources 142 (2005) 169–176, https://doi.org/10.1016/j.jpowsour.2004.10.023.

[49] K.T. Jeng, C.C. Chien, N.Y. Hsu, S.C. Yen, S. Du Chiou, S.H. Lin, W.M. Huang, Performance of direct methanol fuel cell using carbon nanotube-supported Pt-Ru anode catalyst with controlled composition, J. Power Sources 160 (2006) 97–104, https://doi.org/10.1016/j.jpowsour.2006.01.057.

[50] C. Wang, M. Waje, X. Wang, J.M. Tang, R.C. Haddon, Yan, proton exchange membrane fuel cells with carbon nanotube based electrodes, Nano Lett. 4 (2004) 345–348, https://doi.org/10.1021/nl034952p.

[51] X. Sun, R. Li, B. Stansfield, J.P. Dodelet, S. Désilets, 3D carbon nanotube network based on a hierarchical structure grown on carbon paper backing, Chem. Phys. Lett. 394 (2004) 266–270, https://doi.org/10.1016/j.cplett.2004.07.014.

[52] X. Sun, R. Li, G. Lebrun, B. Stansfield, J.P. Dodelet, S. Désilets, Formation of carbon nanotubes on carbon paper and stainless steel screen by ohmically heating catalytic sites, Int. J. Nanosci. 01 (2002) 223–234, https://doi.org/10.1142/s0219581x02000309.

[53] W. Li, X. Wang, Z. Chen, M. Waje, Y. Yan, Carbon nanotube film by filtration as cathode catalyst support for proton-exchange membrane fuel cell, Langmuir 21 (2005) 9386–9389, https://doi.org/10.1021/la051124y.

[54] E.S. Steigerwalt, G.A. Deluga, D.E. Cliffel, C.M. Lukehart, A Pt-Ru/graphitic carbon nanofiber nanocomposite exhibiting high relative performance as a direct-methanol fuel cell anode catalyst, J. Phys. Chem. B 105 (2001) 8097–8101, https://doi.org/10.1021/jp011633i.

[55] K. Vinodgopal, M. Haria, D. Meisel, P. Kamat, Fullerene-based carbon nanostructures for methanol oxidation, Nano Lett. 4 (2004) 415–418, https://doi.org/10.1021/nl035028y.

[56] M. Endo, Y.A. Kim, T. Hayashi, T. Yanagisawa, H. Muramatsu, M. Ezaka, H. Terrones, M. Terrones, M.S. Dresselhaus, Microstructural changes induced in "stacked cup" carbon nanofibers by heat treatment, Carbon 41 (2003) 1941–1947, https://doi.org/10.1016/S0008-6223(03)00171-4.

[57] L. Feng, N. Xie, J. Zhong, Carbon nanofibers and their composites: a review of synthesizing, properties and applications, Materials (Basel) 7 (5) (2014) 3919–3945, https://doi.org/10.3390/ma7053919.

[58] D.A. Bulushev, Noble metal nanoparticles on carbon fibers special issue, in: Advanced Catalysis in Hydrogen Production from Formic Acid and Methanol, 2009, pp. 3195–3205, https://doi.org/10.1081/E-ENN2-120034032. View project.

[59] C.A. Bessel, K. Laubernds, N.M. Rodriguez, R.T.K. Baker, Graphite nanofibers as an electrode for fuel cell applications, J. Phys. Chem. B 105 (2001) 1121–1122, https://doi.org/10.1021/jp003280d.

[60] E.S. Steigerwalt, G.A. Deluga, C.M. Lukehart, Pt-Ru/carbon fiber nanocomposites: synthesis, characterization, and performance as anode catalysts of direct methanol fuel cells. A search for exceptional performance, J. Phys. Chem. B 106 (2002) 760–766, https://doi.org/10.1021/jp012707t.

[61] J. Guo, G. Sun, Q. Wang, G. Wang, Z. Zhou, S. Tang, L. Jiang, B. Zhou, Q. Xin, Carbon nanofibers supported Pt-Ru electrocatalysts for direct methanol fuel cells, Carbon 44 (2006) 152–157, https://doi.org/10.1016/j.carbon.2005.06.047.

[62] D.L.V.K. Prasad, E.D. Jemmis, Stuffing improves the stability of fullerenelike boron clusters, Phys. Rev. Lett. 100 (2008), https://doi.org/10.1103/PhysRevLett.100.165504.

[63] E. Oberdörster, Nanotechnology: environmental implications and solutions, Environ. Health Perspect. 113 (2005) 70–79, https://doi.org/10.1289/ehp.113-1257682.

[64] J.N. Ding, Y. Fan, C.X. Zhao, Y.B. Liu, C.T. Yu, N.Y. Yuan, Electrical conductivity of waterborne polyurethane/graphene composites prepared by solution mixing, J. Compos. Mater. 46 (2012) 747–752, https://doi.org/10.1177/0021998311413835.

[65] J.Y. Huang, H. Yasuda, H. Mori, Highly curved carbon nanostructures produced by ball-milling, Chem. Phys. Lett. 303 (1999) 130–134, https://doi.org/10.1016/S0009-2614(99)00131-1.

[66] C. Wang, Z.X. Guo, S. Fu, W. Wu, D. Zhu, Polymers containing fullerene or carbon nanotube structures, Prog. Polym. Sci. 29 (2004) 1079–1141, https://doi.org/10.1016/j.progpolymsci.2004.08.001.

[67] R.L. Murry, D.L. Strout, G.E. Scuseria, Theoretical studies of fullerene annealing and fragmentation, Int. J. Mass Spectrom. 138 (1994) 113–131, https://doi.org/10.1016/0168-1176(94)04037-0.

[68] H.-D. Beckhaus, S. Verevkin, C. Rüchardt, F. Diederich, C. Thilgen, H.-U. Ter Meer, H. Mohn, W. Müller, C70 ist stabiler als C60: experimentelle Bestimmung der Bildungswärme von C70, Angew. Chem. 106 (1994) 1033–1035, https://doi.org/10.1002/ange.19941060916.

Chapter 16 Carbonaceous nanomaterials for alcohol fuel cells **317**

[69] A. Hirsch, I. Lamparth, H.R. Karfunkel, Fullerene chemistry in three dimensions: isolation of seven regioisomeric bisadducts and chiral trisadducts of C60 and di(ethoxycarbonyl)methylene, Angew. Chem. Int. Ed. Eng. 33 (1994) 437–438, https://doi.org/10.1002/anie.199404371.

[70] H.W. Kroto, The stability of the fullerenes Cn, with n = 24, 28, 32, 36, 50, 60 and 70, Nature 329 (1987) 529–531, https://doi.org/10.1038/329529a0.

[71] N. Sivaraman, R. Dhamodaran, I. Kaliappan, T.G. Srinivasan, P.R.V. Rao, C.K. Mathews, Solubility of C60 in organic solvents, J. Organomet. Chem. 57 (1992) 6077–6079, https://doi.org/10.1021/jo00048a056.

[72] R.S. Ruoff, D.S. Tse, R. Malhotra, D.C. Lorents, Solubility of C60 in a variety of solvents, J. Phys. Chem. 97 (1993) 3379–3383, https://doi.org/10.1021/j100115a049.

[73] Z. Da Meng, L. Zhu, S. Ye, Q. Sun, K. Ullah, K.Y. Cho, W.C. Oh, Fullerene modification CdSe/TiO_2 and modification of photocatalytic activity under visible light, Nanoscale Res. Lett. 8 (2013) 1–10, https://doi.org/10.1186/1556-276X-8-189.

17

Carbon-based nanomaterials for alcohol fuel cells

Merve Akin[a,b], Ramazan Bayat[a,b], Vildan Erduran[a,b], Muhammed Bekmezci[a,b], Iskender Isik[b], and Fatih Şen[a]

[a]Şen Research Group, Department of Biochemistry, Dumlupinar University, Kütahya, Turkey. [b]Department of Materials Science and Engineering, Faculty of Engineering, Dumlupinar University, Kütahya, Turkey

1 Introduction

With the increase in the world population, the need for energy is increasing day by day. Renewable energy resources are needed to solve problems such as environmental pollution caused by energy resources used today (especially fossil fuels), and the depletion of resources. Fuel cells are promising as a renewable energy source. Many types of research are carried out on the effective use of fuel cells. According to these studies, the materials to be used in fuel cells must be performance-enhancing, cheap, and environmentally friendly. Fuel cells consist of an electrolyte sandwiched between an anode and a cathode. A fuel cell is an electrochemical device in which electrocatalysts oxidize fuel at the anode, reducing oxygen at the cathode [1]. Electricity is produced by the reaction of hydrogen and oxygen.

The use of pure hydrogen in fuel cells brings difficulties in storage. As an alternative, alcohol as a fuel is one of the types that are more environmentally friendly and easier to manufacture, transport, and store than pure hydrogen. However, they need catalyst in order to activate alcohol oxidation reaction. Catalysts are components that play an important role in fuel cells and accelerate the reactions. According to studies, the most used catalyst in fuel cells is platinum (Pt). Because Pt is expensive and also is affected by CO in alcohol fuel cells, various catalyst support materials have been investigated for a solution. It is important to increase the electro-oxidation reaction rate of alcohol by increasing the temperature or using an active electrocatalyst [2]. At this point, the use of carbon-

Nanomaterials for Direct Alcohol Fuel Cells. https://doi.org/10.1016/B978-0-12-821713-9.00025-1
Copyright © 2021 Elsevier Inc. All rights reserved.

based nanomaterials in fuel cells holds much promise. Recently, carbon-assisted electrocatalysts have been used in direct studies such as methanol fuel cells [3]. Carbon-based nanomaterials with high surface area and good electrical conductivity have attracted the attention of many researchers, and their use in fuel cells has increased as a result.

This chapter describes fuel cells and their components. The advantages of using carbon-based nanomaterials in alcohol fuel cells are demonstrated and information about their application areas is given.

2 Fuel cells

For the effective use of renewable energy sources, it is important to synthesize low-cost, high-performance, and environmentally friendly materials [4]. Fuel cells, discovered by William Grove, is a technology that creates energy by using various sources, including hydrogen and methanol [5].

The fuel cell is an environmentally friendly device that converts chemical energy directly into electrical energy. It has a high energy conversion efficiency and low emissions [6]. Fuel cells are essential for promising electrochemical energy conversion and storage [7].

Fuel cells consist of an anode, a cathode, and a separation membrane [7]. In fuel cells, the electrolyte is placed between the anode and the cathode, preventing the short circuit that may occur between the anode and the cathode [8]. It also creates heat as a by-product by directly converting the chemical energy of fuel and oxygen into electricity [9]. Unlike batteries, fuel cells do not run out, so long as fuel and oxidants are supplied, and they do not need to be recharged [10]. Fuel cells must be able to feed reagents, obtain reaction products, and maintain proper operating conditions [11]. Depending on the types of fuel cells, there are differences in efficiency rates. As the availability of the heat released from the fuel cell increases, its efficiency increases by up to 80%.

Fuel cell types can be determined according to the area to be used. Fuel cell types include direct methanol fuel cells (DMFCs), proton exchange membrane fuel cells (PEMFCs), solid oxide fuel cells (SOFCs), phosphoric acid fuel cells (PAFCs), and alkaline fuel cells (AFCs) [12]. Fuel cells can be a source of energy for many devices. It can be used in many areas, such as electronic devices, electric vehicles, mobile phones, computers, spacecraft, and in the home [13]. Many researchers think that the use of fuel cells will increase soon.

2.1 Basic working principle of fuel cells

The material used in the fuel cell (hydrogen) is given by the anode and oxygen (air) by the cathode. At the anode, hydrogen atoms split into their electrons and protons. The electrolytes (membranes) in the anode and cathode do not allow electrons to pass. Protons pass through the membrane and reach the cathode. Electrons, on the other hand, reach the cathode in a different way (Fig. 1). In this external path, electricity is produced, and heat is released. Water is released when oxygen and hydrogen are fed to the cathode. Because a simple cell will produce a small amount of power, sufficient electrical current can be provided by combining unit cells in bulk [14]. By applying cogeneration of waste heat to fuel cells, fuel efficiency can increase by 90% [15]. Catalysts are used to increase the performance of fuel cells; they accelerate reactions at the anode and cathode [12]. Pt is the most used material as an electrocatalyst. However, different materials have been investigated for use as electrocatalysts due to the cost and limited supply of Pt. Hydration of the membranes in the fuel cell is required for proton conductivity, but the excess water in the electrodes reduces the performance by carrying the electrode [16]. It is important to maintain this balance in fuel cells.

2.2 Fuel cell components

2.2.1 Membranes

Membranes are the layer sandwiched between the anode and the cathode. It separates the anode and cathode reactants. It has ionic conductivity and a structure that does not allow electron

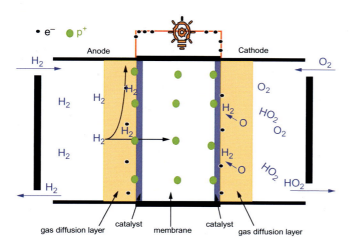

Fig. 1 The basic working principle of a fuel cell [17].

transfer. The operation of the fuel cell also depends on the moistening of the membrane and is important for ensuring ionic conductivity [18].

2.2.2 Gas diffusion layers

Gas diffusion layers (GDLs) are obtained from conductive and porous materials such as carbon cloth with a thickness in the range of 100–400 μm [19]. These layers are located between flow channels and catalyst layers [20]. It helps to reduce contact resistance by being between these layers, and it is effective in spreading GDL reactants into the catalyst [21]. It plays a role in the transport of gas and electrons to the required places in the membrane electrode assembly (MEA) [22]. The GDL must carry water and electrons to the oxygen and hydrogen catalyst layers and conduct heat as well [23].

2.2.3 Bipolar plates

Bipolar (BP) plates are positively charged from one side to the other based on the anode and cathode. They distribute fuel and air homogeneously and ensure the transfer of electric current between cells [24]. BP plates need to remove heat from active areas and prevent gas leaks [25]. Another function of the plates is to take products out of the cell. BPs must be resistant to low field contact resistance, high conductivity, and corrosion in the GDL/BP interface [26].

2.2.4 Catalysts

A catalyst is a substance that changes the reaction rate. For good performance, a catalyst must be used in the fuel cell. Catalysts are not consumed during the reaction and cannot change the reaction balance [27]. The cathode and anode catalyst layers are the basic elements of the fuel cell and are necessary for reactions to take place [28]. Electro-oxidation reaction or electroreduction should increase the rate of the reaction [22]. Catalysts are important for the oxygen reduction reaction (ORR) in fuel cells [29]. For the fuel cell cycle, the interaction of catalyst support materials with the catalyst, width of surface area, electrical conductivity, and corrosion resistance are important [30]. The durability and stability of the MEA in the fuel cell system are affected by the catalysts used. Electrocatalytic activities of catalysts depend on the support materials, distribution, and size [31].

Studies show that the most used catalysts for fuel cells are Pt and PtRu based. Specific mass activities, instabilities, and high costs make PtRu-based electrocatalysts insufficient for fuel cell applications [32]. Therefore, new catalysts are being investigated to improve the fuel electro-oxidation kinetic performance used in fuel cells [33]. The purposes of new catalysts are to increase the surface area and to minimize the use of support materials and metal. Carbon-supported catalysts support material distribution and electron transport [9]. The pore size of the carbon support plays an important role in mass transport by providing high catalytic [34]. The use of nanoparticle-supported materials as catalysts holds promise for fuel cells [33]. Carbon nanotubes (CNTs) are considered more attractive due to their outstanding mechanical properties, such as high tensile strength combined with high surface area [35].

3 Alcohol fuel cells

The use of pure hydrogen in fuel cells increases the cost of the system. A range of fuels has been the focus of research because of the difficult processes and storage problems connected with hydrogen. Liquid fuels are alternatives that can be used instead of pure hydrogen. Of these fuel cell types, alcohol fuel cells use ethylene glycol, ethanol, and methanol and have low costs. They provide improved storage and distribution of alcohol fuels compared to pure hydrogen. According to research, methanol is better to use than other types of alcohol due to its high electrochemical kinetics. Alcohols show corrosion resistance to acidic and basic environments, as well as low density [34]. It is important to develop poison-resistant electrocatalysts for alcohol fuel cells due to poisoning of alcohol fuel cells, catalysts [36]. Recent research in DMFCs has focused on creating an inexpensive, high-performance, and stable fuel cell for commercialization [37].

3.1 Working principle of alcohol fuel cells

Alcohol fuel cells consist of anode, cathode layers, and membranes, as shown in Fig. 2. Alcohol fuel is given to the oxidized anode layer. Electrons, protons, and carbon dioxide (CO_2) are formed at the anode. Protons reach the cathode by passing through the proton-permeable membrane. Electrons reach the cathode using a different route, and electrical generation occurs as a result. Water is produced by the interaction of protons with the air or oxygen given to the catalyst. The water and CO_2 formed as a result of the reactions are discharged from the cell.

If the alcohol used is ethanol, the cell generates 12 protons and 12 electrons simultaneously per ethanol molecule [39]. If methanol

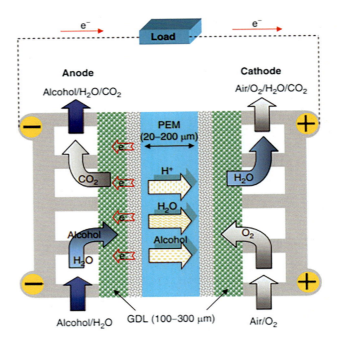

Fig. 2 A DAFC [38].

is used as fuel, the anode fuel is dilute methanol in water, and CO_2 formation is observed as the methanol is consumed [40]. To maintain the concentration of methanol, mechanisms using water from the cathode are being developed [40].

3.2 Direct ethanol fuel cells

Ethanol is an alternative fuel that can be used in fuel cells. It is seen as the ideal fuel due to its low toxicity compared to methanol and hydrogen and the fact that does not pose a hazard during transportation [41]. Ethanol is a biofuel that can be produced from agricultural products in large quantities, and it is suitable for use in low-temperature fuel cells [42]. To ensure the oxidation of ethanol to CO_2, it is necessary to change the structure of the anode catalyst, which is possible with a multifunctional electrocatalyst [43]. Problems such as catalyst poisoning and slow anode kinetics at low temperatures occur directly in ethanol fuel cells [44]. The major disadvantage is the lack of catalyst for full oxidation at high speeds [45]. Electrochemical oxidation and ORR of ethanol are easier in alkaline environments than in the acid medium [46].

Because ethanol contains two bonded carbon atoms, it must break the C—C bond according to Eq. (1) to generate electrical energy [47]. Although this process is not easy compared to methanol, it requires higher temperatures [48].

$$CH_3CH_2OH + 3O_2 \rightarrow 2CO_2 + 3H_2O \tag{1}$$

When we look at Eqs. (2), (3) using ethanol fuel, it is seen that ethanol is oxidized to CO_2 at the anode, protons and electrons are formed, and protons and electrons reach the cathode and react with oxygen at the cathode to form water.

$$Anode: CH_3CH_2OH + 3H_2O \rightarrow 2CO_2 + 12H^+ + 12e^- \tag{2}$$

$$Cathode: 3O_2 + 12H^+ + 12e^- \rightarrow 6H_2O \tag{3}$$

Electricity generation is provided by the ethanol fuel cell within the framework of these equations. It is promising for many electronic devices. In an application, Nissan stated that the first prototypes of the automotive industry working with ethanol fuel cells were built in the e-NV200 electric van [49]. The positive results of the applications and experiments show that the use of ethanol fuel cells will increase.

3.3 Direct methanol fuel cells

The DMFC is a device that generates electrical energy using methanol and oxygen. These cells have advantages such as high density, environmentally friendly products, and lightweight [50]. In DMFCs, the slow reaction of the anode at the cathode creates a problem with energy conversion [51]. Pt is the most used catalyst, but it undergoes CO poisoning due to methanol passing through the anode and provides low activity for the oxygen download reaction [52]. In DMFCs, it is important to adsorb methanol to the surface of the electrode in the fuel cell [53]. When methanol is fed with water in the anode, it prevents thermal management problems better than other fuel cells [54]. Electrocatalysts with high CO tolerance and high durability are needed to improve the use of DMFCs [55]. Studies have found that in particular, the high surface area of carbon-based nanomaterials shows promise in methanol fuel cells. The use of carbon-supported catalysts in methanol alcohol fuel cells can be used to increase catalytic activity, as well as chemical reactions to maintain or change surface properties [56].

Carbon supports such as CNTs and graphene are being investigated to increase the activity and stability of the electrodes in DMFCs [57].

According to the reaction shown in Eq. (4), one molecule of water is required to oxidize one molecule of methanol [58]. On the cathode side, a reaction is observed in which oxygen is reduced to oxygen and the protons to water, as shown in Eq. (5). It is the electro-oxidation of methanol to water and carbon dioxide in the general reaction, as shown in Eq. (6) [59].

$$Anode : CH_3OH + H_2O \rightarrow CO_2 + 6H^+ + 6e^- \qquad (4)$$

$$Cathode : 3/2O_2 + 6H^+ + 6e^- \rightarrow 3H_2O \qquad (5)$$

$$General\ reaction : CH_3OH + 3/2O_2 \rightarrow CO_2 + 2H_2O \qquad (6)$$

DMFCs are divided into active and passive. Passive DMFCs are preferred in portable applications because their structures are simple and they use diffusion to create products [60]. DMFCs are expected to be used in portable electronic devices such as mobile phones and computers as the portable power market grows [61]. In addition, these fuel cells can be used in electric vehicles. Roland Gumpert announced that they are presenting mass production vehicles with methanol fuel cells called Nathalie [62]. In the near future, methanol fuel cells will be used increasingly in many electronic devices and vehicles.

4 Nanomaterials

Nanomaterials are materials that are studied for their atomic size. After many analyses, it has been observed that materials being in nanoscale increases the efficiency, performance, and functionality of the surfaces and adds new properties. By combining these new features, new featured materials are created [63]. Nanomaterials can occur in both artificial and natural forms. Nanostructured materials are also available in natural environments, such as airborne particles or biological media such as proteins, deoxyribonucleic acid (DNA), and viruses [64]. Nanomaterials can be used in many fields, including semiconductors, sensors, fuel cells, drug delivery, and cosmetics [65].

Electric, optical, and magnetic changes in nanosized materials allow them to be widely used in the field of electronics [63]. Carbon-based, composite-based, organic-based, and inorganic-based nanomaterials can be classified based on their composition [66]. They can be classified as zero, one-, two- and three-dimensional nanomaterials, as seen in Fig. 3. Zero-dimensional (0D) materials contain quantum dots, and these materials can be used as fillers in a solid matrix [67]. One-dimensional (1D) nanomaterials contain nanotubes, nanofibers, and nanowires and have large specific surface areas. These materials are useful in electrical transport as

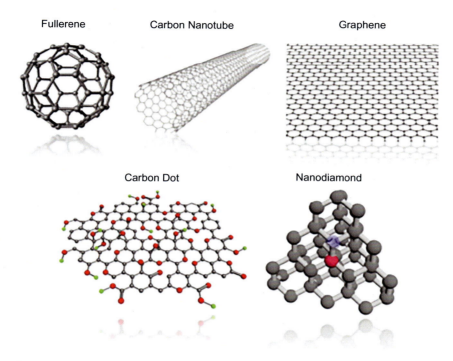

Fig. 3 Carbon-based nanomaterials [70].

electrodes [66]. Two-dimensional (2D) nanomaterials include graphene and nano clay. They have high optical and electronic properties, and their high aspect ratio has increased their use in the pharmaceutical distribution and nanomedicine fields [68]. Three-dimensional (3D) materials include diamond and graphite. They are multilayered materials. Among all these structures, graphene and CNTs are the most researched and studied [69].

Today, nanomaterials are used in many fields, including medicine, electronics, energy, automotive, and renewable energy; and they continue to be studied [71]. The development of nanomaterials for these resources is important for renewable energy, and their use has increased [72].

4.1 Carbon-based nanomaterials for alcohol fuel cells

Carbon is the fourth richly found element in the universe and the fifteenth-most abundant on Earth [66]. Carbon-based nanomaterials exhibit superior chemical and physical properties such as electrical conductivity, high strength, thermal conductivity, and low toxicity. With these properties, they can be used as support catalysts in fuel cells. These materials are important for environmental and energy applications [73]. The catalyst support is widely used in carbon, graphene, CNT, and other carbon-based materials [74]. The properties of carbon supports affect the properties of catalysts used in fuel cells [31]. As seen in Fig. 3, CNTs, graphene, fullerenes, carbon-based quantum dots, graphene oxide, and carbon black (C-black) can be given as examples of carbon-based nanomaterials.

4.2 Carbon nanotubes

CNTs are graphitic cylinders with nanometric diameters [75]. CNTs, a type of carbon fiber made of 2–50 layers, have superior mechanical, electrical and structural properties [75]. They have been the focus of attention because of their electrical and thermal properties and potential for hydrogen storage and fuel cell development [76]. CNTs can be used for catalyst support in fuel cells. The advantages of using CNT in fuel cells include increased catalyst performance, high corrosion resistance, stability, and decreased production costs [77]. With the use of multiwalled carbon nanotubes (MWNTs) in the fuel cell, the loss of Pt surface area decreases and shows ORR activity [78]. Research shows that CNTs positively affect energy storage and their use will provide environmentally friendly products [79]. It has been observed that CNTs are selected to facilitate liquid feeding and electrode transport for DMFCs, showing excellent performance for direct electro-oxidation of $Pt/RuO_2 \times H_2O/CNT$ catalyst methanol [80].

4.3 Graphene

Graphene is a 2D, honeycomb crystal structure made of carbon atoms. The latest member of the 2D materials family, it contains an unusual band structure [81]. Graphene has a large surface area ($2630 \text{ m}^2/\text{g}$) [82], high mechanical strength [83], outstanding thermal conductivity [84], good electrical conductivity, and high intrinsic mobility [85]. The fact that graphene has these properties has encouraged its use in many devices, including solar cells, sensors, and electrochemical cells [86]. Another advantage of graphene is its low cost [87]. Heteroatom-doped graphene was found to show high electrocatalytic activity for ORRs [88]. Yang et al. used a sandwich structure similar to PtCo-graphene/CNPs/graphene catalyst for methanol oxidation [89]. PtRu- and Pt-dispersed graphene-CNT nanocomposites are predicted to

Chapter 17 Carbon-based nanomaterials for alcohol fuel cells **329**

show high electrocatalytic activity against methanol oxidation and ORRs in DMFCs [90].

4.4 Carbon black

C-black consists of carbon particles linked by covalent bonds [91]. It is produced by incomplete combustion of petroleum products and can be used as a support material for metal catalysts due to its high electrical conductivity, high surface area, and stability [92]. Carbon black, which has a good gas diffusion and water treatment capacity, is used to give high porosity to the catalyst layer in DMFCs [93]. Berber et al. stated that a catalyst produced with the double polymer coating of C-black produces high performance, is durable, and has in the wide temperature range (room temperature to 80°C), C-black/PyPBI/Pt-NPs/Nafion in fuel cells [94]. Takahashi et al. stated that the Pt-CeO$_2$/CB anode produced for use in DMFCs is promising [95].

4.5 Nanodiamond

Nanodiamonds (ND) are sp^3 hybrid carbon nanoparticles with diameters between 1 and 20 nm [96]. Nanodiamonds are an attractive material because of their hardness, biocompatibility, and high thermal conductivity [97]. Its electrochemical and chemical stability and high surface area make it very promising as an electrocatalyst support material [98]. Nanodiamonds are widely used because they form a thin cathode catalyst layer, have good methanol tolerance, and are scalable [99]. One study observed that Pt/nanodiamond nanoparticles exhibit good electrocatalytic properties for methanol oxidation, and the initial current of Pt/nanodiamond catalyst, with a Pt percentage range of 50.1%–51.5%, is high [100].

4.6 Fullerene

Fullerenes consist of 60 carbon atoms rolled into a ball [101]. Fullerene molecules, with sp^2 vs sp^3 hybridized carbon atoms, have high electron affinity [102]. Fullerenes have high electrochemical stability and nanometric dimensions, which make them suitable for energy conversion systems [103]. The high surface area of fullerene films makes them well suited for use in fuel cells.

As a support material for methanol oxidation, fullerenes are among the materials of interest due to their electrochemical activities [104], but they are less studied in fuel cells due to the poor solubility of fuller [105]. Nanostructured fullerenes on the conductive surface provide electrochemical properties for oxidation reactions [106]. Zhank et al. observed that Pd nanoparticles supported on fullerene modified with HP-β-CD provide good electrocatalytic activity against ethanol oxidation, and it was predicted that this nanocomposite catalyst could make this a good candidate for ethanol fuel cells [103].

5 Conclusions

It is important to increase the performance characteristics to develop fuel cells and encourage their use. In addition to performance, it is preferred that the materials to be used in its production are cost-effective and highly available. It is a very good alternative to use alcohol, which is safer and easier to produce and store, instead of pure hydrogen fuel in the fuel cell. Many catalysts are produced and tested to increase the performance of fuel cells. It has been observed that the use of carbon-based nanomaterials as catalysts increases the performance of fuel cells. Carbon-based nanomaterials in alcohol fuel cells used as a renewable energy source hold promise in many environments.

References

[1] B. Braunchweig, D. Hibbitts, M. Neurock, A. Wieckowski, Electrocatalysis: a direct alcohol fuel cell and surface science perspective, Catal. Today 202 (1) (2013) 197–209, https://doi.org/10.1016/j.cattod.2012.08.013.

[2] W.J. Zhou, et al., Performance comparison of low-temperature direct alcohol fuel cells with different anode catalysts, J. Power Sources 126 (1–2) (2004) 16–22, https://doi.org/10.1016/j.jpowsour.2003.08.009.

[3] S. Sen Gupta, S. Mahapatra, J. Datta, A potential anode material for the direct alcohol fuel cell, J. Power Sources 131 (1–2) (2004) 169–174, https://doi.org/10.1016/j.jpowsour.2004.01.009.

[4] R. Kumar, et al., Microwave-assisted synthesis of palladium nanoparticles intercalated nitrogen doped reduced graphene oxide and their electrocatalytic activity for direct-ethanol fuel cells, J. Colloid Interface Sci. 515 (2018) 160–171, https://doi.org/10.1016/j.jcis.2018.01.028.

[5] S. Dharmalingam, V. Kugarajah, M. Sugumar, Membranes for microbial fuel cells, in: Microbial Electrochemical Technology, Elsevier, 2019, pp. 143–194.

[6] C. Luo, H. Xie, Q. Wang, G. Luo, C. Liu, A review of the application and performance of carbon nanotubes in fuel cells, J. Nanomater. (2015), https://doi.org/10.1155/2015/560392.

[7] H.-J. Choi, S.-M. Jung, J.-M. Seo, D.W. Chang, L. Dai, J.-B. Baek, Graphene for energy conversion and storage in fuel cells and supercapacitors, Nano Energy 1 (2012) 534–551, https://doi.org/10.1016/j.nanoen.2012.05.001.

[8] M.S. Whittingham, R.F. Savinell, T. Zawodzinski, Introduction: batteries and fuel cells, Chem. Rev. 104 (10) (2004) 4243–4244, https://doi.org/10.1021/cr020705e.

[9] B. Yarar Kaplan, N. Haghmoradi, E. Biçer, C. Merino, S. Alkan Gürsel, High performance electrocatalysts supported on graphene based hybrids for polymer electrolyte membrane fuel cells, Int. J. Hydrog. Energy 43 (52) (2018) 23221–23230, https://doi.org/10.1016/j.ijhydene.2018.10.222.

[10] M.C. Williams, Fuel cells, in: Fuel Cells: Technologies for Fuel Processing, Elsevier, 2011, pp. 11–27.

[11] A. Coralli, B.J.M. Sarruf, P.E.V. De Miranda, L. Osmieri, S. Specchia, N.Q. Minh, Fuel cells, in: Science and Engineering of Hydrogen-Based Energy Technologies: Hydrogen Production and Practical Applications in Energy Generation, Elsevier, 2019, pp. 39–122.

[12] A.M. Abdalla, et al., Nanomaterials for solid oxide fuel cells: a review, Renew. Sust. Energ. Rev. 82 (2018) 353–368, https://doi.org/10.1016/j.rser.2017.09.046.

[13] E. Urbańczyk, A. Jaroń, W. Simka, Ni–Pt sinter as a promising electrode for methanol electrocatalytic oxidation, Int. J. Hydrog. Energy 43 (36) (2018) 17156–17163, https://doi.org/10.1016/j.ijhydene.2018.07.109.

[14] B. Sundén, Fuel cell types—overview, in: Hydrogen, Batteries and Fuel Cells, Elsevier, 2019, pp. 123–144.

[15] W. Wiyaratn, Review on fuel cell technology for valuable chemicals and energy co-generation, Eng. J. 14 (3) (2010) 1–14, https://doi.org/10.4186/ej.2010. 14.3.1.

[16] J. Park, H. Oh, T. Ha, Y. Il Lee, K. Min, A review of the gas diffusion layer in proton exchange membrane fuel cells: durability and degradation, Appl. Energy 155 (2015) 866–880, https://doi.org/10.1016/j.apenergy.2015.06.068.

[17] Y.N. Sudhakar, M. Selvakumar, D.K. Bhat, Biopolymer electrolytes for fuel cell applications, in: Biopolymer Electrolytes, Elsevier, 2018, pp. 151–166.

[18] T. Berning, N. Djilali, A 3D, multiphase, multicomponent model of the cathode and anode of a PEM fuel cell, J. Electrochem. Soc. 150 (2003) 12, https://doi.org/10.1149/1.1621412.

[19] V.K. Mathur, J. Crawford, Fundamentals of gas diffusion layers in PEM fuel cells, in: Recent Trends in Fuel Cell Science and Technology, Springer, New York, 2007, pp. 116–128.

[20] F.Y. Zhang, S.G. Advani, A.K. Prasad, Advanced high resolution characterization techniques for degradation studies in fuel cells, in: Polymer Electrolyte Fuel Cell Degradation, Elsevier Inc., 2012, pp. 365–421.

[21] B.G. Pollet, A.A. Franco, H. Su, H. Liang, S. Pasupathi, Proton exchange membrane fuel cells, in: Compendium of Hydrogen Energy, Elsevier, 2016, pp. 3–56.

[22] S.M. Haile, Fuel cell materials and components, Acta Mater. 51 (19) (2003) 5981–6000, https://doi.org/10.1016/j.actamat.2003.08.004.

[23] J. Andrews, A.K. Doddathimmaiah, Regenerative fuel cells, in: Materials for Fuel Cells, Elsevier Inc., 2008, pp. 344–385.

[24] A. Hermann, T. Chaudhuri, P. Spagnol, Bipolar plates for PEM fuel cells: a review, Int. J. Hydrog. Energy 30 (12) (2005) 1297–1302, https://doi.org/10.1016/j.ijhydene.2005.04.016.

[25] E. Middelman, W. Kout, B. Vogelaar, J. Lenssen, E. De Waal, Bipolar plates for PEM fuel cells, J. Power Sources 118 (1–2) (2003) 44–46, https://doi.org/10.1016/S0378-7753(03)00070-3.

[26] T. Stein, Y. Ein-Eli, Proton exchange membrane (PEM) fuel cell bipolar plates prepared from a physical vapor deposition (PVD) titanium nitride (TiN) coated AISI416 stainless-steel, SN Appl. Sci. 1 (11) (2019) 1–12, https://doi.org/10.1007/s42452-019-1475-3.

[27] L. Feng, X. Sun, S. Yao, C. Liu, W. Xing, J. Zhang, Electrocatalysts and catalyst layers for oxygen reduction reaction, in: Rotating Electrode Methods and Oxygen Reduction Electrocatalysts, Elsevier B.V., 2014, pp. 67–132.

[28] A.A. Kulikovsky, Catalyst layer performance, in: Analytical Modelling of Fuel Cells, Elsevier, 2010, pp. 39–82.

[29] J. Zhang, Z. Xia, L. Dai, Carbon-based electrocatalysts for advanced energy conversion and storage, Sci. Adv. (2015) 1–7, https://doi.org/10.1126/sciadv.1500564.

[30] P.M. Ejikeme, K. Makgopa, K.I. Ozoemena, Effects of Catalyst-Support Materials on the Performance of Fuel Cells, Springer, Cham, 2016, pp. 517–550.

[31] M.W. Xu, Z. Su, Z.W. Weng, Z.C. Wang, B. Dong, An approach for synthesizing PtRu/MWCNT nanocomposite for methanol electro-oxidation, Mater. Chem. Phys. 124 (1) (2010) 785–790, https://doi.org/10.1016/j.matchemphys.2010.07.061.

[32] K. Wang, et al., Ordered mesoporous tungsten carbide/carbon composites promoted Pt catalyst with high activity and stability for methanol electrooxidation, Appl. Catal. B Environ. 147 (2014) 518–525, https://doi.org/10.1016/j.apcatb.2013.09.020.

[33] H. Liu, C. Song, L. Zhang, J. Zhang, H. Wang, D.P. Wilkinson, A review of anode catalysis in the direct methanol fuel cell, J. Power Sources 155 (2) (2006) 95–110, https://doi.org/10.1016/j.jpowsour.2006.01.030.

[34] H.R. Corti, E.R. Gonzalez, Direct Alcohol Fuel Cells, vol. 9789400777, Springer Netherlands, Dordrecht, 2014.

[35] Y. Mu, H. Liang, J. Hu, L. Jiang, L. Wan, Controllable Pt nanoparticle deposition on carbon nanotubes as an anode catalyst for direct methanol fuel cells, J. Phys. Chem. B 109 (47) (2005) 22212–22216, https://doi.org/10.1021/jp0555448.

[36] N. Kakati, J. Maiti, S.H. Lee, S.H. Jee, B. Viswanathan, Y.S. Yoon, Anode catalysts for direct methanol fuel cells in acidic media: do we have any alternative for Pt or Pt-Ru? Am. Chem. Soc. 114 (24) (2014) 12397–12429, https://doi.org/10.1021/cr400389f.

[37] S.S. Siwal, S. Thakur, Q.B. Zhang, V.K. Thakur, Electrocatalysts for electrooxidation of direct alcohol fuel cell: chemistry and applications, Mater. Today Chem. 14 (2019) 100182, https://doi.org/10.1016/j.mtchem.2019.06.004.

[38] T.S. Zhao, R. Chen, Fuel cells—direct alcohol fuel cells | experimental systems, in: Encyclopedia of Electrochemical Power Sources, Elsevier, 2009, pp. 428–435.

[39] L. Jiang, G. Sun, Fuel cells—direct alcohol fuel cells | Direct ethanol fuel cells, in: Encyclopedia of Electrochemical Power Sources, Elsevier, 2009, pp. 390–401.

[40] D. Gervasio, Fuel cells—direct alcohol fuel cells | new materials, in: Encyclopedia of Electrochemical Power Sources, Elsevier, 2009, pp. 420–427.

[41] X. Xue, C. Liu, T. Lu, W. Xing, Synthesis and characterization of Pt/C nanocatalysts using room temperature ionic liquids for fuel cell applications, Fuel Cells 6 (5) (2006) 347–355, https://doi.org/10.1002/fuce.200500105.

[42] E. Antolini, Catalysts for direct ethanol fuel cells, J. Power Sources 170 (1) (2007) 1–12, https://doi.org/10.1016/j.jpowsour.2007.04.009.

[43] C. Lamy, S. Rousseau, E.M. Belgsir, C. Coutanceau, J.M. Léger, Recent progress in the direct ethanol fuel cell: development of new platinum-tin electrocatalysts, Electrochim. Acta 49 (22–23) (2004) 3901–3908, https://doi.org/10.1016/j.electacta.2004.01.078.

[44] W.J. Zhou, et al., Direct ethanol fuel cells based on PtSn anodes: the effect of Sn content on the fuel cell performance, J. Power Sources 140 (1) (2005) 50–58, https://doi.org/10.1016/j.jpowsour.2004.08.003.

[45] J. Mann, N. Yao, A.B. Bocarsly, Characterization and analysis of new catalysts for a direct ethanol fuel cell, Langmuir 22 (25) (2006) 10432–10436, https://doi.org/10.1021/la061200c.

[46] H. Hou, G. Sun, R. He, Z. Wu, B. Sun, Alkali doped polybenzimidazole membrane for high performance alkaline direct ethanol fuel cell, J. Power Sources 182 (1) (2008) 95–99, https://doi.org/10.1016/j.jpowsour.2008.04.010.

[47] S. Rousseau, C. Coutanceau, C. Lamy, J.M. Léger, Direct ethanol fuel cell (DEFC): electrical performances and reaction products distribution under operating conditions with different platinum-based anodes, J. Power Sources 158 (1) (2006) 18–24, https://doi.org/10.1016/j.jpowsour.2005.08.027.

[48] F. Joensen, J.R. Rostrup-Nielsen, Conversion of hydrocarbons and alcohols for fuel cells, J. Power Sources 105 (2) (2002) 195–201, https://doi.org/10.1016/S0378-7753(01)00939-9.

[49] F. Calmon, Nissan Brezilya etanol yakıt hücresinin ilk testlerini bitirdi | Otomotiv Endüstrisi Haberleri | sadece otomatik, Just auto, 2017.

[50] Z.B. Wang, C.Z. Li, D.M. Gu, G.P. Yin, Carbon riveted PtRu/C catalyst from glucose in-situ carbonization through hydrothermal method for direct methanol fuel cell, J. Power Sources 238 (2013) 283–289, https://doi.org/10.1016/j.jpowsour.2013.03.082.

[51] Y. Qiao, C.M. Li, Nanostructured catalysts in fuel cells, J. Mater. Chem. (2010), https://doi.org/10.1039/c0jm02871a.

[52] M. Yaldagard, M. Jahanshahi, N. Seghatoleslami, Pt catalysts on PANI coated WC/C nanocomposites for methanol electro-oxidation and oxygen electro-reduction in DMFC, Appl. Surf. Sci. 317 (2014) 496–504, https://doi.org/10.1016/j.apsusc.2014.08.148.

[53] M.D. Esrafili, R. Nurazar, Potential of C-doped boron nitride fullerene as a catalyst for methanol dehydrogenation, Comput. Mater. Sci. 92 (2014) 172–177, https://doi.org/10.1016/j.commatsci.2014.05.043.

[54] R. Dillon, S. Srinivasan, A.S. Aricò, V. Antonucci, International activities in DMFC R&D: status of technologies and potential applications, J. Power Sources 127 (1–2) (2004) 112–126, https://doi.org/10.1016/j.jpowsour.2003.09.032.

[55] Z. Yang, M.R. Berber, N. Nakashima, A polymer-coated carbon black-based fuel cell electrocatalyst with high CO-tolerance and durability in direct methanol oxidation, J. Mater. Chem. A 2 (44) (2014) 18875–18880, https://doi.org/10.1039/c4ta03185g.

[56] S. Eris, Z. Daşdelen, F. Sen, Enhanced electrocatalytic activity and stability of monodisperse Pt nanocomposites for direct methanol fuel cells, J. Colloid Interface Sci. 513 (2018) 767–773, https://doi.org/10.1016/j.jcis.2017.11.085.

[57] M. Borghei, et al., Enhanced performance of a silicon microfabricated direct methanol fuel cell with PtRu catalysts supported on few-walled carbon nanotubes, Energy 65 (2014) 612–620, https://doi.org/10.1016/j.energy.2013.11.067.

[58] C. Coutanceau, et al., Development of materials for mini DMFC working at room temperature for portable applications, J. Power Sources 160 (1) (2006) 334–339, https://doi.org/10.1016/j.jpowsour.2006.01.073.

[59] C.Y. Chen, P. Yang, Performance of an air-breathing direct methanol fuel cell, J. Power Sources 123 (1) (2003) 37–42, https://doi.org/10.1016/S0378-7753(03)00434-8.

[60] L. Wang, Z. Yuan, F. Wen, Y. Cheng, Y. Zhang, G. Wang, A bipolar passive DMFC stack for portable applications, Energy 144 (2018) 587–593, https://doi.org/10.1016/j.energy.2017.12.039.

[61] S.K. Kamarudin, F. Achmad, W.R.W. Daud, Overview on the application of direct methanol fuel cell (DMFC) for portable electronic devices, Int. J. Hydrog. Energy 34 (16) (2009) 6902–6916, https://doi.org/10.1016/j.ijhydene.2009.06.013.

[62] A.D. Steffen, Nathalie: 745 Mil Menzilli Bir Metanol Yakıt Hücreli Süper Otomobil, İntelligent Living, 2020.

[63] Z. Tüylek, Küçük Şeylerin Hikayesi: nanomalzeme, Nevşehir Bilim Teknol. Derg. 5 (2) (2016) 130–141, https://doi.org/10.17100/nevbiltek.284737.

[64] D. Brabazon, Nanostructured materials, in: Reference Module in Materials Science and Materials Engineering, Elsevier, 2016.

[65] H. Ghadimi, S. Ab Ghani, I. Amiri, Introduction, in: Electrochemistry of Dihydroxybenzene Compounds, Elsevier, 2017, pp. 1–30.

[66] T. Jin, Q. Han, Y. Wang, L. Jiao, 1D nanomaterials: design, synthesis, and applications in sodium–ion batteries, Small 14 (2018) 2, https://doi.org/10.1002/smll.201703086.

[67] P.I. Dolez, Nanomaterials definitions, classifications, and applications, in: Nanoengineering: Global Approaches to Health and Safety Issues, Elsevier, 2015, pp. 3–40.

[68] S. Barua, D. Sahu, N. Shahnaz, R. Khan, Chemistry of two-dimensional nanomaterials, in: Two-Dimensional Nanostructures for Biomedical Technology, Elsevier, 2020, pp. 1–33.

[69] X. Cheng, Nanostructures: fabrication and applications, in: Nanolithography: The Art of Fabricating Nanoelectronic and Nanophotonic Devices and Systems, Elsevier Ltd, 2013, pp. 348–375.

[70] E. Kai-Hua Chow, M. Gu, J. Xu, Carbon nanomaterials: fundamental concepts, biological interactions, and clinical applications, in: Nanoparticles for Biomedical Applications, Elsevier, 2020, pp. 223–242.

[71] O. Alptekin, et al., Use of silica-based homogeneously distributed gold nickel nanohybrid as a stable nanocatalyst for the hydrogen production from the dimethylamine borane, Sci. Rep. 10 (1) (2020) 1–12, https://doi.org/10.1038/s41598-020-64221-y.

[72] R. Kumar, A. Khan, A.M. Asiri, H. Dzudzevic-Cancar, Preparation methods of hydrogen storage materials and nanomaterials, in: Nanomaterials for Hydrogen Storage Applications, Elsevier, 2021, pp. 1–16.

[73] C. Cha, S.R. Shin, N. Annabi, M.R. Dokmeci, A. Khademhosseini, Carbon-based nanomaterials: multifunctional materials for biomedical engineering, ACS Nano 7 (4) (2013) 2891–2897, https://doi.org/10.1021/nn401196a.

[74] Y. Meng, X. Huang, H. Lin, P. Zhang, Q. Gao, W. Li, Carbon-based nanomaterials as sustainable noble-metal-free electrocatalysts, Front. Chem. 7 (2019) 759, https://doi.org/10.3389/fchem.2019.00759.

[75] R.B. Rakhi, Preparation and properties of manipulated carbon nanotube composites and applications, in: Nanocarbon and its Composites: Preparation, Properties and Applications, Elsevier, 2018, pp. 489–520.

[76] S.M. Andersen, et al., Durability of carbon nanofiber (CNF) & carbon nanotube (CNT) as catalyst support for proton exchange membrane fuel cells, Solid State Ionics 231 (2013) 94–101, https://doi.org/10.1016/j.ssi.2012.11.020.

[77] E. Akbari, Z. Buntat, Benefits of using carbon nanotubes in fuel cells: a review, Int. J. Energy Res. (2016), https://doi.org/10.1002/er.3600.

[78] X. Wang, W. Li, Z. Chen, M. Waje, Y. Yan, Durability investigation of carbon nanotube as catalyst support for proton exchange membrane fuel cell, J. Power Sources 158 (1) (2006) 154–159, https://doi.org/10.1016/j.jpowsour.2005.09.039.

[79] R. Kumar, R.K. Singh, D.P. Singh, Natural and waste hydrocarbon precursors for the synthesis of carbon based nanomaterials: graphene and CNTs, Renew. Sust. Energ. Rev. 58 (2016) 976–1006, https://doi.org/10.1016/j.rser.2015.12.120.

[80] L. Cao, et al., Novel nanocomposite $Pt/RuO_2 \cdot xH_2O$/carbon nanotube catalysts for direct methanol fuel cells, Angew. Chem. Int. Ed. 45 (32) (2006) 5315–5319, https://doi.org/10.1002/anie.200601301.

Chapter 17 Carbon-based nanomaterials for alcohol fuel cells **335**

[81] K.I. Bolotin, et al., Ultrahigh electron mobility in suspended graphene, Solid State Commun. 146 (2008) 351–355, https://doi.org/10.1016/j.ssc.2008.02.024.

[82] M.D. Stoller, S. Park, Z. Yanwu, J. An, R.S. Ruoff, Graphene-based ultracapacitors, Nano Lett. 8 (10) (2008) 3498–3502, https://doi.org/10.1021/nl802558y.

[83] M.H. Kang, D. Lee, J. Sung, J. Kim, B.H. Kim, J. Park, Structure and chemistry of 2D materials, in: Comprehensive Nanoscience and Nanotechnology, Elsevier, 2019, pp. 55–90.

[84] V. Singh, D. Joung, L. Zhai, S. Das, S.I. Khondaker, S. Seal, Graphene based materials: past, present and future, Prog. Mater. Sci. (2011) 1178–1271, https://doi.org/10.1016/j.pmatsci.2011.03.003.

[85] Y. Zhu, et al., Graphene and graphene oxide: synthesis, properties, and applications, Adv. Mater. 22 (35) (2010) 3906–3924, https://doi.org/10.1002/adma.201001068.

[86] P. Bazylewski, G. Fanchini, Graphene: properties and applications, in: Comprehensive Nanoscience and Nanotechnology, Elsevier, 2019, pp. 287–304.

[87] D. Verma, K.L. Goh, Functionalized graphene-based nanocomposites for energy applications, in: Functionalized Graphene Nanocomposites and Their Derivatives: Synthesis, Processing and Applications, Elsevier, 2019, pp. 219–243.

[88] M. Liu, R. Zhang, W. Chen, Graphene-supported nanoelectrocatalysts for fuel cells: synthesis, properties, and applications, Chem. Rev. 114 (10) (2014) 5117–5160, https://doi.org/10.1021/cr400523y.

[89] C. Testa, A. Zammataro, A. Pappalardo, G. Trusso Sfrazzetto, Catalysis with carbon nanoparticles, RSC Adv. 9 (47) (2019) 27659–27664, https://doi.org/10.1039/c9ra05689k.

[90] N. Jha, R.I. Jafri, N. Rajalakshmi, S. Ramaprabhu, Graphene-multi walled carbon nanotube hybrid electrocatalyst support material for direct methanol fuel cell, Int. J. Hydrog. Energy 36 (12) (2011) 7284–7290, https://doi.org/10.1016/j.ijhydene.2011.03.008.

[91] A. Mohammad, G.P. Simon, Rubber-clay nanocomposites, in: Polymer Nanocomposites, Elsevier, 2006, pp. 297–325.

[92] R.K. Gautam, A. Verma, Electrocatalyst materials for oxygen reduction reaction in microbial fuel cell, in: Microbial Electrochemical Technology, Elsevier, 2019, pp. 451–483.

[93] G. Wang, G. Sun, Q. Wang, S. Wang, H. Sun, Q. Xin, Effect of carbon black additive in Pt black cathode catalyst layer on direct methanol fuel cell performance, Int. J. Hydrog. Energy 35 (20) (2010) 11245–11253, https://doi.org/10.1016/j.ijhydene.2010.07.045.

[94] M.R. Berber, T. Fujigaya, N. Nakashima, A potential polymer formulation of a durable carbon-black catalyst with a significant fuel cell performance over a wide operating temperature range, Mater. Today Energy 10 (2018) 161–168, https://doi.org/10.1016/j.mtener.2018.08.016.

[95] M. Takahashi, T. Mori, F. Ye, A. Vinu, H. Kobayashi, J. Drennan, Design of high-quality Pt?CeO$_2$ composite anodes supported by carbon black for direct methanol fuel cell application, J. Am. Ceram. Soc. 90 (4) (2007) 1291–1294, https://doi.org/10.1111/j.1551-2916.2006.01483.x.

[96] V. Georgakilas, J.A. Perman, J. Tucek, R. Zboril, Broad family of carbon nanoallotropes: classification, chemistry, and applications of fullerenes, carbon dots, nanotubes, graphene, nanodiamonds, and combined superstructures, Chem. Rev. 115 (11) (2015) 4744–4822, https://doi.org/10.1021/cr500304f.

[97] B.T. Zhang, X. Zheng, H.F. Li, J.M. Lin, Application of carbon-based nanomaterials in sample preparation: a review, Anal. Chim. Acta 784 (2013) 1–17, https://doi.org/10.1016/j.aca.2013.03.054.

336 Chapter 17 Carbon-based nanomaterials for alcohol fuel cells

[98] L. La-Torre-Riveros, R. Guzman-Blas, A.E. Méndez-Torres, M. Prelas, D.A. Tryk, C.R. Cabrera, Diamond nanoparticles as a support for Pt and PtRu catalysts for direct methanol fuel cells, ACS Appl. Mater. Interfaces 4 (2) (2012) 1134–1147, https://doi.org/10.1021/am2018628.

[99] A. Rifai, E. Pirogova, K. Fox, Diamond, carbon nanotubes and graphene for biomedical applications, in: Encyclopedia of Biomedical Engineering, Elsevier, 2019, pp. 97–107.

[100] L.Y. Bian, Y.H. Wang, J.B. Zang, F.W. Meng, Y.L. Zhao, Microwave synthesis and characterization of Pt nanoparticles supported on undoped nanodiamond for methanol electrooxidation, Int. J. Hydrog. Energy 37 (2) (2012) 1220–1225, https://doi.org/10.1016/j.ijhydene.2011.09.118.

[101] N.P. Shetti, D.S. Nayak, K.R. Reddy, T.M. Aminabhvi, Graphene-clay-based hybrid nanostructures for electrochemical sensors and biosensors, in: Graphene-Based Electrochemical Sensors for Biomolecules: A Volume in Micro and Nano Technologies, Elsevier, 2019, pp. 235–274.

[102] A. Kausar, R. Taherian, Electrical conductivity behavior of polymer nanocomposite with carbon nanofillers, in: Electrical Conductivity in Polymer-Based Composites: Experiments, Modelling, and Applications, Elsevier, 2019, pp. 41–72.

[103] J. Coro, M. Suárez, L.S.R. Silva, K.I.B. Eguiluz, G.R. Salazar-Banda, Fullerene applications in fuel cells: a review, Int. J. Hydrog. Energy (2016) 17944–17959, https://doi.org/10.1016/j.ijhydene.2016.08.043.

[104] K.S. Bhavani, T. Anusha, P.K. Brahman, Fabrication and characterization of gold nanoparticles and fullerene-C60 nanocomposite film at glassy carbon electrode as potential electro-catalyst towards the methanol oxidation, Int. J. Hydrog. Energy 44 (47) (2019) 25863–25873, https://doi.org/10.1016/j.ijhydene.2019.08.005.

[105] X. Zhang, J.W. Zhang, P.H. Xiang, J. Qiao, Fabrication of graphene-fullerene hybrid by self-assembly and its application as support material for methanol electrocatalytic oxidation reaction, Appl. Surf. Sci. 440 (2018) 477–483, https://doi.org/10.1016/j.apsusc.2018.01.150.

[106] K. Vinodgopal, M. Haria, D. Meisel, P. Kamat, Fullerene-based carbon nanostructures for methanol oxidation, Nano Lett. 4 (3) (2004) 415–418, https://doi.org/10.1021/nl035028y.

18

Dendrimer-based nanocomposites for alcohol fuel cells

Elif Esra Altuner[a], Muhammed Bekmezci[a,b], Ramazan Bayat[a,b], Merve Akin[a,b], Iskender Isik[b], and Fatih Şen[a]

[a]Şen Research Group, Department of Biochemistry, Dumlupinar University, Kütahya, Turkey. [b]Department of Materials Science and Engineering, Faculty of Engineering, Dumlupinar University, Kütahya, Turkey

1 Introduction

The word *nano* means "dwarf" in Latin. Nanotechnological research includes studies in nanoscale as much as possible. These studies are applied in numerous fields, such as nanoscale polymer chemistry, materials, and synthesis. Today, an important part of nanotechnological studies are nanocomposites, which are within the field of nanomaterials [1]. It is widely believed that nanocomposites are filled with additional filling materials in nanosize and have a certain time value [2]. The carbon black of the elastomers is reinforced with nanocomposites, making it even stronger. Such colloidal silica modification and various materials are reinforced with nanocomposite supports [3]. As supporting materials, there are many examples such as carbon nanotubes (CNTs), carbon nanofibers (CNFs), graphene, some organic materials, polymers, dendrimers, etc. [3]. Dendrimer-based nanocomposites, on the other hand, have a dendrimer structure. *Dendrimer* means "tree" and "part" in Latin [4]. Dendrimer-based nanocomposites get their name due to their molecular structure, divided into trees with a branched structure and partitioned structure. Many syntheses and studies of dendrimers in polymer chemistry have been reported in academic studies [5]. They have been also used for direct alcohol fuel cell applications. Today, direct alcohol fuel cells (DAFCs) are in demand due to their solubility in high-aqueous electrolytes, high-performance energy generation, easy handling,

Nanomaterials for Direct Alcohol Fuel Cells. https://doi.org/10.1016/B978-0-12-821713-9.00007-X
Copyright © 2021 Elsevier Inc. All rights reserved.

and low cost. The most widely used DAFC that is electrooxidized contains methanol. Other alcohols are ethanol (the second most common), ethylene glycol, and propanol [6]. However, these alcohols need catalysts to activate them. Platinum, which is one of the most used as an electrocatalyst in DAFCs, can benefit from supporting nanocomposites because it causes poison in reactions from the anode to the cathode and decreases performance. Platinum is also an expensive metal, and due to its rapid corrosion, it has been necessary for studies into alternatives [7]. As alternative materials, nanohybrid materials composed of palladium and cobalt have also nano catalytic properties [8]. For these reasons, it is also preferred to use palladium and cobalt in platinum alloys and directly in alcohol fuel cells. Similarly, studies have found that ruthenium and copper alloys also show good catalytic effects [9]. This chapter discusses dendrimer-based nanocomposites directly providing electrocatalytic oxidation in alcohol fuel cells, the role of electron transfer in the environment, the advantages and disadvantages of this approach, and the application areas.

2 Alcohol fuel cells

William Grove was the first to come up with the idea of fuel cells, in 1839 [10]. Fuel cells are a valuable resource for economy [10–12]. Researchers have sought to discover devices that will minimize pollution and at the same time, take the fuel performance to the highest level. Therefore, fuel cells are very important for keeping the power density at the highest levels, being environmentally friendly and having a low cost. However, one of the biggest problem with fuel cells is the transport and storage of hydrogen. For this reason, special types in fuel cells, including DAFCs, have been investigated [13]. There are so many studies related to these types of fuel cells and their catalytic systems [14]. The fuel cells are categorized according to their temperature, pressure, and basic materials used, including DAFCs, alkaline carbonate fuel cells, molten carbonate fuel cells, solid oxide fuel cells, phosphoric acid fuel cells, and regenerative fuel cells [15]. In DAFCs, especially methanol fuel cells, the high performance of methanol below 100°C provides great convenience and has led to the production of approximately 6000 devices that can be easily carried and used [16–19]. Due to these advantages, DAFCs take up a very important place.

2.1 Direct methanol fuel cells

Methyl alcohol is used as the main energy source in direct methanol fuel cells. Here, methyl alcohol oxidizes to CO_2, and this constitutes the anode. O_2 is reduced to water, and this constitutes the cathode. Direct methanol fuel cells are used in technological devices because they generate very intense power [20]. Cell phones, laptops, and renewable power devices are the main applications. Many world-renowned electronic companies have introduced the benefits of using methanol fuel cells in laptop production and the features of these devices (Fig. 1) [21]. Direct methyl alcohol fuel cells are preferred because they produce much more energy than the major available energy sources [20, 22]. Fig. 2 shows the ordering of the power densities of these devices.

As a result of the processes of reduction and oxidation reactions taking place at the anode with oxygen (usually air oxygen and methyl alcohol) at the cathode, the flow of fuel is provided

Companies	Announcing date (month, Year)	Power output (W)	Specification (W: weight) (V: volume)	Concentration of fuel solution (wt.% of McOH solution)	Impressive technologies
Toshiba	3, 2003	12 (ave) 20 (max)	W: 900g (without fuel solution) V: 825 ml (275 x 75 x 40mm)	3–6	Developing a system for fuel dilution using produced water
NEC	6, 2003	14 (ave) 24 (max)	W: 893 g (including 298 g of fuel solution) V: 2916 ml (270 x 270 x 40mm)	10	Fuel cell's size reduction due to higher power density
Fujitsu	1, 2004	15	n.g.	30	Their new membrane help lead to smaller and more efficient fuel cells
SAIT (Samsung)	6, 2004	20	n.g.	n.g.	Applying nanomaterials technology
Antig	3, 2005	10	W: 435 g V: 730ml (190 x 128 x 30mm)	10–15	DMFC module fit into a standard laptop optical drive bay
Sanyo Electric and IBM	5, 2005	12 (DMFC only) 72 (combined with Li-polymer battery)	W: 2.2kg V: 1218–4111 (270 x 282 x 16 to 54mm).	n.g.	Docking bay type and hybrid with Li-polymer battery
LG Chem.	9, 2005	25	W: less than 1kg V: less than 1 l in the core volume	n.g.	Lifetime of more than 4000h
Panasonic	1, 2006	13 (DMFC only) 20 (combined with Li-ion battery)	W: 450g (without fuel) V: 400ml	n.g.	Hybrid with Li-ion battery
Antig and their partner	9, 2006	16W (80 Wh)	W: 800g (without fuel) V: 1527ml (218 x 68 x 103)	n.g.	Power supply for a wide range of portable applications
Samsung Electronics	12, 2006	20 (max)		n.g.	Docking station type and high power output

Fig. 1 Major technological devices utilizing direct methanol fuel cells [21].

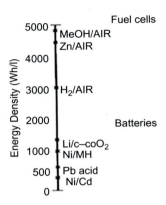

Fig. 2 The sequence of power devices to energy ratios [20].

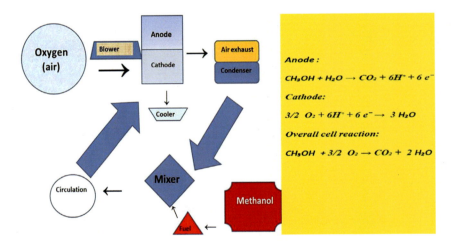

Fig. 3 The process and anode-cathode reactions of direct methanol fuel cells [24].

in the flow area channels directly in the methanol fuel cells. In this way, energy production is provided at a high quality [23]. The process and reactions of anode-cathode of direct methanol fuel cells are shown in Fig. 3.

2.2 Direct ethanol fuel cells

Scientists have generally worked in the area of developing direct methanol fuel cells. However, some undesirable conditions of direct methanol fuel cells occur during the reactions, including corrosion of the platinum electrode, slow oxidation of methanol, reaction, and anode poisoning. It causes poisoning by releasing CO. Because ethanol gives as much energy as methanol and produces less poisoning products, direct ethanol fuel cells have become more popular than direct methanol fuel cells [25–27].

Studies have also examined how to create the platinum electrocatalyst used in direct alcohol cells [28]. Carbon-supported platinum is commonly used as an anode catalyst in low-temperature fuel cells. Because catalysis has a surface effect, the catalyst needs to have the highest possible surface area. Thus, the active phase is dispersed on conductive support as carbon. Pure Pt, however, is not the most efficient anodic catalyst for direct ethanol fuel cells. Indeed, the electro-oxidation of a partially oxygenated organic molecule, such as a primary alcohol, can be performed only with a multifunctional electrocatalyst. Platinum itself is known to be rapidly poisoned by strongly adsorbed species on its surface coming from the dissociative adsorption of ethanol [12]. Efforts to mitigate the poisoning of Pt have been concentrated on the addition of cocatalysts (particularly ruthenium and tin) to platinum. As we have seen in direct methanol fuel cells, platinum is not a very durable metal for anodic reactions. The use of platinum metal in direct ethanol fuel cells requires great stability and durability because it causes corrosion and poisoning. For this reason, it is not preferred to use pure platinum in direct ethanol fuel cells.

Platinum is supported with nanocomposite materials to enable anodic reactions [29]. In various trials on platinum, academic studies have established that platinum-based materials have a high catalytic effect, and they have mostly supported by carbon materials in methanol and ethanol, creating an electrocatalytic reaction, and have a high catalytic effect with the help of second and third metals such as cobalt and ruthenium [30–33]. Researchers have also reported how platinum metal acts as an electrocatalyst directly in alcohol cells [34]. The total oxidation reactions of the direct ethanol fuel cell are as follows [29]:

$$CH_3CH_2OH \rightarrow [CH_3CH_2OH]_{ad} \rightarrow C1_{ad}, C2_{ad}$$
$$\rightarrow CO_2 \text{ (total oxidation)}$$

The scheme of Nafion membrane-supported direct ethanol fuel cells has been reported by Xu et al. [35]. It is shown in Fig. 4.

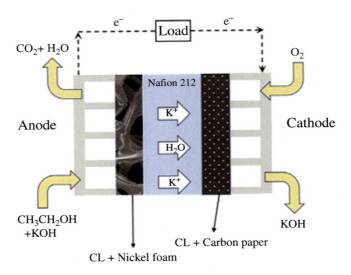

Fig. 4 The anode-cathode reaction process of direct ethanol fuel cells [35].

2.3 Direct ethylene glycol fuel cells

There are studies that have looked at directly creating ethylene glycol fuel cells by electrocatalytic reactions based on the reduction of oxygen with ethylene glycol. As mentioned previously, platinum and platinum alloys are used in catalytic reactions in direct methanol fuel cells, direct ethanol fuel cells, and other types of DAFCs, as well as in direct ethylene glycol fuel cells [36–42]. The total reaction that occurs during electro-oxidation is [36]:

$$(CH_2OH)_2 + 2H_2O \rightarrow 2CO_2 + 10H^+ + 10e^-$$

The reaction and system schemes of the direct ethylene glycol fuel cell are given in Fig. 5. Oxidation reaction takes place at the anode, and electron accumulation occurs in the environment with hydroxyl ions.

Due to the disadvantages of direct ethylene glycol fuel cells and other types of DAFCs, such as CO production due to platinum corrosion and poisoning effect, pure platinum is not used, but platinum alloys are. These alloys are supported by nanocomposite materials [44].

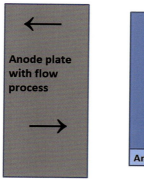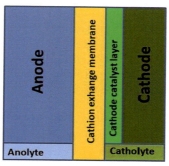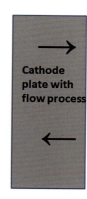

Fig. 5 The process and anode-cathode reactions of direct ethylene glycol fuel cells [43].

3 Nanocomposites types and applications of direct alcohol fuel cells

Nanocomposite materials form in three ways: polymerization, melting materials, and homogenized solutions. Polymerization is a method based on mixing and melting materials to form a homogenized solution [45, 46]. Hundreds of layered nanofilms not exceeding 1 nm in thickness between silicates were formed by interconnecting them with van der Waals bonds and polymerizing them to create nanocomposites. There are spaces between these layers. The desired nanocomposite is formed by placing fillers or silicates between these intermediate layers [46, 47]. Researchers have succeeded in producing a new nanocomposite material that causes the oxidation of methyl alcohol with the help of graphene oxide and polyaniline [48]. There have been studies of the oxidation of graphene oxide with ligand-based monodisperse nanomaterials via the reduction process [49]. Research has also been carried out to oxidize both methyl alcohol and dimethylborane in electrocatalysts formed by the reduction of graphene oxide-supporting platinum [48, 50]. Platinum has also been reported to have a catalytic effect on alcohols with different surfactant materials [51, 52]. The use of Nafion polymers in DAFCs as a nanocomposite increases the quality of energy generation and results in efficiency due to its conductivity. However, because Nafion nanocomposite polymers are expensive, this membrane can be improved with additional materials, or another nanocomposite can be used instead [53]. In experiments with quaternized polyvinyl alcohol/fumed silica (QPVA/FS) nanocomposite materials, the results obtained directly with alkali fuel cells were positive, and this material was promising for DAFCs [54]. PVA membranes can be used in many polymeric nanocomposite membranes due to their high chemical resistance, flexibility, good film-forming capabilities, low cost, and environmental friendliness [54–56]. Likewise, PVA nanocomposites were formed with the help of uncoordinated ZnO nanofilms and directly tested on alcohol fuel cells by Selvi et al. [57].

Likewise, thanks to its modification feature and its ability to react with high density, it is used directly in alcohol fuel cells as nanocomposite materials in polyhydroxy polymers [54, 58]. It has been reported that by producing hexagonal 20%, 30%, and 40% Ni-Co-N-based honeycomb carbon nanocomposites, catalytic reactions directly occur in alcohol fuel cells [57, 59]. Studies have established the permeability of alcohol through membranes by using silica, zeolite, zinc, and various metals as filling materials to form nafion nanocomposites in DAFCs, especially methanol [60–64]. Materials such as CNTs, fullerenes, graphenes, and carbon wires from nanocomposite materials are used directly in alcohol fuel cells to accelerate electrocatalytic reactions [60]. In general, graphene oxide is mostly used in carbon-based nanocomposites. The graph of the use of electrocatalyst carbon nanocomposite materials in direct fuel cells is shown in Fig. 6. Other carbon nanocomposite materials are quantum dots, fullerenes, CNFs, and mesoporous carbons [60]. Thus, nanocomposite

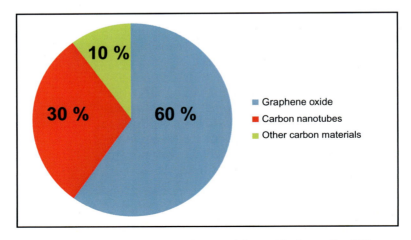

Fig. 6 Percentage rates of carbon-based nanocomposite materials used in the studies [60].

support materials that facilitate passages in the proton exchange membrane in direct fuel cells contribute to catalytic activity, prevent platinum-induced toxicity, increase performance, increase the durability of the cell, and make the electrolytic reaction efficient. Studies on C1 and C3 alcohol oxidation with CNTs have been reported [65].

While nanocomposite materials are a preferred alternative to platinum-induced CO toxicity because they increase energy efficiency, some researchers also use nanocomposites in double, triple, and quadruple systems to further increase energy efficiency and reduce the disadvantages of platinum, and produce various alternatives. These alternatives include the use of dendrimer-based nanocomposites for DAFCs. In this chapter, we will report on the usability of dendrimer-based nanocomposites in direct fuel cells, the extent to which they affect cell performance, their quality, and cellular applications.

3.1 Dendrimer nanocomposite materials

The word *dendrimer* is derived from ancient Latin. The word *dendri* means "tree," and *meros* means "part," so this term literally means "the wooded part" [66]. The first dendrimer macromolecule was synthesized by Vögtle in 1978. This striking dendrimer had a very large size. For this reason, the dendrimer synthesized by Tomelia and Nexkome in nano applicable form has attracted more attention [66–69]. Dendrimers form their structure by connecting functional groups around the molecule from the core located in the center and gradually branching out [66, 70]. The increasing polymerization cycle of dendrimers is called the *degree of generation*. The symbol for the generation degree is the letter *G*. In a dendrimer if there is no binding point, it is called *zeroth generation*, and if its binding point is one, it is called *first-generation* [66, 70]. Fig. 7 shows the structure of a dendrimer and its growth according to the degree of generation.

Dendrimers are used in cancer drugs, wastewater cleaners, textiles, and many other fields. This section provides information on the usability of dendrimers in DAFCs.

The end group numbers of dendrimers are expressed by the following formula [72]:

$$Z = n_c \cdot n_m{}^G$$

$Z =$ the end of group number
$n_c =$ the connection numbers of the main molecule
$n_m =$ the branching connection numbers

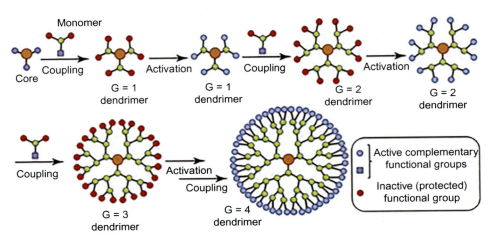

Fig. 7 The structure of a dendrimer and the formation of dendrimers [71].

$G =$ the degree of generation [72]

The solubility of dendrimers depends on the structure of the attached functional groups. While dendrimers formed with functional groups having a hydroscopic structure are hydrolyzed in water, dendrimers with hydrophobic functional groups dissolve mostly in organic solvents. However, it is possible to alter the structure and solubility of dendrimers by changing the structure of the three functional groups [66, 73]. As a result of the branching of the dendrimers, the target point can be coiled and encapsulated. Therefore, it occupies an important place in pharmaceutical chemistry. Studies using dendrimers in the production of cancer drugs have been reported [74, 75]. Reviews on dendrimers stating that end points will undergo catalytic activity have been reported in the research literature [76, 77].

3.1.1 Dendrimer-based nanocomposites for alcohol fuel cells

The material that reduces oxygen in DAFCs and these reduction reactions is platinum. However, the use of platinum alone causes undesirable reactions, reducing the quality and energy efficiency of the cells. In addition, platinum is not durable, expensive, and causes toxicity [78–80]. For these reasons, alternative electrocatalysts are needed. Experiments on alcohol fuel cells of platinum alloys with transition metals have been reported in the literature [80–82]. These supplementary materials work in the form of binary, ternary, and quaternary systems. The catalytic properties of dendrimers in alcohol fuel cells are discussed in this section. Dendrimers can be used as supports in the preparation of monometallic and bimetallic materials used as platinum support materials. Thus, this support by dendrimers is used to carry out electrocatalytic reactions [79, 83]. The internal spaces of branched functions found in the skeleton of dendrimers interact with metallic nanoparticles that aid catalytic reactions in addition to platinum, acting as an electrocatalytic nanoreactor [79, 84]. The branched structures of the dendrimers act as a micelle because they contain a space that surrounds the target point almost entirely. Functional groups of dendrimers are immobilized on the conductive surface of additional materials such as carbon-based nanomaterials that alloy with platinum or transition metals [79, 85–88]. Various types of dendrimers are formed according to their functional groups, and they are used directly in fuel cells by supporting platinum and additional electrocatalysts. Polyamidoamine was coated with zwitterion with the help of ammonium and sulfonic acids, and a dendrimer was formed based on this method. This synthesized dendrimer was used as a support material in direct methyl alcohol fuel cells [89]. With cystamine encapsulated in polyamidoamine dendrimers, electrocatalytic reactions were created in DAFCs supported by platinum [86]. Yu et al. reported on the study of encapsulated platinum nanoparticles for three-dimensional (3D) visualization of organic-inorganic hybrids [90]. CO_2 membrane is also prepared with dendrimers [91], which have the capability of encapsulation because it is very branched and thus contains spaces. At the same time, it is called a *nanoreactor* because it contains this space [92].

4 Conclusions

Dendrimers are part of the class of macromolecules and have a treelike structure with spaces. It highlights their ability to encapsulate target points, as they contain these gaps by branching out. For this reason, dendrimers are used in pharmaceutical chemistry, wastewater treatment, industrial areas, and various fields. This chapter gives descriptions of the use of dendrimers in alcohol fuel cells. Because undesirable side reactions occur in a faster process in direct alcohol fuel cells. Besides, the poisoning of Pt and/or monometallic nanomaterials are very important issue. For this reason, the poisoning effect of platinum can be decreased with the help of bimetallic, trimetallic composite formation. However, to further decrease this poisoning effect in direct alcohol cells, dendrimer-based materials have been extensively used as supporting materials. Studies have been started to look into this; research on the electrocatalytic activation of dendrimers can be carried out by attaching only to transition metals or carbon-derived materials, and by discovering and attaching other materials that may have a catalytic effect along with their electrocatalytic property. These studies are not limited to methyl alcohol fuel cells and direct ethanol fuel cells, but the results should be tested in other types of fuel cells.

References

[1] E.T. Thostenson, C. Li, T.-W. Chou, Nanocomposites in context, Compos. Sci. Technol. 65 (3–4) (2005) 491–516.

[2] D.W. Schaefer, R.S. Justice, How nano are nanocomposites? Macromolecules 40 (24) (2007) 8501–8517.

[3] D.R. Paul, L.M. Robeson, Polymer nanotechnology: nanocomposites, Polymer 49 (15) (2008) 3187–3204.

[4] F. Vögtle, et al., Functional dendrimers, Prog. Polym. Sci. 25 (7) (2000) 987–1041.

[5] L. Balogh, et al., Formation and characterization of dendrimer-based water soluble inorganic nanocomposites, Polym. Mater. Sci. Eng. Wash. 77 (1997) 118–119.

[6] C. Lamy, et al., Recent advances in the development of direct alcohol fuel cells (DAFC), J. Power Sources 105 (2) (2002) 283–296.

[7] L. Xiong, A. Manthiram, Synthesis and characterization of methanol tolerant Pt/TiOx/C nanocomposites for oxygen reduction in direct methanol fuel cells, Electrochim. Acta 49 (24) (2004) 4163–4170.

[8] B. Şen, et al., Silica-based monodisperse PdCo nanohybrids as highly efficient and stable nanocatalyst for hydrogen evolution reaction, Int. J. Hydrog. Energy 43 (44) (2018) 20234–20242.

[9] B. Sen, et al., Monodisperse rutheniumcopper alloy nanoparticles decorated on reduced graphene oxide for dehydrogenation of DMAB, Int. J. Hydrog. Energy 44 (21) (2019) 10744–10751.

[10] E. Antolini, E. Gonzalez, Alkaline direct alcohol fuel cells, J. Power Sources 195 (11) (2010) 3431–3450.

[11] K. Kleiner, Assault on batteries, Nature 441 (7097) (2006) 1046–1047.

[12] T. Zhao, et al., Towards operating direct methanol fuel cells with highly concentrated fuel, J. Power Sources 195 (11) (2010) 3451–3462.

[13] A. Çavuşoğlu, Yakıt pilleri ve kullanım alanları, Uludağ Üniversitesi, 2006.

[14] Ö. Karatepe, et al., Enhanced electrocatalytic activity and durability of highly monodisperse Pt@ PPy–PANI nanocomposites as a novel catalyst for the electro-oxidation of methanol, RSC Adv. 6 (56) (2016) 50851–50857.

[15] I.G.B.N. Makertihartha, M. Zunita, Z. Rizki, P.T. Dharmawijaya, Recent advances on Zeolite modification for direct alcohol fuel cells (DAFCs), In AIP Conference Proceedings, 1818 (2017), https://doi.org/10.1063/1.4976894.

[16] B.G. Güneştekin, Metil alkol yakıt hücrelerinde kullanılacak karbon temelli kompozitlerin üretimi ve petrol mühendisliğinde kullanımı, İskenderun Teknik Üniversitesi/Mühendislik ve Fen Bilimleri Enstitüsü/Petrol, 2020.

[17] M. Kamaruddin, et al., An overview of fuel management in direct methanol fuel cells, Renew. Sust. Energ. Rev. 24 (2013) 557–565.

[18] W. Yuan, et al., Overview on the developments of vapor-feed direct methanol fuel cells, Int. J. Hydrog. Energy 39 (12) (2014) 6689–6704.

Chapter 18 Dendrimer-based nanocomposites for alcohol fuel cells **349**

[19] H. Wu, et al., Novel sulfonated poly (ether ether ketone)/phosphonic acid-functionalized titania nanohybrid membrane by an in situ method for direct methanol fuel cells, J. Power Sources 273 (2015) 544–553.

[20] S.K. Kamarudin, F. Achmad, W.R.W. Daud, Overview on the application of direct methanol fuel cell (DMFC) for portable electronic devices, Int. J. Hydrog. Energy 34 (16) (2009) 6902–6916.

[21] J.-H. Wee, A feasibility study on direct methanol fuel cells for laptop computers based on a cost comparison with lithium-ion batteries, J. Power Sources 173 (1) (2007) 424–436.

[22] A. Hamnett, Mechanism and electrocatalysis in the direct methanol fuel cell, Catal. Today 38 (4) (1997) 445–457.

[23] A.C. Ince, M.F. Serincan, C.O. Colpan, A pseudo two-phase model to study effects of non-condensable gases on the water autonomy of a direct methanol fuel cell system, Int. J. Heat Mass Transf. 163 (2020) 120441.

[24] Y. Na, F. Zenith, U. Krewer, Highly integrated direct methanol fuel cell systems minimizing fuel loss with dynamic concentration control for portable applications, J. Process Control 57 (2017) 140–147.

[25] M. Kamarudin, et al., Direct ethanol fuel cells, Int. J. Hydrog. Energy 38 (22) (2013) 9438–9453.

[26] N. Wongyao, A. Therdthianwong, S. Therdthianwong, Performance of direct alcohol fuel cells fed with mixed methanol/ethanol solutions, Energy Convers. Manag. 52 (7) (2011) 2676–2681.

[27] Q. Xu, et al., A flow field enabling operating direct methanol fuel cells with highly concentrated methanol, Int. J. Hydrog. Energy 36 (1) (2011) 830–838.

[28] F. Şen, G. Gökağaç, Different sized platinum nanoparticles supported on carbon: an XPS study on these methanol oxidation catalysts, J. Phys. Chem. C 111 (15) (2007) 5715–5720.

[29] E. Antolini, Catalysts for direct ethanol fuel cells, J. Power Sources 170 (1) (2007) 1–12.

[30] Z. Daşdelen, et al., Enhanced electrocatalytic activity and durability of Pt nanoparticles decorated on GO-PVP hybrid material for methanol oxidation reaction, Appl. Catal. B Environ. 219 (2017) 511–516.

[31] S. Şen, F. Şen, G. Gökağaç, Preparation and characterization of nano-sized Pt–Ru/C catalysts and their superior catalytic activities for methanol and ethanol oxidation, Phys. Chem. Chem. Phys. 13 (15) (2011) 6784–6792.

[32] F. Şen, G. Gökağaç, Activity of carbon-supported platinum nanoparticles toward methanol oxidation reaction: role of metal precursor and a new surfactant, tert-octanethiol, J. Phys. Chem. C 111 (3) (2007) 1467–1473.

[33] B. Çelik, et al., Monodisperse Pt (0)/DPA@ GO nanoparticles as highly active catalysts for alcohol oxidation and dehydrogenation of DMAB, Int. J. Hydrog. Energy 41 (13) (2016) 5661–5669.

[34] S. Eris, Z. Daşdelen, F. Sen, Enhanced electrocatalytic activity and stability of monodisperse Pt nanocomposites for direct methanol fuel cells, J. Colloid Interface Sci. 513 (2018) 767–773.

[35] J. Zhang, et al., Balancing the electron conduction and mass transfer: effect of nickel foam thickness on the performance of an alkaline direct ethanol fuel cell (ADEFC) with 3D porous anode, Int. J. Hydrog. Energy 45 (38) (2020) 19801–19812.

[36] A. Serov, C. Kwak, Recent achievements in direct ethylene glycol fuel cells (DEGFC), Appl. Catal. B Environ. 97 (1–2) (2010) 1–12.

[37] E. Peled, V. Livshits, T. Duvdevani, High-power direct ethylene glycol fuel cell (DEGFC) based on nanoporous proton-conducting membrane (NP-PCM), J. Power Sources 106 (1–2) (2002) 245–248.

350 Chapter 18 Dendrimer-based nanocomposites for alcohol fuel cells

[38] R. De Lima, et al., On the electrocatalysis of ethylene glycol oxidation, Electrochim. Acta 49 (1) (2003) 85–91.

[39] A. Neto, et al., Electro-oxidation of ethylene glycol on PtRu/C and PtSn/C electrocatalysts prepared by alcohol-reduction process, J. Appl. Electrochem. 35 (2) (2005) 193–198.

[40] R. Chetty, K. Scott, Catalysed titanium mesh electrodes for ethylene glycol fuel cells, J. Appl. Electrochem. 37 (9) (2007) 1077–1084.

[41] V. Selvaraj, M. Vinoba, M. Alagar, Electrocatalytic oxidation of ethylene glycol on Pt and Pt–Ru nanoparticles modified multi-walled carbon nanotubes, J. Colloid Interface Sci. 322 (2) (2008) 537–544.

[42] N.W. Maxakato, C.J. Arendse, K.I. Ozoemena, Insights into the electro-oxidation of ethylene glycol at Pt/Ru nanocatalysts supported on MWCNTs: adsorption-controlled electrode kinetics, Electrochem. Commun. 11 (3) (2009) 534–537.

[43] Z. Pan, Y. Bi, L. An, A cost-effective and chemically stable electrode binder for alkaline-acid direct ethylene glycol fuel cells, Appl. Energy 258 (2020) 114060.

[44] R. Baronia, et al., Electro-oxidation of ethylene glycol on PtCo metal synergy for direct ethylene glycol fuel cells: reduced graphene oxide imparting a notable surface of action, Int. J. Hydrog. Energy 44 (20) (2019) 10023–10032.

[45] C.C. Okpala, Nanocomposites—an overview, Int. J. Eng. Res 8 (2013) 17–23.

[46] P. Zapata, et al., Preparation of nanocomposites by in situ polimerization, J. Chil. Chem. Soc. 53 (1) (2008) 1359–1360.

[47] M. Alexandre, P. Dubois, Polymer-layered silicate nanocomposites: preparation, properties and uses of a new class of materials, Mater. Sci. Eng. R. Rep. 28 (1–2) (2000) 1–63.

[48] S. Eris, et al., Nanostructured polyaniline-rGO decorated platinum catalyst with enhanced activity and durability for methanol oxidation, Int. J. Hydrog. Energy 43 (3) (2018) 1337–1343.

[49] Y. Yıldız, et al., Different ligand based monodispersed Pt nanoparticles decorated with rGO as highly active and reusable catalysts for the methanol oxidation, Int. J. Hydrog. Energy 42 (18) (2017) 13061–13069.

[50] Y. Yıldız, et al., Monodisperse Pt nanoparticles assembled on reduced graphene oxide: highly efficient and reusable catalyst for methanol oxidation and dehydrocoupling of dimethylamine-borane (DMAB), J. Nanosci. Nanotechnol. 16 (6) (2016) 5951–5958.

[51] F. Şen, S. Şen, G. Gökağaç, Efficiency enhancement of methanol/ethanol oxidation reactions on Pt nanoparticles prepared using a new surfactant, 1, 1-dimethyl heptanethiol, Phys. Chem. Chem. Phys. 13 (4) (2011) 1676–1684.

[52] Y. Li, et al., Recent advances in the fabrication of advanced composite membranes, J. Mater. Chem. A 1 (35) (2013) 10058–10077.

[53] F.A. Zakil, S. Kamarudin, S. Basri, Modified Nafion membranes for direct alcohol fuel cells: an overview, Renew. Sust. Energ. Rev. 65 (2016) 841–852.

[54] S. Rajesh Kumar, et al., Fumed silica nanoparticles incorporated in quaternized poly (vinyl alcohol) nanocomposite membrane for enhanced power densities in direct alcohol alkaline fuel cells, Energies 9 (1) (2016) 15.

[55] J.-M. Yang, N.-C. Wang, H.-C. Chiu, Preparation and characterization of poly (vinyl alcohol)/sodium alginate blended membrane for alkaline solid polymer electrolytes membrane, J. Membr. Sci. 457 (2014) 139–148.

[56] A. de Souza Gomes, J.C.D. Filho, Hybrid membranes of PVA for direct ethanol fuel cells (DEFCs) applications, Int. J. Hydrog. Energy 37 (7) (2012) 6246–6252.

[57] J. Selvi, et al., Optical, electrical, mechanical, and thermal properties and non-isothermal decomposition behavior of poly (vinyl alcohol)–ZnO nanocomposites, Iran. Polym. J. (2020) 1–12.

Chapter 18 Dendrimer-based nanocomposites for alcohol fuel cells **351**

[58] J. Qiao, et al., Alkaline solid polymer electrolyte membranes based on structurally modified PVA/PVP with improved alkali stability, Polymer 51 (21) (2010) 4850–4859.

[59] G. Li, et al., Ni-co-N doped honeycomb carbon nano-composites as cathodic catalysts of membrane-less direct alcohol fuel cell, Carbon 140 (2018) 557–568.

[60] G. Rambabu, S.D. Bhat, F.M. Figueiredo, Carbon nanocomposite membrane electrolytes for direct methanol fuel cells—a concise review, Nanomaterials 9 (9) (2019) 1292.

[61] B.P. Tripathi, V.K. Shahi, Organic–inorganic nanocomposite polymer electrolyte membranes for fuel cell applications, Prog. Polym. Sci. 36 (7) (2011) 945–979.

[62] R. Jiang, H.R. Kunz, J.M. Fenton, Composite silica/Nafion® membranes prepared by tetraethylorthosilicate sol–gel reaction and solution casting for direct methanol fuel cells, J. Membr. Sci. 272 (1–2) (2006) 116–124.

[63] P. Dimitrova, et al., Modified Nafion®-based membranes for use in direct methanol fuel cells, Solid State Ionics 150 (1–2) (2002) 115–122.

[64] V. Tricoli, F. Nannetti, Zeolite–Nafion composites as ion conducting membrane materials, Electrochim. Acta 48 (18) (2003) 2625–2633.

[65] B. Çelik, et al., Nearly monodisperse carbon nanotube furnished nanocatalysts as highly efficient and reusable catalyst for dehydrocoupling of DMAB and C1 to C3 alcohol oxidation, Int. J. Hydrog. Energy 41 (4) (2016) 3093–3101.

[66] M. Bulut, A. Ezgi, Dendrimerlerin önemi ve kullanım alanları, Teknik Bilimler Dergisi 2 (1) (2012) 5–11.

[67] d.A. Bosman, H. Janssen, E. Meijer, About dendrimers: structure, physical properties, and applications, Chem. Rev. 99 (7) (1999) 1665–1688.

[68] J.M. Frechet, Functional polymers and dendrimers: reactivity, molecular architecture, and interfacial energy, Science 263 (5154) (1994) 1710–1715.

[69] M. Van Genderen, E. De Brabander, E. Meijer, Advances in Dendritic Macromolecules, 4, JAI Press, 1999, pp. 61–105.

[70] A. Yücel, Dendrimerlerin Önemi ve Uygulama Alanları, Bitirme Ödevi, Süleyman Demirel Üniversitesi. Tekstil Mühendisliği Bölümü, Isparta, 2011.

[71] D. Medina-Cruz, et al., Drug-delivery nanocarriers for skin wound-healing applications, in: Wound Healing, Tissue Repair, and Regeneration in Diabetes, Elsevier, 2020, pp. 439–488.

[72] O. Namirti, R. Atav, Tekstilde yeni bir konsept olan dendrimerlerin tarihçesi, sınıflandırılması, molekül yapısı ve özellikleri, Pamukkale Üniversitesi Mühendislik Bilimleri Dergisi 17 (2) (2011) 109–115.

[73] P.E. Froehling, Dendrimers and dyes—a review, Dyes Pigments 48 (3) (2001) 187–195.

[74] M. Carvalho, R. Reis, J.M. Oliveira, Dendrimer nanoparticles for colorectal cancer applications, J. Mater. Chem. B 8 (6) (2020) 1128–1138.

[75] S.J. Lee, et al., Enzyme-responsive doxorubicin release from dendrimer nanoparticles for anticancer drug delivery, Int. J. Nanomedicine 10 (2015) 5489.

[76] Y. Niu, R.M. Crooks, Dendrimer-encapsulated metal nanoparticles and their applications to catalysis, C.R. Chim. 6 (8–10) (2003) 1049–1059.

[77] G.E. Oosterom, et al., Transition metal catalysis using functionalized dendrimers, Angew. Chem. Int. Ed. 40 (10) (2001) 1828–1849.

[78] H.A. Gasteiger, et al., Methanol electrooxidation on well-characterized platinum-ruthenium bulk alloys, J. Phys. Chem. 97 (46) (1993) 12020–12029.

[79] J. Ledesma-Garcia, et al., Pt dendrimer nanocomposites for oxygen reduction reaction in direct methanol fuel cells, J. Solid State Electrochem. 14 (5) (2010) 835–840.

[80] W. Vielstich, A. Lamm, H. Gasteiger, Handbook of Fuel Cells. Fundamentals, Technology, Applications, 2003.

[81] V. Baglio, et al., Investigation of Pt–Fe catalysts for oxygen reduction in low temperature direct methanol fuel cells, J. Power Sources 159 (2) (2006) 900–904.

[82] V. Baglio, et al., Development of Pt and Pt–Fe catalysts supported on multi-walled carbon nanotubes for oxygen reduction in direct methanol fuel cells, J. Electrochem. Soc. 155 (8) (2008) B829.

[83] R.W. Scott, O.M. Wilson, R.M. Crooks, Synthesis, characterization, and applications of dendrimer-encapsulated nanoparticles, J. Phys. Chem. B 109 (2) (2005) 692–704.

[84] M.R. Knecht, D.W. Wright, Dendrimer-mediated formation of multicomponent nanospheres, Chem. Mater. 16 (24) (2004) 4890–4895.

[85] G. Vijayaraghavan, K.J. Stevenson, Synergistic assembly of dendrimer-templated platinum catalysts on nitrogen-doped carbon nanotube electrodes for oxygen reduction, Langmuir 23 (10) (2007) 5279–5282.

[86] S. Raghu, et al., Platinum–dendrimer nanocomposite films on gold surfaces for electrocatalysis, Catal. Lett. 119 (1) (2007) 40–49.

[87] J. Ledesma-Garcia, et al., Evaluation of assemblies based on carbon materials modified with dendrimers containing platinum nanoparticles for PEM-fuel cells, Int. J. Hydrog. Energy 34 (4) (2009) 2008–2014.

[88] T. Maiyalagan, Pt–Ru nanoparticles supported PAMAM dendrimer functionalized carbon nanofiber composite catalysts and their application to methanol oxidation, J. Solid State Electrochem. 13 (10) (2009) 1561–1566.

[89] Z. Gu, et al., Polybenzimidazole/zwitterion-coated polyamidoamine dendrimer composite membranes for direct methanol fuel cell applications, Int. J. Hydrog. Energy 38 (36) (2013) 16410–16417.

[90] Y. Ju, et al., Three-dimensional TEM study of dendrimer-encapsulated Pt nanoparticles for visualizing structural characteristics of the whole organic–inorganic hybrid nanostructure, Anal. Chem. (2021).

[91] F. Ito, et al., Development of high-performance polymer membranes for CO_2 separation by combining functionalities of polyvinyl alcohol (PVA) and sodium polyacrylate (PAANa), J. Polym. Res. 26 (5) (2019) 1–9.

[92] S. Sadjadi, Dendrimers as nanoreactors, in: Organic Nanoreactors, Elsevier, 2016, pp. 159–201.

19

Metal organic framework-based nanocomposites for alcohol fuel cells

Bahar Yilmaz[a], Ramazan Bayat[b,c], Muhammed Bekmezci[b,c], and Fatih Şen[b]

[a]*Karamanoglu Mehmetbey University, Faculty of Engineering, Department of Bioengineering, Karaman, Turkey.* [b]*Şen Research Group, Department of Biochemistry, Dumlupinar University, Kütahya, Turkey.* [c]*Department of Materials Science and Engineering, Faculty of Engineering, Dumlupinar University, Kütahya, Turkey*

1   Introduction

Because of the growing energy needs of the world, energy storage and conversion technologies such as fuel cells, metal-air batteries, and water-splitting systems have drawn tremendous research interest [1, 2]. Oxygen evolution reactions (OERs) play a vital role in these next generations of energy technology [3]. To transform the energy within fuel into electricity, traditional electricity systems first use a combustion reaction. Fuel and oxygen oxidizers need to be thoroughly combined in order for the combustion reaction to occur efficiently. After that, before electrical energy is produced, several intermediate processes are needed. Every intermediate approach leads to energy loss, decelerating efficiency [3]. It is possible to transform electricity from gasoline directly into electrical energy in a fuel cell. The fuel and oxidizer are located in different compartments; unlike in conventional production systems, they do not mix. Their fusion occurs only by the transfer of ions and electrons between these compartments. A fuel cell transforms fuel energy directly into an electrochemical reaction that creates electrical energy. It produces electricity through fuel supplied externally (on the anode side) and oxidizer (on the cathode side). In an electrolyte/electrode unit, they react. As a rule, the cell is reached by those who answer, while the reaction products exit the cell [4, 5]. So long as the needed flow of fuel and oxidizer is given, fuel cells will operate. Substances that can react are continuously absorbed in fuel

Nanomaterials for Direct Alcohol Fuel Cells. https://doi.org/10.1016/B978-0-12-821713-9.00006-8
Copyright © 2021 Elsevier Inc. All rights reserved.

cells, the battery electrodes react and alter as the battery is filled and discharged, the fuel cell electrodes are catalytic and relatively stable [6, 7].

2 Fuel cells

Fuel cells are one of the most popular energy-generating tools of our age that can produce energy from organic materials by means of chemicals [5]. Fuel cells are composed of an electrolyte with ionic conductivity and an anode and cathode where various reactions take place. Fuel cells are electrochemical cells, just like batteries, but they do have some differences. The most important difference between fuel cells and batteries is the need to constantly be fed fuel from the outside. In other words, an electrochemical cell that can work continuously if fuel is fed to it from an external source is called a *fuel cell*.

Fuel cells present a very important alternative that is on the brink of commercialization today for the energy systems of the future. Fuel cells have many advantages, especially if they are used in portable energy systems. These energy systems are often considered alternatives to batteries and internal combustion engines. Fuel cell systems do not need to be charged, as they can run continuously if fuel is fed to them; in addition, they achieve a higher energy density than batteries [8, 9].

2.1 Basic principles of fuel cells

The first fuel cell was designed by William Grove in 1842. It produced hydrogen and oxygen by immersing platinum electrodes in sulfuric acid with the presence of electric current. By observing that a current in the opposite direction is formed by cutting the current passing through the external circuit, Grove created a gas battery. The first serious use of the fuel cell was by the National Aeronautics and Space Administration (NASA), in its *Gemini* spacecraft in 1962. The fuel cell used was manufactured by General Electric, and a polymer membrane that acted as an electrolyte was used for the first time. Because the use of polymer membranes eliminated the use of heavy and nonportable liquid electrolytes, it significantly reduced the size of the fuel cell. Currently, polyelectrolyte membrane fuel cells are still being developed and are making significant progress. For this purpose, research has been initiated in many areas in the world, from fixed applications of fuel cells that require a large amount of energy to their use in portable systems such as cars, bicycles, computers, and mobile phones [5, 8, 10].

2.2 Applications of fuel cells

Fuel cells do not harm the environment and are very efficient at electricity production. They work very quietly, producing only water, electric current, and heat as waste. For these reasons, fuel cells seem close to taking their place in the transport sector. There are also vehicle applications such as buses, trains, cars, and submarines that are still being tested around the world and run on fuel cells [11, 12]. Among the fuel cell types, which are named for the type of electrolyte used, PEMFC, which is the highest amount of power produced per unit volume, is used often. Cars powered by fuel cells are getting more attention than other battery-powered vehicles. In addition to offering the advantages of battery-powered vehicles, fuel-cell vehicles refuel faster and increase the range of replenishment. Fuel cell-powered cars require less maintenance and are quieter than internal combustion engine-powered cars. It also has fewer moving parts than a standard engine. The operating life of PEMFC is actually longer than the life of the car itself. PEMFC is operational and available when the fuel cell-powered car is scrapped [10, 13, 14].

Also, fuel cells can provide zero emissions in vehicles with the use of hydrogen, and close to zero emissions can also be achieved with the use of other fuels. Fuel cells can provide more effective operation than a vehicle with a powerful battery [15]. Considering environmental factors, fuel cells are in good condition compared to energy efficiency and harmful substances. Some railway companies and locomotive manufacturers plan to produce commercial fuel cell-powered locomotives in the next 10–15 years. Fuel cells, especially between long distances (desert or wide plain) and given the increasing cost of electric train transportation and energy cables, appear to be an alternative as an energy-generating system [8].

In submarine applications, fuel cell-powered submarines are very attractive to manufacturers due to the increasing costs of nuclear-powered submarines and environmental threats, the noise of diesel-powered submarines, and the difficulty of surfacing at certain intervals [16].

Studies on the spacecraft that saw the first applications of fuel cells are continuing. Such applications are quite interesting: the spacecraft provides its energy with solar cells. And with some of that increased energy, the water contained in the vehicle decomposes into its components, hydrogen and oxygen, by electrolysis. At the same time, when the sun is not visible, the hydrogen, and oxygen produced are combined into fuel cells, providing water and electric current [17, 18].

2.3 Operating principle of fuel cells

Fuel cells consist of an anode, a cathode, and an electrolyte. As a result of reactions occurring at the anode and cathode, voltage, ions, and electrons are formed. The resulting electron moves from one conductive line, and the ion moves from the ion-permeable electrolyte to the other electrode (Fig. 1) [9, 19–21].

The movement of electrons creates a current. The difference in voltage between electrodes and current in this system creates the power. For these reactions to be continuous, substances that constantly react to the fuel cell must be sent, and the reaction products must be removed from the environment [8, 18].

2.4 Types of fuel cells

Currently, many types of fuel cells are available, with different uses, operating conditions, and fuels. Fuel cells are usually named according to their electrolytes or the fuel they use:

Proton exchange membrane fuel cells (PEMFCs): The PEMFC is a category of fuel cells designed specifically for transportation, stationary fuel cells, and portable fuel cells [14, 22, 23].

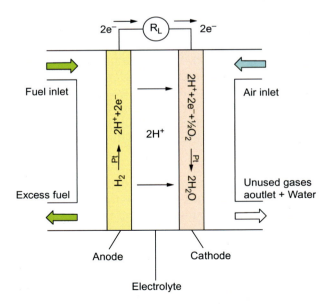

Fig. 1 Simple schematic representation of the fuel cells [19]. Reprinted (adapted) with permission from G. Slaughter, T. Kulkarni, Enzymatic glucose biofuel cell and its application, J. Biochips Tissue Chips. 05 (2015). https://doi.org/10.4172/2153-0777.1000110. Copyright (2015) *Journal of Bioengineering and Bioelectronics.*

Alkaline (basic) fuel cells (AFCs): The AFC is one of the most advanced and highly efficient fuel cell technologies. It generates energy by taking advantage of the redox reaction between hydrogen and oxygen [24–26].

Phosphoric acid fuel cells (PAFCs): The PAFC is a fuel cell that uses phosphoric acid as the electrolyte fluid [27, 28].

Molten carbonate fuel cells (MCFCs): The MCFC is a high-temperature fuel cell that runs at 600°C and above. The efficiency is the highest of all fuel cells [29, 30].

Solid oxide fuel cells (SOFCs): The SOFC is a device that generates electricity directly from the fuel. Solid oxide or ceramic is the electrolyte material in these cells. Ceramic fuel cells work at significantly higher temperatures than polymer-based ones [31–33].

Direct borohydride fuel cell (DBFCs): A DBFC uses sodium borohydride as fuel. The value of sodium borohydride over standard hydrogen in AFCs is that it stops carbon dioxide in the air from poisoning the fuel cell with alkaline fuel and waste [34–36].

Direct carbon fuel cells (DCFCs): A DCFC uses fuel rich in carbon, including coal or a biomass. The cell produces electricity by combining carbon and oxygen, which emits carbon dioxide as a by-product [31, 37, 38].

Direct glucose fuel cells (DGFC): The DGFC has been investigated as an important application of biofuel cells to provide the required energy for implanted medical devices [39–41].

In addition to the fuel cells mentioned here, other types of fuel cells are not commonly used. Some of these fuel cells can operate at low temperatures, while others can operate at high temperatures. Some fuel cells can work with hydrogen, some with carbon monoxide, some with alcohol, some with coal, some with borohydride, and some with glucose. Some are solid, while others are in the form of electrodes immersed in liquid [31].

Direct alcohol fuel cells (DAFCs): DAFCs can be used in electrical equipment for transportation, as well as mobile applications that turn the chemical energy in liquid alcohol into electrical energy [42]. DAFCs can be used as power sources for portable electronic devices, including computers for laptops, mobile phones, and others. As DAFCs have comparatively higher energy by mass density, they use low-molecular-weight alcohols and are more manageable, storable and transportable. They have a range of advantages compared to PMFCs. Methanol, ethanol, ethylene glycol, glycerol, 1-propanol, and 2-propanol are alcohols that can be used in DAFCs [43–48].

3 Nanocomposites

Nanocomposites are materials formed by the dispersion of nanometer-sized particles in a matrix. The essence of nanotechnology is to work in the molecular dimension, obtaining large structures whose molecular structures have been renovated. The nanometric-sized properties of materials vary according to the macro-sized properties of the same material. The advantages that nanocomposites bring to the material include increasing the module, strengthening it, increasing its heat resistance, preventing gas leakage into the material, and reducing its flammability [49–51]. Powders in nanoparticle form are used as support in materials such as ceramics, metals, or polymers to form nanocomposites with advanced properties.

Nanocomposite materials combine the various properties of their components, have at least one component with dimensions of less than 100 nm. For structural applications, this definition is used as fiber- or granular-reinforced materials supported by a binder or matrix phase. Polymer nanocomposites form an alternative class of materials to traditionally filled polymers, while its main components consist of polymers that act as carrier matrices [52, 53].

The filling in the nanoparticles is dispersed into the polymer matrix by various methods. Because nanoparticles have a high surface area compared to the volume of polymers and, accordingly, more atoms can interact with different bonds and they show tremendous changes compared to their equivalents with larger grain sizes. In particular, significant improvement is observed in tensile strength, thermal and chemical resistance, glassy transition temperature, thermal decomposition, and viscoelasticity [54, 55]. In recent years, polymer nanocomposites have formed a precursor class among multicomponent polymer systems. As such, nanocomposites have been the most studied materials in research development and commercialization worldwide. It has been used in many applications, from food packaging to gas and oxygen barriers, from electrochromic tools for high-resolution displays to drug release systems, from thin-film capacitors to sports equipment [56–58].

3.1 Properties of nanocomposites

Composite materials are hybrids produced by combining two or more materials with different properties. The aim of the production of composite materials is to create a substance with new and improved characteristics from insoluble ingredients.

These fundamental properties include durability, flexibility, dimensional stability, thermomechanical properties, and water permeability. The basic structure of composites is composed of the main material called a *matrix* and a siding material called *reinforcement*. These materials produced in nanoparticles maximize intersurface adhesion due to their size characteristics, significantly increasing the interface interactions between the polymer matrix and decompiler nanomaterial [59, 60]. The unique design of nanocomposites allows them to adopt the desired properties and have superior properties to conventional composites. Nanocomposites basically feature three-class differences: ceramic matrix nanocomposites [Al_2O_3/TiO_2, Al_2O_3/SiO_2, Al_2O_3/Sic, Al_2O_3/carbon nanotube (CNT)]; metal matrix nanocomposites (Co/Cr, Fe-Cr/Al_2O_3, Fe-MgO); and polymer matrix nanocomposites (polyester/TiO_2, polymer/CNT) [61, 62].

4 Metal organic frameworks

Metal organic frameworks (MOFs) are one-, two-, or three-dimensional structures of one type composed of metal ions or clusters that are coordinated through organic ligands (Fig. 2) [63]. MOFs are unified groups that are a subset of polymers of coordination that are typically porous. The organic ligands used, such as 1,4-benzendi carboxylic acid, are often called *buttresses* or *binders*. It is clearly seen that a metal-organic system is a coordination network with potential organic ligand spaces. A coordination network expands in one dimension with repetitions. However, crosslinks include two or more independent chains consisting of loop or spiral links or repetitive coordination compounds that stretch across two or three entities [64–69].

Crystal MOFs fall into the category of high-potential porous materials that can be used in magnetic, sensor design and drug

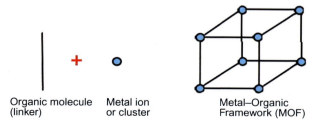

Fig. 2 Metal organic frameworks (MOFs) [63]. Reprinted (adapted) with permission from V. Borovkov, Book review of "Lanthanide metal-organic frameworks", Front. Chem. 3 (2015) 50. https://doi.org/10.3389/fchem.2015.00050. Copyright (2015) *Frontiers in Chemistry*.

delivery, adsorbents, gas storage, retention/separation, and catalysis. MOFs are typically formed by self-assembly, where secondary structure units relate to organic spacers (ligands) to form complex networks. Organic spacers or metallic sibilants can be modified to control the porosity of the MOF, which is crucial to its functions and benefits for specific applications.

Over the past few years, MOFs have been developed as an attractive group of crystalline materials at the interface between coordination chemistry and material science due to their unusual catalytic functionalities. They are self-assembled hybrid organic and/or inorganic units that form a porous and periodic metal node and/or polynuclear secondary construction unit structure. These units have a tendency to coordinate the metal ions, rigidity, and length of organic ligands. Their special geometry topologies, the outstanding amount of surface area, high porosity, and various applications make these materials promising for applied research, catalysis resolution, proton piping, gas recovery, and drug distribution, identification, and separation.

For metal-organic structures, such as postsynthetic alteration, many functionalization methodologies are currently being studied by researchers. And these modifications may be achieved by using functional ligands to create metal clusters or organic ligands and adding some useful molecules inside the pores, such as metal nanoparticles, organometallics, heteropoly acids, and enzymes. The properties of MOF materials can be controlled by organic ligands or inorganic metal ions. To generate functional sites in the MOF architecture, several well-known methodologies are used, including organic ligands, the addition of various guest elements, including metal and organic ligands, and the creation of multiple functional sites with other guest molecules. These methods make MOFs an ideal medium for the design and processing of materials that are functionally manufactured [70, 71].

From the point of view of catalysis, MOF materials provide a number of interesting advantages over conventional catalytic like clay, zeolites, and/or mesoporous materials such as silica. Some can be highlighted among many noteworthy, multifunctional catalytic potentialities of MOFs. This study highlights the extraordinary multicatalytic ability and recent MOF implementations, with special attention paid to the view of catalytic chemistry. A focus is put on the photocatalysis, biocatalysts, and electrocatalysis characteristics of MOFs with relevant examples. In the original section, the electrically catalytic activity was explored by various MOFs, such as oxidation and reduction of candidates for various types of chemical reactions. The second portion of the report focuses on photocatalyst processes for different MOFs, including multiple photocatalytic degradation and reductions in CO_2 and water separation. The third portion of this study looks at the usage of MOF-based porous materials for biomedical use of biological species such as sensing and biosensing, pharmaceutical goods, biomimetic systems, and antibacterial agents. Finally, it also stresses the findings, hindrances, and future catalytic potential for MOFs [34, 64].

4.1 Structural classification and synthesis of metal organic frameworks

A new class of porous materials is characterized by MOFs. They have a modular structure that offers tremendous structural diversity and large opportunities for materials with tailored properties to be developed. Two components can be differentiated in MOFs: secondary building units (clusters or metal ions) and organic molecules connecting the former to give basically periodic porous structures. A large number of MOFs arise from variations of these elements of the structure. In particular, by 2007, the Cambridge Crystallographic Data Centre had deposited 131 variants of the geometry of secondary building units. In addition, the central metal atom may be replaced in each geometry variant. In particular, removing secondary building units, while the linker is terephthalic acid, leads to very different MOFs. One of the first synthesized MOFs and MOF-5 is the tetrahedral ZnO_4 moiety, found at the sites of a cubic lattice and connected by terephthalate dianions. Elongating the linker's carbon chain and maintaining the initial topology allow materials that have identical structure and symmetry but vary in pore size to be synthesized. As a rule, the members of such a structural sequence are denoted with the isoreticular metal-organic framework (IRMOF) code followed by a number. Via different linkers, secondary building units can be combined into a system structure [31, 39].

4.2 MOF applications

Many applications that utilize the cage structure of MOFs, such as gas storage and isolation, fluid separation and purifying, electrochemical energy storage, catalysis, and sensing are being developed in various fields [72, 73]. To manufacture inorganic functional materials that possess unmatched design capabilities, such as carbons, metal mixing, and composites, MOFs were used as basic precursors in direct applications [74]. Carbonaceous materials are currently gaining tremendous interest for their comprehensive applications, such as adsorption, catalysis, batteries, fuel cells, overcapacitors, drug distribution, and imagery [75, 76]. In addition, sensors are an important material for carbon-based materials and one of the important applications for human health.

There are various approaches to preparing such carbon materials, but direct carbonization by organic precursors is the most common method used to prepare nanoporous carbons due to its versatility and simplicity. However, these textiles have some limitations, such as low surface area, unordered styles, and cohesive measurements, which greatly limit their use. There is a great opportunity for air purification from toxic gases in the MOF systems. Ammonia absorption, for example, may occur either by chemisorption at coordinated, unsaturated, active sites, causing the frame to collapse in most cases; or by the formation of hydrogen bonds with functional groups of organic binders. For hydrogen sulfide, certain impurity reduction mechanisms are also suitable. It is more difficult to extract all sulfur and nitrogen dioxides and chlorine because they may not adhere to coordinately unsaturated active sites. In this case, for each mission, the MOF pores should be functionalized. MOFs should be humidity-resistant for the purification of aqueous solutions [67, 70, 77].

4.3 Structural features of metal organic frameworks

The topology of metal-organic structures is influenced by their structural characteristics. The nature and synthesis of MOFs greatly help produce the geometric and chemical characteristics of the structural elements of MOFs and organic linkers. In favorable environments, it has been found that multidentate ligands can aggregate, lock, and shape the structural features of MOFs in some positions. In MOFs, two components can be distinguished: secondary building units and organic molecules that connect them to provide porous structures that are essentially periodic. Multiple combinations of elements of the structure contribute to an overwhelming number of MOFs [39, 78, 79].

4.4 Functionalization of metal-organic frameworks

MOFs with a wide porous surface area provide a convenient forum for developing usable materials designed to solve specific catalysis, gas separation, and power engineering problems. For this reason, MOF functionalization is still one of the most topical

Chapter 19 Metal organic framework-based nanocomposites **363**

issues attracting researchers' interest. This problem is even more significant given the number of publications than the optimization of existing synthesis methods and the design of novel structures [39, 80, 81]. There is no more than one coordination vacancy for the metal sites of MOFs, which restricts the applicability of such materials (e.g., in catalysis, where at least two coordination vacancies are necessary).

Two fundamental approaches to the functionalization of MOFs exist:
- The usage of linkers that contain functional groups in MOF synthesis
- Postsynthetic modification of MOFs [82]

4.5 Metal organic frameworks as promising materials in fuel cells

MOFs are a new type of product that has gained enormous interest in the last decade. A number of MOFs are used in industrial manufacturing [83, 84]. MOFs are open networks of secondary construction units and organic connectors with a metal base. And these architectures can have one, two, or three dimensions. The structures are crystalline and long-lasting.

The system contains highly uniform pores or channels. The difference between MOFs and other porous materials is that MOFs can be made functional and have distinctive properties. This versatility helps us to build MOFs to meet particular requirements, making them extremely versatile and adaptable. Their toughness varies considerably from that of conventional materials such as zeolites, which are porous. The chemical structure of MOFs influences the properties and applications of these compounds. In addition, because MOFs are designed to create close links between metal centers and organic connectors, they form networks. They typically demonstrate good thermal stability of up to very high temperatures [82, 85]. Also, for some MOFs, strong chemical stability has been found [86]. However, it remains difficult to produce chemically stable MOFs due to their high susceptibility to ligand exchange.

Two primary components are made of MOFs: a metal ion, or metal ion cluster; and an organic molecule is known as the *binder*. For this reason, these materials are also known as *organic-inorganic composites*. Lately, however, this language was not explicitly recommended. Organic units are typically ligands that are monovalent, divalent, trivalent, or tetravalent. The metal

364 Chapter 19 Metal organic framework-based nanocomposites

and binder are chosen to determine the composition (and thus the properties) of the MOFs. To determine how many ligands attach to the metal and in which direction, the metal coordination defines the pore's proportions and types. To define and organize MOF systems, a nomenclature scheme has been developed. Subunit MOPs can be defined as secondary structure units common to various structures. Each topology is given a symbol of three bold letters, also known as a *network* [67].

4.5.1 Ligand exchange

To replace an existing organic binding group with a new binder through ligand exchange or partial ligand exchange inside prefabricated MOFs, postsynthetic alteration techniques may be used. This move makes it easier to change the overall configuration of the pores (and, in some situations, the MOFs) for various purposes, such as selective adsorption, gas storage, and fine-tuning of the material for catalysis. The MOF crystals were prefabricated, washed with solvent for ligand exchange, and then dipped in a new binder solution. Also postsynthetic ligand exchange allows functional groups to be used in MOFs that would otherwise not be able to tolerate MOF synthesis due to temperature, pH, or other conditions of reaction, or that could inhibit the synthesis itself by competing in the lending ligand with donor groups [34, 87, 88].

4.5.2 Metal exchange

Some synthetic techniques can also be used to replace the MOF ion. Solvent washes precast MOF crystals similar to post-common ligand exchange and then soak the crystals in a new metal solution produced by the postsynthetic metal exchange. Postsynthetic metal replacement allows the exchange of MOFs in a similar way, but ion exchange is easier [10, 34, 89].

4.5.3 Stratified synthesis

Postsynthetic modifications may be added to extend the structure of the MOF, as well as changing the function of ligands and metals themselves. MOFs may be transformed from a highly ordered crystalline material into a heterogeneously porous material with postsynthetic modification. The installation of domains in MOF crystals can be handled using postsynthetic techniques, which represent unique structural and functional features. Various MOFs were created, where layer functions are difficult but typically crystallographically stable from layer to layer [34, 89].

5 Conclusion

The combining of MOFs with various functional materials is a technique to further improve MOF performance. Adding new functions to the MOFs is also a highly powerful and feasible way to use them. Many MOF nanocomposites, including graphene, CNTs, metal nanoparticulate and nanorod materials, metal oxides, complexes, and even enzymes, have already been successfully prepared. Currently, MOFs are being manufactured. In MOF nanocomposites, the various functions of the MOF and functional materials combine synergistically to establish new physical and chemical properties, not just in the individual components. In addition, they have multifunctional properties. Their uses have been enhanced by the combination of MOFs, and other functional materials. Furthermore, parallel with the MOF enrichment, studies on high-performance MOF variations with innovative designs have led to new design strategies for MOF composites.

References

[1] M.H. Mostafa, S.H.E. Abdel Aleem, S.G. Ali, A.Y. Abdelaziz, Energy-management solutions for microgrids, in: Distrib. Energy Resour. Microgrids Integr. Challenges Optim, Elsevier, 2019, pp. 483–515, https://doi.org/10.1016/B978-0-12-817774-7.00020-X.

[2] T. Zhang, H. Yang, High efficiency plants and building integrated renewable energy systems: building-integrated photovoltaics (BIPV), in: Handb. Energy Effic. Build. A Life Cycle Approach, Elsevier, 2018, pp. 441–595, https://doi.org/10.1016/B978-0-12-812817-6.00040-1.

[3] F. Yin, X. Zhang, X. He, H. Wang, Metal-organic frameworks and their applications in hydrogen and oxygen evolution reactions, in: Met. Fram, InTech, 2016, https://doi.org/10.5772/64657.

[4] A. Kirubakaran, S. Jain, R.K. Nema, A review on fuel cell technologies and power electronic interface, Renew. Sust. Energ. Rev. 13 (2009) 2430–2440, https://doi.org/10.1016/j.rser.2009.04.004.

[5] N. Sazali, W.N.W. Salleh, A.S. Jamaludin, M.N.M. Razali, New perspectives on fuel cell technology: a brief review, Membranes (Basel) 10 (2020), https://doi.org/10.3390/membranes10050099.

[6] X. Cao, C. Tan, M. Sindoro, H. Zhang, Hybrid micro-/nano-structures derived from metal-organic frameworks: preparation and applications in energy storage and conversion, Chem. Soc. Rev. 46 (2017) 2660–2677, https://doi.org/10.1039/c6cs00426a.

[7] A.A. Olajire, Synthesis chemistry of metal-organic frameworks for CO_2 capture and conversion for sustainable energy future, Renew. Sust. Energ. Rev. 92 (2018) 570–607, https://doi.org/10.1016/j.rser.2018.04.073.

[8] M.Z.F. Kamarudin, S.K. Kamarudin, M.S. Masdar, W.R.W. Daud, Review: direct ethanol fuel cells, Int. J. Hydrog. Energy 38 (2013) 9438–9453, https://doi.org/10.1016/j.ijhydene.2012.07.059.

366 Chapter 19 Metal organic framework-based nanocomposites

[9] M. Muthuvel, X. Jin, G.G. Botte, Fuel cells—exploratory fuel cells | direct carbon fuel cells, in: Encycl. Electrochem. Power Sources, Elsevier, 2009, pp. 158–171, https://doi.org/10.1016/B978-044452745-5.00912-6.

[10] Z. Zakaria, S.K. Kamarudin, S.N. Timmiati, Membranes for direct ethanol fuel cells: an overview, Appl. Energy 163 (2016) 334–342, https://doi.org/10.1016/j.apenergy.2015.10.124.

[11] C. Santoro, C. Arbizzani, B. Erable, I. Ieropoulos, Microbial fuel cells: from fundamentals to applications. A review, J. Power Sources 356 (2017) 225–244, https://doi.org/10.1016/j.jpowsour.2017.03.109.

[12] Y.A.S.J.P. Kumari, K. Padmaja, P.R. Kumari, A fuel cell and its applications, AIP Conf. Proc. 1992 (2018) 6–10, https://doi.org/10.1063/1.5047980.

[13] D. Murzin, Engineering Catalysis, De Gruyter, 2013, https://doi.org/10.1515/9783110283372.

[14] C.O. Colpan, I. Dincer, F. Hamdullahpur, Portable Fuel Cells—Fundamentals, Technologies and Applications, Springer, Dordrecht, 2008, pp. 87–101, https://doi.org/10.1007/978-1-4020-8295-5_6.

[15] K. Asazawa, K. Yamada, H. Tanaka, A. Oka, M. Taniguchi, T. Kobayashi, A platinum-free zero-carbon-emission easy fuelling direct hydrazine fuel cell for vehicles, Angew. Chem. 119 (2007) 8170–8173, https://doi.org/10.1002/ange.200701334.

[16] S. Krummrich, J. Llabrés, Methanol reformer—the next milestone for fuel cell powered submarines, Int. J. Hydrog. Energy 40 (2015) 5482–5486, https://doi.org/10.1016/j.ijhydene.2015.01.179.

[17] E.A. Rapley, G.P. Crockford, D.F. Easton, M.R. Stratton, D.T. Bishop, The genetics of testicular germ cell tumours, in: Germ Cell Tumours V, Springer, London, 2002, pp. 3–22, https://doi.org/10.1007/978-1-4471-3281-3_1.

[18] M.A. Deyab, Corrosion protection of aluminum bipolar plates with polyaniline coating containing carbon nanotubes in acidic medium inside the polymer electrolyte membrane fuel cell, J. Power Sources 268 (2014) 50–55, https://doi.org/10.1016/j.jpowsour.2014.06.021.

[19] G. Slaughter, T. Kulkarni, Enzymatic glucose biofuel cell and its application, J. Biochips Tissue Chips 05 (2015), https://doi.org/10.4172/2153-0777.1000110.

[20] X. Li, Principles of Fuel Cells, CRC Press, 2005, https://doi.org/10.1201/9780203942338.

[21] S.T. Revankar, P. Majumdar, Fuel Cells, CRC Press, 2016, https://doi.org/10.1201/b15965.

[22] D. Gautam, S. Anjum, S. Ikram, Proton exchange membrane (PEM) in fuel cells : a review, IUP J. Chem. III (2010) 51–81.

[23] S.J. Peighambardoust, S. Rowshanzamir, M. Amjadi, Review of the proton exchange membranes for fuel cell applications, Int. J. Hydrog. Energy (2010) 9349–9384, https://doi.org/10.1016/j.ijhydene.2010.05.017.

[24] R.C.T. Slade, J.P. Kizewski, S.D. Poynton, R. Zeng, J.R. Varcoe, Alkaline membrane fuel cells, in: Fuel Cells, Springer, New York, 2013, pp. 9–29, https://doi.org/10.1007/978-1-4614-5785-5_2.

[25] F. Bidault, D.J.L. Brett, P.H. Middleton, N.P. Brandon, Review of gas diffusion cathodes for alkaline fuel cells, J. Power Sources 187 (2009) 39–48, https://doi.org/10.1016/j.jpowsour.2008.10.106.

[26] G.F. McLean, T. Niet, S. Prince-Richard, N. Djilali, An assessment of alkaline fuel cell technology, Int. J. Hydrog. Energy 27 (2002) 507–526, https://doi.org/10.1016/S0360-3199(01)00181-1.

[27] N. Sammes, R. Bove, K. Stahl, Phosphoric acid fuel cells: fundamentals and applications, Curr. Opin. Solid State Mater. Sci. 8 (2004) 372–378, https://doi.org/10.1016/j.cossms.2005.01.001.

Chapter 19 Metal organic framework-based nanocomposites **367**

[28] Phosphoric acid fuel cells (PAFCs), in: Assess. Res. Needs Adv. Fuel Cells, Elsevier, 1986, pp. 13–94, https://doi.org/10.1016/B978-0-08-033990-0.50007-1.

[29] A. Mehmeti, F. Santoni, M. Della Pietra, S.J. McPhail, Life cycle assessment of molten carbonate fuel cells: state of the art and strategies for the future, J. Power Sources 308 (2016) 97–108, https://doi.org/10.1016/j.jpowsour.2015.12.023.

[30] J. Milewski, T. Wejrzanowski, Ł. Szabłowski, R. Baron, A. Szczęśniak, K. Ćwieka, Development of molten carbonate fuel cells at warsaw university of technology, Energy Procedia (2017) 1496–1501, https://doi.org/10.1016/j.egypro.2017.12.598.

[31] J. Chen, C.X. Zhao, M.M. Zhi, K. Wang, L. Deng, G. Xu, Alkaline direct oxidation glucose fuel cell system using silver/nickel foams as electrodes, Electrochim. Acta 66 (2012) 133–138, https://doi.org/10.1016/j.electacta.2012.01.071.

[32] A.B. Stambouli, E. Traversa, Solid oxide fuel cells (SOFCs): a review of an environmentally clean and efficient source of energy, Renew. Sust. Energ. Rev. 6 (2002) 433–455, https://doi.org/10.1016/S1364-0321(02)00014-X.

[33] S. Hussain, L. Yangping, Review of solid oxide fuel cell materials: cathode, anode, and electrolyte, Energy Trans. 4 (2020) 113–126, https://doi.org/10.1007/s41825-020-00029-8.

[34] F.A. Zakil, S.K. Kamarudin, S. Basri, Modified Nafion membranes for direct alcohol fuel cells: an overview, Renew. Sust. Energ. Rev. 65 (2016) 841–852, https://doi.org/10.1016/j.rser.2016.07.040.

[35] J. Ma, N.A. Choudhury, Y. Sahai, A comprehensive review of direct borohydride fuel cells, Renew. Sust. Energ. Rev. 14 (2010) 183–199, https://doi.org/10.1016/j.rser.2009.08.002.

[36] B. Sen, B. Demirkan, B. Şimşek, A. Savk, F. Sen, Monodisperse palladium nanocatalysts for dehydrocoupling of dimethylamineborane, Nano-Struct. Nano-Obj. 16 (2018) 209–214, https://doi.org/10.1016/j.nanoso.2018.07.008.

[37] S. Giddey, S.P.S. Badwal, A. Kulkarni, C. Munnings, A comprehensive review of direct carbon fuel cell technology, Prog. Energy Combust. Sci. 38 (2012) 360–399, https://doi.org/10.1016/j.pecs.2012.01.003.

[38] A.C. Rady, S. Giddey, S.P.S. Badwal, B.P. Ladewig, S. Bhattacharya, Review of fuels for direct carbon fuel cells, Energy Fuel 26 (2012) 1471–1488, https://doi.org/10.1021/ef201694y.

[39] J. Chen, H. Zheng, J. Kang, F. Yang, Y. Cao, M. Xiang, An alkaline direct oxidation glucose fuel cell using three-dimensional structural Au/Ni-foam as catalytic electrodes, RSC Adv. 7 (2017) 3035–3042, https://doi.org/10.1039/C6RA27586A.

[40] D. Basu, S. Basu, A study on direct glucose and fructose alkaline fuel cell, Electrochim. Acta 55 (2010) 5775–5779, https://doi.org/10.1016/j.electacta.2010.05.016.

[41] M.H. de Sá, L. Brandão, Non-enzymatic direct glucose fuel cells (DGFC): a novel principle towards autonomous electrochemical biosensors, Int. J. Hydrog. Energy 45 (2020) 29749–29762, https://doi.org/10.1016/j.ijhydene.2019.09.105.

[42] A.M. Sheikh, K. Ebn-Alwaled Abd-Alftah, C.F. Malfatti, On reviewing the catalyst materials for direct alcohol fuel cells (DAFCs), J. Multidiscip. Eng. Sci. Technol. 1 (2014). ISSN: 3159-0040.

[43] H.B. McClean, M. Hare, An Explorative Study of the Perspectives of Professionals Working With Young Deaf Children and of Their Families Prior to the Transition of Their Children to Full Time Early Years and Foundation Stage Educational Placements, 2019.

368 Chapter 19 Metal organic framework-based nanocomposites

[44] E.H. Yu, U. Krewer, K. Scott, Principles and materials aspects of direct alkaline alcohol fuel cells, Energies 3 (2010) 1499–1528, https://doi.org/10.3390/en3081499.

[45] F. Şen, G. Gökagaç, Activity of carbon-supported platinum nanoparticles toward methanol oxidation reaction: role of metal precursor and a new surfactant, tert-octanethiol, J. Phys. Chem. C 111 (2007) 1467–1473, https://doi.org/10.1021/jp065809y.

[46] Z. Daşdelen, Y. Yıldız, S. Eriş, F. Şen, Enhanced electrocatalytic activity and durability of Pt nanoparticles decorated on GO-PVP hybride material for methanol oxidation reaction, Appl. Catal. B Environ. 219 (2017) 511–516, https://doi.org/10.1016/j.apcatb.2017.08.014.

[47] B. Çelik, G. Başkaya, H. Sert, Ö. Karatepe, E. Erken, F. Şen, Monodisperse Pt(0)/DPA@GO nanoparticles as highly active catalysts for alcohol oxidation and dehydrogenation of DMAB, Int. J. Hydrog. Energy 41 (2016) 5661–5669, https://doi.org/10.1016/j.ijhydene.2016.02.061.

[48] S. Eris, Z. Daşdelen, F. Sen, Enhanced electrocatalytic activity and stability of monodisperse Pt nanocomposites for direct methanol fuel cells, J. Colloid Interface Sci. 513 (2018) 767–773, https://doi.org/10.1016/j.jcis.2017.11.085.

[49] S.H. Din, Nano-Composites and Their Applications, En Press, 2018, https://doi.org/10.24294/can.v0i0.875.

[50] S. Şen, F. Şen, G. Gökağaç, Preparation and characterization of nano-sized Pt-Ru/C catalysts and their superior catalytic activities for methanol and ethanol oxidation, Phys. Chem. Chem. Phys. 13 (2011) 6784–6792, https://doi.org/10.1039/c1cp20064j.

[51] F. Şen, G. Gökağaç, Different sized platinum nanoparticles supported on carbon: an XPS study on these methanol oxidation catalysts, J. Phys. Chem. C 111 (2007) 5715–5720, https://doi.org/10.1021/jp068381b.

[52] M. Safdari, M.S. Al-Haik, A review on polymeric nanocomposites: effect of hybridization and synergy on electrical properties, in: Carbon-Based Polym. Nanocomposites Environ. Energy Appl, Elsevier Inc., 2018, pp. 113–146, https://doi.org/10.1016/B978-0-12-813574-7.00005-8.

[53] S. Eris, Z. Daşdelen, Y. Yıldız, F. Sen, Nanostructured polyaniline-rGO decorated platinum catalyst with enhanced activity and durability for methanol oxidation, Int. J. Hydrog. Energy 43 (2018) 1337–1343, https://doi.org/10.1016/j.ijhydene.2017.11.051.

[54] S.S. Siwal, Q. Zhang, N. Devi, V.K. Thakur, Carbon-based polymer nanocomposite for high-performance energy storage applications, Polymers (Basel) 12 (2020) 1–30, https://doi.org/10.3390/polym12030505.

[55] S.K. Kumar, R. Krishnamoorti, Nanocomposites: structure, phase behavior, and properties, Annu. Rev. Chem. Biomol. Eng. 1 (2010) 37–58, https://doi.org/10.1146/annurev-chembioeng-073009-100856.

[56] S.H. Din, Nano-composites and their applications: a review, Charact. Appl. Nanomater. 2 (2019), https://doi.org/10.24294/can.v2i1.875.

[57] R. Bogue, Nanocomposites: a review of technology and applications, Assem. Autom. 31 (2011) 106–112, https://doi.org/10.1108/01445151111117683.

[58] Ö. Karatepe, Y. Yildiz, H. Pamuk, S. Eris, Z. Dasdelen, F. Sen, Enhanced electrocatalytic activity and durability of highly monodisperse Pt@PPy-PANI nanocomposites as a novel catalyst for the electro-oxidation of methanol, RSC Adv. 6 (2016) 50851–50857, https://doi.org/10.1039/c6ra06210e.

[59] K. Müller, E. Bugnicourt, M. Latorre, M. Jorda, Y. Echegoyen Sanz, J. Lagaron, O. Miesbauer, A. Bianchin, S. Hankin, U. Bölz, G. Pérez, M. Jesdinszki, M. Lindner, Z. Scheuerer, S. Castelló, M. Schmid, Review on the processing and properties of polymer nanocomposites and nanocoatings

and their applications in the packaging, automotive and solar energy fields, Nano 7 (2017) 74, https://doi.org/10.3390/nano7040074.

[60] B. Le, J. Khaliq, D. Huo, X. Teng, I. Shyha, A review on nanocomposites. Part 1: mechanical properties, J. Manuf. Sci. Eng. Trans. ASME 142 (2020), https://doi.org/10.1115/1.4047047.

[61] R.A.M. Said, M.A. Hasan, A.M. Abdelzaher, A.M. Abdel-Raoof, Review—insights into the developments of nanocomposites for its processing and application as sensing Materials, J. Electrochem. Soc. 167 (2020), https://doi.org/10.1149/1945-7111/ab697b, 037549.

[62] C.C. Okpala, Nanocomposites—an overview, 2013. www.ijerd.com. (Accessed 3 February 2021).

[63] V. Borovkov, Book review of "Lanthanide metal-organic frameworks", Front. Chem. 3 (2015) 50, https://doi.org/10.3389/fchem.2015.00050.

[64] T. Rasheed, K. Rizwan, M. Bilal, H.M.N. Iqbal, Metal-organic framework-based engineered materials—fundamentals and applications, Molecules 25 (2020), https://doi.org/10.3390/molecules25071598.

[65] S. Soni, P.K. Bajpai, C. Arora, A review on metal-organic framework: synthesis, properties and application, Charact. Appl. Nanomater. 2 (2018), https://doi.org/10.24294/can.v2i2.551.

[66] Q. Wang, D. Astruc, State of the art and prospects in metal-organic framework (MOF)-based and MOF-derived nanocatalysis, Chem. Rev. 120 (2020) 1438–1511, https://doi.org/10.1021/acs.chemrev.9b00223.

[67] H.C.J. Zhou, S. Kitagawa, Metal-organic frameworks (MOFs), Chem. Soc. Rev. 43 (2014) 5415–5418, https://doi.org/10.1039/c4cs90059f.

[68] F. Sen, Metal organic frameworks (MOF's) for biosensing and bioimaging applications, in: Mater. Res. Found, Materials Research Forum LLC, 2019, pp. 308–360, https://doi.org/10.21741/9781644900437-11.

[69] Metal-Organic Framework Composites, Materials Research Forum LLC, 2019, https://doi.org/10.21741/9781644900437.

[70] M.K. Karnena, M. Konni, V. Saritha, Nano-Catalysis Process for Treatment of Industrial Wastewater, 2020, pp. 229–251, https://doi.org/10.4018/978-1-7998-1241-8.ch011.

[71] B.K. Kang, M.H. Woo, J. Lee, Y.H. Song, Z. Wang, Y. Guo, Y. Yamauchi, J.H. Kim, B. Lim, D.H. Yoon, Mesoporous Ni-Fe oxide multi-composite hollow nanocages for efficient electrocatalytic water oxidation reactions, J. Mater. Chem. A 5 (2017) 4320–4324, https://doi.org/10.1039/c6ta10094e.

[72] V.F. Samanidou, E.A. Deliyanni, Metal organic frameworks: synthesis and application, Molecules 25 (2020) 960, https://doi.org/10.3390/molecules25040960.

[73] A. Khan, M. Jawaid, A.M.A. Asiri, W. Ni, M.M. Rahman, N. Karaman, K. Cellat, H. Acıdereli, A. Khan, F. Şen, Metal-organic framework with immobilized nanoparticles, in: Met. Framew. Nanocomposites, CRC Press, 2020, pp. 275–290, https://doi.org/10.1201/9780429346262-11.

[74] F.-Y. Yi, R. Zhang, H. Wang, L.-F. Chen, L. Han, H.-L. Jiang, Q. Xu, Metal-organic frameworks and their composites: synthesis and electrochemical applications, Small Methods 1 (2017) 1700187, https://doi.org/10.1002/smtd.201700187.

[75] R. Kötz, M. Hahn, R. Gallay, Temperature behavior and impedance fundamentals of supercapacitors, J. Power Sources 154 (2006) 550–555, https://doi.org/10.1016/j.jpowsour.2005.10.048.

[76] Z.Z. Benabithe, Calorimetry characterization of carbonaceous materials for energy applications: review, in: Calorim. - Des. Theory Appl. Porous Solids, InTech, 2018, https://doi.org/10.5772/intechopen.71310.

[77] B. Şen, A. Aygün, A. Şavk, C. Yenikaya, S. Cevik, F. Şen, Metal-organic frameworks based on monodisperse palladium–cobalt nanohybrids as highly active and reusable nanocatalysts for hydrogen generation, Int. J. Hydrog. Energy 44 (2019) 2988–2996, https://doi.org/10.1016/j.ijhydene.2018.12.051.

[78] X. Gao, A.Y. Fu, B. Liu, J.C. Jin, L.T. Dou, L.X. Chen, Unique topology analysis by ToposPro for a metal-organic framework with multiple coordination centers, Inorg. Chem. 58 (2019) 3099–3106, https://doi.org/10.1021/acs.inorgchem.8b03104.

[79] L.R. MacGillivray, Metal-Organic Frameworks: Design and Application, John Wiley and Sons, 2010, https://doi.org/10.1002/9780470606858.

[80] J. Hao, X. Xu, H. Fei, L. Li, B. Yan, Functionalization of metal-organic frameworks for photoactive materials, Adv. Mater. 30 (2018) 1705634, https://doi.org/10.1002/adma.201705634.

[81] S.J. Garibay, Functionalization of Metal-Organic Frameworks with Metalloligands and Postsynthetic Modification, UC San Diego, 2011. UC San Diego Electronic Theses and Dissertations https://escholarship.org/uc/item/4fk887bn. (Accessed 3 February 2021).

[82] S. Mandal, S. Natarajan, P. Mani, A. Pankajakshan, Post-synthetic modification of metal–organic frameworks toward applications, Adv. Funct. Mater. 31 (2021) 2006291, https://doi.org/10.1002/adfm.202006291.

[83] A.U. Czaja, N. Trukhan, U. Müller, Industrial applications of metal–organic frameworks, Chem. Soc. Rev. 38 (2009) 1284–1293, https://doi.org/10.1039/b804680h.

[84] J.L.C. Rowsell, O.M. Yaghi, Metal-organic frameworks: a new class of porous materials, Microporous Mesoporous Mater. 73 (2004) 3–14, https://doi.org/10.1016/j.micromeso.2004.03.034.

[85] G. Mouchaham, S. Wang, C. Serre, Metal-Organic Frameworks: Applications in Separations and Catalysis, first ed., 2018.

[86] S. Yuan, L. Feng, K. Wang, J. Pang, M. Bosch, C. Lollar, Y. Sun, J. Qin, X. Yang, P. Zhang, Q. Wang, L. Zou, Y. Zhang, L. Zhang, Y. Fang, J. Li, H.C. Zhou, Stable metal–organic frameworks: design, synthesis, and applications, Adv. Mater. 30 (2018), https://doi.org/10.1002/adma.201704303.

[87] M. Kim, J.F. Cahill, Y. Su, K.A. Prather, S.M. Cohen, Postsynthetic ligand exchange as a route to functionalization of "inert" metal-organic frameworks, Chem. Sci. 3 (2012) 126–130, https://doi.org/10.1039/c1sc00394a.

[88] C.C. Chiu, F.K. Shieh, H.H.G. Tsai, Ligand exchange in the synthesis of metal-organic frameworks occurs through acid-catalyzed associative substitution, Inorg. Chem. 58 (2019) 14457–14466, https://doi.org/10.1021/acs.inorgchem.9b01947.

[89] Y. Noori, K. Akhbari, Post-synthetic ion-exchange process in nanoporous metal-organic frameworks; an effective way for modulating their structures and properties, RSC Adv. 7 (2017) 1782–1808, https://doi.org/10.1039/c6ra24958b.

20

Carbon-polymer hybrid-supported nanomaterials for alcohol fuel cells

Ramazan Bayat[a,b], Nimeti Doner[c], and Fatih Şen[a]

[a]Şen Research Group, Department of Biochemistry, Dumlupinar University, Kütahya, Turkey. [b]Department of Materials Science and Engineering, Faculty of Engineering, Dumlupinar University, Kütahya, Turkey. [c]Department of Mechanical Engineering, Faculty of Engineering, Gazi University, Ankara, Turkey

1 Introduction

The population is increasing on planet Earth, which has caused a number of problems. Climate change and increasing energy needs are very important global issues that are often in conflict, and climate change is being researched. Widely used fossil fuels, which meet 85% of the energy needs on the planet [1], cause climate change. Researchers estimate that fossil fuel reserves will be depleted in 50–60 years—a rather limited period—and so alternative energy systems must be established. In addition, considering the environmental damage caused by fossil fuels, there is an increasing need for alternative fuels generated through cleaner and safer procedures. The rapid depletion of fossil fuels, as well as their unstable and unsustainable prices, make them less good for meeting these increasing energy needs. Solutions to these problems are being sought in this century, including the efficient use of fuels and the use of different renewable energy sources [2, 3]. Examples of renewable energy resources being researched include thermal energy, solar cells, ocean energy, biomass, wind energy, fuel cells, and geothermal energy [4–6].

Energy consumption increased by 2.9% in 2019 and 1.5% annually over the last 10 years. This increase in energy consumption increased the global carbon emission rate by 2% [7]. The use of fossil fuels has been found to cause global warming [8, 9]. Many countries around the world accept that permanent solutions to climate change and to their energy needs must be found. To address these problems, many countries have signed the Paris

Nanomaterials for Direct Alcohol Fuel Cells. https://doi.org/10.1016/B978-0-12-821713-9.00013-5
Copyright © 2021 Elsevier Inc. All rights reserved.

and Kyoto climate agreements, but thus far, these agreements have not found definitive solutions. Rapid population growth and rising energy demand are the biggest obstacles to achieving energy use and transformation targets. Providing clean and efficient energy sources seem to be the biggest problem that humans face in the 21st century [10].

Recently, there has been rapid progress in the production and development of alcohol fuel cells (AFCs) from various renewable energy sources. Fuel cells are electrochemical devices used to directly and efficiently convert chemical energy (fuels) into electricity. Fuel cells have many advantages over other energy sources [11]: they are much smaller, more powerful, quieter, and have few or no negative environmental effects. Fuel cells can cover a broad range of energy needs, from a few watts to several hundred megawatts [12]. Storage methods are required to store sufficient energy for use in electronic devices and electrical appliances. Fuel cells have superior energy conversion systems and meet the requirements for electric power generation. Electrochemical storage systems usually contain probes, electrolytic fluids, and power receivers. These components are manufactured using carbon-based nanomaterials, carbon-polymer hybrid materials, and metal oxide materials. However, these materials cannot be used efficiently in fuel cells due to the difficulties of commercialization and development. Besides, conductive polymers have unique properties, such as being low cost, environmentally friendly, and easy to use. Polymer materials can be used in a hybrid manner with carbon-based nanomaterials. Carbon-polymer hybrid nanomaterials are increasingly used in aviation, packaging, automotive industry, and fuel cells [13, 14]. With the use of carbon-polymer hybrid materials, it is possible to produce high-performance fuel cells [15].

The carbon-based products graphite, carbon nanotubes (CNTs), graphene, and fullerenes are widely used as electrocatalysts in fuel cells. Conductive polymers, with their high conductors and high stability, are used today with carbon-based nanomaterials in fuel cells [16].

In this chapter, carbon-polymer hybrid materials and their use in alcohol fuel cells will be discussed, and information about the results obtained using carbon-polymer hybrid-supported nanomaterials in alcohol fuel cells will be presented.

2 Fuel cells

Fuel cells are the general name given to devices that can obtain electrical energy as a result of reactions with electrical chemicals from the chemical fuel that they contain [17, 18]. No mechanical or thermal processing is required for the chemical reaction to take place [19–21]. There must be oxygen for the reaction to start. Unlike traditional power sources, fuel cells are systems that convert chemical energy directly into electrical energy so long as fuel and oxidants are supplied [22]. A fuel cell consists of an anode (positive end), a cathode (negative end), and an electrolyte that provides the transport of charges between the two ends of the fuel cell (Fig. 1) [23, 24].

The British scientists Anthony Carlisle and Willam Nicolas reported in 1800 that water can be separated into hydrogen and oxygen by using electricity. In 1838, William Grove developed a hydrogen fuel cell, with the constant current being obtained by reverse electrolysis of water. The basis of fuel cells was established. Grove connected several systems made of two platinum electrodes immersed in sulfuric acid with a series of circuits, using oxygen as the oxidizer and hydrogen as the fuel, and prepared a mechanism known as a gas battery, which was the first fuel cell [25, 26]. Fuel cells can operate on substances such as methanol, ethanol, hydrogen, and phosphoric acid.

Fuel cells can be classified according to their operating temperature or pressure. Generally, it is possible to classify systems operating with high, medium, and low temperature or at high, medium, and low (atmospheric) pressure. Also, fuel cells can be distinguished by the oxidants and/or fuels they use. For practical reasons, fuel cell systems can be divided only according to the type of electrolyte used. Nanomaterials with large surface areas are used for electrocatalysis in the cathode and anode in fuel cells. Today, carbon-based materials, especially alcohol fuel cells, are used for support in fuel cells [27]. The following names and

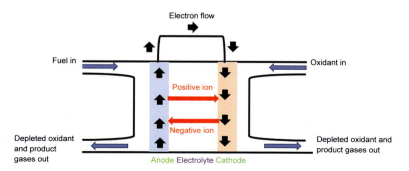

Fig. 1 Simple schematic representation of a fuel cell [23]. Reprinted (adapted) with permission from Y. Manoharan, S.E. Hosseini, B. Butler, H. Alzhahrani, B.T.F. Senior, T. Ashuri, J. Krohn, Hydrogen fuel cell vehicles; Current status and future prospect, Appl. Sci. 9 (2019). https://doi.org/10.3390/app9112296. Copyright (2020) Applied Sciences (MDPI).

abbreviations are frequently used: proton exchange membrane fuel cell (PEMFC), phosphoric acid fuel cell (PAFC), direct methanol fuel cell (DMFC), alkaline fuel cell (AFC), molten carbonate fuel cell (MCFC), and solid oxide classified as fuel cells (SOFC) [28, 29].

2.1 Alcohol fuel cells

Because of the costly use of hydrogen in fuel cells and difficulties involved with their use and storage, alternative fuels to hydrogen have been investigated. It has been found that alcohol can be an alternative fuel to hydrogen. Alcohols are more easily obtained, transported, and stored as fuel than hydrogen. Direct alcohol fuel cells (DAFCs) are located under PEMFCs [30–32]. DAFCs oxidize the alcohol at the anode using various alcohols and generate electrical energy (Fig. 2) [33]. Methanol is widely used in DAFCs because it can produce more energy than other types of alcohol [34].

Platinum (Pt) and ruthenium (Ru) are the most commonly used catalysts in DAFCs. However, these metals show low catalytic activity in the presence of carbon monoxide (CO) [35–38]. As a result of research, it has been established that the use of carbon-polymer hybrid nanomaterials in DAFCs can provide more efficient fuel cells that are CO resistant and can operate at high temperatures [30, 34].

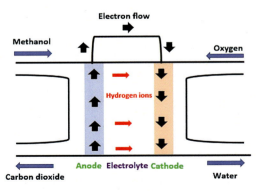

Fig. 2 Simple schematic diagram of a direct methanol fuel cell [33]. Reprinted (adapted) with permission from R. Govindarasu, S. Somasundaram, Studies on influence of cell temperature in direct methanol fuel cell operation, Processes 8 (2020). https://doi.org/10.3390/PR8030353. Copyright (2020) Processes (MDPI).

3 Carbon nanomaterials

Carbon-based nanomaterials (CBNs) were discovered about 150 years ago as a result of studies carried out with the understanding that these important materials hold great potential. Sumio Iijima created multiwalled carbon nanotubes (MWCNTs) in 1991, and Novoselov made single-layer graphene layers in 2004 [39]. As a result of the discovery and the comprehensive characterization of such materials, high-strength materials are being developed for use in areas such as electronic devices, efficient electron emitters, fuel cells, biosensors, and drug delivery systems. CBNs are frequently used in these areas as a result of their mechanical, thermal, optical, and electrical properties. They often have an important application in fuel cells due to their high electrical conductivity, large surface area, and porous amorphous structure [40]. For example, CNTs can carry high currents without heating due to their length. When CBNs are used as an electrode material in fuel cells, their active surface area increases energy conversion efficiency by allowing the metals to reach a faster reaction density to the transport of fuels. It has been observed that composite structures formed with graphene-doped or codoped metal, metal oxides, and conductive polymers are more efficient compared to traditional Pt/C structures. Carbon-reinforced materials are developed and frequently used in alcohol fuel cells [41–43].

Carbon-based polymer nanohybrid structures have applications in the field of energy storage. These materials are frequently used and developed due to their easy processing, good conductivity, applicability, and lightness. The basic materials of the carbon family include graphene, CNTs, and fullerenes. Nanomaterials such as carbon quantum dots, nanohorns, nanofibers, and nanoribbons are synthesized from these materials. The common features of carbon-based materials are their small size and their unique morphological structure. Nanocarbons with sp^2 hybridization have a smooth and long structure. Having unique structures gives carbon-reinforced materials high electrical conductivity, thermal conductivity, and mechanical properties [44].

3.1 Graphene

Graphene is a structure in which carbon atoms are formed in a honeycomb form. Graphene, unlike graphite, has a single-layer structure that is regular, with good electrical conductivity [45, 46]. Therefore, a lot of research today is being done on its use and application in every aspect of life.

Graphene was first described by Novoselov in 2004 [47]. Today, due to its mechanical, thermal, electrical, and optical properties, it is used in many areas. Due to the unique properties of graphene, it has been the subject of much research [48–50]. Graphene is a promising material for the future, with its excellent thermal conductivity, high electrical conductivity, semimetalness, zero band energy, flexibility, lightness, transparency, and high strength, which can be used in many fields. Graphene has different forms with different properties, such as zigzags, chair spreads, and thin sheets. The number of layers gives different properties to graphene. For example, single-layer graphene is a semiconductor, while multilayer graphene is conductive. Graphene has a promising structure for fuel cells due to its excellent electron-carrying capacity and other properties [44, 48, 51].

Qin et al. synthesized nanocomposite materials using nitrogen-added graphene with titanium metal-organic frames. Nitrogen-doped graphene and titanium metal-organic frames are used as cathodic materials in fuel cells and used as oxygen-reducing catalysts. In their study, they found that nitrogen-added graphene and titanium metal-organic frames can be a solution to the high costs of catalysts that use precious metals and the low durability of conventional catalysts [52].

Hussain et al. used nitrobenzene-modified platinum-graphene nanoparticles as oxygen-reducing catalysts in alkaline solutions. According to the data from their study, it was shown that a nitrobenzene-modified, platinum-graphene hybrid can be used as a promising cathode catalyst in fuel cells by promoting four-electron reduction from oxygen [53].

You et al. used a graphene oxide-poly(N-vinyl-2-pyrrolidone)-ruthenium hybrid nanomaterial as a catalyst in the fuel cell. The hybrid material that was produced offers a new method of greatly improving catalytic dehydrogenation in the catalyst [54].

Today, graphene-based graphite, CNTs, and fullerene are produced using two-dimensional (2D) graphene.

3.2 Carbon nanotubes

Smalley et al. developed the hexagonal and pentagonal structure of carbon atoms [55], which was called a *fullerene*. The first structure that was created was the C_{60} structure. Later, studies on fullerenes led to the production of CNTs, which are elongated fullerenes in the form of a graphene cylinder or tube. CNTs are available in the form of MWCNTs and single-layer carbon nanotubes (SWCNT). SWCNTs are found in the form of tubes consisting of a single plate that is 1–3 nm in diameter and approximately 1 µm in size. MWCNTs take the form of tubes with a diameter of 10–40 nm, approximately 1 µm in diameter, and rolled within each other. Although SWCNTs and MWCNTs are in tubular form, SWCNTs have higher surface area and lower conductivity [56–58]. CNTs show extraordinary behavior. It has an elastic modulus of 1 TPa and a strength that is 100 times greater than steel [59]. CNTs have many areas of use due to their electrical, mechanical, and thermal properties. It can be used in many areas, such as nanosensors, hydrogen absorption, energy storage, heat barriers, and electronic devices. Today, CNTs are used in AFCs and are used in reduction and oxidation reactions [60, 61].

CNTs are promising for their use in fuel cells with their electrical properties. Hanif et al. used zeolitic imidazolate frameworks-nitrogen-added CNTs in alkali fuel cells as an oxygen reducer in the catalyst, and their performance in fuel cells was observed. Their study established that zeolitic imidazolate frameworks-nitrogen doped CNTs performed better than platinum-carbon cathodes [62].

A composite membrane based on CNTs and sulfonated polysulfide functionalized with polydopamine was produced. It has been observed that the composite membrane has more efficient methanol permeability than Nafion membranes [63].

Sharma et al. prepared electrodes using CNTs and used them in a microbial fuel cell. CNTs are used as an electron carrier in fuel cells. It has been observed that the CNT-based electrodes that the researchers obtained are six times more efficient than graphite electrodes [64].

3.3 Fullerene

Smalley and his team obtained spherical lattice formations called fullerenes as a result of laser vaporization of graffiti in 1984 [55]. These researchers won the Nobel Prize in Chemistry in 1996 for this work [65]. Fullerene has a closed lattice structure formed by carbon atoms. These structures have a large specific surface area, unique three-dimensional (3D) form, chemical stability, and good electrical conductivity [66]. These structures have the formula $C_{20} + m$ [67]. Fullerenes are 1 nm in diameter and are composed of various numbers of carbon atoms [68, 69]. C_{60}, $C_{70,}$ and C_{80} are the most commonly known forms [65]. The smallest fullerene achievable today, C_{20}, consists of 20 carbon atoms; and the largest fullerene, C_{540}, consists of 540 carbon atoms. Among the fullerenes, C_{60} is the most stable and the best known.

It consists of 12 pentagons and 20 hexagonal honeycomb forms as a result of the sp^2 hybridization of C_{60} carbon atoms [65].

The lattice structures make fullerenes extremely strong. They are resistant to 3000 atm of pressure. The high pressure does not break the structure, and they easily return to their original form after pressure is removed from the fullerene. Although C_{60} is soft like graphite, it turns into a hard diamond when compressed. C_{60} molecules have the lowest energy range, at 1.6 EV [70]. C_{60} crystals become superconducting when they reach critical temperatures [71].

Due to their excellent properties, fullerenes are used in many fields, such as cosmetics, astronomy, pharmacy, physics, fuel cells, electronics, energy. They are increasingly important in electrochemistry, especially in AFCs, due to their good performance in the electron transfer process and their effects on chemical reactions [72].

Rambabu et al. designed a new membrane to be used in AFCs as an alternative to the Nafion-117 membrane. A sulfonated polyether ether ketone composite membrane was formed using sulfonated fullerenes, and 4-benzene diazonium sulfonic acid was used for the functionalization of fullerene. In the study, it was found as a result of the measurements that the sulfanod polyether ether ketone composite membrane exhibits similar properties as the Nafion-117 membrane [73].

Bhavani et al. performed a study on the production of bimetallic nanoparticles that showed high electro-oxidation in the catalyst in AFCs. For this purpose, they developed a fullerene C_{60} catalyst supported by palladium-tungsten bimetallic nanoparticles. The electrochemical, cyclic voltage, and electrochemical properties of the developed catalyst were investigated. It was observed that the developed catalyst was more tolerant of poisoning than platinum-carbon-supported catalysts, and it works in a stable structure for a long time. The results also showed that the composite material that was developed is promising for AFCs [74].

4 Polymer materials

Polymer materials are synthetic or natural structures where monomers come together to form a long chain. While deoxyribonucleic acid (DNA), protein, and enzymes are natural polymers, polyester, polyethylene, and polyvinyl chloride are examples of synthetic polymers. Synthetic polymers are increasingly important due to the interest and daily increase in polymer use [15, 75].

Polymer materials are resistant to breakage and feature flexible structures, low densities, easy processing, and low cost [76–78].

Organic polymers that can conduct electricity are called *conductive polymers (CPs)*. CPs provide conductivity alone or by forming a hybrid with metals. They exhibit unique conductivity and optical properties and can be produced using simple, inexpensive methods [79–82] and are used in solar cells, energy storage [83], sensors, and electronics.

Polyaniline (PANI), poly(3,4-ethylenedioxythiophene) (PEDOT), poly-3-methyl thiophene, and polypyrrole (PPy) have desirable properties such as chemical resistance, electrical conductivity, suitable structure, accessibility, environmental resistance, redox activity, and biocompatibility. These structures can be used as a support material for electrocatalysts in CP fuel cells [84–89].

4.1 Carbon-polymer hybrid-supported nanomaterials

Materials consisting of two or more components with superior properties are called *composite materials*. Composites exhibit new properties in harmony with each other. If the materials that make up the composite materials are in the nanoscale regime, such materials are called *nanocomposite materials* [15, 90]. These materials contribute greatly to the technological progress with their properties. Nanocomposite materials are classified as polymeric matrices, ceramics, and metals [91].

Polymer composites are developed and used in many areas today. Carbon-polymer hybrid nanomaterials are produced by placing carbon-based nanomaterials into the polymer matrix [15]. Carbon-based materials are used with CPs especially. By using polymer and carbon-based nanomaterials, high-conductivity carbon-polymer hybrid structures are formed [92], and they are used in energy storage, sensors, and electronics [93]. They can be used to develop fuel cells by combining various materials and obtaining carbon-polymer hybrid nanomaterials of the desired properties. Many studies have been done in this area [94–96].

Sen et al. aimed to develop an effective catalyst in a fuel cell operating at low temperatures in a direct methanol fuel cell. In the study, the carbon-polymer hybrid structure was synthesized using carbon black, platinum, and propylamine microwave. The data obtained showed that the carbon black-platinum/

propylamine polymer hybrid structure works five times more effectively than conventional platinum-carbon catalysts [97].

PVA-sulfonated CNTs (PVA-SCNTs) were synthesized for use in PEMFCs. It has been established from previous studies, that SCNTs increase the conductivity and mechanical strength of the membranes in fuel cells. Vani et al. showed that the membrane they prepared using PVA-SCNT was as effective and cost the same as a Nafion 115 membrane [98].

A carbon paste electrode (Cu_2O/PPy/CPE) modified with polypyrrole-copper oxide was prepared to measure the performance of ethanol oxidation in an alkaline environment. With this hybrid prepared structure, the oxidation of ethanol was measured. The work showed that the Cu_2O/PPy/CPE hybrid structure was a good catalyst for the oxidation of ethanol [99].

A Pt-CNT-PANI carbon-polymer hybrid structure gives important results in the oxidation of ethanol in acidic environments. A Pt-CNT-PANI structure enables the AFC to operate successfully in an acidic environment by increasing the conductivity and accelerating the reaction [100].

In AFCs, the enhanced catalytic activity of the catalysts requires stability and durability. Mozafari et al., designed a new electrode for use with a Pd catalyst. They synthesized PANI-MWCNTs-SnO_2 on a titanium (Ti) network substrate and synthesized a new electrode. They showed that Pd/PANI-MWCNTs-SnO_2/Ti electrodes with unique properties are promising for the oxidation of ethanol in AFCs [101].

5 Conclusion

Today, many studies are being done to look for ways to solve the energy crisis. Many of these are on AFCs, which are renewable energy sources. Carbon-polymer hybrid nanomaterials created using carbon-based materials and polymer materials can provide efficient, long-term operation of alcohol fuel batteries. AFCs, which are formed by the combination of superior properties of carbon and polymer materials, are predicted to meet the world's needs for clean, sustainable energy today and in the future.

References

[1] C.J. Cleveland, C. Morris, Climate change, in: Handb. Energy, Elsevier, 2014, pp. 805–820, https://doi.org/10.1016/B978-0-12-417013-1.00045-5.

[2] D.S. Su, G. Centi, A perspective on carbon materials for future energy application, J. Energy Chem. 22 (2013) 151–173, https://doi.org/10.1016/S2095-4956(13)60022-4.

Chapter 20 Carbon-polymer hybrid-supported nanomaterials **381**

[3] B.W.L. Jang, R. Gläser, M. Dong, C.J. Liu, Fuels of the future, Energy Environ. Sci. 3 (2010) 253, https://doi.org/10.1039/c003390c.

[4] A. Inayat, A.M. Nassef, H. Rezk, E.T. Sayed, M.A. Abdelkareem, A.G. Olabi, Fuzzy modeling and parameters optimization for the enhancement of biodiesel production from waste frying oil over montmorillonite clay K-30, Sci. Total Environ. 666 (2019) 821–827, https://doi.org/10.1016/j.scitotenv.2019.02.321.

[5] T. Wilberforce, A. Baroutaji, Z. El Hassan, J. Thompson, B. Soudan, A.G. Olabi, Prospects and challenges of concentrated solar photovoltaics and enhanced geothermal energy technologies, Sci. Total Environ. 659 (2019) 851–861, https://doi.org/10.1016/j.scitotenv.2018.12.257.

[6] M.S. Nazir, A.J. Mahdi, M. Bilal, H.M. Sohail, N. Ali, H.M.N. Iqbal, Environmental impact and pollution-related challenges of renewable wind energy paradigm—a review, Sci. Total Environ. 683 (2019) 436–444, https://doi.org/10.1016/j.scitotenv.2019.05.274.

[7] R. Adib Ren, M. Eckhart Mohamed El-Ashry David Hales Kirsty Hamilton Peter Rae, F. Bariloche, Renewables 2019 global status report, n.d.

[8] R. Turconi, A. Boldrin, T. Astrup, Life cycle assessment (LCA) of electricity generation technologies: overview, comparability and limitations, Renew. Sust. Energ. Rev. 28 (2013) 555–565, https://doi.org/10.1016/j.rser.2013.08.013.

[9] K. Elsaid, M. Kamil, E.T. Sayed, M.A. Abdelkareem, T. Wilberforce, A. Olabi, Environmental impact of desalination technologies: a review, Sci. Total Environ. 748 (2020) 141528, https://doi.org/10.1016/j.scitotenv.2020.141528.

[10] M. Liu, R. Zhang, W. Chen, Graphene-supported nanoelectrocatalysts for fuel cells: synthesis, properties, and applications, Chem. Rev. 114 (2014) 5117–5160, https://doi.org/10.1021/cr400523y.

[11] M.A. Abdelkareem, K. Elsaid, T. Wilberforce, M. Kamil, E.T. Sayed, A. Olabi, Environmental aspects of fuel cells: a review, Sci. Total Environ. 752 (2021) 141803, https://doi.org/10.1016/j.scitotenv.2020.141803.

[12] M.A. Abdelkareem, A. Allagui, E.T. Sayed, M. El Haj Assad, Z. Said, K. Elsaid, Comparative analysis of liquid versus vapor-feed passive direct methanol fuel cells, Renew. Energy 131 (2019) 563–584, https://doi.org/10.1016/j.renene.2018.07.055.

[13] V.B. Mohan, K. tak Lau, D. Hui, D. Bhattacharyya, Graphene-based materials and their composites: a review on production, applications and product limitations, Compos. Part B Eng. 142 (2018) 200–220, https://doi.org/10.1016/j.compositesb.2018.01.013.

[14] H. Abbasi, M. Antunes, J.I. Velasco, Recent advances in carbon-based polymer nanocomposites for electromagnetic interference shielding, Prog. Mater. Sci. 103 (2019) 319–373, https://doi.org/10.1016/j.pmatsci.2019.02.003.

[15] S.S. Siwal, Q. Zhang, N. Devi, V.K. Thakur, Carbon-based polymer nanocomposite for high-performance energy storage applications, Polymers (Basel) 12 (2020) 1–30, https://doi.org/10.3390/polym12030505.

[16] M.M. Bruno, F.A. Viva, Carbon materials for fuel cells, in: Direct Alcohol Fuel Cells Mater. Performance, Durab. Appl, Springer Netherlands, 2013, pp. 231–270, https://doi.org/10.1007/978-94-007-7708-8_7.

[17] F. Şen, Fuel cell electrochemistry, in: Mater. Res. Found, Materials Research Forum LLC, 2019, pp. 73–108, https://doi.org/10.21741/9781644900079-4.

[18] R. Kumar, E.T.S.G. da Silva, R.K. Singh, R. Savu, A.V. Alaferdov, L.C. Fonseca, L.C. Carossi, A. Singh, S. Khandka, K.K. Kar, O.L. Alves, L.T. Kubota, S.A. Moshkalev, Microwave-assisted synthesis of palladium

382 Chapter 20 Carbon-polymer hybrid-supported nanomaterials

nanoparticles intercalated nitrogen doped reduced graphene oxide and their electrocatalytic activity for direct-ethanol fuel cells, J. Colloid Interface Sci. 515 (2018) 160–171, https://doi.org/10.1016/j.jcis.2018.01.028.

[19] J.P. Lemmon, Energy: reimagine fuel cells, Nature 525 (2015) 447–449, https://doi.org/10.1038/525447a.

[20] I. Staffell, D. Scamman, A. Velazquez Abad, P. Balcombe, P.E. Dodds, P. Ekins, N. Shah, K.R. Ward, The role of hydrogen and fuel cells in the global energy system, Energy Environ. Sci. 12 (2019) 463–491, https://doi.org/10.1039/c8ee01157e.

[21] H.A. Gasteiger, N.M. Markovic, Just a dream or future reality? Science 324 (2009) 48–49, https://doi.org/10.1126/science.1172083.

[22] R.N. Singh, Madhu, R. Awasthi, Alcohol fuel cells, in: New Futur. Dev. Catal. Batter. Hydrog. Storage Fuel Cells, Elsevier B.V., 2013, pp. 453–478, https://doi.org/10.1016/B978-0-444-53880-2.00021-1.

[23] Y. Manoharan, S.E. Hosseini, B. Butler, H. Alzhahrani, B.T.F. Senior, T. Ashuri, J. Krohn, Hydrogen fuel cell vehicles; Current status and future prospect, Appl. Sci. 9 (2019), https://doi.org/10.3390/app9112296.

[24] S.M. Haile, Fuel cell materials and components, Acta Mater. 51 (2003) 5981–6000, https://doi.org/10.1016/j.actamat.2003.08.004.

[25] J. Wisniak, Historical notes: electrochemistry and fuel cells: the Contribution of William Robert Grove, Indian J. Hist. Sci. 50 (2015), https://doi.org/10.16943/ijhs/2015/v50i4/48318.

[26] A. Serov, I.V. Zenyuk, C.G. Arges, M. Chatenet, Hot topics in alkaline exchange membrane fuel cells, J. Power Sources 375 (2018) 149–157, https://doi.org/10.1016/j.jpowsour.2017.09.068.

[27] A. Lavacchi, H. Miller, F. Vizza, Carbon-Based Nanomaterials, 2013, pp. 115–144, https://doi.org/10.1007/978-1-4899-8059-5_5.

[28] A.M. Abdalla, S. Hossain, A.T. Azad, P.M.I. Petra, F. Begum, S.G. Eriksson, A.K. Azad, Nanomaterials for solid oxide fuel cells: a review, Renew. Sust. Energ. Rev. 82 (2018) 353–368, https://doi.org/10.1016/j.rser.2017.09.046.

[29] C. Coutanceau, S. Brimaud, C. Lamy, J.M. Léger, L. Dubau, S. Rousseau, F. Vigier, Review of different methods for developing nanoelectrocatalysts for the oxidation of organic compounds, Electrochim. Acta 53 (2008) 6865–6880, https://doi.org/10.1016/j.electacta.2007.12.043.

[30] C. Lamy, E.M. Belgsir, J.M. Léger, Electrocatalytic oxidation of aliphatic alcohols: application to the direct alcohol fuel cell (DAFC), J. Appl. Electrochem. 31 (2001) 799–809, https://doi.org/10.1023/A:1017587310150.

[31] N.A. Karim, S.K. Kamarudin, Introduction to direct alcohol fuel cells (DAFCs), in: Direct Liq. Fuel Cells, Elsevier, 2021, pp. 49–70, https://doi.org/10.1016/b978-0-12-818624-4.00002-9.

[32] H.R. Corti, E.R. Gonzalez, Direct Alcohol Fuel Cells: Materials, Performance, Durability and Applications, 2014, https://doi.org/10.1007/978-94-007-7708-8.

[33] R. Govindarasu, S. Somasundaram, Studies on influence of cell temperature in direct methanol fuel cell operation, Processes 8 (2020), https://doi.org/10.3390/PR8030353.

[34] D. Gervasio, Fuel cells—direct alcohol fuel cells | new materials, in: Encycl. Electrochem. Power Sources, Elsevier, 2009, pp. 420–427, https://doi.org/10.1016/B978-044452745-5.00247-1.

[35] M. Yaldagard, M. Jahanshahi, N. Seghatoleslami, Pt catalysts on PANI coated WC/C nanocomposites for methanol electro-oxidation and oxygen electroreduction in DMFC, Appl. Surf. Sci. 317 (2014) 496–504, https://doi.org/10.1016/j.apsusc.2014.08.148.

Chapter 20 Carbon-polymer hybrid-supported nanomaterials **383**

[36] H. Burhan, K. Cellat, G. Yılmaz, F. Şen, Direct methanol fuel cells (DMFCs), in: Direct Liq. Fuel Cells, Elsevier, 2021, pp. 71–94, https://doi.org/10.1016/B978-0-12-818624-4.00003-0.

[37] F. Şen, G. Gökağaç, Different sized platinum nanoparticles supported on carbon: an XPS study on these methanol oxidation catalysts, J. Phys. Chem. C 111 (2007) 5715–5720, https://doi.org/10.1021/jp068381b.

[38] F. Şen, S. Şen, G. Gökağaç, Efficiency enhancement of methanol/ethanol oxidation reactions on Pt nanoparticles prepared using a new surfactant, 1,1-dimethyl heptanethiol, Phys. Chem. Chem. Phys. 13 (2011) 1676–1684, https://doi.org/10.1039/c0cp01212b.

[39] S. Tang, G. Sun, J. Qi, S. Sun, J. Guo, Q. Xin, G.M. Haarberg, Review of new carbon materials as catalyst supports in direct alcohol fuel cells, Chin. J. Catal. 31 (2010) 12–17, https://doi.org/10.1016/s1872-2067(09)60034-6.

[40] L.Q. Hoa, M.C. Vestergaard, E. Tamiya, Carbon-based nanomaterials in biomass-based fuel-fed fuel cells, Sensors 17 (2017) 1–21, https://doi.org/10.3390/s17112587.

[41] S. Şen, F. Şen, G. Gökağaç, Preparation and characterization of nano-sized Pt-Ru/C catalysts and their superior catalytic activities for methanol and ethanol oxidation, Phys. Chem. Chem. Phys. 13 (2011) 6784–6792, https://doi.org/10.1039/c1cp20064j.

[42] S. Eris, Z. Daşdelen, F. Sen, Investigation of electrocatalytic activity and stability of Pt@f-VC catalyst prepared by in-situ synthesis for methanol electrooxidation, Int. J. Hydrog. Energy 43 (2018) 385–390, https://doi.org/10.1016/j.ijhydene.2017.11.063.

[43] Y. Yıldız, S. Kuzu, B. Sen, A. Savk, S. Akocak, F. Şen, Different ligand based monodispersed Pt nanoparticles decorated with rGO as highly active and reusable catalysts for the methanol oxidation, Int. J. Hydrog. Energy 42 (2017) 13061–13069, https://doi.org/10.1016/j.ijhydene.2017.03.230.

[44] K.P. Loh, Q. Bao, P.K. Ang, J. Yang, The chemistry of graphene, J. Mater. Chem. 20 (2010) 2277–2289, https://doi.org/10.1039/b920539j.

[45] A.K. Geim, K.S. Novoselov, The rise of graphene, in: Nanosci. Technol. A Collect. Rev. From Nat. Journals, World Scientific Publishing Co., 2009, pp. 11–19, https://doi.org/10.1142/9789814287005_0002.

[46] W. Gao, The chemistry of graphene oxide, in: Graphene Oxide Reduct. Recipes, Spectrosc. Appl, Springer International Publishing, 2015, pp. 61–95, https://doi.org/10.1007/978-3-319-15500-5_3.

[47] K.S. Novoselov, A.K. Geim, S.V. Morozov, D. Jiang, Y. Zhang, S.V. Dubonos, I.V. Grigorieva, A.A. Firsov, Electric field in atomically thin carbon films, Science 306 (2004) 666–669, https://doi.org/10.1126/science.1102896.

[48] D. Chen, L. Tang, J. Li, Graphene-based materials in electrochemistry, Chem. Soc. Rev. 39 (2010) 3157–3180, https://doi.org/10.1039/b923596e.

[49] P. Martin, Electrochemistry of gaphene: new horizons for sensing and energy storage, Chem. Rec. 9 (2009) 211–223, https://doi.org/10.1002/tcr.200900008.

[50] D.A.C. Brownson, C.E. Banks, Graphene electrochemistry: an overview of potential applications, Analyst 135 (2010) 2768–2778, https://doi.org/10.1039/c0an00590h.

[51] U.R. Farooqui, A.L. Ahmad, N.A. Hamid, Graphene oxide: a promising membrane material for fuel cells, Renew. Sust. Energ. Rev. 82 (2018) 714–733, https://doi.org/10.1016/j.rser.2017.09.081.

[52] X. Cao, C. Tan, M. Sindoro, H. Zhang, Hybrid micro-/nano-structures derived from metal-organic frameworks: preparation and applications in energy

storage and conversion, Chem. Soc. Rev. 46 (2017) 2660–2677, https://doi.org/10.1039/c6cs00426a.

[53] S. Hussain, N. Kongi, L. Matisen, J. Kozlova, V. Sammelselg, K. Tammeveski, Platinum nanoparticles supported on nitrobenzene-functionalised graphene nanosheets as electrocatalysts for oxygen reduction reaction in alkaline media, Electrochem. Commun. 81 (2017) 79–83, https://doi.org/10.1016/j.elecom.2017.06.009.

[54] B. Şen, A. Aygün, A. Şavk, S. Duman, M.H. Calimli, E. Bulut, F. Şen, Polymer-graphene hybrid stabilized ruthenium nanocatalysts for the dimethylamine-borane dehydrogenation at ambient conditions, J. Mol. Liq. 279 (2019) 578–583, https://doi.org/10.1016/j.molliq.2019.02.003.

[55] Kroto H W, Heath J R, O'Brien S C, Curl R F & Smalley R E. C, This Week's Citation Classic ® C 60-The Third Man, n.d.

[56] G. Che, B.B. Lakshmi, E.R. Fisher, C.R. Martin, Carbon nanotubule membranes for electrochemical energy storage and production, Nature 393 (1998) 346–349, https://doi.org/10.1038/30694.

[57] S. Maass, F. Finsterwalder, G. Frank, R. Hartmann, C. Merten, Carbon support oxidation in PEM fuel cell cathodes, J. Power Sources 176 (2008) 444–451, https://doi.org/10.1016/j.jpowsour.2007.08.053.

[58] D.B. Mawhinney, V. Naumenko, A. Kuznetsova, J.T. Yates, J. Liu, R.E. Smalley, Infrared spectral evidence for the etching of carbon nanotubes: ozone oxidation at 298 K [7], J. Am. Chem. Soc. 122 (2000) 2383–2384, https://doi.org/10.1021/ja994094s.

[59] R. Zhang, Y. Zhang, F. Wei, Synthesis and properties of ultralong carbon nanotubes, in: Nanotub. Superfiber Mater. Chang. Eng. Des, Elsevier Inc., 2013, pp. 87–136, https://doi.org/10.1016/B978-1-4557-7863-8.00004-9.

[60] W. Li, C. Liang, W. Zhou, J. Qiu, Z. Zhou, G. Sun, Q. Xin, Preparation and characterization of multiwalled carbon nanotube-supported platinum for cathode catalysts of direct methanol fuel cells, J. Phys. Chem. B 107 (2003) 6292–6299, https://doi.org/10.1021/jp022505c.

[61] B. Rajesh, K. Ravindranathan Thampi, J.M. Bonard, N. Xanthopoulos, H.J. Mathieu, B. Viswanathan, Carbon nanotubes generated from template carbonization of polyphenyl acetylene as the support for electrooxidation of methanol, J. Phys. Chem. B 107 (2003) 2701–2708, https://doi.org/10.1021/jp0219350.

[62] S. Hanif, N. Iqbal, X. Shi, T. Noor, G. Ali, A.M. Kannan, NiCo–N-doped carbon nanotubes based cathode catalyst for alkaline membrane fuel cell, Renew. Energy 154 (2020) 508–516, https://doi.org/10.1016/j.renene.2020.03.060.

[63] F. Altaf, R. Gill, R. Batool, Zohaib-Ur-Rehman, H. Majeed, G. Abbas, K. Jacob, Synthesis and applicability study of novel poly(dopamine)-modified carbon nanotubes based polymer electrolyte membranes for direct methanol fuel cell, J. Environ. Chem. Eng. 8 (2020), https://doi.org/10.1016/j.jece.2020.104118, 104118.

[64] T. Sharma, A.L. Mohana Reddy, T.S. Chandra, S. Ramaprabhu, Development of carbon nanotubes and nanofluids based microbial fuel cell, Int. J. Hydrog. Energy 33 (2008) 6749–6754, https://doi.org/10.1016/j.ijhydene.2008.05.112.

[65] E. Ulloa, Fullerenes and their applications in science and technology, in: Introd. to Nanotechnol. Conf, 2013, pp. 1–5.

[66] Y. Pan, X. Liu, W. Zhang, Z. Liu, G. Zeng, B. Shao, Q. Liang, Q. He, X. Yuan, D. Huang, M. Chen, Advances in photocatalysis based on fullerene C60 and its derivatives: properties, mechanism, synthesis, and applications, Appl. Catal. B Environ. 265 (2020) 118579, https://doi.org/10.1016/j.apcatb.2019.118579.

[67] A. Hirsch, I. Lamparth, H.R. Karfunkel, Fullerene chemistry in three dimensions: isolation of seven regioisomeric bisadducts and chiral trisadducts of C60 and di(ethoxycarbonyl)methylene, Angew. Chem. Int. Ed. Eng. 33 (1994) 437–438, https://doi.org/10.1002/anie.199404371.

[68] M.S. Mauter, M. Elimelech, Environmental applications of carbon-based nanomaterials, Environ. Sci. Technol. 42 (2008) 5843–5859, https://doi.org/10.1021/es8006904.

[69] M.S. Dresselhaus, G. Dresselhaus, P.C. Eklund, Structure of fullerenes, in: Sci. Fullerenes Carbon Nanotub, Elsevier, 1996, pp. 60–79, https://doi.org/10.1016/b978-012221820-0/50003-4.

[70] Y. Quo, N. Karasawa, W.A. Goddard, Prediction of fullerene packing in C60 and C70 crystals, Nature 351 (1991) 464–467, https://doi.org/10.1038/351464a0.

[71] V. Martínez-Agramunt, E. Peris, Photocatalytic properties of a palladium metallosquare with encapsulated fullerenes via singlet oxygen generation, Inorg. Chem. 58 (2019) 11836–11842, https://doi.org/10.1021/acs.inorgchem.9b02097.

[72] Z. Da Meng, L. Zhu, S. Ye, Q. Sun, K. Ullah, K.Y. Cho, W.C. Oh, Fullerene modification CdSe/TiO$_2$ and modification of photocatalytic activity under visible light, Nanoscale Res. Lett. 8 (2013) 1–10, https://doi.org/10.1186/1556-276X-8-189.

[73] G. Rambabu, S.D. Bhat, Sulfonated fullerene in SPEEK matrix and its impact on the membrane electrolyte properties in direct methanol fuel cells, Electrochim. Acta 176 (2015) 657–669, https://doi.org/10.1016/j.electacta.2015.07.045.

[74] K.S. Bhavani, T. Anusha, J.V.S. Kumar, P.K. Brahman, Enhanced electrocatalytic activity of methanol and ethanol oxidation in alkaline medium at bimetallic nanoparticles electrochemically decorated fullerene-C$_{60}$ nanocomposite electrocatalyst: an efficient anode material for alcohol fuel cell applications, Electroanalysis 33 (2021) 97–110, https://doi.org/10.1002/elan.202060154.

[75] H. Luo, X. Zhou, C. Ellingford, Y. Zhang, S. Chen, K. Zhou, D. Zhang, C.R. Bowen, C. Wan, Interface design for high energy density polymer nanocomposites, Chem. Soc. Rev. 48 (2019) 4424–4465, https://doi.org/10.1039/c9cs00043g.

[76] Q. Li, L. Chen, M.R. Gadinski, S. Zhang, G. Zhang, H. Li, A. Haque, L.Q. Chen, T. Jackson, Q. Wang, Flexible high-temperature dielectric materials from polymer nanocomposites, Nature 523 (2015) 576–579, https://doi.org/10.1038/nature14647.

[77] X. Huang, B. Sun, Y. Zhu, S. Li, P. Jiang, High-k polymer nanocomposites with 1D filler for dielectric and energy storage applications, Prog. Mater. Sci. 100 (2019) 187–225, https://doi.org/10.1016/j.pmatsci.2018.10.003.

[78] S. Luo, J. Yu, S. Yu, R. Sun, L. Cao, W.-H. Liao, C.-P. Wong, Significantly enhanced electrostatic energy storage performance of flexible polymer composites by introducing highly insulating-ferroelectric microhybrids as fillers, Adv. Energy Mater. 9 (2019) 1803204, https://doi.org/10.1002/aenm.201803204.

[79] G. Inzelt, Introduction, in: Conduct. Polym. A New Era Electrochem, 2012, pp. 1–6, https://doi.org/10.1007/978-3-642-27621-7_1.

[80] H. Pang, L. Xu, D.X. Yan, Z.M. Li, Conductive polymer composites with segregated structures, Prog. Polym. Sci. 39 (2014) 1908–1933, https://doi.org/10.1016/j.progpolymsci.2014.07.007.

[81] W. Zhang, A.A. Dehghani-Sanij, R.S. Blackburn, Carbon based conductive polymer composites, J. Mater. Sci. (2007) 3408–3418, https://doi.org/10.1007/s10853-007-1688-5.

[82] T. Nezakati, A. Seifalian, A. Tan, A.M. Seifalian, Conductive polymers: opportunities and challenges in biomedical applications, Chem. Rev. 118 (2018) 6766–6843, https://doi.org/10.1021/acs.chemrev.6b00275.

[83] K. Singh, S. Kumar, K. Agarwal, K. Soni, V. Ramana Gedela, K. Ghosh, Three-dimensional graphene with MoS2 nanohybrid as potential energy storage/transfer device, Sci. Rep. 7 (2017) 1–12, https://doi.org/10.1038/s41598-017-09266-2.

[84] Z. Qi, J. Shan, P.G. Pickup, Conducting polymer-supported fuel cell catalysts, ACS Symp. Ser. 832 (2002) 166–183, https://doi.org/10.1021/bk-2003-0832.ch013.

[85] L.-P. Fan, T. Gao, Applications of nanoscale polypyrrole proton exchange membrane in microbial fuel cells, Int. J. Electrochem. Sci. 14 (2019) 470–480, https://doi.org/10.20964/2019.01.41.

[86] K. Dutta, S. Das, D. Rana, P.P. Kundu, Enhancements of catalyst distribution and functioning upon utilization of conducting polymers as supporting matrices in DMFCs: a review, Polym. Rev. 55 (2015) 1–56, https://doi.org/10.1080/15583724.2014.958771.

[87] S. Ghosh, T. Maiyalagan, R.N. Basu, Nanostructured conducting polymers for energy applications: towards a sustainable platform, Nanoscale 8 (2016) 6921–6947, https://doi.org/10.1039/c5nr08803h.

[88] Ö. Karatepe, Y. Yildiz, H. Pamuk, S. Eris, Z. Dasdelen, F. Sen, Enhanced electrocatalytic activity and durability of highly monodisperse Pt@PPy-PANI nanocomposites as a novel catalyst for the electro-oxidation of methanol, RSC Adv. 6 (2016) 50851–50857, https://doi.org/10.1039/c6ra06210e.

[89] S. Eris, Z. Daşdelen, Y. Yıldız, F. Sen, Nanostructured polyaniline-rGO decorated platinum catalyst with enhanced activity and durability for methanol oxidation, Int. J. Hydrog. Energy 43 (2018) 1337–1343, https://doi.org/10.1016/j.ijhydene.2017.11.051.

[90] X. Liu, C. Lai, Z. Xiao, S. Zou, K. Liu, Y. Yin, T. Liang, Z. Wu, Superb electrolyte penetration/absorption of three-dimensional porous carbon nanosheets for multifunctional supercapacitor, ACS Appl. Energy Mater. 2 (2019) 3185–3193, https://doi.org/10.1021/acsaem.9b00002.

[91] N. Hamadneh, W. Khan, Polymer Nanocomposites—Synthesis Techniques, Classification and Properties, 2016.

[92] V.K. Prateek, R.K. Thakur, Gupta, recent progress on ferroelectric polymer-based nanocomposites for high energy density capacitors: synthesis, dielectric properties, and future aspects, Chem. Rev. 116 (2016) 4260–4317, https://doi.org/10.1021/acs.chemrev.5b00495.

[93] C.I. Idumah, A. Hassan, Emerging trends in graphene carbon based polymer nanocomposites and applications, Rev. Chem. Eng. 32 (2016) 223–264, https://doi.org/10.1515/revce-2015-0038.

[94] B. Çelik, G. Başkaya, H. Sert, Ö. Karatepe, E. Erken, F. Şen, Monodisperse Pt(0)/DPA@GO nanoparticles as highly active catalysts for alcohol oxidation and dehydrogenation of DMAB, Int. J. Hydrog. Energy 41 (2016) 5661–5669, https://doi.org/10.1016/j.ijhydene.2016.02.061.

[95] F. Şen, G. Gökagaç, Activity of carbon-supported platinum nanoparticles toward methanol oxidation reaction: role of metal precursor and a new surfactant, tert-octanethiol, J. Phys. Chem. C 111 (2007) 1467–1473, https://doi.org/10.1021/jp065809y.

Chapter 20 Carbon-polymer hybrid-supported nanomaterials **387**

[96] Z. Daşdelen, Y. Yıldız, S. Eriş, F. Şen, Enhanced electrocatalytic activity and durability of Pt nanoparticles decorated on GO-PVP hybride material for methanol oxidation reaction, Appl. Catal. B Environ. 219 (2017) 511–516, https://doi.org/10.1016/j.apcatb.2017.08.014.

[97] S. Eris, Z. Daşdelen, F. Sen, Enhanced electrocatalytic activity and stability of monodisperse Pt nanocomposites for direct methanol fuel cells, J. Colloid Interface Sci. 513 (2018) 767–773, https://doi.org/10.1016/j.jcis.2017.11.085.

[98] R. Vani, S. Ramaprabhu, P. Haridoss, Mechanically stable and economically viable polyvinyl alcohol-based membranes with sulfonated carbon nanotubes for proton exchange membrane fuel cells, Sustain. Energy Fuels 4 (2020) 1372–1382, https://doi.org/10.1039/c9se01031a.

[99] D. Rajesh, C. Mahendiran, C. Suresh, The promotional effect of ag in Pd-Ag/carbon nanotube-graphene electrocatalysts for alcohol and formic acid oxidation reactions, ChemElectroChem. 7 (2020) 2629–2636, https://doi.org/10.1002/celc.202000642.

[100] A. De, R. Adhikary, J. Datta, Proactive role of carbon nanotube-polyaniline conjugate support for Pt nano-particles toward electro-catalysis of ethanol in fuel cell, Int. J. Hydrog. Energy 42 (2017) 25316–25325, https://doi.org/10.1016/j.ijhydene.2017.08.073.

[101] V. Mozafari, J. Basiri Parsa, Electrochemical synthesis of Pd supported on PANI-MWCNTs-SnO_2 nanocomposite as a novel catalyst towards ethanol oxidation in alkaline media, Synth. Met. 259 (2020) 116214, https://doi.org/10.1016/j.synthmet.2019.116214.

21

Polymer-based nanocatalyts for alcohol fuel cells

Ilyas Ilker Isler[a], Haydar Goksu[a], Vildan Erduran[b,c], Iskender Isik[c], and Fatih Şen[b]

[a]Kaynasli Vocational College, Duzce University, Duzce, Turkey. [b]Şen Research Group, Department of Biochemistry, Dumlupinar University, Kütahya, Turkey. [c]Department of Materials Science and Engineering, Faculty of Engineering, Dumlupinar University, Kütahya, Turkey

1 Introduction

The need for energy corresponding to higher levels of consumption is increasing day by day. The demand for fossil fuels increased after international trade developed within the scope of globalization accelerated industrialization and population growth. However, the limited use of natural resources in the world, as well as the detrimental effect of fossil fuel use on the environment, have pushed scientists to search for new, alternative energy resources. For this reason, it has been observed that the interest in and number of studies on alternative energy sources to fossil fuels have increased [1].

Today, alternatives to fossil fuels include solar, geothermal, wind, hydroelectric, and nuclear energy, all of which are natural resources. Fuel cells are among the principal energy sources of the future, thanks to their high efficiency, ecological friendliness, and safety. The energy production potential of solar cells, which are widespread in the world, depends on seasonal effects, sunrise and sunset. In wind turbines, it depends on the flowing direction of air and wind. Fuel cells can produce energy continuously, regardless of cyclical effects, and so interest in them has increased. In recent years, there has been a considerable number of scientific studies on fuel cells. It is expected that its use in Turkey, as well as the rest of the world, will become widespread in the future [2].

Fuel cells, which produce chemical energy directly from electricity using hydrogen, are part of conversion technology. These

Nanomaterials for Direct Alcohol Fuel Cells. https://doi.org/10.1016/B978-0-12-821713-9.00004-4
Copyright © 2021 Elsevier Inc. All rights reserved.

batteries, referred to as *electrochemical converters*, decouple chemical energy directly into electrical energy, with no intermediate steps. Because there are no intense heat treatments such as in internal combustion engines, these batteries are environmentally friendly and ensure that electrical energy is produced with high efficiency [1–3]. A wide range of catalysts and nanoparticles are used in the studies of fuel cell and hydrogen storage systems [4–9]. Effective nanocatalysts have been discovered for use in reactions for hydrogen production [10–19].

Fuel cells have a batterylike structure and contain electrochemical cells, such as batteries. They consist of an anode and a cathode, along with an ionic conductive electrolyte. Fuel cells have some important differences from batteries. Whereas batteries do not need to be fed from outside, fuel cells are continuously fed from an exterior source. If this cycle can be achieved, there is a continuity of energy within the fuel cells [20].

In additional, hydrogen, natural gas, and liquid petroleum gas (LPG) can be utilized directly inside the battery. The most common fuel cell is hydrogen-based. Ethanol and methanol are at the forefront due to their low molecular weight and potential for producing high energy. Due to these properties of alcohol, intensive research and development studies continue [21].

In the future, fuel cells will have advantages over existing energy systems in terms of mobility and portability and may present a viable alternative to batteries and internal combustion engines. If continuous feeding is possible due to high energy density, this solves problems such as the need for charging and termination [22, 23].

2 Alcohol fuel cells

Fuel cells are being used in increasingly wider areas due to their efficient and silent technology that is compatible with the environment. They can convert the chemical energy of the existing fuel into electricity and heat without any need for combustion and intermediate input. Although the operating system of fuel cells has been known since 1889, it has not been applied very much in other areas except for the aerospace industry [24].

The fuel cell system generates electrical energy via electrochemical reactions of suitable fuel and oxidizers. In other words, a fuel cell is a generator that converts fuel chemical energy directly into electrical energy as a result of the electrochemical interaction of fuel and air [25]. Energy stored in batteries and accumulators is converted into electrical energy by an electrochemical reaction. The energy provided in this way is limited to the energy stored within these devices. Fuel cells, on the other hand, are energy generation systems that can make this conversion without interruption if fuel and air are supplied from an external source [26].

Hydrogen can be used as the primary energy source directly in the fuel cell, or fuels containing hydrogen, such as natural gas, methanol, LPG, gasoline, or naphtha can be used if they are converted [24].

The transformation that occurs in the fuel cell (Fig. 1) is similar to that in the transformation in the battery or accumulator. The main separation between the fuel cell and other traditional systems is that fuel cells can transform energy as long as fuel and oxidizers are supplied. In others, this conversion is restricted to the energy stored in them.

2.1 The history of alcohol fuel cells

The improvement of fuel cell technology is crucial to the development of alcohol fuel cell technology and its contribution to the economy [28, 29]. Fuel cell technology was invented by William Grove in 1839 and began to evolve in the 20th century. Alcohol fuel cells in the beginning of the 20th century could be used in practical applications [30]. The carbonate cell, which is produced only by liquid contamination in the alcohol fuel cell, has prevented it

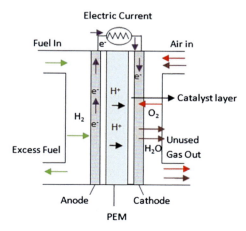

Fig. 1 Diagram of a simple alcohol fuel cell [27].

from being adapted to pure oxygen-only systems. This system was first used by the National Aeronautics and Space Administration (NASA) in the *Apollo* space program in 1960, and then other spacecraft [31]. In later years, different fuel cells have emerged, the main purpose of which is to transmit current within them.

The rapidly developing fuel cell industry also brings with it some uncertain sales and application development situations caused by its new nature. It also requires discussion of the transition from research and development activities, which are important in the beginning, to production, marketing, and sales activities [21].

Phosphoric acid fuel cells (PAFCs), which emerged in the 20th century, have become one of the most significant indicators of technological progress in alcohol fuel cells, with them showing resistance to heat of approximately 200°C. However, since the cost per kilowatt of these fuel cells is higher than what engineers and developers target for electricity generation compared to energy from fossil fuels, their use is limited to military use in the United States and developed European countries. In recent years, molten carbonates and alcohol fuel cells using solid oxide electrolytes have become more common because these cells are capable of using low-purity hydrogen and also provide greater efficiency in transferring heat [32–34]. Another common example is proton exchange membrane fuel cells (PEMFCs), fuel cells with a working potential below 100°C that are typically used as electrolytes [35].

While alkaline fuel cell technology became widespread during the early period of space exploration, the demand for alcohol fuel cells, which is the focus of low-cost proton-exchange membrane fuel cell (PEMFC) technology, increased due to the expense of fossil fuels, especially after the 1979 energy crisis. Due to this demand, a number of studies have explored PEMFCs in order to achieve lower costs. After oil prices dropped again in the 1980s and 1990s, PEMFC technology were directed exclusively to military applications [35–37]. Then, efforts to prevent environmental pollution beginning in the late 1990s set the stage for scientists to focus on PEMFC technology for general energy use again [38].

2.2 Types and functions of alcohol fuel cells

A fuel cell usually consists of two electrodes with a compressed electrode between them.

These cells are divided into five subcategories based on the material used as the electrolyte, as described in Table 1.

Table 1 Fuel cell types and properties [39].

Fuel cell	Electrolyte	In the electrolyte carrier	Cell material	Fuel type	Power production efficiency (%)	Scope of application
PAFC	Phosphoric acid	H^+	Carbon	H_2, hydrocarbon, fossil fuels	37–42	Commercial applications
Molten carbonate fuel cell	Carbonate	CO_3^{-2}	Nickel, stainless steel	H_2, hydrocarbon	45–60	Power stations
PEFC	Polymer ion change movie	H^+	Carbon	H_2, hydrocarbon	60	Transport vehicles, military
Solid oxide fuel cell	On zinc pinned yttria	O_2^{-2}	Ceramic	H_2, hydrocarbon	60–70	Commercial applications
Alkaline fuel cell	Potassium hydroxide	OH^-	Carbon	H_2	42–73	Space studies

2.2.1 Phosphoric acid fuel cells

Phosphoric acid is used in fuel cells as an electrolyte, but also as a catalyst. Platinum is often used as well. PAFCs are used in many areas, including schools, hotels, and hospitals [40–42]. The efficiency of PAFCs is 40%, and the operating temperature is about 400°C. The power density of a PAFC is less than that of other fuel cells [43], which causes them to be heavier and take up more space. When we need to examine the working system, the hydrogen ions that form by separating the electrons from the hydrogen molecules sent to the anode are phosphoric [44]. The acid reaches the cathode from the electrolyte and electrons from the outer circuit (Fig. 2). Here, the circuit is expected to be completed as a result of combining hydrogen ions and electrons with oxygen [45].

2.2.2 Molten carbonate fuel cells

In molten carbonate fuel cells, the electrolyte material is alkali metal carbonate. For this carbonate to work properly in the fuel cell, it must be in liquid and fluid form (Fig. 3). The operating temperature of these fuel cells is 700°C. If this temperature is high, the materials inside the cell are quickly deformed [46–48].

394 Chapter 21 Polymer-based nanocatalyts for alcohol fuel cells

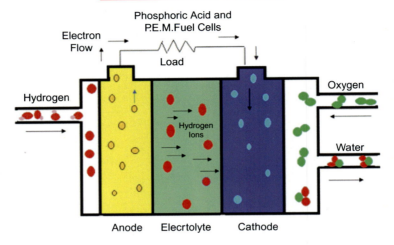

Fig. 2 Diagram of a PAFC [27].

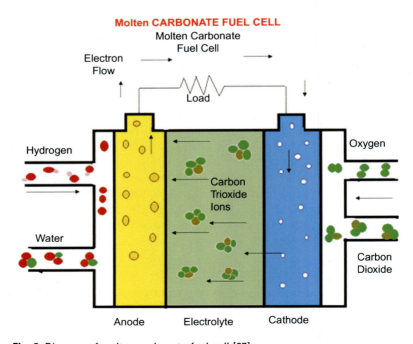

Fig. 3 Diagram of molten carbonate fuel cell [27].

While the molten carbonate fuel cell technology aims to increase the durability and development of the batteries, it continues its efforts to increase its marketing capacity. Molten carbonate fuel cell prototype facilities, whose market is under development, are still lagging behind PAFCs [49].

2.2.3 Alkaline fuel cells

Studies on alkaline fuel cells have a long history. Work on alkaline fuel cells was applied by NASA to their space program and studies [50]. In these batteries, alkaline potassium hydroxide is used instead of electrolytes [51–53]. While the battery is running, the average temperature is 70°C, and the efficiency is 70%. However, the power generation of these batteries is very low. Another disadvantage in alkaline batteries is the corrosion created by the substances set out within them. Carbon dioxide in the fuel can react chemically with the potassium hydroxide in the electrolyte. This may cause corrosion [24, 54].

The development of highly efficient, nonnoble metal-based catalysts for the oxygen reduction reaction is important for alkaline fuel cells [55]. Studies on such fuel cells have fallen behind others. This is the fact that it is less preferred due to the negativities in its content [56, 57].

2.2.4 Solid oxide fuel cells

Solid oxide fuel cells have attracted a great deal of attention in industry due to their unique benefits. This type of fuel cell can contain high temperatures without using expensive catalysts. The extreme tolerance level of this fuel cell against high temperatures and damage caused by fuel gives it a greater efficiency than other fuel cells. The energy obtained facilitates the conversion of hydrocarbon into hydrogen energy [58].

The technologies used to generate energy by cleaner types of fuel (such as methane, hydrogen, and ammonia) were considered to be more promising than those for the use of power generators [59].

Solid oxide fuel cell technology could change the course of today's world because of its optimum power generation capabilities for automobiles and home appliances with maximum electric efficiency [60]. Due to its structure and the content of the material used, as well as its potential, this type of fuel cell is under development [61–64].

Owing to the high temperature level, the cell electrolyte is made of ceramic. Because the elements in the structure are solid, there is no evaporation or reduction in the electrolytes. The fuel cells should be enlarged and production methods streamlined in order to be further developed [65].

2.2.5 Polymer electrolyte fuel cells

Polymer electrolyte fuel cells (PEFCs) are commonly used in the transportation industry [66, 67]. There is a polymer electrolyte plate in the middle of these fuel cells. As a conductive mechanism, it contains positive ions that break through the membrane. These positive ions are either hydrogen or proton ions (Fig. 4).

PEFCs are electrochemical devices that transform the chemical energy of hydrogen fuel into electric energy and heat. Because of their high efficiency, zero emissions at the point of use, large power volume, and quick reaction to load change, these fuel cells replaced the internal combustion engine in vehicles [68]. They are recognized as ecofriendly energy converters for automobiles and residential energy systems [69].

Hydrogen ions move only in one direction within the plate. This is the basic working principle of the PEMFC. PEM electrons, which are naturally electrolyte insulators, are impermeable and very thin [70].

The operating temperature of PEM is below 100°C. It uses an acid polymer as electrolyte. The structure of the PEM includes a catalyst, bipolar plates, current collectors, membranes, gaskets, and other equipment. The anode side is powered with fuel and

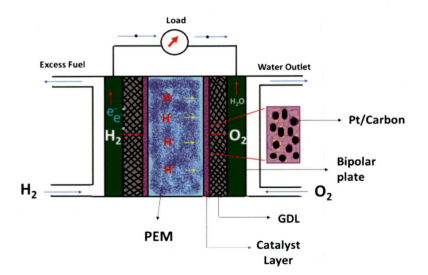

Fig. 4 Diagram of a PEM [27].

the cathode side with oxidant [71]. In this process, water is created by the chemical reaction of hydrogen and oxygen, and two independent electrochemical reactions occur in the electrodes [25, 64].

3 Using nanocatalyst polymer fuel cells, which reduces costs and increases conductivity

The role of catalysts in increasing the distribution of fuel cells is significant. It allows electrocatalysts to increase the electrochemically active surface area, become uniformly disperse, and provide the optimal environment for fuel cell reactions [72].

The most important role of catalysts in alcohol fuel cells is to serve as an electron source with increased conductivity. Having a larger surface area to hold metal nanoparticles in energy production, providing better stability, ensuring porosity, and having higher corrosion resistance in fuel cell operating conditions are among their most important roles [73].

There are many polymer material forms, which vary according to the structure, including conductivity and thermal stability. The surface preparation area increases its efficiency with various physiochemical approaches [74].

The reason for the high costs of using PEFCs is the presence of metal catalysts, perfluoro sulfonic acid polymer electrolyte membranes, bipolar plates, and auxiliary components [75]. However, the development of improved electrocatalysts, high-performance polymer electrolyte membranes, and inexpensive bipolar plates that meet the quality product development target requirements is a major challenge. Topics that researchers focus on today include low-cost materials, such as new hydrocarbon membranes [76].

Over the past few years, the use of carbon materials has increased with the development of fuel cells. Catalysts of additives for the fuel cell offer conductivity, high surface area, high stability, and suitable porosity to provide good reactants [77]. Of the nanocatalites used to increase fuel consumption in the fuel cell, the number of carbon-containing ones is quite high. It is among the favorites due to its high compatibility degree and low cost. The disadvantage of these carbons is that they contain high amounts of micropores. This situation can prevent the flow of reactants [78, 79].

The reason for the initiation of efforts to increase the efficiency of modern fuel cells is the materials that reduce the energy efficiency in the content of carbon-containing substances. The increased use of porous carbon materials and carbon gels enhances the thermal conductivity and efficiency of fuel cells [80].

CNT- and molten carbonate—containing materials have been used recently in fuel cells instead of carbon-containing materials. Using materials containing CNTs and molten carbonates, the complexity of synthesis methods and the efficiency of fuel cell catalysts increase with the versatility of the pore size and adaptation of the pore distribution. New types of catalysts used in these fuel cells replace the previous mesocarbons and increase the cell additive and energy level, thereby reducing the cost [81].

At the same time, CNTs are unique because of their high electronic conductivity. Their superior stability when tested with acids is compatible with metal and carbon interactions. Catalyst-CNTs in the fuel cell are better than molten carbonates used as a ground in the fuel cell. In molten carbonate fuel cell, the structure of stable metals resembles that of carbon black-based catalysts. Moreover, the efficiency of meso can be increased by graphitizing porous carbons in molten carbonate fuel cells [82–84].

4 Conclusion

In recent years, many countries in the world have increased their investments in fuel cells to prevent silent operation and environmental pollution within the framework of technological developments. The whole world has begun to recognize the importance of hydrogen, the energy of the future. The aim is to make vehicles environmentally friendly with fuel cells, which provide quieter operation compared to other electric vehicles.

A fuel cell qualifies as a system in which fuel can be converted directly into energy without any intermediate processing. In recent years, studies on the efficiency of these cells have focused on the type of fuel and the catalyst material used. A number of studies have reevaluated the effect on the efficiency of fuel cells for various catalysts used.

This chapter introduced polymer-based nanocatalyts for alcohol fuel cells. It began with a descriptive introduction about fuel cells and their types. Then, polymer-based nanocatalyts for alcohol fuel cell types were examined individually, and various membrane properties and application methods were discussed. The basic principles of the application of these fuel cells were demonstrated.

Important steps have been taken by the world's largest companies to develop fuel cell technologies. Fuel cells have the potential to operate at low temperatures and produce high levels of power,

Chapter 21 Polymer-based nanocatalyts for alcohol fuel cells **399**

thus increasing the importance of these fuel cells in the automotive and energy sectors.

Short assembly time, high energy production efficiency, waste heat reuse, system reliability, and ability to work in different fuel types are the most important features of fuel cells. They are more advantageous than traditional power systems. Alcohol fuel cells, with the potential to solve the energy deficit problems of a wide audience ranging from space research to production facilities and individual vehicles, are the technology of the future.

References

[1] D. Hart, Sustainable energy conversion: fuel cells—the competitive option? J. Power Sources 86 (2000) 23–27, https://doi.org/10.1016/S0378-7753(99)00455-3.

[2] J. Romm, The car and fuel of the future, Energy Policy 34 (2006) 2609–2614, https://doi.org/10.1016/j.enpol.2005.06.025.

[3] A. Bauen, D. Hart, Assessment of the environmental benefits of transport and stationary fuel cells, J. Power Sources 86 (2000) 482–494, https://doi.org/10.1016/S0378-7753(99)00445-0.

[4] B. Çelik, Y. Yildiz, H. Sert, E. Erken, Y. Koşkun, F. Şen, Monodispersed palladium-cobalt alloy nanoparticles assembled on poly(N-vinyl-pyrrolidone) (PVP) as a highly effective catalyst for dimethylamine borane (DMAB) dehydrocoupling, RSC Adv. 6 (2016) 24097–24102, https://doi.org/10.1039/c6ra00536e.

[5] B. Sen, S. Kuzu, E. Demir, S. Akocak, F. Sen, Highly monodisperse RuCo nanoparticles decorated on functionalized multiwalled carbon nanotube with the highest observed catalytic activity in the dehydrogenation of dimethylamine-borane, Int. J. Hydrog. Energy 42 (2017) 23292–23298, https://doi.org/10.1016/j.ijhydene.2017.06.032.

[6] B. Sen, B. Demirkan, B. Şimşek, A. Savk, F. Sen, Monodisperse palladium nanocatalysts for dehydrocoupling of dimethylamineborane, Nano-Struct. Nano-Obj. 16 (2018) 209–214, https://doi.org/10.1016/j.nanoso.2018.07.008.

[7] F. Şen, G. Gökagaç, Activity of carbon-supported platinum nanoparticles toward methanol oxidation reaction: role of metal precursor and a new surfactant, tert-octanethiol, J. Phys. Chem. C 111 (2007) 1467–1473, https://doi.org/10.1021/jp065809y.

[8] S. Şen, F. Şen, G. Gökağaç, Preparation and characterization of nano-sized Pt-Ru/C catalysts and their superior catalytic activities for methanol and ethanol oxidation, Phys. Chem. Chem. Phys. 13 (2011) 6784–6792, https://doi.org/10.1039/c1cp20064j.

[9] Z. Daşdelen, Y. Yıldız, S. Eriş, F. Şen, Enhanced electrocatalytic activity and durability of Pt nanoparticles decorated on GO-PVP hybride material for methanol oxidation reaction, Appl. Catal. B Environ. 219 (2017) 511–516, https://doi.org/10.1016/j.apcatb.2017.08.014.

[10] B. Sen, E. Kuyuldar, A. Şavk, H. Calimli, S. Duman, F. Sen, Monodisperse ruthenium–copper alloy nanoparticles decorated on reduced graphene oxide for dehydrogenation of DMAB, Int. J. Hydrog. Energy 44 (2019) 10744–10751, https://doi.org/10.1016/j.ijhydene.2019.02.176.

[11] B. Şen, B. Demirkan, M. Levent, A. Şavk, F. Şen, Silica-based monodisperse PdCo nanohybrids as highly efficient and stable nanocatalyst for hydrogen

evolution reaction, Int. J. Hydrog. Energy 43 (2018) 20234–20242, https://doi.org/10.1016/j.ijhydene.2018.07.080.

[12] B. Şen, B. Demirkan, A. Savk, R. Kartop, M.S. Nas, M.H. Alma, S. Sürdem, F. Şen, High-performance graphite-supported ruthenium nanocatalyst for hydrogen evolution reaction, J. Mol. Liq. 268 (2018) 807–812, https://doi.org/10.1016/j.molliq.2018.07.117.

[13] S. Eris, Z. Daşdelen, F. Sen, Investigation of electrocatalytic activity and stability of Pt@f-VC catalyst prepared by in-situ synthesis for methanol electrooxidation, Int. J. Hydrog. Energy 43 (2018) 385–390, https://doi.org/10.1016/j.ijhydene.2017.11.063.

[14] Ö. Karatepe, Y. Yildiz, H. Pamuk, S. Eris, Z. Dasdelen, F. Sen, Enhanced electrocatalytic activity and durability of highly monodisperse Pt@PPy-PANI nanocomposites as a novel catalyst for the electro-oxidation of methanol, RSC Adv. 6 (2016) 50851–50857, https://doi.org/10.1039/c6ra06210e.

[15] Y. Yildiz, E. Erken, H. Pamuk, H. Sert, F. Şen, Monodisperse Pt nanoparticles assembled on reduced graphene oxide: highly efficient and reusable catalyst for methanol oxidation and dehydrocoupling of dimethylamine-borane (DMAB), J. Nanosci. Nanotechnol. 16 (2016) 5951–5958, https://doi.org/10.1166/jnn.2016.11710.

[16] Y. Yıldız, S. Kuzu, B. Sen, A. Savk, S. Akocak, F. Şen, Different ligand based monodispersed Pt nanoparticles decorated with rGO as highly active and reusable catalysts for the methanol oxidation, Int. J. Hydrog. Energy 42 (2017) 13061–13069, https://doi.org/10.1016/j.ijhydene.2017.03.230.

[17] F. Şen, S. Şen, G. Gökağaç, Efficiency enhancement of methanol/ethanol oxidation reactions on Pt nanoparticles prepared using a new surfactant, 1,1-dimethyl heptanethiol, Phys. Chem. Chem. Phys. 13 (2011) 1676–1684, https://doi.org/10.1039/c0cp01212b.

[18] S. Eris, Z. Daşdelen, Y. Yıldız, F. Sen, Nanostructured polyaniline-rGO decorated platinum catalyst with enhanced activity and durability for methanol oxidation, Int. J. Hydrog. Energy 43 (2018) 1337–1343, https://doi.org/10.1016/j.ijhydene.2017.11.051.

[19] B. Çelik, S. Kuzu, E. Erken, H. Sert, Y. Koşkun, F. Şen, Nearly monodisperse carbon nanotube furnished nanocatalysts as highly efficient and reusable catalyst for dehydrocoupling of DMAB and C1 to C3 alcohol oxidation, Int. J. Hydrog. Energy 41 (2016) 3093–3101, https://doi.org/10.1016/j.ijhydene.2015.12.138.

[20] S. Muench, A. Wild, C. Friebe, B. Häupler, T. Janoschka, U.S. Schubert, Polymer-based organic batteries, Chem. Rev. 116 (2016) 9438–9484, https://doi.org/10.1021/acs.chemrev.6b00070.

[21] S. Giddey, S.P.S. Badwal, A. Kulkarni, C. Munnings, A comprehensive review of direct carbon fuel cell technology, Prog. Energy Combust. Sci. 38 (2012) 360–399, https://doi.org/10.1016/j.pecs.2012.01.003.

[22] S. Sharma, B.G. Pollet, Support materials for PEMFC and DMFC electrocatalysts—a review, J. Power Sources 208 (2012) 96–119, https://doi.org/10.1016/J.JPOWSOUR.2012.02.011.

[23] A. Chen, P. Holt-Hindle, Platinum-based nanostructured materials: synthesis, properties, and applications, Chem. Rev. 110 (2010) 3767–3804, https://doi.org/10.1021/cr9003902.

[24] X. Li, Principles of Fuel Cells, Taylor & Francis, 2006.

[25] L. Carrette, K.A. Friedrich, U. Stimming, Fuel cells—fundamentals and applications, Fuel Cells 1 (2001) 5–39, https://doi.org/10.1002/1615-6854(200105)1:1<5::AID-FUCE5>3.0.CO;2-G.

[26] K. Kordesch, J. Gsellmann, M. Cifrain, S. Voss, V. Hacker, R.R. Aronson, C. Fabjan, T. Hejze, J. Daniel-Ivad, Intermittent use of a low-cost alkaline fuel

Chapter 21 Polymer-based nanocatalyts for alcohol fuel cells **401**

cell-hybrid system for electric vehicles, J. Power Sources 80 (1999) 190–197, https://doi.org/10.1016/S0378-7753(98)00261-4.

[27] S. Dharmalingam, V. Kugarajah, M. Sugumar, Membranes for microbial fuel cells, in: Microb. Electrochem. Technol, Elsevier, 2019, pp. 143–194, https://doi.org/10.1016/b978-0-444-64052-9.00007-8.

[28] G.A. Olah, A. Goeppert, G.K.S. Prakash, Beyond Oil and Gas: The Methanol Economy: Second Edition, Wiley-VCH, 2009, https://doi.org/10.1002/9783527627806.

[29] C.-J. Winter, On Energies-of-Change, the Hydrogen Solution: Policy, Business and Technology Decisions Ahead, Gerling Akademie Verlag, Munich, 2000 (G. International Hydrogen Energy Forum 2000: Policy-Business-Technology 2000).

[30] J.H. Reid, Process of Generating Electricity, 1902.

[31] K.V. Kordesch, G.R. Simader, Environmental impact of fuel cell technology, Chem. Rev. 95 (1995) 191–207, https://doi.org/10.1021/cr00033a007.

[32] K. Strickland, R. Pavlicek, E. Miner, Q. Jia, I. Zoller, S. Ghoshal, W. Liang, S. Mukerjee, Anion resistant oxygen reduction electrocatalyst in phosphoric acid fuel cell, ACS Catal. 8 (2018) 3833–3843, https://doi.org/10.1021/acscatal.8b00390.

[33] K.M. Skupov, I.I. Ponomarev, D.Y. Razorenov, V.G. Zhigalina, O.M. Zhigalina, I.I. Ponomarev, Y.A. Volkova, Y.M. Volfkovich, V.E. Sosenkin, Carbon nanofiber paper cathode modification for higher performance of phosphoric acid fuel cells on polybenzimidazole membrane, Russ. J. Electrochem. 53 (2017) 728–733, https://doi.org/10.1134/S1023193517070114.

[34] H.S. Wang, C.P. Chang, Y.J. Huang, Y.C. Su, F.G. Tseng, A high-yield and ultra-low-temperature methanol reformer integratable with phosphoric acid fuel cell (PAFC), Energy 133 (2017) 1142–1152, https://doi.org/10.1016/j.energy.2017.05.140.

[35] A. Öztürk, R.G. Akay, S. Erkan, A.B. Yurtcan, Introduction to fuel cells, in: Direct Liq. Fuel Cells, Elsevier, 2021, pp. 1–47, https://doi.org/10.1016/b978-0-12-818624-4.00001-7.

[36] Y. Wang, B. Seo, B. Wang, N. Zamel, K. Jiao, X.C. Adroher, Fundamentals, materials, and machine learning of polymer electrolyte membrane fuel cell technology, Energy AI. 1 (2020) 100014, https://doi.org/10.1016/j.egyai.2020.100014.

[37] N.F. Thomas, R. Jain, N. Sharma, S. Jaichandar, Advancements in automotive applications of fuel cells—a comprehensive review, in: Lect. Notes Mech. Eng, Springer, 2020, pp. 51–64, https://doi.org/10.1007/978-981-15-3631-1_6.

[38] C. Lamy, J.-M. Léger, S. Srinivasan, Direct methanol fuel cells: from a twentieth century electrochemist's dream to a twenty-first century emerging technology, in: Mod. Asp. Electrochem, Kluwer Academic Publishers, 2005, pp. 53–118, https://doi.org/10.1007/0-306-46923-5_3.

[39] M.V. Lototskyy, I. Tolj, L. Pickering, C. Sita, F. Barbir, V. Yartys, The use of metal hydrides in fuel cell applications, Prog. Nat. Sci. Mater. Int. 27 (2017) 3–20, https://doi.org/10.1016/j.pnsc.2017.01.008.

[40] S. Trentadue, D. de Mesquita Sousa, M. Santarelli, L.F.M. Mendes, D. de Mesquita Sousa, R.P. da Costa Neto, Solid Oxide Fuel Cell Integration in Combined Heat and Power Systems: Analysis of the Business Case in the Commercial Sector, Energy and Management Engineering Examination Committee, Universitat Politècnica de Catalunya, 2019. https://upcommons.upc.edu/handle/2117/332158. (Accessed 30 December 2020).

[41] X. Guo, H. Zhang, Z. Hu, S. Hou, M. Ni, T. Liao, Energetic, exergetic and ecological evaluations of a hybrid system based on a phosphoric acid fuel cell and an organic Rankine cycle, Energy (2020), https://doi.org/10.1016/j.energy.2020.119365, 119365.

402 Chapter 21 Polymer-based nanocatalyts for alcohol fuel cells

[42] W. Chen, C. Xu, H. Wu, Y. Bai, Z. Li, B. Zhang, Energy and exergy analyses of a novel hybrid system consisting of a phosphoric acid fuel cell and a triple-effect compression–absorption refrigerator with [mmim]DMP/CH3OH as working fluid, Energy 195 (2020) 116951, https://doi.org/10.1016/j.energy.2020.116951.

[43] T. Brenscheidt, K. Janowitz, H.J. Salge, H. Wendt, F. Brammer, Performance of onsi PC25 PAFC cogeneration plant, Int. J. Hydrog. Energy 23 (1998) 53–56, https://doi.org/10.1016/s0360-3199(97)00029-3.

[44] D.S. Chan, C.C. Wan, Influence of PTFE dispersion in the catalyst layer of porous gas-diffusion electrodes for phosphoric acid fuel cells, J. Power Sources 50 (1994) 163–176, https://doi.org/10.1016/0378-7753(94)01897-9.

[45] M. Watanabe, T. Mizukami, K. Tsurumi, T. Nakamura, P. Stoneharf, Activity and stability of ordered and disordered Co-Pt alloys for phosphoric acid fuel cells, J. Electrochem. Soc. (1994), https://doi.org/10.1149/1.2059162.

[46] J.M. Young, A. Mondal, T.A. Barckholtz, G. Kiss, L. Koziol, A.Z. Panagiotopoulos, Predicting chemical reaction equilibria in molten carbonate fuel cells via molecular simulations, AICHE J. (2020), https://doi.org/10.1002/aic.16988.

[47] E. Audasso, B. Bosio, D. Bove, E. Arato, T. Barckholtz, G. Kiss, J. Rosen, H. Elsen, R. Blanco Gutierrez, L. Han, T. Geary, C. Willman, A. Hilmi, C.Y. Yuh, H. Ghezel-Ayagh, The effects of gas diffusion in molten carbonate fuel cells working as carbon capture devices, J. Electrochem. Soc. 167 (2020) 114515, https://doi.org/10.1149/1945-7111/aba8b6.

[48] A. Lysik, T. Wejrzanowski, K. Cwieka, J. Skibinski, J. Milewski, F.M.B. Marques, T. Norby, W. Xing, Silver coated cathode for molten carbonate fuel cells, Int. J. Hydrog. Energy 45 (2020) 19847–19857, https://doi.org/10.1016/j.ijhydene.2020.05.112.

[49] B. Zohuri, B. Zohuri, Hydrogen energy technology, renewable source of energy, in: Hybrid Energy Syst, Springer International Publishing, 2018, pp. 135–179, https://doi.org/10.1007/978-3-319-70721-1_5.

[50] S.C. Price, X. Ren, A.C. Jackson, Y. Ye, Y.A. Elabd, F.L. Beyer, Bicontinuous alkaline fuel cell membranes from strongly self-segregating block copolymers, Macromolecules 46 (2013) 7332–7340, https://doi.org/10.1021/ma400995n.

[51] S.J. Lue, W.H. Pan, C.M. Chang, Y.L. Liu, High-performance direct methanol alkaline fuel cells using potassium hydroxide-impregnated polyvinyl alcohol/carbon nano-tube electrolytes, J. Power Sources 202 (2012) 1–10, https://doi.org/10.1016/j.jpowsour.2011.10.091.

[52] Z. Zhang, C. Zuo, Z. Liu, Y. Yu, Y. Zuo, Y. Song, All-solid-state Al-air batteries with polymer alkaline gel electrolyte, J. Power Sources 251 (2014) 470–475, https://doi.org/10.1016/j.jpowsour.2013.11.020.

[53] M.S. Naughton, F.R. Brushett, P.J.A. Kenis, Carbonate resilience of flowing electrolyte-based alkaline fuel cells, J. Power Sources 196 (2011) 1762–1768, https://doi.org/10.1016/j.jpowsour.2010.09.114.

[54] E.H. Yu, X. Wang, U. Krewer, L. Li, K. Scott, Direct oxidation alkaline fuelcells: from materials to systems, Energy Environ. Sci. 5 (2012) 5668–5680, https://doi.org/10.1039/C2EE02552C.

[55] S. Hanif, N. Iqbal, X. Shi, T. Noor, G. Ali, A.M. Kannan, NiCo–N-doped carbon nanotubes based cathode catalyst for alkaline membrane fuel cell, Renew. Energy 154 (2020) 508–516, https://doi.org/10.1016/j.renene.2020.03.060.

[56] L. Wang, Y. Han, X. Feng, J. Zhou, P. Qi, B. Wang, Metal-organic frameworks for energy storage: batteries and supercapacitors, Coord. Chem. Rev. 307 (2016) 361–381, https://doi.org/10.1016/j.ccr.2015.09.002.

[57] S. Singh Parihar, R.D. Roseman, R.Y. Lin, High Temperature Seals for Solid Oxide Fuel Cells, 2007.

[58] S.C. Singhal, Advances in solid oxide fuel cell technology, Solid State Ionics 135 (2000) 305–313, https://doi.org/10.1016/S0167-2738(00)00452-5.

[59] K.H.M. Al-Hamed, I. Dincer, A novel integrated solid-oxide fuel cell powering system for clean rail applications, Energy Convers. Manag. 205 (2020) 112327, https://doi.org/10.1016/j.enconman.2019.112327.

[60] S. Dwivedi, Solid oxide fuel cell: materials for anode, cathode and electrolyte, Int. J. Hydrog. Energy 45 (2020) 23988–24013, https://doi.org/10.1016/j.ijhydene.2019.11.234.

[61] L. Gubler, G.G. Scherer, Trends for fuel cell membrane development, Desalination 250 (2010) 1034–1037, https://doi.org/10.1016/j.desal.2009.09.101.

[62] S.J. Skinner, Recent advances in perovskite-type materials for solid oxide fuel cell cathodes, Int. J. Inorg. Mater. 3 (2001) 113–121, https://doi.org/10.1016/S1466-6049(01)00004-6.

[63] C. Buzea, I.I. Pacheco, K. Robbie, Nanomaterials and nanoparticles: sources and toxicity, Biointerphases 2 (2007) MR17–MR71, https://doi.org/10.1116/1.2815690.

[64] B.C.H. Steele, A. Heinzel, Materials for fuel-cell technologies, Nature 414 (2001) 345–352, https://doi.org/10.1038/35104620.

[65] N.Q. Minh, Solid oxide fuel cell technology—features and applications, Solid State Ionics 174 (2004) 271–277, https://doi.org/10.1016/j.ssi.2004.07.042.

[66] J. Marcinkoski, J.P. Kopasz, T.G. Benjamin, Progress in the US DOE fuel cell subprogram efforts in polymer electrolyte fuel cells, Int. J. Hydrog. Energy 33 (2008) 3894–3902, https://doi.org/10.1016/j.ijhydene.2007.12.068.

[67] K.B. Prater, Polymer electrolyte fuel cells: a review of recent developments, J. Power Sources 51 (1994) 129–144, https://doi.org/10.1016/0378-7753(94)01934-7.

[68] A. Ibrahim, O. Hossain, J. Chaggar, R. Steinberger-Wilckens, A. El-Kharouf, GO-nafion composite membrane development for enabling intermediate temperature operation of polymer electrolyte fuel cell, Int. J. Hydrog. Energy 45 (2020) 5526–5534, https://doi.org/10.1016/j.ijhydene.2019.05.210.

[69] I. Bae, B. Kim, D.Y. Kim, H. Kim, K.H. Oh, In-plane 2-D patterning of microporous layer by inkjet printing for water management of polymer electrolyte fuel cell, Renew. Energy 146 (2020) 960–967, https://doi.org/10.1016/j.renene.2019.07.003.

[70] Y. Wang, D.F. Ruiz Diaz, K.S. Chen, Z. Wang, X.C. Adroher, Materials, technological status, and fundamentals of PEM fuel cells—a review, Mater. Today 32 (2020) 178–203, https://doi.org/10.1016/j.mattod.2019.06.005.

[71] S.J. Hamrock, M.A. Yandrasits, Proton exchange membranes for fuel cell applications, J. Macromol. Sci. C Polym. Rev. 46 (2006) 219–244, https://doi.org/10.1080/15583720600796474.

[72] M. Shao, Electrocatalysis in fuel cells, Catalysts 5 (2015) 2115–2121, https://doi.org/10.3390/catal5042115.

[73] F. Vigier, S. Rousseau, C. Coutanceau, J.M. Leger, C. Lamy, Electrocatalysis for the direct alcohol fuel cell, in: Top. Catal, Springer, 2006, pp. 111–121, https://doi.org/10.1007/s11244-006-0113-7.

[74] Y. Wang, Nanomaterials for Direct Alcohol Fuel Cell, Pan Stanford Publishing, Singapore, 2016, https://doi.org/10.1201/9781315364902.

[75] R. Borup, J. Meyers, B. Pivovar, Y.S. Kim, R. Mukundan, N. Garland, D. Myers, M. Wilson, F. Garzon, D. Wood, P. Zelenay, K. More, K. Stroh, T. Zawodzinski, J. Boncella, J.E. McGrath, M. Inaba, K. Miyatake, M. Hori, K. Ota, Z. Ogumi, S. Miyata, A. Nishikata, Z. Siroma, Y. Uchimoto, K. Yasuda, K.I. Kimijima, N. Iwashita, Scientific aspects of polymer electrolyte fuel cell durability and degradation, Chem. Rev. 107 (2007) 3904–3951, https://doi.org/10.1021/cr050182l.

404 Chapter 21 Polymer-based nanocatalyts for alcohol fuel cells

[76] J. Xie, D.L. Wood, D.M. Wayne, T.A. Zawodzinski, P. Atanassov, R.L. Borup, Durability of PEFCs at high humidity conditions, J. Electrochem. Soc. 152 (2005) A104, https://doi.org/10.1149/1.1830355.

[77] L. Yang, J. Shui, L. Du, Y. Shao, J. Liu, L. Dai, Z. Hu, Carbon-based metal-free ORR electrocatalysts for fuel cells: past, present, and future, Adv. Mater. 31 (2019), https://doi.org/10.1002/adma.201804799.

[78] J. Garche, C.K. Dyer, Encyclopedia of Electrochemical Power Sources, Academic Press, 2009.

[79] E. Antolini, Carbon supports for low-temperature fuel cell catalysts, Appl. Catal. B Environ. (2009), https://doi.org/10.1016/j.apcatb.2008.09.030.

[80] I. Nitta, O. Himanen, M. Mikkola, Thermal conductivity and contact resistance of compressed gas diffusion layer of PEM fuel cell, Fuel Cells 8 (2008) 111–119, https://doi.org/10.1002/fuce.200700054.

[81] A.L. Dicks, Molten carbonate fuel cells, Curr. Opin. Solid State Mater. Sci. 8 (2004) 379–383, https://doi.org/10.1016/j.cossms.2004.12.005.

[82] M. Bischoff, Molten carbonate fuel cells: a high temperature fuel cell on the edge to commercialization, J. Power Sources 160 (2006) 842–845, https://doi.org/10.1016/j.jpowsour.2006.04.118.

[83] G. Maggio, S. Freni, S. Cavallaro, Light alcohols/methane fuelled molten carbonate fuel cells: a comparative study, J. Power Sources 74 (1998) 17–23, https://doi.org/10.1016/S0378-7753(98)00003-2.

[84] G. Wilemski, Simple porous electrode models for molten carbonate fuel cells, J. Electrochem. Soc. 130 (1983) 117–121, https://doi.org/10.1149/1.2119635.

22

Different synthesis methods of nanomaterials for direct alcohol fuel cells

Vildan Erduran[a,b], Muhammed Bekmezci[a,b], Iskender Isik[b], and Fatih Şen[a]

[a]Şen Research Group, Department of Biochemistry, Dumlupinar University, Kütahya, Turkey. [b]Department of Materials Science and Engineering, Faculty of Engineering, Dumlupinar University, Kütahya, Turkey

1 Introduction

Energy, generally defined as the most basic resource necessary for the production, is the most important factor in many activities in the world. The tremendous growth in population throughout the world, especially in the 20th century, has increased the need for energy to fuel industrialization, urbanization, trade, and production, so energy has become the most important issue for developed and developing countries alike. Today, energy needs are met mainly from fossil fuels. The major disadvantages of fossil fuels are that they are harmful to the environment. One of the main causes of global warming, which is one of the most important problems threatening our world, is toxic gases caused by the use of fossil fuels [1,2]. Due to these negative features, it is very important to minimize the demand for fossil fuels. For this purpose, the use of alternative energy sources such as wind energy [3], solar energy [4], and fuel cells [5] instead of fossil fuels is being explored by scientists. One of the alternative energy sources studied is fuel cells. Fuel cells are prominent in this effort because they are efficient, economical, silent, and environmentally friendly.

Fuel cells can convert chemical energy directly into electrical energy and contribute to creating renewable energy. The simplest way to explain the principle of the operation of fuel cells is to say that it is the opposite of electrolysis. By applying the reverse of the electrolysis method, hydrogen and oxygen gases are combined and water and electric current are obtained [6]. The fuel cell

Nanomaterials for Direct Alcohol Fuel Cells. https://doi.org/10.1016/B978-0-12-821713-9.00026-3
Copyright © 2021 Elsevier Inc. All rights reserved.

was discovered by the Swiss chemist Christian F. Schoenbien in 1839 and was put into practical use by the English scientist William Grove in 1845.

Research and application studies have shown that a wide variety of fuels can be used in fuel cells. However, when choosing fuel, the price, energy content, environmental effects, storage, transportation, and ease of use should be taken into consideration. Fuel supply to fuel cells is carried out either directly or indirectly. Alcohols, natural gas, liquid petroleum gas (LPG), and hydrogen can be used as fuel. Regardless of what fuel is used, it is essential to obtain hydrogen-rich gas from fuels. Alcohols such as methanol, ethanol, ethylene glycol, glycerol, and propanol have become suitable sources for fuel cells due to their high energy density, high solubility in aqueous electrolytes, safe storage and ease of use, and high reaction ability. The main problem with alcohols used in fuel cells is poor reaction kinetics, which requires the use of catalysts.

Generally, catalysts are substances that increase the reaction rate by decreasing the activation energy of a chemical reaction and do not change its structure afterward. A catalyst needs to have long-lasting and stable activity against a reaction, and should also be resistant to poisoning and deterioration. Preparation methods of catalysts can be divided into two main categories: chemical and electrochemical. Chemical synthesis of the catalysts is carried out by methods such as precipitation, impregnation, and thermal cracking. The electrochemical preparation method is based on electron exchange.

Synthesizing the catalysts to be used in fuel cells to solve the problem of activity of the alcohols at nanometer size is a very important field of study. Solving the problem of the activity of alcohols with the nanocatalysts to be prepared will cause the spread of fuel cells and increase the amount of clean energy obtained.

1.1 History of fuel cells

The first fuel cell was developed in 1839 by Schönbein. But it was Grove who brought the fuel cell into the implementation stage. That is why Grove is considered the inventor of the fuel cell [7]. During his studies on the H_2O_2 battery, Grove realized that constant current and power were produced as a result of the reverse reaction of the electrolysis of the water, thus making a great invention by pure chance [8]. In the following years, he succeeded in generating more electric current by connecting fifty of the system is used in its previous work in series [9].

Mond and Langer are the first to develop Grove's cell. Unlike Grove's work, they designed porous, three-dimensional (3D) electrodes used in the fuel cell. Thus, they designed the general components of the new generation of fuel cells. Mond and Langer proved the opposite of Grove's theory (that only pure hydrogen could be used as fuel in this system) by using coal instead of hydrogen in fuel cells [10]. In the late 19th and early 20th centuries, William W. Jacques and Emil Baur did important research on fuel cells. Baur created the first molten carbonate fuel cell in 1921 and implemented studies on solid oxide electrolytes at high temperatures in the 1930s. Jacques has created high-power systems with 1.5- and 30-kW fuel cell components. The operating principle of a fuel cell is shown in Fig. 1 [11].

In 1933, Francis Thomas Bacon developed the first fuel cell, consisting of hydrogen and oxygen, for practical use. It managed to convert air and hydrogen directly into electrical energy through electrochemical processes in the fuel cell. During World War II, he developed a fuel cell for use in the Royal Navy's submarines [12]. The use of a material called polytetrafluoroethylene (popularly known as Teflon) in 1950 and Nafion as fuel in 1955 contributed significantly to the development of fuel cells. In 1959, a team led by Harry Ihrig designed a 15-kW fuel cell, which is still used on a tractor on display at the Smithsonian Institute in Washington, DC [13]. As a result of the studies carried out in laboratories at the National Aeronautics and Space Administration (NASA) in 1990, a direct alcohol fuel cell (DAFC) was developed using methanol as fuel [14].

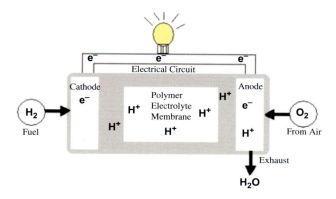

Fig. 1 The operating principle of a fuel cell [11].

1.2 Usage areas of fuel cells

The usage applications of fuel cells, which have a very wide variety, are generally divided into two areas: portable and fixed. While the main purpose of portable applications is to produce solutions for transportation systems, fixed applications mostly include solutions for residential and commercial needs [15].

The fact that fuel cells work silently and create energy more effectively without causing environmental pollution increases the interest in vehicles that can employ this technology. Because cars running on gasoline or diesel cause air pollution, vehicles powered by fuel cells have attracted attention. Fuel cell vehicles are expected to replace diesel or gasoline-powered vehicles over time. Many car manufacturers in the world are examining fuel cell systems and developing prototypes. Fuel cells can be used in large vehicles such as tractors and trucks, as well as in cars and buses. In this way, the cost of petroleum-based fuels will be saved and emissions that cause environmental pollution will be prevented.

One of the interesting research and application areas of fuel cells is spacecraft. The electrical energy required for spacecraft is produced by solar panels, batteries, or a system known as *radio-isotope thermal generator (RTG)*. But solar systems are expensive and bulky. The fact that the batteries are heavy and have a short life caused them to be found unsuitable for space exploration. Nuclear-powered fuel cells were preferred to meet the electrical needs of spacecraft.

The biggest advantage of using fuel cells as a power source in space flights is their low weight. For this reason, the electrical energy of the *Gemini* and *Apollo* space shuttles was supplied with solid polymer and alkaline fuel cells (Fig. 2). Currently, the energy of the space shuttles is supplied by 12-kW fuel cells [15].

Large-scale energy production is not the only potential use for fuel cells. There are studies looking at replacing the batteries used in mobile phones, computers, and cameras with fuel cells. Microprocessor fuel cells can make a big contribution to computer chips as well. In addition, promising studies are being carried out on the topic of using very small fuel cells in hearing aids and pacemakers [15].

1.3 The energy of the new century: Hydrogen

Hydrogen, which is a colorless, odorless, tasteless, and transparent element, is the lightest in nature. It is the perfect complement to electrons in future clean energy systems. Because it is very active chemically, it is very difficult to find it as an element in pure nature.

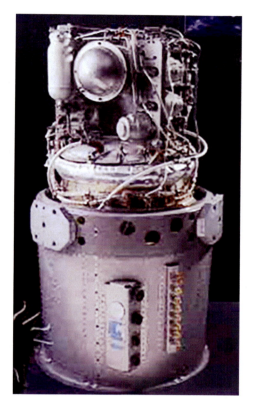

Fig. 2 The fuel cell used in the *Gemini* spacecraft [15].

Hydrogen is not natural fuel; rather, it is a synthetic fuel that can be produced from various raw materials such as water, fossil fuels, and biomass by utilizing primary energy sources. After hydrogen is produced, it can be easily transported via pipelines or tankers. The absence of obstacles in the transportation of hydrogen is one of the main factors that will make the use of hydrogen widespread.

Hydrogen has a number of very interesting properties. Although it is the lightest element, it has the highest energy content among all known fuels, with a value of 120.7 kJ/kg per unit. When hydrogen is liquefied at −252.77°C, its volume changes tremendously. The volume of liquid hydrogen is only 1/700 of its gaseous volume. The thermal value of the unit mass of liquid hydrogen is 141.9 MJ/kg, 3.2 times higher than petroleum. The thermal value of the unit mass of the gas hydrogen is the same as liquid hydrogen, and 2.8 times that of natural gas.

Compared to fossil fuels, hydrogen has many advantages. Fossil fuels are only suitable for burning with fire, while hydrogen is a fuel suitable for catalytic combustion, direct steam production, chemical conversion by hybridization, and fuel cell and electricity production.

Its flame combustion feature makes hydrogen suitable for use as fuel in internal combustion engines, gas turbines, and furnaces. The direct conversion of hydrogen into steam makes its application in steam turbines easier. The catalytic combustion feature of hydrogen can be used in kitchen stoves, water heaters, and stoves.

Hydrogen has a more efficient and environmentally friendly combustion feature compared to other fuels. Because there is no carbon, nitrogen, or sulfur in the structure of hydrogen, substances such as CO, CO_2, NO_x, SO_x are not released when it is burned.

For the combustion of hydrogen, less air is needed than for other types of fuel. This is an important advantage that enables its reaction with a lower fuel/air mixture.

It can be used in hydrogen, liquid, or gas form. The path to be chosen for storage depends on the use of hydrogen and the expected performance of hydrogen. Today, hydrogen can be stored in a liquid and gaseous state under high pressure [16]. Examining the fuel properties of hydrogen shows that if hydrogen is used as a fuel in engines, it has many advantages over petroleum-based fuels. The water vapor released as a combustion product does no harm to the environment and has high flame velocity and ignition ability and a high-octane number. Another important feature of hydrogen is that it can be used as fuel in various mixing ratios with gasoline and diesel. This feature will make it possible to use hydrogen without significant changes to existing engines during the transition period [17,18].

The most suitable use for hydrogen, which is considered as the energy source of the future, is in hydrogen fuel cells. The main problem encountered in this area is the lack of hydrogen in its pure form in nature. Studies to solve this problem have led to the search for an alternative fuel to pure hydrogen. Among these sources, alcohols such as methanol, ethanol, propanetriol especially stand out due to their high energy density, easy production, and cheapness [19,20].

2 Direct alcohol fuel cells

A *fuel cell* can be defined as a system that converts the energy stored in fuel into electricity via a chemical reaction. Although many products are used in fuel cells, the most common is hydrogen gas. However, there are some obstacles to using hydrogen as a fuel, including the difficulties connected with storage and transportation. In addition, as stated previously, hydrogen does not exist in its pure form in nature. For all these reasons, the search for alternative fuels directed researchers to alcohols such as methanol, ethanol, ethylene glycol. The advantages of these alcohols are as follows:

- Because they are easy to produce, they are very cheap compared to their alternatives.
- They have a very high solubility ratio in aqueous electrolytes.
- They do not cause environmental pollution.
- They are very easy to store and transport.
- They have a higher power density than hydrogen gas.

In DAFCs, the polymer electrolyte membrane operates with fuel diluted alcohol solutions. When the working principle of alcohol fuel cells is directly examined, the first step is to add an aqueous alcohol solution to the anode side. Carbon dioxide (CO_2) is formed as a result of the oxidation of this solution. In the next stage, electrons are carried to the cathode through an external circuit. At the cathode, oxygen reacts with protons and combines with electrons from the external circuit to form water. Because DAFCs have high application potential compared to the available clean energy technologies, they have been the subject of an increasing number of studies in recent years. Alcohols such as methanol, ethanol, propanol, and ethylene glycol are generally used in these fuel cells [21–23].

2.1 Direct methanol fuel cells

Direct methanol fuel cells were first developed in 1965. The principle of operation is similar to that of polymer electrolyte membrane fuel cells (PEMFCs). For this reason, direct methanol fuel cells have been categorized by some sources as a type of PEMFC. The main difference between direct methanol fuel cells and PEMFCs, though, is that methanol can be used directly as a fuel without requiring any converters. Methanol, which is used as fuel, can be mixed with water and given directly to the system. Mixing methanol with water is an important way to moisten the membrane.

Direct methanol fuel cells consist of four main components: the anode, cathode, liquid electrolytes, and a membrane with selective permeability. The most critical process of direct methanol fuel cell that is studied is the choice of a membrane with a catalyst and selective permeability.

There are many reasons why methanol is used as fuel in fuel cells, including easy storage, high energy density, and conversion speed [24]. The working principle of methanol fuel cells, which are very simple energy systems, is based on the oxidation of methanol on a catalyst layer to form carbon dioxide. Hydrogen ions, electrons, and CO_2 are formed as a result of the breakdown of methanol by an electrochemical reaction at the anode. As hydrogen ions pass from the proton-selective membrane to the cathode, the electrons pass from the outer circuit to the cathode and start the electricity generation process. Hydrogen ions coming to the cathode provide the formation of oxygen and water [25].

Direct methanol fuel cells stand out as a highly promising system for the future. Their main advantages are listed as follows [26,27]:

- They can be easily and cheaply produced.
- They are easy to move and store.
- Available distribution channels can be used for fuel distribution.
- It does not cause environmental pollution.
- They operate very quietly because they do not contain any moving parts.
- They are able to produce high-quality electrical energy required by high technology systems.
- They can work with larger-molecule alcohols, such as ethanol or propanol.
- The fuel can be used directly.
- There is no need for a fuel treatment system.
- It is possible to use these cells not only in high energy systems, but also in small devices such as computers, mobile phones, and cameras.
- If the use of electronic devices becomes widespread, the charging problem will disappear [28].

2.2 Direct ethanol fuel cells

The working principle behind the direct ethanol fuel cell, which has become widespread recently, is generally similar to that of the direct methanol fuel cell. However, in the ethanol fuel cell, ethanol is used as fuel instead of hydrogen gas.

Ethanol has important advantages over methanol. First, it has a higher energy density and lower toxicity than methanol. Ethanol, which is rich in hydrogen, is a less flammable liquid than pure hydrogen. It is easier and more economical to produce ethanol. In addition, it is very environmentally friendly and less dangerous to use than methanol. Ethanol, which is in liquid form under atmospheric pressure, is very easy to store, distribute, and transport. In these aspects, ethanol stands out as a promising liquid fuel that can be used instead of hydrogen, natural gas, or other alcohols in fuel cells. Despite all the advantages mentioned here, ethanol has a significant disadvantage. This disadvantage is that the structure of ethanol is more complex than methanol. Because ethanol has a more complex structure than methanol, many intermediate products are formed during the reaction. In studies with ethanol, CO poisoning is more likely. In addition, in the reactions using ethanol, very strong carbon-carbon bonds should be broken. This is not the case because methanol has only one carbon element in its structure. Therefore, much more energy is needed during the conversion of ethanol [29–32].

It is also different in intermediates observed in methanol and ethanol fuel cells. While the first intermediates formed when using methanol are formaldehyde and formate types, acetaldehyde and acetate are obtained as intermediates when ethanol is used [33].

The working principle observed directly in the ethanol fuel cell can be summarized as follows: An ethanol-water mixture added directly to the anode side turns into proton, electron, and CO_2 as a result of an electrochemical reaction. The protons formed to pass through the polymer membrane and reach the cathode side. The CO_2 formed by the oxidation of ethanol cannot pass through the membrane and leaves the fuel cell by the anode. The electrons formed as a result of the reaction pass to the cathode side. During this transition, electricity production occurs. Hydrogen ions passing to the cathode side combine with the electrons coming to the cathode and the oxygen added to the cell to form water. As a result of the reaction at the anode, 12 mol of electrons are released for each mole of ethanol. This causes the energy density obtained by the use of ethanol to be greater than that of methanol or hydrogen [34].

2.3 Direct ethylene glycol fuel cells

The direct ethylene glycol fuel cell generally operates on the same principle as the direct methanol fuel cell, but instead of hydrogen gas, ethylene glycol is used as fuel directly in the anode of the ethylene glycol fuel cell. It is directly considered an important energy carrier for alcohol fuel cells. Ethylene glycol has many advantages. First, ethylene glycol originates from biomass. Ethylene glycol, which has a high energy density, is less toxic and more environmentally friendly than methanol. Although the energy density of ethylene glycol is lower than that of ethanol, it is close to that of methanol. These properties of ethylene glycol have been directly examined in the category of alcohols used in alcohol fuel cells [35].

Different intermediate products are seen in fuel cells using ethylene glycol in other fuel cells. These intermediate products can be listed as glycolaldehyde, glycolic acid, and oxalic acid [35,36].

2.4 Direct propanol fuel cells

The most studied alcohol for the purpose of alcohol fuel cells is methyl alcohol. However, other alcohols that have been researched in recent years include alcohols such as 1-propanol and 2-propanol. Studies have found that some second-degree alcohols, such as 2-propanol, can outperform methanol. The most important advantage of 2-propanol over methanol is that it has faster reaction kinetics. In addition, it is less toxic and more environmentally friendly than methanol. The main oxidation product of 2-propanol is acetone at low potential. Wang et al. have conducted studies on fuel cells using 2-propanol as fuel and reported high yield rates. In the study, methanol, ethanol, 1-propanol, and 2-propanol were used as fuel, and their performances were compared. According to the findings, 2-propanol showed the highest cell performance among the tested alcohols [37]. In a study with alcohol fuel cells, Qi et al. found that 2-propanol in the low-current-density region may actually be a better fuel than methanol [38]. They also examined in detail the advantages and disadvantages of using 2-propanol as fuel in the fuel cell [39]. In another study with 2-propanol, the performances of 2-propanol diluted with anhydrous 2-propanol were compared. It was determined that anhydrous 2-propanol showed higher performance in direct oxidation fuel cells than did dilute 2-propanol solutions. In this study, the performance achieved by the anhydrous 2-propanol at 60°C cell temperature was the best reported for liquid-feed direct fuel cells to date [14]. Bergens et al. also directly studied the performance of 2-propanol fuel cells. In these studies, they decided that 2-propanol is a promising fuel that can be used directly in alcohol fuel cells, in line with the results of Qi et al. studies with 2-propanol showed a lower level of cathode poisoning than cathode poisoning in methanol [40]. Bergens et al. found that in the reactions using 2-propanol, different intermediates were formed depending on the potential. According to this study, acetone as an intermediate product at low potential and intermediate products that prevent oxidation at high potential are formed [41].

3 Studies related to nanocatalyst synthesis

Catalysts can be defined as substances that reduce activation energy and accelerate the reaction. The catalysts are classed into two groups: heterogeneous and homogeneous. Since the heterogeneous catalyst does not dissolve in solution, the catalysis occurs in a phase separate from the solution. A homogeneous catalyst can be easily dissolved in a solution. Therefore, the catalysis process is followed by spectroscopic and chromatographic techniques.

An important part of the work done to synthesize high performance, environmentally friendly, and cheap catalysts is carried out in nanoscale. Nanocatalysts, which constitute an important group of nanosized materials, have been a topic of interest for researchers in recent years due to their interesting properties. In studies conducted with metal nanocatalysts, it was found that the surface area-to-volume ratio increased with the reduction of particle size. This means that the surface area per unit volume reaches a very high rate. Increased surface area causes both the physical and chemical properties of the metal nanocatalyst to change. These changes are considered a critical step in achieving the goals of the studies on metal nanocatalysts [42].

The most important part of metal nanocatalyst synthesis is providing particle size control and distribution. However, it has been observed in studies with metal nanocatalysts that even in environments where the best stabilizers are used, metal can agglomerate and become metal ingot [43]. This metal agglomeration results in reduced catalytic activity and catalyst life. In Fig. 3, transmission electron microscopy (TEM) images of freshly prepared rhodium nanoparticles stabilized with laurate and after

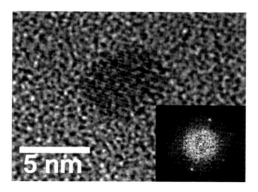

Fig. 3 TEM image of rhodium nanoparticles [15].

hydrolysis for the fifth time in the hydrolysis of ammonia-borane are shown. As can be seen from these images, rhodium nanoscopes clumped after being used for the fifth time in the hydrolysis reaction of ammonia-bora and lost their initial catalytic activity to a large extent [15].

Spaced macromolecules such as zeolites, carbon-based materials, metal oxides, polymers, minerals, and metal-organic frameworks (MOFs) have been used to address this agglomeration problem, which reduces the catalytic activity of metal nanocatalysts [44,45]. Today, advances in the synthesis of carbon nanomaterials are facilitating the synthesis of more stable and active catalysts. Carbon materials are particularly promising materials to increase catalytic activity in the fuel cell. Carbon nanomaterials have many advantages, including being stable in both acidic and basic environments and providing a high surface area to dissolve metal catalysts [46].

Carbon black (a generic name for the family of carbon materials) is one of the catalyst support materials used in fuel cells. Its biggest advantage over other nanostructures is that its cost is very low. Nanostructures such as carbon nanotubes (CNTs) and graphene are much more expensive than carbon black. The main disadvantage of carbon black is its poor corrosion resistance [47,48].

The advantages of using graphene as a catalyst support material in fuel cells include its noteworthy electrical conductivity, mechanical strength and thermal conductivity. Graphene has a surface area that can be considered perfect for a chemical reaction (theoretically >2600 m^2/g) [47].

CNTs are considered promising support materials in fuel cells due to their electrical and structural properties. Their most interesting feature is their tubular structure. High surface area, excellent electronic conductivity, and high chemical stability are other important features of CNTs. Studies have shown that these materials have a positive effect on many catalysts, which causes carbon black to provide higher catalytic activity and stability [49].

Today, a number of different catalyst production methods, including impregnation, colloidal, sol-gel, hydrothermal, microemulsion, coprecipitation, polyol, etc., are used. An important element of these methods involves the production of nanosize catalysts [50].

3.1 Impregnation

The first step in the impregnation method is to dissolve the catalyst salts in a suitable solvent such as water or alcohol. Then the macromolecule, which is intended to be used as a support, is added into the solution and the catalyst particles are absorbed onto the molecule surface. This is the reason why this method is called *impregnation*. In the impregnation method, the high surface area support material is mixed with a solution containing catalyst precursor, and then the solvent is removed from the impregnated catalyst. Impregnation is a fast and economical method, but its main disadvantage is the difficulties encountered during the uniform distribution of catalyst components [51].

Since the impregnation method is easy to use, it is the most common method of preparing a DAFC catalyst. Hexachloroplatinic acid, with the formula H_2PtCl_6, and ruthenium chloride [52–58], with the formula $RuCl_3$, are used as the catalyst salt. Ethyl alcohol [52,53] was used in some studies to dissolve these catalyst salts, and other studies used water [54–58]. In these studies, the catalyst solutions were precipitated by means of a reducing agent. Different reducing materials were used in each study; Wang and colleagues and Zhou and colleagues used formaldehyde (HCHO) as a reductant, Choi and colleagues and Yang and colleagues used hydrogen, and Park et al. and Choi et al. used $NaBH_4$.

In the first stage of the study carried out by Wang et al. on HCHO, the catalyst salts at 80°C were dissolved in a suitable solvent medium and mixed for a certain time in a mixture of carbon black water and isopropyl alcohol. In the next step, pH was adjusted to 8 with the help of NaOH, and then HCHO was added slowly to the solution. Pure water has been used several times to remove chlorine ions in the solution. The catalyst obtained after washing with pure water was dried at 80°C for 4 h.

Choi et al. used HCHO as a reducing agent as well. In the first step, they missed a sufficient amount of carbon black and chloroplatinic acid solution together. Later, 37% HCHO was added. The pH of the solution was brought to 14 with the help of 0.5 M NaOH. After impregnation, the mixture was taken to the oil bath and heated to 363 K. In the final stage, it was added to a solution containing compounds such as $NaNO_3$, NaCl, and HCl. The chlorine ions in the obtained material were washed with pure water until finished and then the catalysts were dried. Another study conducted by Choi et al. used hydrogen gas as the reducer. After the catalyst salts were dissolved in water, they were mixed with Vulcan XC-72. The dried solution was then treated with 5% H_2/N_2 gas. A similar study was carried out by Yang et al. In yet another study by Choi et al., $NaBH_4$ was used and metal salts were dissolved in water in the first stage. In the next step, 0.2 M $NaBH_4$ was added, and the solid obtained was purified. Park et al. used 0.1 g $NaBH_4$ per 1 mL of water after dissolving metal salts in water [52–58].

3.2 Colloid method

In the first stage of the colloid method, the catalyst salts are dissolved in a suitable solvent. The main purpose of this method is to limit the size of hydrophilic catalysts that will become solid when reduced. For this purpose, molecules called *surfactants* are left in the solution. When the catalyst is reduced, the surfactant with hydrophilic and hydrophobic portions stabilizes the nanoparticles and prevents agglomeration. In this way, it is possible to produce nanocatalysts.

In the colloid method, various catalyst salts are used for platinum (Pt) and ruthenium (Ru). Many studies used chloroplatinic acid as the catalyst salt [59–61], and Dubau et al. used $PtCl_2$ [62]. In all these studies, ruthenium chloride was used as a source of Ru [63].

In a study by Liu et al., chloroplatinic acid was dissolved in ethylene glycol. The solution was stirred in the microwave reactor for 60 s and then mixed with a decantiol-toluene solution. In the new mixture, the process was terminated by allowing Pt to pass to the toluene phase. After the catalyst particles were placed on the support, toluene was blown off and the resulting solid was dried [59].

In another study conducted by Kim et al., hexachloroplatinic acid and ruthenium chloride salts were dissolved in a water-propanol mixture and then mixed with poly(*N*-vinyl-2-pyrrolidone). After stirring at room temperature, the solution was refluxed. In this study, the effect of the poly(*N*-vinyl-2-pyrrolidone) ratio and the catalyst ratio on the particle size of catalysts was investigated [60].

In another study, Xue et al., H_2PtCl_6, $RuCl_3$, and XC-72 carbon black were mixed for 30 min at room temperature in acetone. In the next step, the solution was kept at 80°C for half an hour, and acetone was dried. The resulting Pt-Ru/C catalyst was washed with acetone and dichloromethane and dried [64].

3.3 Microemulsion

Microemulsion is a technique that has become widespread for catalyst synthesis in recent years. It can be defined as a nanosized drop of aqueous liquid containing a noble metal precursor. The first step in this method is to form nanoparticles by water-oil microemulsion reaction. In the next step, the reduction is carried out. The reduction step is accomplished by adding substances such as N_2H_4, HCHO, and $NaBH_4$ to the microemulsion system. As a result, the reaction is kept in nanoscale microemulsion and the size of the particles formed is easily controlled by the microemulsion size. Removal of surfactant molecules can be easily accomplished by heat treatment of high-surface-area carbon support nanoparticles.

This method has many advantages, including metallic composition and ease of controlling particle size. The major disadvantage of this method is the use of expensive surfactant molecules. Also, the many separations and washing steps make the process very difficult to perform [65].

The difference between microemulsion from other methods is that the catalyst is put into a new solvent that will form a second phase after dissolution. As in other methods, the surfactant is used in microemulsion to prevent collapse and agglomeration. As a result, the reaction is carried out in a nanoscale microemulsion. Therefore, the size of the particles obtained can be easily determined by the microemulsion size [65].

In the study of Xu et al., sodium dodecylbenzenesulfonate ($C_{18}H_{29}NaO_3S$, NaDDBS) was used to decrease the surface tension. In this study, where n-butanol was used as an auxiliary surfactant, cyclohexane was used during the oil phase. Chloroplatinic acid and ruthenium chloride were used as a metal salt. Sodium boron hydride and formaldehyde were used as reducers in these studies [66].

3.4 Coprecipitation

In coprecipitation, the metal salt solution and the alkaline solution are mixed at the same time and the precipitation process is carried out. In the next step, the precipitated compounds are filtered and dried. In this method, the nanoscale structure of the catalyst depends upon many factors such as solution concentration, pH, and temperature affect. The coprecipitation method has many advantages over other methods, including the catalyst component distribution being homogeneous, a relatively low reaction temperature, and low cost. Despite these advantages, coprecipitation is not a controlled method in terms of reaction kinetics, which is a considerable disadvantage because the uncontrolled reaction kinetics causes the synthesized catalysts to have uncontrolled particle morphology. This is undesirable in terms of catalyst performance and reusability [51].

3.5 Sol-gel method

The first purpose of the sol-gel method is to convert the materials into a colloidal structure. The next goal is the hydrolysis of the starting materials in solution and then concentrating them as a gel. After condensation, the precipitated compounds are washed and dried, and then crystalline products are obtained. The disadvantage of the sol-gel method is the difficulty and length of the application time. The most important advantage is that high-purity products can be obtained even at low temperatures. In addition, the synthesis of multicomponent materials is possible with this method [67,68].

3.6 Hydrothermal method

The hydrothermal method is based on the synthesis of crystalline materials. For this purpose, the starting materials and solvent are heated in an autoclave or reactor at a constant temperature (generally, higher than 200°C). With this method, the term *hydrothermal* is used when water is the solvent, whereas when an organic solvent is used instead of water, the name is *solvothermal*.

The structure of the crystal material targeted to be synthesized depends on many factors, including the type of starting materials, the ratio of starting materials to solvent, reaction time, temperature, and pressure. The most important advantage of the hydrothermal method is that highly dispersed products can be synthesized. In addition, this method is fairly simple, not needing long processing time or high energy consumption [69,70].

3.7 Polyol method

In the polyol method, polyalcohols such as diethylene glycol, ethylene glycol, triethylene glycol, butanediol, and propanediol are used. This method has emerged as a very simple and economical option. In the first stage of the polyol method, metal salts are dissolved by heating in solution. In the next step, metal particles are grown with homogeneous nucleation. The biggest advantage of this method is that the particle size and shape properties can be well controlled [71].

4 Nanomaterial synthesis

As the catalyst in the anode layer, alcohol is one of the most important factors that directly determine the performance of fuel cells. It is a very important area because fuel cells have the potential of producing more economical and environmentally friendly noble metals that will give higher performance than commonly used Pt and Ru alloys. For this reason, many research and development (R&D) studies are conducted on the alloying and modification of metals. Rapid developments in the field of nanotechnology increase the hopes for reaching the targets of this area of study. The development of a catalyst with the desired properties can directly turn alcohol fuel cells into an extremely important commercial product.

Pt and Ru are used as the anode catalysts in a significant proportion of alcohol fuel cells. These two metals have various functions. While Pt facilitates the separation of hydrogen from the alcohol surface, Ru helps to break the CO bond retained on the Pt surface and to complete the reaction. Pt is mostly used as a catalyst in the cathode layer.

Many studies have been carried out for the synthesis of nano-catalysts used directly in alcohol fuel cells. Pt has been used in the majority of the studies on the synthesis of materials to be used as a catalyst in alcohol oxidation, but because it is expensive and easily poisoned by intermediates, it is difficult to use in alcohol fuel cells. In studies with Pt, the main goal is to increase catalytic activity and decrease the cost of platinum. Two methods are generally used to achieve this goal. In the first, while Pt nanoparticles are distributed homogeneously on conductive supports with wide surface support area; in the second, alloys of Pt with different metals are prepared. There are also studies in which both methods are applied simultaneously. Nanoparticle materials are used in the preparation of conductive supports with a large surface support area. Active surface areas of nanoparticle materials are quite wide. Also, many studies are carried out to enlarge these surface areas. Compared with pure metals, nanocatalysts have been found to reduce costs, increase catalyst surface area, and demonstrate higher catalytic activity [72–74].

Many studies on direct methanol fuel cells were carried out in England in the 1960s and 1970s. After the main problems related to Pt anode catalysts were identified, studies were conducted to develop more active and durable catalysts, and Pt-Ru and Pt-Rh catalysts have been reported to show highly effective catalytic activity [75,76].

Different ratios have been tried in a significant part of studies on Pt-Ru nanomaterials. $Pt_{66}Ru_{33}$ catalysts for oxidation have been shown to perform best in oxidized Pt-Ru alloy studies that can be well characterized. In studies without support, the best activity was generally found in the ratio of $Pt_{50}:Ru_{50}$ [77,78].

Another study claimed that Ru could be used as both a supportive and proton conductor. In other words, it has been suggested to increase activity by helping proton transport. However, EXAFS results do not support this idea. According to these results, the Ru present in Pt-Ru catalysts at low potentials is usually metallic. These results show that aqueous, oxidized bulk

Ru clusters do not play an important role in the oxidation of methanol [79–81].

One of the main problems encountered during the alcohol oxidation reaction is the formation of undesirable intermediates such as CO. These formed intermediates cover the surface of the catalyst and adversely affect their catalytic performance. This phenomenon is often referred to as *catalyst poisoning*. Bianchini and colleagues found that using two or three metal alloys is an effective method to solve the problem of poisoning in the catalyst, and determined that the synthesized two or three metal alloy catalysts increase the activity of alcohol oxidation. In this study, where the performances of palladium (Pd) catalysts were examined, the activities of Pd and Pt catalysts at different temperatures and pressure values also were compared. It was found that the Pd catalysts produced more power density at room temperature than Pt catalysts [82–84]. It has been reported that Pd, which shows high catalytic activity, especially in the oxidation of small organic molecules, can be used directly in alcohol fuel cells [50,85–87].

Pt-Ru catalysts are the primary nanomaterials used in methanol oxidation. These catalysts have been used for a long time, but the desired performance and durability values have not been achieved [88]. For this reason, many studies are being conducted to look at the synthesis of new catalysts that will show superior catalytic activity and durability. Combination methods are used in a significant number of these studies. Initial studies focused on triple and quadruple combinations depending on Pt-Ru alloys. Many studies have been carried out for the Pt-Ru-Os-Ir quaternary composition. $Pt_{47}Ru_{29}Os_{20}Ir_4$ composition developed as a result of these studies showed better oxidation performance than $Pt_{50}Ru_{50}$ combination. Studies have shown that nanomaterial activity is highly dependent on the structure of the alloy. Despite the superior performance of the $Pt_{47}Ru_{29}Os_{20}Ir_4$ alloy, the $Pt_{56}Ru_{20}Os_{20}Ir_4$ composition had a rather low efficiency. Although the ratios of Os and Ir elements remain the same, changing Pt and Ru ratios completely changed the performance of the nanomaterial [89].

In their study, Takeshi et al. aimed to synthesize more economical electrocatalysts for DAFCs. For this purpose, they proposed using Ni-based catalysts instead of Pt-based catalysts. However, Ni/C performance was lower than Pt/C performance. In addition, the findings determined that Ni-based nanocatalysts are exposed to more CO poisoning than Pt-based electrocatalysts. The same study found that Ni-based nanocatalysts show more tolerance to acetone contamination caused by 2-propanol [90].

Lopes et al. compared the performance of ethanol fuel cells directly with that of methanol fuel cells. In this study, they reported that alcohol transfer to the cathode electrode is less in ethanol fuel cells, which is an important advantage. To develop this beneficial characteristic, studies have been carried out with different nanocatalysts. As a result, they proposed using Pt-Co (3:1)/C catalysts to accelerate the reaction required in the cathode relative to the Pt/C binary. The proposed Pt-Co (3:1)/C catalyst was also found to have higher ethanol tolerance [91].

Liu et al. aimed to obtain a more effective Pt nanocatalyst in their study. They compared the performances of Pt nanoparticles directly in the alcohol fuel cell with unsupported Pt-black and carbon-supported Pt. It has been determined that the performance of Nafion Pt nanoparticles is higher in the presence and absence of methanol than the Pt black and Pt/C pairs [92].

In another study by Shen et al., the performances of Pd-C catalysts supported with nanocrystalline oxides such as CeO_2, Co_3O_4, Mn_3O_4, and NiO were investigated. The effects of these nanocatalysts on the oxidation of alcohols such as methyl alcohol, ethyl alcohol, glycerol, and ethylene glycol in the alkaline medium were investigated. The study reported that these nanocatalysts showed higher performance compared to Pt-based nanocatalysts. In addition, it has been determined that the same nanocatalysts have a higher tolerance of contamination. The most obvious difference was observed in Pd-NiO-C electrocatalysts and ethanol oxidation [93].

Xua et al. have done many studies on nanocatalysts used directly in alcohol fuel cells. CeO_2 was used as a basis in this research. The effect of adding CeO_2 to Pt-C catalysts on alcohol oxidation was examined, and optimized studies were carried out to give the best performance. It was determined that the $Pt-CeO_2$ duo performs quite well. The reason for the high performance of the $Pt-CeO_2$ nanocatalyst has been linked to its tolerance of poisoning [93].

The research literature shows that Pt is used as a catalyst in a significant part of the studies related to ethanol electrooxidation. For ethanol electrooxidation, PtSn/C, Pt-RuSn/C, and Pt-Ru/C catalysts and their performance were also examined [94]. Studies have shown that Pt-Sn-Ni or Pt-Sn-Rh triple alloys perform better than Pt-Sn binary alloys. This performance advantage is created by the fact that the tin (Sn) and nickel (Ni) oxide particles in the Pt-Sn-Ni alloy are close to the Pt nanoparticle. Also, the addition of Rh to the Pt-Sn alloy has been reported to change the Pt-Pt bond distance and the electrochemical structure of the alloy [95].

A triple metal nanocatalyst containing Pd, Cu, and Co was synthesized on graphene oxide by Yang et al., and 2 mg of the synthesized catalyst was taken and dispersed in water, ethanol, and 5% Nafion solution. The prepared mixture was dried on the glassy carbon electrode surface. Nitrogen gas was used for 30 min to remove dissolved oxygen before measurements were taken. The crystal structure of the prepared catalyst was analyzed by X-ray diffraction (XRD), and the size and morphology of the nanoparticles by TEM. The results of the study showed that the prepared catalyst has high activity and stability against methanol electrooxidation [96].

Xu et al. conducted studies on Pd-Ru nanocatalysts in the ratios of 1:1, 2:1, and 1:2. Each of the synthesized materials was applied to the glassy carbon electrode surface and their performances were examined. The electrochemical active surface area was measured to interpret the electrochemical activity of the nanocatalyst. Then the durability of the catalyst was determined. The data obtained from the study showed that Pd-Ru nanoparticles have higher activity in the alkaline environment than Pd and commercial Pd/C. It has also been reported that the prepared nanocatalysts have high stability against methanol electrooxidation [97].

Studies aiming to support the nanocatalysts with metal-organic cages were conducted by Mehek et al. In the study using the hydrothermal method, Co-MOF was synthesized. In the next step, graphene oxide synthesized by the Hummer method was added to dimethylformamide. Then the solution was heated in an autoclave. The crystals obtained were mixed with 5% Nafion and ethanol. The prepared solution was applied to the glassy, carbon electrode surface. The data obtained from the study revealed that Co-MOF catalysts synthesized by the hydrothermal method showed lower performance than Pt catalysts. Despite this low performance, the prepared catalyst has a low cost and acceptable catalytic activity [98].

Pt/TiO$_2$-C catalysts were prepared by Fan et al. Titanium isopropoxide was mixed with urea in the first stage of the synthesis process of the nanocatalyst. In the next step, the mixture obtained was gradually added to a 95% ethanol solution. Then the mixture was dried under vacuum at 80°C, and titanium urea was synthesized. The final mixture was heated at 900°C under argon gas. H$_2$PtCl$_4$ was then added to the mixture. NaBH$_4$ was used for the reduction of the mixture, which was mixed for a sufficient time. At the last stage, the mixture was filtered, washed, and dried. It was reported that the nanomaterial obtained gave a performance that showed promise for its use in the future [99].

Wasmus and Küver focused on the addition of the second component on Pt in the development of strong Pt-based catalysts to be used directly in alcohol fuel cells. According to this study, adding Ru, Sn, W, or Re to Pt can significantly improve the performance of the nanomaterial by reducing the poisoning problem [100].

5 Conclusion

DAFCs present a promising option to meet the increasing energy demands of the world. For this reason, many research and application studies have been carried out to overcome the deficiencies of this material. A very important part of the research in this area deals with nanomaterials. The ability to produce nanomaterials with desired properties will be a critical turning point in terms of energy history. The production of inexpensive, durable, and high-performance nanomaterials will directly facilitate the use of alcohol fuel cells to meet the world's increased energy needs.

References

[1] H. Fayaz, R. Saidur, N. Razali, F.S. Anuar, A.R. Saleman, M.R. Islam, An overview of hydrogen as a vehicle fuel, Renew. Sustain. Energy Rev. 16 (2012) 5511–5528, https://doi.org/10.1016/j.rser.2012.06.012.

[2] M. Balat, Potential importance of hydrogen as a future solution to environmental and transportation problems, Int. J. Hydrogen Energy 33 (2008) 4013–4029, https://doi.org/10.1016/j.ijhydene.2008.05.047.

[3] C.W. Zheng, Z.N. Xiao, Y.H. Peng, C.Y. Li, Z.B. Du, Rezoning global offshore wind energy resources, Renew. Energy 129 (2018) 1–11, https://doi.org/10.1016/j.renene.2018.05.090.

[4] M. Masip, High energy neutrinos from the sun, Astropart. Phys. 97 (2018) 63–68, https://doi.org/10.1016/j.astropartphys.2017.11.003.

[5] S. Hardman, G. Tal, Who are the early adopters of fuel cell vehicles? Int. J. Hydrogen Energy 43 (2018) 17857–17866, https://doi.org/10.1016/j.ijhydene.2018.08.006.

[6] F. Gonzatti, M. Miotto, F.A. Farret, Proposal for automation and control of a PEM fuel cell stack, J. Control. Autom. Electr. Syst. 28 (2017) 493–501, https://doi.org/10.1007/s40313-017-0322-2.

[7] H. Oman, Fuel cells for "personal electricity", IEEE Aerosp. Electron. Syst. Mag. 15 (2000) 43–45, https://doi.org/10.1109/62.873475.

[8] V. Utgikar, T. Thiesen, Life cycle assessment of high temperature electrolysis for hydrogen production via nuclear energy, Int. J. Hydrogen Energy 31 (2006) 939–944, https://doi.org/10.1016/j.ijhydene.2005.07.001.

[9] H.T. Bui, N.K. Shrestha, N. Cho, C. Bathula, H. Opoku, Y.Y. Noh, S.H. Han, Oxygen reduction reaction on nickel-based Prussian blue analog frameworks synthesized via electrochemical anodization route, J. Electroanal. Chem. 828 (2018) 80–85, https://doi.org/10.1016/j.jelechem.2018.09.033.

[10] M.L. Perry, T.F. Fuller, An ECS Centennial Series Article. A Historical Perspective of Fuel Cell Technology in the 20th Century, 2002, https://doi.org/10.1149/1.1488651.

[11] S. Mekhilef, R. Saidur, A. Safari, Comparative study of different fuel cell technologies, Renew. Sustain. Energy Rev. 16 (2012) 981–989, https://doi.org/10.1016/j.rser.2011.09.020.

[12] C. Stone, A.E. Morrison, From curiosity to "power to change the world®", Solid State Ion. (2002) 1–13, https://doi.org/10.1016/S0167-2738(02)00315-6.

[13] R. von Helmolt, U. Eberle, Fuel cell vehicles: status 2007, J. Power Sources 165 (2007) 833–843, https://doi.org/10.1016/j.jpowsour.2006.12.073.

[14] B. Viswanathan, Fuel cells, Energy Source. (2017) 329–356, https://doi.org/10.1016/B978-0-444-56353-8.00014-9.

[15] P. Kurzweil, History | fuel cells, Encycl. Electrochem. Power Sources (2009) 579–595, https://doi.org/10.1016/B978-044452745-5.00005-8.

[16] A. Boretti, Comparison of fuel economies of high efficiency diesel and hydrogen engines powering a compact car with a flywheel based kinetic energy recovery systems, Int. J. Hydrogen Energy 35 (2010) 8417–8424, https://doi.org/10.1016/j.ijhydene.2010.05.031.

[17] F. Alavi, E. Park Lee, N. van de Wouw, B. De Schutter, Z. Lukszo, Fuel cell cars in a microgrid for synergies between hydrogen and electricity networks, Appl. Energy 192 (2017) 296–304, https://doi.org/10.1016/j.apenergy.2016.10.084.

[18] T. Wilberforce, Z. El-Hassan, F.N. Khatib, A. Al Makky, A. Baroutaji, J.G. Carton, A.G. Olabi, Developments of electric cars and fuel cell hydrogen electric cars, Int. J. Hydrogen Energy 42 (2017) 25695–25734, https://doi.org/10.1016/j.ijhydene.2017.07.054.

[19] X.H. Yan, P. Gao, G. Zhao, L. Shi, J.B. Xu, T.S. Zhao, Transport of highly concentrated fuel in direct methanol fuel cells, Appl. Therm. Eng. 126 (2017) 290–295, https://doi.org/10.1016/j.applthermaleng.2017.07.186.

[20] A. Eftekhari, B. Fang, Electrochemical hydrogen storage: opportunities for fuel storage, batteries, fuel cells, and supercapacitors, Int. J. Hydrogen Energy 42 (2017) 25143–25165, https://doi.org/10.1016/j.ijhydene.2017.08.103.

[21] E. Antolini, E.R. Gonzalez, Alkaline direct alcohol fuel cells, J. Power Sources 195 (2010) 3431–3450, https://doi.org/10.1016/j.jpowsour.2009.11.145.

[22] A. Caglar, H. Kivrak, Highly active carbon nanotube supported PdAu alloy catalysts for ethanol electrooxidation in alkaline environment, Int. J. Hydrogen Energy 44 (2019) 11734–11743, https://doi.org/10.1016/j.ijhydene.2019.03.118.

[23] M. Nacef, A.M. Affoune, Comparison between direct small molecular weight alcohols fuel cells' and hydrogen fuel cell's parameters at low and high temperature. Thermodynamic study, Int. J. Hydrogen Energy 36 (2011) 4208–4219, https://doi.org/10.1016/j.ijhydene.2010.06.075.

[24] B.C. Ong, S.K. Kamarudin, S. Basri, Direct liquid fuel cells: a review, Int. J. Hydrogen Energy 42 (2017) 10142–10157, https://doi.org/10.1016/j.ijhydene.2017.01.117.

[25] M.Z.F. Kamaruddin, S.K. Kamarudin, W.R.W. Daud, M.S. Masdar, An overview of fuel management in direct methanol fuel cells, Renew. Sustain. Energy Rev. 24 (2013) 557–565, https://doi.org/10.1016/j.rser.2013.03.013.

[26] A.L. Dicks, D.A.J. Rand, Fuel Cell Systems Explained, Wiley, 2018, https://doi.org/10.1002/9781118706992.

[27] F. Dalena, A. Senatore, A. Marino, A. Gordano, M. Basile, A. Basile, Methanol production and applications: an overview, Methanol Sci. Eng. (2018) 3–28, https://doi.org/10.1016/B978-0-444-63903-5.00001-7.

[28] Intelligent Energy, Microqual hook up in telecom collaboration, Fuel Cells Bull. 2013 (2013) 5–6, https://doi.org/10.1016/s1464-2859(13)70253-0.

[29] C. Du, J. Mo, H. Li, Renewable hydrogen production by alcohols reforming using plasma and plasma-catalytic technologies: challenges and opportunities, Chem. Rev. 115 (2015) 1503–1542, https://doi.org/10.1021/cr5003744.

[30] Y. Xin, B. Sun, X. Zhu, Z. Yan, H. Liu, Y. Liu, Effects of plate electrode materials on hydrogen production by pulsed discharge in ethanol solution, Appl. Energy 181 (2016) 75–82, https://doi.org/10.1016/j.apenergy.2016.08.047.

[31] M. Bertau, H. Offermanns, L. Plass, F. Schmidt, H.J. Wernicke, Methanol: The Basic Chemical and Energy Feedstock of the Future: Asinger's Vision Today, Springer Berlin Heidelberg, 2014, https://doi.org/10.1007/978-3-642-39709-7.

[32] J. Thepkaew, S. Therdthianwong, A. Kucernak, A. Therdthianwong, Electrocatalytic activity of mesoporous binary/ternary PtSn-based catalysts for ethanol oxidation, J. Electroanal. Chem. 685 (2012) 41–46, https://doi.org/10.1016/j.jelechem.2012.09.006.

[33] B. Ruiz-Camacho, A. Medina-Ramírez, M. Villicaña Aguilera, J.I. Minchaca-Mojica, Pt supported on mesoporous material for methanol and ethanol oxidation in alkaline medium, Int. J. Hydrogen Energy 44 (2019) 12365–12373, https://doi.org/10.1016/j.ijhydene.2019.01.180.

[34] U.B. Demirci, Theoretical means for searching bimetallic alloys as anode electrocatalysts for direct liquid-feed fuel cells, J. Power Sources 173 (2007) 11–18, https://doi.org/10.1016/j.jpowsour.2007.04.069.

[35] L.L. de Souza, A.O. Neto, C.A.L.G.d.O. Forbicini, Direct oxidation of ethylene glycol on PtSn/C for application in alkaline fuel cell, Int. J. Electrochem. Sci. 12 (2017) 11855–11874, https://doi.org/10.20964/2017.12.57.

[36] S. Zhang, L. Liu, J. Yang, Y. Zhang, Z. Wan, L. Zhou, Pd-Ru-Bi nanoalloys modified three-dimensional reduced graphene oxide/MOF-199 composites as a highly efficient electrocatalyst for ethylene glycol electrooxidation, Appl. Surf. Sci. 492 (2019) 617–625, https://doi.org/10.1016/j.apsusc.2019.06.228.

[37] J. Wang, S. Wasmus, R.F. Savinell, Evaluation of ethanol, 1-propanol, and 2-propanol in a direct oxidation polymer-electrolyte fuel cell: a real-time mass spectrometry study, J. Electrochem. Soc. 142 (1995) 4218–4224, https://doi.org/10.1149/1.2048487.

[38] I.D.A. Rodrigues, J.P.I. De Souza, E. Pastor, F.C. Nart, Cleavage of the C-C bond during the electrooxidation of 1-propanol and 2-propanol: effect of the Pt morphology and of codeposited Ru, Langmuir 13 (1997) 6829–6835, https://doi.org/10.1021/la9704415.

[39] Z. Qi, M. Hollett, A. Attia, A. Kaufman, Low temperature direct 2-propanol fuel cells, Electrochem. Solid St. 5 (2002) A129, https://doi.org/10.1149/1.1475197.

[40] Z. Qi, A. Kaufman, Liquid-feed direct oxidation fuel cells using neat 2-propanol as fuel, J. Power Sources (2003) 54–60, https://doi.org/10.1016/S0378-7753(03)00061-2.

[41] M.E.P. Markiewicz, D.M. Hebert, S.H. Bergens, Electro-oxidation of 2-propanol on platinum in alkaline electrolytes, J. Power Sources 161 (2006) 761–767, https://doi.org/10.1016/j.jpowsour.2006.05.002.

[42] E.S. Papazoglou, A. Parthasarathy, BioNanotechnology, Synth. Lect. Biomed. Eng. 7 (2007) 1–141, https://doi.org/10.2200/S00051ED1V01Y200610BME007.

428 Chapter 22 Different synthesis methods of nanomaterials for DAFCs

[43] S. Özkar, R.G. Finke, Transition-metal nanocluster stabilization fundamental studies: hydrogen phosphate as a simple, effective, readily available, robust, and previously unappreciated stabilizer for well-formed, isolable, and redissolvable Ir(0) and other transition-metal nanoclusters, Langmuir 19 (2003) 6247–6260, https://doi.org/10.1021/la0207522.

[44] Z. Peng, J. Wu, H. Yang, Synthesis and oxygen reduction electrocatalytic property of platinum hollow and platinum-on-silver nanoparticles, Chem. Mater. 22 (2010) 1098–1106, https://doi.org/10.1021/cm902218j.

[45] V. Mazumder, S. Sun, Oleylamine-mediated synthesis of Pd nanoparticles for catalytic formic acid oxidation, J. Am. Chem. Soc. 131 (2009) 4588–4589, https://doi.org/10.1021/ja9004915.

[46] Z.A.C. Ramli, S.K. Kamarudin, Platinum-based catalysts on various carbon supports and conducting polymers for direct methanol fuel cell applications: a review, Nanoscale Res. Lett. (2018), https://doi.org/10.1186/s11671-018-2799-4.

[47] A. Riese, Nanostructured Carbon Materials for Active and Durable Electrocatalysts and Supports in Fuel Cells, 2015 (Electron. Thesis Diss. Repos.) https://ir.lib.uwo.ca/etd/3256. (erişim 31 Ocak 2021).

[48] B. Çelik, S. Kuzu, E. Erken, H. Sert, Y. Koşkun, F. Şen, Nearly monodisperse carbon nanotube furnished nanocatalysts as highly efficient and reusable catalyst for dehydrocoupling of DMAB and C1 to C3 alcohol oxidation, Int. J. Hydrogen Energy 41 (2016) 3093–3101, https://doi.org/10.1016/j.ijhydene.2015.12.138.

[49] J. Qi, N. Benipal, C. Liang, W. Li, PdAg/CNT catalyzed alcohol oxidation reaction for high-performance anion exchange membrane direct alcohol fuel cell (alcohol = methanol, ethanol, ethylene glycol and glycerol), Appl. Catal. Environ. 199 (2016) 494–503, https://doi.org/10.1016/j.apcatb.2016.06.055.

[50] C. Bianchini, P.K. Shen, Palladium-based electrocatalysts for alcohol oxidation in half cells and in direct alcohol fuel cells, Chem. Rev. 109 (2009) 4183–4206, https://doi.org/10.1021/cr9000995.

[51] N.M. Deraz, The comparative jurisprudence of catalysts preparation methods: I. Precipitation and impregnation methods, J. Ind. Environ. Chem. (2018). http://www.alliedacademies.org/journal-industrial-environmental-chemistry/. (erişim 31 Ocak 2021).

[52] M. Carmo, V.A. Paganin, J.M. Rosolen, E.R. Gonzalez, Alternative supports for the preparation of catalysts for low-temperature fuel cells: the use of carbon nanotubes, J. Power Sources 142 (2005) 169–176, https://doi.org/10.1016/j.jpowsour.2004.10.023.

[53] Z.B. Wang, G.P. Yin, P.F. Shi, Effects of ozone treatment of carbon support on Pt-Ru/C catalysts performance for direct methanol fuel cell, Carbon N. Y. 44 (2006) 133–140, https://doi.org/10.1016/j.carbon.2005.06.043.

[54] K.W. Park, J.H. Choi, S.A. Lee, C. Pak, H. Chang, Y.E. Sung, PtRuRhNi nanoparticle electrocatalyst for methanol electrooxidation in direct methanol fuel cell, J. Catal. 224 (2004) 236–242, https://doi.org/10.1016/j.jcat.2004.02.010.

[55] J.S. Choi, W.S. Chung, H.Y. Ha, T.H. Lim, I.H. Oh, S.A. Hong, H.I. Lee, Nanostructured Pt-Cr anode catalyst over carbon support, for direct methanol fuel cell, J. Power Sources 156 (2006) 466–471, https://doi.org/10.1016/j.jpowsour.2005.05.075.

[56] G.A. Camara, M.J. Giz, V.A. Paganin, E.A. Ticianelli, Correlation of electrochemical and physical properties of PtRu alloy electrocatalysts for PEM fuel cells, J. Electroanal. Chem. 537 (2002) 21–29, https://doi.org/10.1016/S0022-0728(02)01223-8.

Chapter 22 Different synthesis methods of nanomaterials for DAFCs **429**

[57] B. Yang, Q. Lu, Y. Wang, L. Zhuang, J. Lu, P. Liu, J. Wang, R. Wang, Simple and low-cost preparation method for highly dispersed PtRu/C catalysts, Chem. Mater. 15 (2003) 3552–3557, https://doi.org/10.1021/cm034306r.

[58] K. Hola, Z. Markova, G. Zoppellaro, J. Tucek, R. Zboril, Tailored functionalization of iron oxide nanoparticles for MRI, drug delivery, magnetic separation and immobilization of biosubstances, Biotechnol. Adv. 33 (2015) 1162–1176, https://doi.org/10.1016/j.biotechadv.2015.02.003.

[59] Z. Liu, X.Y. Ling, X. Su, J.Y. Lee, Carbon-supported Pt and PtRu nanoparticles as catalysts for a direct methanol fuel cell, J. Phys. Chem. B 108 (2004) 8234–8240, https://doi.org/10.1021/jp049422b.

[60] T. Kim, M. Takahashi, M. Nagai, K. Kobayashi, Preparation and characterization of carbon supported Pt and PtRu alloy catalysts reduced by alcohol for polymer electrolyte fuel cell, Electrochim. Acta (2004) 817–821, https://doi.org/10.1016/j.electacta.2004.01.124.

[61] X. Xue, T. Lu, C. Liu, W. Xu, Y. Su, Y. Lv, W. Xing, Novel preparation method of Pt-Ru/C catalyst using imidazolium ionic liquid as solvent, Electrochim. Acta 50 (2005) 3470–3478, https://doi.org/10.1016/j.electacta.2004.12.034.

[62] L. Dubau, F. Hahn, C. Coutanceau, J.M. Léger, C. Lamy, On the structure effects of bimetallic PtRu electrocatalysts towards methanol oxidation, J. Electroanal. Chem. 554–555 (2003) 407–415, https://doi.org/10.1016/S0022-0728(03)00308-5.

[63] B. Sen, S. Kuzu, E. Demir, S. Akocak, F. Sen, Highly monodisperse RuCo nanoparticles decorated on functionalized multiwalled carbon nanotube with the highest observed catalytic activity in the dehydrogenation of dimethylamine–borane, Int. J. Hydrogen Energy 42 (2017) 23292–23298, https://doi.org/10.1016/j.ijhydene.2017.06.032.

[64] A. Brouzgou, S.Q. Song, P. Tsiakaras, Low and non-platinum electrocatalysts for PEMFCs: current status, challenges and prospects, Appl. Catal. Environ. 127 (2012) 371–388, https://doi.org/10.1016/j.apcatb.2012.08.031.

[65] A. Bangisa, Electrochemical Study of Electrode Support Material for Direct Methanol Fuel Cell Applications, 2013. http://etd.uwc.ac.za/xmlui/handle/11394/3319. (erişim 31 Ocak 2021).

[66] C. Xu, T.S. Zhao, Q. Ye, Effect of anode backing layer on the cell performance of a direct methanol fuel cell, Electrochim. Acta 51 (2006) 5524–5531, https://doi.org/10.1016/j.electacta.2006.02.030.

[67] M. Bhatt, N.K. Mishra, R. Singh, A. Kumar, N. Yadav, N.K. Mishra, P. Chaudhary, 2015, www.isca.me (erişim 31 Ocak 2021).

[68] A.H.M. Yusoff, M.N. Salimi, M.F. Jamlos, A review: synthetic strategy control of magnetite nanoparticles production, Adv. Nano Res. 6 (2018) 1–19, https://doi.org/10.12989/anr.2018.6.1.001.

[69] L.Y. Meng, B. Wang, M.G. Ma, K.L. Lin, The progress of microwave-assisted hydrothermal method in the synthesis of functional nanomaterials, Mater. Today Chem. 1–2 (2016) 63–83, https://doi.org/10.1016/j.mtchem.2016.11.003.

[70] N. Ye, T. Yan, Z. Jiang, W. Wu, T. Fang, A review: conventional and supercritical hydro/solvothermal synthesis of ultrafine particles as cathode in lithium battery, Ceram. Int. 44 (2018) 4521–4537, https://doi.org/10.1016/j.ceramint.2017.12.236.

[71] F. Fievet, S. Ammar-Merah, R. Brayner, F. Chau, M. Giraud, F. Mammeri, J. Peron, J.Y. Piquemal, L. Sicard, G. Viau, The polyol process: a unique method for easy access to metal nanoparticles with tailored sizes, shapes and compositions, Chem. Soc. Rev. 47 (2018) 5187–5233, https://doi.org/10.1039/c7cs00777a.

[72] L. Ning, X. Liu, M. Deng, Z. Huang, A. Zhu, Q. Zhang, Q. Liu, Palladium-based nanocatalysts anchored on CNT with high activity and durability for ethanol electro-oxidation, Electrochim. Acta 297 (2019) 206–214, https://doi.org/10.1016/j.electacta.2018.11.188.

[73] F. Şen, G. Gökagaç, Activity of carbon-supported platinum nanoparticles toward methanol oxidation reaction: role of metal precursor and a new surfactant, tert-octanethiol, J. Phys. Chem. C 111 (2007) 1467–1473, https://doi.org/10.1021/jp065809y.

[74] F. Sen, Y. Karatas, M. Gulcan, M. Zahmakiran, Amylamine stabilized platinum (0) nanoparticles: active and reusable nanocatalyst in the room temperature dehydrogenation of dimethylamine-borane, RSC Adv. 4 (2014) 1526–1531, https://doi.org/10.1039/c3ra43701a.

[75] B.D. McNicol, D.A.J. Rand, K.R. Williams, Direct methanol-air fuel cells for road transportation, J. Power Sources 83 (1999) 15–31, https://doi.org/10.1016/S0378-7753(99)00244-X.

[76] B. Çelik, G. Başkaya, H. Sert, Ö. Karatepe, E. Erken, F. Şen, Monodisperse Pt(0)/DPA@GO nanoparticles as highly active catalysts for alcohol oxidation and dehydrogenation of DMAB, Int. J. Hydrogen Energy 41 (2016) 5661–5669, https://doi.org/10.1016/j.ijhydene.2016.02.061.

[77] D. Chu, S. Gilman, Methanol electro-oxidation on unsupported Pt-Ru alloys at different temperatures, J. Electrochem. Soc. 143 (1996) 1685–1690, https://doi.org/10.1149/1.1836700.

[78] Y. Takasu, T. Fujiwara, Y. Murakami, K. Sasaki, M. Oguri, T. Asaki, W. Sugimoto, Effect of structure of carbon-supported PtRu electrocatalysts on the electrochemical oxidation of methanol, J. Electrochem. Soc. 147 (2000) 4421, https://doi.org/10.1149/1.1394080.

[79] D.R. Rolison, P.L. Hagans, K.E. Swider, J.W. Long, Role of hydrous ruthenium oxide in Pt-Ru direct methanol fuel cell anode electrocatalysts: the importance of mixed electron/proton conductivity, Langmuir 15 (1999) 774–779, https://doi.org/10.1021/la9807863.

[80] J.W. Long, R.M. Stroud, K.E. Swider-Lyons, D.R. Rolison, How to make electrocatalysts more active for direct methanol oxidation—avoid PtRu bimetallic alloys! J. Phys. Chem. B 104 (2000) 9772–9776, https://doi.org/10.1021/jp001954e.

[81] S. Şen, F. Şen, G. Gökağaç, Preparation and characterization of nano-sized Pt-Ru/C catalysts and their superior catalytic activities for methanol and ethanol oxidation, Phys. Chem. Chem. Phys. 13 (2011) 6784–6792, https://doi.org/10.1039/c1cp20064j.

[82] S. Eris, Z. Daşdelen, F. Sen, Enhanced electrocatalytic activity and stability of monodisperse Pt nanocomposites for direct methanol fuel cells, J. Colloid Interface Sci. 513 (2018) 767–773, https://doi.org/10.1016/j.jcis.2017.11.085.

[83] Y. Yildiz, E. Erken, H. Pamuk, H. Sert, F. Şen, Monodisperse Pt nanoparticles assembled on reduced graphene oxide: highly efficient and reusable catalyst for methanol oxidation and dehydrocoupling of dimethylamine-borane (DMAB), J. Nanosci. Nanotechnol. 16 (2016) 5951–5958, https://doi.org/10.1166/jnn.2016.11710.

[84] Y. Yıldız, S. Kuzu, B. Sen, A. Savk, S. Akocak, F. Şen, Different ligand based monodispersed Pt nanoparticles decorated with rGO as highly active and reusable catalysts for the methanol oxidation, Int. J. Hydrogen Energy 42 (2017) 13061–13069, https://doi.org/10.1016/j.ijhydene.2017.03.230.

[85] B. Çelik, Y. Yildiz, H. Sert, E. Erken, Y. Koşkun, F. Şen, Monodispersed palladium-cobalt alloy nanoparticles assembled on poly(N-vinyl-pyrrolidone) (PVP) as a highly effective catalyst for dimethylamine borane (DMAB) dehydrocoupling, RSC Adv. 6 (2016) 24097–24102, https://doi.org/10.1039/c6ra00536e.

Chapter 22 Different synthesis methods of nanomaterials for DAFCs **431**

[86] Z. Daşdelen, Y. Yıldız, S. Eriş, F. Şen, Enhanced electrocatalytic activity and durability of Pt nanoparticles decorated on GO-PVP hybride material for methanol oxidation reaction, Appl. Catal. Environ. 219 (2017) 511–516, https://doi.org/10.1016/j.apcatb.2017.08.014.

[87] F. Şen, G. Gökağaç, Different sized platinum nanoparticles supported on carbon: an XPS study on these methanol oxidation catalysts, J. Phys. Chem. C 111 (2007) 5715–5720, https://doi.org/10.1021/jp068381b.

[88] B. Şen, B. Demirkan, A. Savk, R. Kartop, M.S. Nas, M.H. Alma, S. Sürdem, F. Şen, High-performance graphite-supported ruthenium nanocatalyst for hydrogen evolution reaction, J. Mol. Liq. 268 (2018) 807–812, https://doi.org/10.1016/j.molliq.2018.07.117.

[89] B. Gurau, R. Viswanathan, R. Liu, T.J. Lafrenz, K.L. Ley, E.S. Smotkin, E. Reddington, A. Sapienza, B.C. Chan, T.E. Mallouk, S. Sarangapani, Structural and electrochemical characterization of binary, ternary, and quaternary platinum alloy catalysts for methanol electro-oxidation, J. Phys. Chem. B 102 (1998) 9997–10003, https://doi.org/10.1021/jp982887f.

[90] T. Kobayashi, J. Otomo, C.J. Wen, H. Takahashi, Direct alcohol fuel cell—relation between the cell performance and the adsorption of intermediate originating in the catalyst-fuel combinations, J. Power Sources 124 (2003) 34–39, https://doi.org/10.1016/S0378-7753(03)00622-0.

[91] T. Lopes, E. Antolini, F. Colmati, E.R. Gonzalez, Carbon supported Pt-Co (3:1) alloy as improved cathode electrocatalyst for direct ethanol fuel cells, J. Power Sources 164 (2007) 111–114, https://doi.org/10.1016/j.jpowsour.2006.10.052.

[92] Z. Liu, Z.Q. Tian, S.P. Jiang, Synthesis and characterization of Nafion-stabilized Pt nanoparticles for polymer electrolyte fuel cells, Electrochim. Acta 52 (2006) 1213–1220, https://doi.org/10.1016/j.electacta.2006.07.027.

[93] P.K. Shen, C. Xu, Alcohol oxidation on nanocrystalline oxide Pd/C promoted electrocatalysts, Electrochem. Commun. 8 (2006) 184–188, https://doi.org/10.1016/j.elecom.2005.11.013.

[94] J.M. Léger, S. Rousseau, C. Coutanceau, F. Hahn, C. Lamy, How bimetallic electrocatalysts does work for reactions involved in fuel cells?: Example of ethanol oxidation and comparison to methanol, Electrochim. Acta (2005), https://doi.org/10.1016/j.electacta.2005.01.051.

[95] F. Colmati, E. Antolini, E.R. Gonzalez, Preparation, structural characterization and activity for ethanol oxidation of carbon supported ternary Pt-Sn-Rh catalysts, J. Alloys Compd. 456 (2008) 264–270, https://doi.org/10.1016/j.jallcom.2007.02.015.

[96] F. Yang, B. Zhang, S. Dong, C. Wang, A. Feng, X. Fan, Y. Li, Reduced graphene oxide supported Pd-Cu-Co trimetallic catalyst: synthesis, characterization and methanol electrooxidation properties, J. Energy Chem. 29 (2019) 72–78, https://doi.org/10.1016/j.jechem.2018.02.007.

[97] H. Xu, B. Yan, K. Zhang, J. Wang, S. Li, C. Wang, Y. Shiraishi, Y. Du, P. Yang, Facile fabrication of novel PdRu nanoflowers as highly active catalysts for the electrooxidation of methanol, J. Colloid Interface Sci. 505 (2017) 1–8, https://doi.org/10.1016/j.jcis.2017.05.067.

[98] R. Mehek, N. Iqbal, T. Noor, H. Nasir, Y. Mehmood, S. Ahmed, Novel Co-MOF/graphene oxide electrocatalyst for methanol oxidation, Electrochim. Acta 255 (2017) 195–204, https://doi.org/10.1016/j.electacta.2017.09.164.

[99] Y. Fan, Z. Yang, P. Huang, X. Zhang, Y.M. Liu, Pt/TiO$_2$-C with hetero interfaces as enhanced catalyst for methanol electrooxidation, Electrochim. Acta 105 (2013) 157–161, https://doi.org/10.1016/j.electacta.2013.04.158.

[100] S. Wasmus, A. Küver, Methanol oxidation and direct methanol fuel cells: a selective review, J. Electroanal. Chem. 461 (1999) 14–31, https://doi.org/10.1016/s0022-0728(98)00197-1.

23

The synthesis and characterization of size-controlled bimetallic nanoparticles

Haydar Goksu[a], Muhammed Bekmezci[b,c], Ramazan Bayat[b,c], Elif Esra Altuner[b], and Fatih Şen[b]

[a]Kaynasli Vocational College, Duzce University, Duzce, Turkey. [b]Şen Research Group, Department of Biochemistry, Dumlupinar University, Kütahya, Turkey. [c]Department of Materials Science and Engineering, Faculty of Engineering, Dumlupinar University, Kütahya, Turkey

1 Introduction

The nanoparticles present in solid atomic or molecular size and have superior physicochemical properties that distinguish them from conventional cast solids. Solid nanoparticles have a unique form that can give them specific features. In nanosized materials, there are certain nano ranges in which each solid has its own characteristics. That's why the resulting solid particle has different names depending on its nanosize. However, the definition of nanoparticles varies according to the type of material used and the area used [1, 2]. Generally, nanoparticles have various surface shapes, which require different methods and catalysts with different nanoscale sizes for controlling particle morphology.

To control the particle morphology, nanocatalysts having different activity and selectivity levels are tested to modify the surface of nanoparticles. Obtaining different-sized nanoparticles can provide different properties and levels of stability. For example, decreasing the nanoscale of particles and changing them into nanoparticles enhance their activity because of increased surface area and a combination of coordinated atoms on the surface. Decreasing nanoscales enhances sintering and aggregation in

Nanomaterials for Direct Alcohol Fuel Cells. https://doi.org/10.1016/B978-0-12-821713-9.00027-5
Copyright © 2021 Elsevier Inc. All rights reserved.

nanoparticles. Defects occurring at times on the surface of particles causes forming of undercoordinated atoms having more activate to reaction with chemicals species compared some particles have a face centered cubic metal sequences [3, 4].

Monometallic nanoparticles are formed by combining a metal and support material containing different functional groups. The features of monometallic depend on the metal used in the nanoparticles, and the electronic, magnetic, and transition states of metals affect catalytic activity and stability [5]. Bimetallic nanoparticles, consisting of two different metals, have recently attracted much interest in technology and scientific research [6].

The most important feature of bimetallic nanoparticles is the type and nanoscale of the metals in nanoparticles. By combining two metals, a new form of the bimetallic nanoparticle is obtained, which exhibits electronic, optical, and catalytic and stability features that are superior to those of individual metals. For the last decade, various studies related to bimetallic nanoparticles have been conducted, and various techniques have been tested to synthesize and examine characteristic features.

Currently, the most current topics related to bimetallic nanoparticles are surface functionalization, contact aggregate, and core shells [1]. As mentioned previously, when bimetallic nanoparticles form, they achieve better catalytic properties than monometallic nanoparticles. For example, in a study conducted on bimetallic and monometallic catalysts of PtNi and Pt nanoparticles, superior properties of PtNi bimetallic than Pt monometallic nanoparticles were found [7]. Similar findings have been revealed by other studies comparing bimetallic nanoparticles to monometallic nanoparticles. Generally, electron uptake by cationic metal ions leads to the formation of elemental metals in the bimetallic nanoparticle. That also can be called the *synergic effect* of metals on support materials [8]. In this situation, changing the electronic properties of nanometals occurs due to electrons transferring among metals and reducing agents. Forming alloys between two metals in nanoparticles is another important effect of bimetallic nanoparticles compared to monometallic nanoparticles [9].

2 Synthesis of bimetallic and monometallic nanoparticles and their differences

The fabrication or synthesis of bimetallic nanoparticles is commonly dependent on the preparation of monometallic nanoparticles or other modification methods used to synthesize these nanoparticles. Therefore, knowing about the preparation and synthesis of monometallic nanoparticles is crucial for the synthesis of bimetallic nanoparticles [10]. The preparation of monometallic nanoparticles is carried out in two ways. In the first way, called the *top-down method*, the bulk particles are divided into nanoscales of atoms or ions. In the second way, known as the *bottom-up method*, monometallic nanoparticles are constituted with atoms and ions. Although both ways have been widely used in the synthesis of monometallic nanoparticles, the bottom-up method is accepted better than the top-down method due to some advantages that it has, like being easy to implement and creating adjustable nanoparticles. In the synthesis of monometallic nanoparticles, metals are used as initial materials [11–13]. Metals used to prepare monometallic nanoparticles generally present in aqueous media, and these metals are in an ionic state. The reduction of metal in monometallic nanoparticles is provided by electron donors as a known reduction agent such as sodium boron hydride, carbohydrates, or polyalcohol [11].

2.1   Bimetallic nanoparticle synthesis methods

The synthesis of bimetallic nanoparticles differs from the synthesis of monometallic nanoparticles in certain approaches, such as the distribution of metals and particle structure [14]. So far, various methods have improved and been tried in the synthesis of bimetallic nanoparticles. Bimetallic nanoparticles in these methods differ from each other due to features like shape, particle size, and composition, which affect the properties of synthesized bimetallic nanoparticles [1].

2.2   Chemical reduction of ionic metals to elemental metallic form in the synthesis of bimetallic nanoparticles

The chemical reduction method basically depends on reducing agents that are the source of electrons. Ionic metals are reduced by taking electrons from agents such as sodium borohydride, citric acid, hydrogen, ammonia borane, and hydrazine. In the synthesis of bimetallic nanoparticles, the reduction of metals to elemental form (i.e., having no charge) is taken place by successive reductions that form a core-shell of nanoparticles. One metal is reduced first and the other metal surrounds the first by deposition [15]. The synthesis of nanoparticles is schematized in Fig. 1 [1].

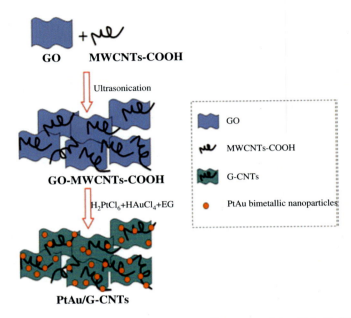

Fig. 1 Reduction of metals in synthesis bimetallic nanoparticles (PtAu/G-CNTs nanocomposites) [1].

Reduction of metals in bimetallic nanoparticles takes place in a suitable solvent. Also, some support materials are used to stabilize the reduced metals. Supporting chemicals prevent agglomeration of the metals and enhance the stabilization of catalysts consisted of nanoparticles. The common support chemicals used in studies to stability the bimetallic nanoparticles are multiwalled carbon nanotubes (MWCNTs) [16, 17], reduced graphene oxide [18], graphene oxide [19], and chitosan, among others. Although the chemical method is widely used to synthesize nanoparticles, it can cause a possible spread of chemicals that can be harmful to the environment and the health of living creatures [13].

2.3 Thermal method

Synthesis of bimetallic nanoparticles using the thermal method is carried out in solvents to pyrolyze the precursors. The isolation of raw nanoparticles becomes difficult to achieve at high temperatures. This process requires high energy to break the bonds of materials used in the synthesis of bimetallic nanoparticles. Photochemical techniques are used to minimize external factors [1].

2.4 Fabrication of bimetallic nanoparticles using microemulsion

Another important method of obtaining bimetallic nanoparticles is microemulsion. In this method, phases in materials of nanoparticles are separated and the reduction of metals is achieved. The separation phases are ensured by a surfactant that exists between two adjacent phases that maintain the stability of phases. The microemulsion is consistent with hydrophilic (polar) and hydrophobic (apolar) structures and a surfactant. Systems containing hydrophilic and hydrophobic structures that commonly take an oil-water form are widely used to fabricate bimetallic nanoparticles. The separation phase is ensured by the surfactant between two phases. The contents, surfactant type, ratio, and concentration compounds are effective features of bimetallic nanoparticles. Bimetallic atoms are arranged with regular sequencing between the boundary of adjacent phases in nanoscale [13, 20]. In the synthesis of bimetallic nanoparticles by microemulsion, metals in ionic form are held in microemulsion phases and reducing chemicals are held in another phase [21].

3 Using electrical current for fabrication of metallic nanoparticles

Electrical current is used to synthesize all the metallic nanoparticles. A sheet consisting of metals and metallic salts is reduced using a cathode electrode to obtain elemental metallic particles. The electrical current creates electrodes that act as a reducing agent in the electrolyte system [22]. Tetra-alkyl ammonium is commonly used as a stabilization salt in this system. The system using electrical current is extensively preferred due to its advantages, such as easy application, low cost, need for highly pure particles, and adjustable particle sample sizes. Generally, the electrolytic current is mainly preferred in industrial areas for the synthesis of nanoparticles [1].

3.1 The use of highly energetic radiations in the synthesis of bimetallic nanoparticles

Radiations such as X-rays, gamma rays, and ultraviolet rays can be used to produce free radicals from various alcohols. The produced free radicals are extremely active due to their high electron content. Hence, these free radicals act as reducing agents in the fabrication of bimetallic nanoparticles and bimetallic catalysts for biosensor applications or other applications of nanotechnology. In addition, nanoparticles having core-shell structures can be synthesized using high energic rays [1]. The number of metals to be used in nanoparticles can be adjusted by radiating doses of gamma rays. The nanoparticles having core shells are fabricated under high-dose radiation of gamma rays, and bimetallic nanoparticles are synthesized under low doses of gamma rays [23].

3.2 Bimetallic nanoparticles in sol-gel form

Preparing bimetallic nanoparticles in sol-gel form is easy, economical, and effective, and the bimetallic products are of good quality. Bimetallic nanoparticles prepared in sol-gel form can be controlled in terms of their chemical contents. The interactions of particles are ensured by physical bonds, especially by Van der Waals bonds, which form a sol-gel structure. A sol-gel contract consists of two structures that are in solid and liquid phases. These two parts form the whole structure of nanoparticles [1, 24]. Indeed, all of the aforementioned methods have shortcomings.

For these reasons, recent studies target to improve novel approaches to fabricate bimetallic nanoparticles that are simple, low cost, and ecofriendly. To achieve that target, there have been many attempts. For example, various tissues extracted from plants have been tried for developing nanoparticles. Samples like flavones, terpenoids, and phenolic extracts have been tested as reducing and stabilizing agents in the development of nanoparticles [25]. Some extracted tissues from plants used in the synthesis of nanoparticles are *Anacardium occidentale* [26], mahogany leaves [27], *Piper pedicellatum* [28], *Punica granatum* [29], and *Azadirachta indica* [30].

3.3 Effect of metal distributions on forming alloy nanoparticles

Bimetallic nanoclusters or nanoparticles with a heterogeneous structure contain different metal distributions in a medium containing wet chemicals. The heterogeneous distribution of these metals is the source of the difference between reducing potentials. This difference leads to a core-shell structure or a heterogeneous chemical forming in the nanocluster. However, homogeneous metallic distributions on support materials lead to the forming of alloy nanoparticles. To create nanoparticles with homogeneous metal distribution (namely, for the production of alloy nanoclusters), controlling the reaction routes or kinetics is strongly recommended. To control the reaction mechanism, various approaches have been undertaken [31–33].

An effective and successful route for producing alloy nanoparticles requires choosing the right reducing agent to reduce metals simultaneously in nanoparticles. So far, many researchers have managed to reduce different metals simultaneously using effective routes. Sodium borohydride has been widely used as a reducing agent in the synthesis of bimetallic nanoparticles [34–38]. Selecting support materials with rich, functional groups and sufficient surface area is an important issue in the fabrication of nanomaterials. Adjusting the reduction of metals in bimetallic materials can be accomplished by selecting an appropriate adsorption technique. Sen et al. used a number of effective routes to produce alloy nanoparticles such as palladium-ruthenium, ruthenium-copper, and palladium-cobalt [39–41]. Some metals facilitate the forming of alloy nanoclusters. One study reported that Cu^{+2} ions have been used in the synthesis of Au-Pd alloy and Cu^{+2} acted as a bridge in the forming of Au-Pu bimetallic nanoparticles [42]. In addition, to synthesize nanomaterials, Wang et al. recently tested many techniques or routes using low-cost metals (e.g., Zn^{2+}, In^{+3}, Fe^{3+}, Ni^{2+}, Co^{2+}, Cu^{2+}) and noble metals (e.g., Pt^{4+}, Au^{3+}, Ru^{+3}, Pd^{2+}, Rh^{3+}). They revealed an interesting chemical reaction of noble and nonnoble metals with octa decylamine (ODA) as the reducing agent. ODA has a low capacity for reduction at high temperatures. Thus, only noble metal ions can be reduced at high temperatures. At high temperatures, metal oxide formation occurs when only nonnoble metals are added to the reaction medium. However, two metal types are simultaneously added to the reaction medium; nonnoble metals are reduced in the presence of noble metals and ODA in the reaction medium. That finding shows that the noble metals reduce nonnoble metals [43–45].

4 Porous form and surface area of bimetallic nanoparticles

Today, the requirements for nanoparticles consist of low particle size and a sufficiently high surface area for synthesizing appropriate nanocatalysts. This is because nanoparticles are used in a number of areas, such as sensing materials, catalytic studies, and other applications. The success of catalytic studies mostly depends on nanoparticles with high surface area and low particle size. Having a porous structure enhances the surface area, low density, and high gas permeability of alloy nanoparticles. The alloy nanoparticles having these features are greatly preferred in catalytic studies and applications. Generally, a metal with low noble features is used in support materials in the preparation of porous bimetallic nanoparticles [46, 47]. Wang et al. [48] developed a procedure for preparing porous nanoparticles; here, it dissolved the metal in concentrated nitric acid, forming a porous structure. By using this method, it is possible to prepare nanoparticles with a high surface area, porous structure, and high catalytic activity that can be used in many catalytic applications. Fig. 2 shows scanning electron microscopy (SEM) images of bimetallic nanoparticles that have been synthesized [1].

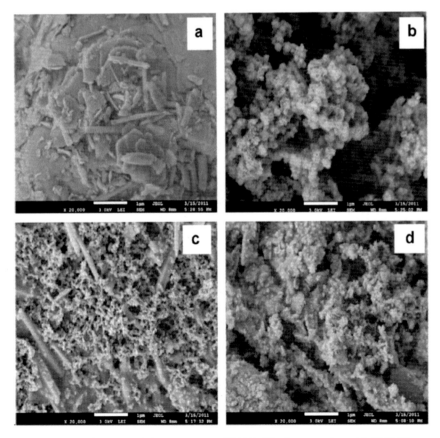

Fig. 2 SEM images: (A) kaolin, (B) Fe/Ni, (C) K-Fe/Ni, and (D) K-Fe/Ni after reaction with DBG [1].

5 Important metals used for the preparation bimetallic nanoparticles

5.1 Iron-copper and iron-nickel bimetallic nanoparticles

Recently, by using iron (Fe) and another metal, several bimetallic nanoparticles have been prepared, and they exhibit a number of interesting features. A recent study related to iron-copper (Fe-Cu) was conducted, and a bimetallic Fe-Cu nanoparticle was prepared via an electrical explosion. This bimetallic nanoparticle exhibited homogeneous distribution through the entire structure of the nanocluster. After annealing Fe and Cu at 200–400°C, the consolidation of a Fe-Cu alloy was achieved. However, the sintering of these metals occurs at low temperatures. Above this temperature range (i.e., at 600°C) were some phase boundaries of iron, copper, or Fe-Cu forms. In some sides of nanoparticles saturated with iron, copper depositions having 1 μm were detected. In addition, at different temperatures, various features of Fe-Cu nanoparticles have been observed [49]. Another important study [50] of iron-nickel (Fe-Ni) bimetallic nanoparticles have been synthesized via chemical reduction using cetyltrimethylammonium bromide and sodium dodecylbenzene sulfonate as supporting elements. The formation of the core-shell shape in bimetallic nanoparticles was detected using ultraviolet (UV) spectra upon decreasing ions of Fe and Ni^{+2} in the reaction medium. The prepared FeNi alloy bimetallic nanoparticles were used in hydrogen evolution from sodium borohydride at different experimental parameters. Also, in this same study, this alloy was tested as an adsorbent in the removal of golden yellow. The catalytic activity of FeNi alloys differs depending on the ratio of the metals in nanoparticles. Various iron-based, bimetallic nanoparticles [51–53] like the ones already mentioned have been improved, but many features of iron-based nanomaterial still need to be explored.

5.2 Effect of nickel on the formation of nanoparticles

Ni-based nanomaterials, especially bimetallic nanoparticles, are used in terms of magnetic and catalytic features. The main advantages of using Ni-based nanoparticles are low cost, enhanced high surface area, and prevention of stability and catalytic activity in catalyst materials [54]. Some bimetallic nanoparticles prepared with a Ni-base include silver-nickel [55], Fe-Ni [50], and Ni-Cu [56].

Ni-based bimetallic nanoparticles are used in reactions like oxidation and catalytic dehydrogenation due to their high catalytic activity and stability. In research on Ni-based nanoparticles, in order to enhance their catalytic activity, various support materials such as graphene oxide, MWCNTs, and graphite have been tested in different reactions [57–59].

5.3 Using platinum in the fabrication of bimetallic nanoparticles

Bimetallic nanoparticles containing platinum (Pt) have fascinating features that are useful in applications such as biomedical, automotive, and catalytic areas. When a bimetallic nanoparticle contains Pt, its surface area and catalytic activity are enhanced. However, sources of Pt are restricted and Pt is rarely found in nature. Various metals such as gold, copper, and palladium have been used to fabricate bimetallic nanoparticles with Pt. Comparison of monometallic nanoparticles to bimetallic nanoparticles showed that bimetallic nanoparticles have better catalytic features than monometallic nanoparticles [60–62].

5.4 Usage of palladium in the fabrication of bimetallic nanoparticles

Using palladium (Pd) instead of Pt in electrocatalytic applications will provide solutions to a number of technical barriers originating from Pt. Recent studies have shown that Pd-based nanoparticles are good alternative to Pt-based materials in terms of catalytic activity [62]. Unlike Pt, Pd is a low-cost metal used in many different areas due to its abundance and easy access. These bimetallic nanoparticles have many applications. Pd bimetallic nanoparticles can be used in catalytic reactions, alcohol oxidation, and the preparation of electrocatalytic and biosensing materials. Although there are many studies on bimetallic nanoparticles originating from Pd, it is a wide area that needs further study.

6 Conclusion

Each metal has positive and negative properties in terms of catalytic performance, which can be changed by being mixed with each other. Changing the ratio of metal in bimetallic nanocomposite changes features of the entire composition, and these new features may be wanted or unwanted. Bimetallic nanoparticles are used in many fields, such as catalytic, sensor, electrosensory, and biodegradable areas. The properties of nanoparticles that can be used in many applications can be changed depending on the proportion of the metals, so nanoparticles exhibiting different features can be used in a number of appropriate areas. The nanoparticles obtained by mixing metals in different proportions can create a brand new feature that is not found in either metal individually.

Bimetallic nanoparticles have some unique properties not seen in monometallic nanoparticles. Thus, we can state that bimetallic nanoparticles are more important than monometallic nanoparticles. These unique properties of bimetallic nanoparticles made them attractive for use in electronic, magnetic, optical, and other applications. These features in bimetallic nanoparticles are dressed as a synergic effect of various metals.

Generally, the obtained properties of bimetallic nanoparticles enhance catalytic activity, reusability, and stability compared to monometallic nanoparticles. However, extensive applicable areas of bimetallic nanoparticles and their various advantages remain unstudied. That is why, today, researchers are intensely investigating the revealed features of bimetallic nanoparticles. Bimetallic nanoparticles, having a controllable crystal particle size, geometry, and magnetic properties, can be utilized in many fields. In addition, support materials can affect the properties of bimetallic nanoparticles due to their porous surface area. The support materials to be used in the framework of nanomaterials are very important to this discussion.

References

[1] G. Sharma, et al., Novel development of nanoparticles to bimetallic nanoparticles and their composites: a review, J. King Saud Univ. Sci. 31 (2) (2019) 257–269, https://doi.org/10.1016/J.JKSUS.2017.06.012.

[2] R. Ferrando, J. Jellinek, R.L. Johnston, Nanoalloys: From theory to applications of alloy clusters and nanoparticles, Chem. Rev. 108 (3) (2008) 845–910, https://doi.org/10.1021/cr040090g.

[3] H. Jung, M.E. King, M.L. Personick, Strategic synergy: advances in the shape control of bimetallic nanoparticles with dilute alloyed surfaces, Curr. Opin. Colloid Interface Sci. 40 (2019) 104–117, https://doi.org/10.1016/J.COCIS.2019.02.004.

[4] D.D. Robertson, M.L. Personick, Growing nanoscale model surfaces to enable correlation of catalytic behavior across dissimilar reaction environments, Chem. Mater. 31 (4) (2019) 1121–1141, https://doi.org/10.1021/acs.chemmater.8b04595.

[5] I. Sondi, B. Salopek-Sondi, Silver nanoparticles as antimicrobial agent: a case study on *E. coli* as a model for Gram-negative bacteria, J. Colloid Interface Sci. 275 (1) (2004) 177–182, https://doi.org/10.1016/J.JCIS.2004.02.012.

444 Chapter 23 The synthesis and characterization of size-controlled bimetallic nanoparticles

[6] G. Sharma, V.K. Gupta, S. Agarwal, A. Kumar, S. Thakur, D. Pathania, Fabrication and characterization of Fe@MoPO nanoparticles: ion exchange behavior and photocatalytic activity against malachite green, J. Mol. Liq. 219 (2016) 1137–1143, https://doi.org/10.1016/J.MOLLIQ.2016.04.046.

[7] S. De, J. Zhang, R. Luque, N. Yan, Ni-based bimetallic heterogeneous catalysts for energy and environmental applications, Energ. Environ. Sci. 9 (11) (2016) 3314–3347, https://doi.org/10.1039/C6EE02002J.

[8] G. Savitha, R. Saha, G. Sekar, Bimetallic chiral nanoparticles as catalysts for asymmetric synthesis, Tetrahedron Lett. 57 (47) (2016) 5168–5178, https://doi.org/10.1016/J.TETLET.2016.10.011.

[9] G. Sharma, M. Naushad, A. Kumar, S. Devi, M.R. Khan, Lanthanum/cadmium/polyaniline bimetallic nanocomposite for the photodegradation of organic pollutant, Iran. Polym. J. (Engl. Ed.) 24 (12) (2015) 1003–1013, https://doi.org/10.1007/s13726-015-0388-2.

[10] N. Toshima, T. Yonezawa, Bimetallic nanoparticles—novel materials for chemical and physical applications, New J. Chem. 22 (11) (1998) 1179–1201, https://doi.org/10.1039/a805753b.

[11] Y. Zhu, P. Chandra, K.-M. Song, C. Ban, Y.-B. Shim, Label-free detection of kanamycin based on the aptamer-functionalized conducting polymer/gold nanocomposite, Biosens. Bioelectron. 36 (1) (2012) 29–34, https://doi.org/10.1016/J.BIOS.2012.03.034.

[12] B. Bhushan, Introduction to Nanotechnology, Springer, Berlin, Heidelberg, 2017, pp. 1–19.

[13] R. Mandal, A. Baranwal, A. Srivastava, P. Chandra, Evolving trends in bio/chemical sensor fabrication incorporating bimetallic nanoparticles, Biosens. Bioelectron. 117 (2018) 546–561, https://doi.org/10.1016/J.BIOS.2018.06.039.

[14] S. Alayoglu, B. Eichhorn, Rh–Pt bimetallic catalysts: synthesis, characterization, and catalysis of core–shell, alloy, and monometallic nanoparticles, J. Am. Chem. Soc. 130 (51) (2008) 17479–17486, https://doi.org/10.1021/ja8061425.

[15] B. Şen, et al., Polymer-graphene hybrid stabilized ruthenium nanocatalysts for the dimethylamine-borane dehydrogenation at ambient conditions, J. Mol. Liq. 279 (2019) 578–583, https://doi.org/10.1016/J.MOLLIQ.2019.02.003.

[16] R. Ghosh Chaudhuri, S. Paria, Core/shell nanoparticles: classes, properties, synthesis mechanisms, characterization, and applications, Chem. Rev. 112 (4) (2012) 2373–2433, https://doi.org/10.1021/cr100449n.

[17] A. Savk, et al., Multiwalled carbon nanotube-based nanosensor for ultrasensitive detection of uric acid, dopamine, and ascorbic acid, Mater. Sci. Eng. C 99 (2019) 248–254, https://doi.org/10.1016/j.msec.2019.01.113.

[18] S. Eris, Z. Daşdelen, Y. Yıldız, F. Sen, Nanostructured polyaniline-rGO decorated platinum catalyst with enhanced activity and durability for methanol oxidation, Int. J. Hydrogen Energy 43 (3) (2018) 1337–1343, https://doi.org/10.1016/j.ijhydene.2017.11.051.

[19] B. Şen, A. Aygün, A. Şavk, M.H. Çalımlı, S.K. Gülbay, F. Şen, Bimetallic palladium-cobalt nanomaterials as highly efficient catalysts for dehydrocoupling of dimethylamine borane, Int. J. Hydrogen Energy (2019), https://doi.org/10.1016/J.IJHYDENE.2019.01.215.

[20] X. Zhang, K.Y. Chan, Water-in-oil microemulsion synthesis of platinum-ruthenium nanoparticles, their characterization and electrocatalytic properties, Chem. Mater. (2003), https://doi.org/10.1021/cm0203868.

[21] R.M. Félix-Navarro, M. Beltrán-Gastélum, E.A. Reynoso-Soto, F. Paraguay-Delgado, G. Alonso-Nuñez, J.R. Flores-Hernández, Bimetallic Pt–Au nanoparticles supported on multi-wall carbon nanotubes as electrocatalysts for

oxygen reduction, Renew. Energy 87 (2016) 31–41, https://doi.org/10.1016/j.renene.2015.09.060.

[22] R. Katwal, H. Kaur, G. Sharma, M. Naushad, D. Pathania, Electrochemical synthesized copper oxide nanoparticles for enhanced photocatalytic and antimicrobial activity, J. Ind. Eng. Chem. 31 (2015) 173–184, https://doi.org/10.1016/J.JIEC.2015.06.021.

[23] J.S. Aaron, J. Oh, T.A. Larson, S. Kumar, T.E. Milner, K.V. Sokolov, Increased optical contrast in imaging of epidermal growth factor receptor using magnetically actuated hybrid gold/iron oxide nanoparticles, Opt. Express 14 (26) (2006) 12930, https://doi.org/10.1364/OE.14.012930.

[24] K.W. Cheah, et al., Monometallic and bimetallic catalysts based on Pd, Cu and Ni for hydrogen transfer deoxygenation of a prototypical fatty acid to diesel range hydrocarbons, Catal. Today (2019), https://doi.org/10.1016/J.CATTOD.2019.03.017.

[25] A. Baranwal, A.K. Chiranjivi, A. Kumar, V.K. Dubey, P. Chandra, Design of commercially comparable nanotherapeutic agent against human disease-causing parasite, Leishmania, Sci. Rep. 8 (1) (2018) 8814, https://doi.org/10.1038/s41598-018-27170-1.

[26] D.S. Sheny, J. Mathew, D. Philip, Phytosynthesis of Au, Ag and Au–Ag bimetallic nanoparticles using aqueous extract and dried leaf of Anacardium occidentale, Spectrochim. Acta A: Mol. Biomol. Spectrosc. 79 (1) (2011) 254–262, https://doi.org/10.1016/J.SAA.2011.02.051.

[27] S. Mondal, et al., Biogenic synthesis of Ag, Au and bimetallic Au/Ag alloy nanoparticles using aqueous extract of mahogany (*Swietenia mahogani* JACQ.) leaves, Colloids Surf. B: Biointerfaces 82 (2) (2011) 497–504, https://doi.org/10.1016/J.COLSURFB.2010.10.007.

[28] C. Tamuly, M. Hazarika, S.C. Borah, M.R. Das, M.P. Boruah, In situ biosynthesis of Ag, Au and bimetallic nanoparticles using *Piper pedicellatum* C.DC: green chemistry approach, Colloids Surf. B: Biointerfaces 102 (2013) 627–634, https://doi.org/10.1016/J.COLSURFB.2012.09.007.

[29] A. Kumari, V.K. Nigam, D.M. Pandey, Regulation of lignin biosynthesis through RNAi in aid of biofuel production, in: Microbial Factories, Springer India, New Delhi, 2015, pp. 185–201.

[30] S.S. Shankar, A. Rai, A. Ahmad, M. Sastry, Rapid synthesis of Au, Ag, and bimetallic Au core–Ag shell nanoparticles using neem (*Azadirachta indica*) leaf broth, J. Colloid Interface Sci. 275 (2) (2004) 496–502, https://doi.org/10.1016/J.JCIS.2004.03.003.

[31] S. Guo, S. Sun, FePt nanoparticles assembled on graphene as enhanced catalyst for oxygen reduction reaction, J. Am. Chem. Soc. 134 (5) (2012) 2492–2495, https://doi.org/10.1021/ja2104334.

[32] Y. Xu, S. Hou, Y. Liu, Y. Zhang, H. Wang, B. Zhang, Facile one-step room-temperature synthesis of Pt_3Ni nanoparticle networks with improved electro-catalytic properties, Chem. Commun. 48 (21) (2012) 2665–2667, https://doi.org/10.1039/C2CC16798K.

[33] X. Gu, Z.-H. Lu, H.-L. Jiang, T. Akita, Q. Xu, Synergistic catalysis of metal–organic framework-immobilized Au–Pd nanoparticles in dehydrogenation of formic acid for chemical hydrogen storage, J. Am. Chem. Soc. 133 (31) (2011) 11822–11825, https://doi.org/10.1021/ja200122f.

[34] B. Sen, E. Kuyuldar, A. Şavk, H. Calimli, S. Duman, F. Sen, Monodisperse ruthenium–copper alloy nanoparticles decorated on reduced graphene oxide for dehydrogenation of DMAB, Int. J. Hydrogen Energy 44 (21) (2019) 10744–10751, https://doi.org/10.1016/j.ijhydene.2019.02.176.

[35] S. Taçyıldız, B. Demirkan, Y. Karataş, M. Gulcan, F. Sen, Monodisperse Ru Rh bimetallic nanocatalyst as highly efficient catalysts for hydrogen generation from hydrolytic dehydrogenation of methylamine-borane, J. Mol. Liq. 285 (2019) 1–8, https://doi.org/10.1016/j.molliq.2019.04.019.

[36] H. Sert, et al., Activated carbon furnished monodisperse Pt nanocomposites as a superior adsorbent for methylene blue removal from aqueous solutions, J. Nanosci. Nanotechnol. 17 (7) (2017) 4799–4804, https://doi.org/10.1166/jnn.2017.13776.

[37] Y. Karataş, A. Aygun, M. Gülcan, F. Şen, A new highly active polymer supported ruthenium nanocatalyst for the hydrolytic dehydrogenation of dimethylamine-borane, J. Taiwan Inst. Chem. Eng. 99 (2019) 60–65.

[38] Y. Karatas, M. Gülcan, F. Sen, Catalytic methanolysis and hydrolysis of hydrazine-borane with MONODISPERSE Ru NPs@Nano-CeO_2 catalyst for hydrogen generation at room temperature, Int. J. *Hydrogen Energy* (2019), https://doi.org/10.1016/j.ijhydene.2019.04.012.

[39] B. Şen, A. Şavk, A. Şavk, S. Akocak, F. Şen, Bimetallic palladium–iridium alloy nanoparticles as highly efficient and stable catalyst for the hydrogen evolution reaction, Int. J. Hydrogen Energy 43 (44) (2018) 20183–20191, https://doi.org/10.1016/j.ijhydene.2018.07.081.

[40] B. Sen, A. Aygün, M. Ferdi Fellah, M. Harbi Calimli, F. Sen, Highly monodispersed palladium-ruthenium alloy nanoparticles assembled on poly(N-vinyl-pyrrolidone) for dehydrocoupling of dimethylamine–borane: an experimental and density functional theory study, J. Colloid Interface Sci. 546 (2019) 83–91, https://doi.org/10.1016/j.jcis.2019.03.057.

[41] B. Şen, A. Aygün, A. Şavk, C. Yenikaya, S. Cevik, F. Şen, Metal-organic frameworks based on monodisperse palladiumcobalt nanohybrids as highly active and reusable nanocatalysts for hydrogen generation, Int. J. Hydrogen Energy 44 (5) (2019) 2988–2996, https://doi.org/10.1016/J.IJHYDENE.2018.12.051.

[42] L. Zhang, et al., Cu^{2+}-assisted synthesis of hexoctahedral Au–Pd alloy nanocrystals with high-index facets, J. Am. Chem. Soc. 133 (43) (2011) 17114–17117, https://doi.org/10.1021/ja2063617.

[43] D. Wang, Q. Peng, Y. Li, Nanocrystalline intermetallics and alloys, Nano Res. 3 (8) (2010) 574–580, https://doi.org/10.1007/s12274-010-0018-4.

[44] Z. Wu, Anti-galvanic reduction of thiolate-protected gold and silver nanoparticles, Angew. Chem. Int. Ed. 51 (12) (2012) 2934–2938, https://doi.org/10.1002/anie.201107822.

[45] D. Wang, T. Xie, Q. Peng, Y. Li, Ag, Ag_2S, and Ag_2Se nanocrystals: synthesis, assembly, and construction of mesoporous structures, J. Am. Chem. Soc. (2008). https://doi.org/10.1021/JA710004H.

[46] J. Rugolo, J. Erlebacher, K. Sieradzki, Length scales in alloy dissolution and measurement of absolute interfacial free energy, Nat. Mater. 5 (12) (2006) 946–949, https://doi.org/10.1038/nmat1780.

[47] J. Snyder, T. Fujita, M.W. Chen, J. Erlebacher, Oxygen reduction in nanoporous metal–ionic liquid composite electrocatalysts, Nat. Mater. 9 (11) (2010) 904–907, https://doi.org/10.1038/nmat2878.

[48] D. Wang, Q. Lv, F. Liu, K. Wang, Quadrotor longitudinal controller based on L1 adaptive control method, J. Project. Rock. Miss. Guid. (2011).

[49] A.S. Lozhkomoev, et al., The formation of FeCu composite based on bimetallic nanoparticles, Vacuum 159 (2019) 441–446, https://doi.org/10.1016/J.VACUUM.2018.10.078.

[50] S.S. Alruqi, S.A. Al-Thabaiti, Z. Khan, Iron-nickel bimetallic nanoparticles: surfactant assisted synthesis and their catalytic activities, J. Mol. Liq. 282 (2019) 448–455, https://doi.org/10.1016/J.MOLLIQ.2019.03.021.

Chapter 23 The synthesis and characterization of size-controlled bimetallic nanoparticles **447**

[51] J. Kwon, X. Mao, J. Lee, Fe-based multifunctional nanoparticles with various physicochemical properties, Curr. Appl. Phys. 17 (8) (2017) 1066–1078, https://doi.org/10.1016/J.CAP.2017.04.018.

[52] C. Lei, Y. Sun, D.C.W. Tsang, D. Lin, Environmental transformations and ecological effects of iron-based nanoparticles, Environ. Pollut. 232 (2018) 10–30, https://doi.org/10.1016/J.ENVPOL.2017.09.052.

[53] A.V.B. Reddy, et al., Recent progress on Fe-based nanoparticles: synthesis, properties, characterization and environmental applications, J. Environ. Chem. Eng. 4 (3) (2016) 3537–3553, https://doi.org/10.1016/J.JECE.2016.07.035.

[54] M. Shah, Q.-X. Guo, Y. Fu, The colloidal synthesis of unsupported nickel-tin bimetallic nanoparticles with tunable composition that have high activity for the reduction of nitroarenes, Catal. Commun. 65 (2015) 85–90, https://doi.org/10.1016/J.CATCOM.2015.02.026.

[55] J. Pinkas, J. Sopoušek, P. Brož, V. Vykoukal, J. Buršík, J. Vřešťál, Synthesis, structure, stability and phase diagrams of selected bimetallic silver- and nickel-based nanoparticles, Calphad 64 (2019) 139–148, https://doi.org/10.1016/J.CALPHAD.2018.11.013.

[56] X. Ma, K. Qi, S. Wei, L. Zhang, X. Cui, In situ encapsulated nickel-copper nanoparticles in metal-organic frameworks for oxygen evolution reaction, J. Alloys Compd. 770 (2019) 236–242, https://doi.org/10.1016/J.JALLCOM.2018.08.096.

[57] R.M. Tesfaye, G. Das, B.J. Park, J. Kim, H.H. Yoon, Ni-Co bimetal decorated carbon nanotube aerogel as an efficient anode catalyst in urea fuel cells, Sci. Rep. 9 (1) (2019) 479, https://doi.org/10.1038/s41598-018-37011-w.

[58] Y. Lv, H. Cui, P. Liu, F. Hao, W. Xiong, H. Luo, Functionalized multi-walled carbon nanotubes supported Ni-based catalysts for adiponitrile selective hydrogenation to 6-aminohexanenitrile and 1,6-hexanediamine: switching selectivity with [Bmim]OH, J. Catal. 372 (2019) 330–351, https://doi.org/10.1016/J.JCAT.2019.03.023.

[59] V. Veeramani, et al., Metal organic framework derived nickel phosphide/graphitic carbon hybrid for electrochemical hydrogen generation reaction, J. Taiwan Inst. Chem. Eng. 96 (2019) 634–638, https://doi.org/10.1016/J.JTICE.2018.12.019.

[60] P. Puja, P. Kumar, A perspective on biogenic synthesis of platinum nanoparticles and their biomedical applications, Spectrochim. Acta A: Mol. Biomol. Spectrosc. 211 (2019) 94–99, https://doi.org/10.1016/J.SAA.2018.11.047.

[61] B. Karthikeyan, M. Murugavelu, Nano bimetallic Ag/Pt system as efficient opto and electrochemical sensing platform towards adenine, Sens. Actuators B 163 (1) (2012) 216–223, https://doi.org/10.1016/J.SNB.2012.01.039.

[62] M. Luo, et al., Palladium-based nanoelectrocatalysts for renewable energy generation and conversion, Mater. Today Nano 1 (2018) 29–40, https://doi.org/10.1016/J.MTNANO.2018.04.008.

24

The synthesis and characterization of size-controlled monometallic nanoparticles

Muhammed Bekmezci[a,b], Vildan Erduran[a,b], Mustafa Ucar[c], and Fatih Şen[a]

[a]Şen Research Group, Department of Biochemistry, Dumlupinar University, Kütahya, Turkey. [b]Department of Materials Science and Engineering, Faculty of Engineering, Dumlupinar University, Kütahya, Turkey. [c]Faculty of Science and Arts, Department of Chemistry, Afyon Kocatepe University, Afyonkarahisar, Turkey

1 Introduction

Fuel cells are technically considered to be a source of power that transforms chemical energy into electricity [1, 2]. When fuel cells are taken into consideration in terms of the hydrogen economy, it is predicted that they can replace conventional internal combustion engines in transportation vehicles [3–5]. It is possible to depict fuel cells with an electrolyte as shown in Fig. 1A. Hydrogen (H_2) is oxidized to the anode by releasing electrons that are then transferred to the cathode by an external circuit (e.g., electrical appliances) to minimize oxygen (O_2) [6–8]. In the context of transportable electronic apparatuses where the energy requirements are low, direct alcohol fuel cells (DAFCs) show the best features—namely, high energy density (in other words, small volume for the amount of power generated), low cost, relatively low working temperature, the safety of operation, and ease of storage and refueling.

The inherently slower half-reaction has a thin layer of electrocatalyst coated on each electrode to maintain the kinetics of the slow half-reaction, and an oxidation reaction and a reduction reaction occur on the anode and cathode, respectively. The most expensive and essential component of fuel cells is the electrocatalyst. The

Nanomaterials for Direct Alcohol Fuel Cells. https://doi.org/10.1016/B978-0-12-821713-9.00022-6
Copyright © 2021 Elsevier Inc. All rights reserved.

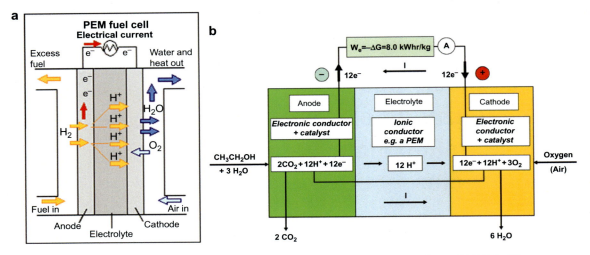

Fig. 1 Schematic illustration of (A) polymer electrolyte membrane fuel cells (PEMFCs) and (B) DAFCs [7].

basis of fuel cell work is to minimize cost and improve catalytic performance. In this chapter, anode and cathode catalysts will be examined in DAFCs. They play an essential role in reducing our dependence on fossil fuels. Thus, it has become one of the most popular power systems in the world. Nanostructured materials dramatically increase not only the electrolytes of DAFCs, but also the efficiency of the anode, cathode, and electrodes. This chapter emphasizes the importance of nanostructured monometallic materials in DAFCs. Low noble metal catalysts staged on various supports for the electrolytic reduction reaction are nanomaterials that increase the transfer of protons in electrolytes [7, 9–22]. Many materials have been synthesized for alcohol fuel cells, and research has been done on their efficiency [23, 24].

2 Direct alcohol fuel cells

Hydrogen is the most widely utilized anode fuel for all types of fuel cells due to the simplicity of the molecule and the reaction. The electro-oxidation of hydrogen is kinetically favorable, which can be effectively catalyzed by platinum (Pt) in both acidic and alkaline electrolytes without exerting any negative influence on the catalysts. Therefore, the academic and industrial focus on hydrogen fuel cells has concentrated on hydrogen production, hydrogen delivery, and hydrogen storage instead of hydrogen oxidation reaction (HOR). Hydrogen may be produced by steam reforming of fossil or alcohol fuels [8]. As a hydrogen carrier, methanol or ethanol can be used on the anode of a fuel cell after a steam reforming process in order to take advantage of the high efficiency of HOR. This type of fuel cell is called *reformed alcohol fuel cells (RAFCs)* or *indirect alcohol fuel cells (IAFCs)*.

In contrast, DAFCs immediately use the alcohol fuel as the electron donor on the anode. The fundamental difference between IAFCs and DAFCs is that IAFCs use hydrogen on the anode, while DAFCs use alcohol. Thereafter, a similar but broader category of fuel cells was defined as *direct liquid fuel cells (DLFCs),* where the liquid compound was directly fed into the anode.

In terms of storage of energy, DLFCs have shown great perspective as the power sources for portable electronic devices. Table 1 lists the theoretical energy densities of several fuels compared with those of gasoline and biodiesel. Methanol and ethanol have been under extensive and intensive study as anode fuels, leading to direct methanol fuel cells (DMFCs) and direct ethanol fuel cells (DEFCs), respectively. The electrode reactions are given by

$$CH_3OH + H_2O \rightarrow CO_2 + 6H^+ + 6e^- \quad E_0 = 0.016 \text{ V}$$
$$C_2H_5OH + 3H_2O \rightarrow 2CO_2 + 12H^+ + 12e^- \quad E_0 = 0.085 \text{ V}$$

Combined with the standard electrode potential of oxygen reduction reaction ($E_0 = 1.229$ V), the ideal potential (E_0^{eq}) for DMFCs and DEFCs are 1.213 and 1.144 V, respectively [8, 26].

Table 1 The energy density of different fuels [25].

Fuel	kWh/L	kWh/kg
Biodisel	9.17	11.72
Butanol	8.11	10.17
Ethanol	6.67	8.33
Ethylene glycol	5.87	5.30
Formic acid	2.10	1.72
Gasoline	9.50	12.89
Glucose	6.64	4.32
Glycerol	6.26	4.96
Hydrazine	5.36	5.42
Hydrogen (@700 bar)	1.56	39.72

2.1 Types of direct alcohol fuel cells

Various alcohol fuels have been used in DAFCs, with typical advantages and disadvantages. The most common fuel sources for DAFCs are methanol, ethanol, propanol, glycerol, and ethylene glycol. Reduction-oxidation reactions, theoretical cell potential energy, and energy intensities for each of the alcohols mentioned are given in Table 2. The given energy density is assumed to be 100% by the weight of the fuel [27].

DMFCs are a type of DAFC. Much research has been done on DMFCs due to their many advantageous properties. These features include that they can be supplied directly as a liquid without any action on the anode, have very good electrochemical activity, are easily biodegradable, and can be easily obtained due to their low costs. In addition, they have a high carbon content, and their easy transport and storage constitute the reason why methanol is preferred [28–31]. In addition, while the energy density is 180 Wh/L in hydrogen, 4820 Wh/L in methanol has a high energy

Table 2 General reactions of DAFCs with different alcohols [27].

Type of fuel	Reaction	Standard theoretical potential, E^0 (V)	Energy density (Wh/L)
Methanol	Anode: $CH_3OH + H_2O \rightarrow CO_2 + 6H^+ + 6e^-$ Cathode: $6H^+ + 6e^- + 3/2O_2 \rightarrow 3H_2O$ Overall: $CH_3OH + 3/2O_2 \rightarrow CO_2 + 2H_2O$	1.213	4820
Ethanol	Anode: $C_2H_5OH + 3H_2O \rightarrow 2CO_2 + 12H^+ + 12e^-$ Cathode: $12H^+ + 12e^- + 3O_2 \rightarrow 6H_2O$ Overall: $C_2H_5OH + 3O_2 \rightarrow 2CO_2 + 3H_2O$	1.145	6280
Propanol	Anode: $C_3H_7OH + 5H_2O \rightarrow 3CO_2 + 18H^+ + 18e^-$ Cathode: $18H^+ + 18e^- + 9/2O_2 \rightarrow 9H_2O$ Overall: $C_3H_7OH + 9/2O_2 \rightarrow 3CO_2 + 4H_2O$	1.122	7080
Ethylene glycol	Anode: $C_2H_6O_2 + 2H_2O \rightarrow 2CO_2 + 10H^+ + 10e^-$ Cathode: $10H^+ + 10e^- + 5/2O_2 \rightarrow 5H_2O$ Overall: $C_2H_6O_2 + 5/2O_2 \rightarrow 2CO_2 + 3H_2O$	1.220	5800
Glycerol	Anode: $C_3H_8O_3 + 3H_2O \rightarrow 3CO_2 + 14H^+ + 14e^-$ Cathode: $14H^+ + 14e^- + 7/2O_2 \rightarrow 7H_2O$ Overall: $C_3H_8O_3 + 7/2O_2 \rightarrow 3CO_2 + 4H_2O$	1.210	6400

Reprinted (adapted) with permission from G.L. Soloveichik, Liquid fuel cells, Beilstein J. Nanotechnol. 5 (2014) 1399–1418, https://doi.org/10.3762/bjnano.5.153. Copyright (2020) Beilstein J. Nanotechnol.

density. However, due to the disadvantage of these cells being toxic, necessary precautions should be taken.

Because it may be produced easily from raw materials, including sugar, it is thought that ethanol has some advantageous compared to the other alcohols. In addition, ethanol is seen as a promising fuel due to its lack of toxicity and high energy density (8030 Wh/kg) [32,33]. Furthermore, the exact reaction of ethanol is difficult to achieve due to the tight C—C bond, which limits the efficiency of the cleavage reaction. While this issue could be overcome by increasing the reaction temperature, Yang et al. observed that there was a decline in yield since the electrolyte membrane could be dehydrated at high temperatures [34]. Ethanol is larger in molecular size than methanol, thus reducing its effect on the cathode. Song et al. found that the permeability rate between methanol and membrane is smaller [35]. Because of the low electrochemical activity of ethanol, the oxidation reaction was slower in the anode. The power density of these DEFCs is about 1/7 of the DMFCs [33].

During the oxidation of DEFCs, by-products such as formaldehyde and acetic acid are produced which reduces yield. Acetic acid cannot be removed from electro-oxidation because it cannot oxidize any longer. It should be removed because it increases the complexity of the system [36]. The ethylene glycol molecule carries two hydroxyl groups. It is also used as antifreeze in automobiles. It is a good and preferred component in making polyethylene terephthalate (PET); PET production is over 7 million tons. The properties are superior to alcohols with a single hydroxyl group. This is because the vapor pressure is higher than that of a single hydroxyl group [37].

Glycerol is a by-product of biodiesel production, and it is thought that the demand for glycerol production will increase by sixfold after 2020 [38,39]. In appropriate methods, such as fuel cells, waste glycerol may be used. The price of raw glycerol is considerably cheaper than methanol and ethanol, as it is considered to be a waste product. This means that the fuel cell can be used more economically [40]. Glycerol is poisonous, reactive, and nonflammable; since it is bio-renewable, it is a promising fuel in fuel cell systems. In comparison, DGFCs have a high energy density of 6.4 kWh/L as a hypothesis [41]. Today, it is still being studied as a type of DAFC [42]. In addition, the current density of DGFCs is 200 mA/cm^2, and the power density was approximately 45 mW/cm^2.

2.2 Catalyst for direct alcohol fuel cells

In the DAFC system, the membrane electrode assembly (MEA) is the most crucial part of the determination of yield. MEAs consist of three parts: electrode, electrolyte, and support matter for electrode catalysts [43–46]. The catalyst is used to reduce the activation energy for the reaction to electro-oxidation. The reduction of the activation energy is important for the breaking of the bonds and a higher rate of reaction. This makes it possible to shorten the reaction time using catalysts. Catalysts are necessary for both the cathode and anode of the electrode in the MEA system. The oxygen reduction rate (ORR) in the cathode and the alcohol oxidation rate (AOR) in the anode are significantly affected by the catalysts [43,45,46]. Because the half-reactions for the anode and cathode are different, their designs must be distinct to ensure optimum efficiency.

2.2.1 Anode catalysts

The system's AOR is related to the anode. Hence, when choosing a catalyst, the ideal choice should be made according to the anode electrodes, because the main problem in DMFCs is usually anode catalysts. Some of these problems include the polarity of the catalyst, problems with the degradation of performance due to difficulties in breaking the C—C bond in alcohol molecules. Palladium (Pd) and Pt are the most common catalysts that are still accepted as anodes in DAFCs [47].

2.2.1.1 Pt-based electrocatalysts for the alcohol oxidation reaction

As the simplest alcohol, methanol has been employed as a model compound and provided insightful information about the mechanism of the electro-oxidation of alcohols. Meanwhile, Pt has been the metal most involved in catalyzing this reaction. Therefore, we will expand the discussion from the methanol-Pt system. Fig. 2 presents the reaction pathways for the methanol oxidation reaction (MOR) on the surface of Pt [48,49].

The strongly adsorbed carbonyl group (COads) and the weakly adsorbed carboxyl group (COOHads) involving reaction routes are

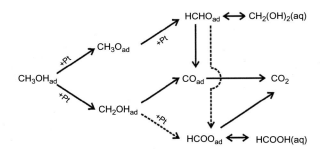

Fig. 2 Reaction pathways for methanol oxidation reaction on Pt [50].

called the *indirect* and *direct pathway*, respectively, for MORs [51]. The two pathways compete with each other from the bifurcation point. It has been proven beyond doubt that COads is the primary poisoning intermediate passivating the surface of Pt, which ultimately results in the sluggish kinetics of the electro-oxidation of methanol in an acidic electrolyte. Therefore, a more competent, Pt-based electrocatalyst for MORs should be able to inhibit the formation of COads on Pt or weaken the bonding between COads and Pt [8].

It is widely used in Pt fuel cell studies that give good results in the DMFC system [44,47,52]. In particular, the oxidative properties of ethanol in an acidic environment have been touted as promising by Ghumman et al. [53], especially in acidic environments. In their study, they used 1.0 mol of H_2SO_4 solution as the solution buffer, and they found that Pt completely oxidizes ethanol to CO_2 even though it is in few quantities. They also found that Pt was better performing Pt with other metals than alone Pt. As a result, on the surface, the Ru alloy is easily contaminated on the Pt surface, while increasing the tolerance for contaminants (COads) only at low temperatures and at the available concentrations. Only Pt at 80°C turns into more CO_2 than other catalysts. The reason for this is the increase in the Pt surface active area when the temperature increases. In addition, the poisoning effect increase [54]. At higher temperatures, it was found that anodic contamination increased and the active surface area was rapidly becoming unusable. The best way to keep the ethanol oxidation reaction (EOR) stable is to use a Pt-containing alloy catalyst at lower temperatures when a bimetallic catalyst is formed [47].

In the fuel cell operations, problems of commercialization of catalysts, especially Pt cost, were encountered. In addition to the high cost, the success of Pt for DMFC studies was not obtained in ethanol studies. In this case, many researchers began to work with Pd in the Pt group metals. Pd is a more electropositive metal than Pt. A number of studies found that the ideal metal for Pt alloying is Pd in DEFCs [47]. In an acidic medium, Pd was the least reactive catalyst and showed better performance in the alkaline environment than Pt. Since the cost of Pd is more affordable than Pt, doing study trials with Pd may reduce the cost of studies. In addition, the pH environment needs to be higher in order for Pd to perform well.

For this reason, there are some auxiliary catalysts for the best efficiency of Pd [55–59]. Therefore, Pd studies are performed in environments with high OH^- ion concentration. In Pd-based catalysts, as in Pt-based catalysts, the Pd-CO usage surface is formed, the COads are dissolved, and oxidation to CO_2 occurs [55–63]. Studies with Pd have been started and bimetallic catalyst studies are usually performed, but trimetallic catalyst studies are rarely performed.

In the ethanol oxidation studies, Chen et al. chose the base catalyst as Au [60]. Although Au is not active at low pH, it is relatively active at high pH, which is similar to Pd. In the basic environment, they found that the initial potency of Au was greater than 0.72, but Pt oxidized ethanol more slowly. Although the medium is added as a cocatalyst, adsorption and desorption on the Au surface are slow. Therefore, the reaction does not accelerate. In addition, when the oxidation of Au with different alcohols was compared, the best performance was given with ethylene glycol, and the worst performance was observed in methanol. Cao et al. chose iridium (Ir) as a catalyst in ethanol oxidation [64] compared to the results of Pt studies; they found that Ir was also sensitive to similar catalytic poisoning, and the value of the linear screening test decreased after a while. In addition, when added to the medium, the activity of Ir increased and the activity of Ir-Sn lend was found to be analogous PtSn activity. It may be thought that Ir is analogous to the surface area of Pt and may replace Pt. They found that Ir-Sn gave a steep gradient similar to the polarization curve of Pt-Sn. This means that IrFC can be trusted for use in DEFCs.

2.2.2 Cathode catalysts

The cathode and anode are very important to determine the performance of DEFCs. However, just being good is not good enough. If a weak anode with a good anode oxidizes ethanol but slowly lowers oxygen, performance is low. Therefore, to increase the efficiency of DEFCs, it is necessary to determine how both the anode and the cathode work. As the cathode surface area, Pt is used in basic and acidic environments [65,66]. This is due to the ability of Pt to efficiently reduce oxygen as a cathode. Although there has been no intensive study on DEFC cathode development, experiments with Pt showed high performance.

We will examine the catalyzed reaction before mentioning cathode catalysts. Because oxygen is in the air, it is used spontaneously as a cathode fuel. Therefore, many OORs that may be useful for the fuel cell have been extensively investigated. The reaction process of OOR can be defined as "direct or 4 electron paths or" series 2 electron processes [67,68].

In an acidic medium:

$$O_2 + 4H^+ + 4e^- \rightarrow 2H_2O \quad E_0 = 1.229 \text{ V}$$

or

$$O_2 + 2H^+ + 2e^- \rightarrow H_2O_2$$
$$H_2O_2 + 2H^+ + 2e^- \rightarrow 2H_2O$$

In an alkaline medium:

$$O_2 + 2H_2O + 4e^- \rightarrow 4OH^- \quad E_0 = 0.40 \text{ V}$$

or

$$O_2 + H_2O + 2e^- \rightarrow HO_2^- + OH^-$$
$$HO_2^- + H_2O + 2e^- \rightarrow 3OH^-$$

Pt/C performs very well in ORR as a cathode alone [47]. However, Lopes et al. added Pd to the structure as an alloy [69]. At the end of their study, they found that the catalyst reduced oxygen better. ORR performance increased due to both the deterioration of the Pt-Pt configuration and the distance between them. It is decreased at the active catalyst surface because of orbital spaces in sp^2 electrons in the oxygen structure. In other words, when the auxiliary catalyst is added to the medium such as Pd, Pt's affinity toward COads and OHads decreases, while the affinity for O_2 species increases. In addition, except for Pd, it has been discovered that Co can be improved in increasing the oxygen reduction ability of Pt [70]. It is improved not by increasing the tolerance of Co-Pt, but by increasing the ORR performance of the active surface. The reason for not increasing ethanol tolerance is that Pt/C and PtCo/C activity occur close to each other. When Pd is added, it increases both the ORR activity of Pt and ethanol tolerance. Generally, when an evaluation is made, the toleration of the catalyst against EOR activity increases, as does the ORR performance.

In a different study, it was observed that ORR performance increased with the addition of catalysts such as Ni and Sn [71]. As shown by Pd, COads reduce the adsorption of species such as OHads, while increasing the absorption of O_2 species. It also weakens the O—O bond to facilitate absorption. PtSn/C bimetallic catalyst formation was seen to demonstrate higher oxygen reactivity than PtSnNi/C trimetallic catalysts. Here, it can be concluded that the trimetallic catalyst overlaps the surface area in the OOR mechanism. Therefore, a comprehensive DOE is required to minimize the possibility of overlap. Thus, the optimum proportion among the base and the auxiliary catalysts is determined.

The PdAu alloy was investigated as a cathode by Xu et al., and the researchers found that catalysts are comparable to Pt cathodes in terms of performance [72]. OOR alkaline environment (KOH) at 50 mV/s up to a maximum of 50 mA/cm^2. In the previous experiments, due to the low ethanol oxidation rate, PdAu alloy showed

458 Chapter 24 The synthesis and characterization of size-controlled monometallic nanoparticles

high performance as the cathode [60,73]. This is thought to be due to the presence of Au. Because Au does not perform well during oxidation, it has an inverse effect on the reduction reaction.

In addition, the development of nanomaterials and fuel cell technology is hoped. There are examples in the literature on the development of nanomaterials [9–11,18,20,22,74–83].

3 Conclusions

Many types of monometallic nanomaterials have been used to improve the efficiency of alcohol fuel cells as electrocatalysts. In general, an oxidation reaction and a reduction reaction occur on the anode and the cathode, respectively, with a thin layer of electrocatalyst coating on each electrode to support the half-reaction, the kinetics of which is intrinsically sluggish. The most expensive and essential component of fuel cells is the electrocatalyst. The basis of fuel cell work is to minimize costs and improve catalytic performance.

For this reason, in this chapter, anode, and cathode monometallic catalysts in alcohol fuel cells were examined. Nanostructured materials increase dramatically, not only the electrolytes of DAFCs, but also the efficiency of the anode, cathode, and electrodes. This chapter emphasized the importance of nanostructured monometallic materials in DAFCs. Low noble metal catalysts staged on various supports for the electrolytic reduction reaction are nanomaterials that increase the transfer of protons in electrolytes. This has increased expectations about the efficiency of DAFCs; however, further improvements are required in order to improve the efficiency of DAFCs and the use of them in daily life.

References

[1] M.S. Dresselhaus, I.L. Thomas, Alternative energy technologies, Nature 414 (2001) 332–337, https://doi.org/10.1038/35104599.

[2] B.C.H. Steele, A. Heinzel, Materials for fuel-cell technologies, Nature 414 (2001) 345–352, https://doi.org/10.1038/35104620.

[3] T. Plan, Fuel Cells (2015) 1–49. 3.4.

[4] J. Tollefson, Hydrogen vehicles: fuel of the future? Nature 464 (2010) 1262–1264, https://doi.org/10.1038/4641262a.

[5] G.D. Berry, A.D. Pasternak, G.D. Rambach, J. Ray Smith, R.N. Schock, Hydrogen as a future transportation fuel, Energy 21 (1996) 289–303, https://doi.org/10.1016/0360-5442(95)00104-2.

[6] C. Lamy, Direct alcohol fuel cells (DAFCs), Encycl. Appl. Electrochem. (2014) 321–330, https://doi.org/10.1007/978-1-4419-6996-5_188.

Chapter 24 The synthesis and characterization of size-controlled monometallic nanoparticles **459**

[7] Y. Wang, K.S. Chen, J. Mishler, S.C. Cho, X.C. Adroher, A review of polymer electrolyte membrane fuel cells: technology, applications, and needs on fundamental research, Appl. Energy 88 (2011) 981–1007, https://doi.org/10.1016/j.apenergy.2010.09.030.

[8] L. Su, Noble Metal Based Nanomaterials in the Application of Direct Alcohol Fuel Cells, 2013.

[9] S. Eris, Z. Daşdelen, F. Sen, Investigation of electrocatalytic activity and stability of Pt@f-VC catalyst prepared by in-situ synthesis for methanol electrooxidation, Int. J. Hydrogen Energy 43 (2018) 385–390, https://doi.org/10.1016/j.ijhydene.2017.11.063.

[10] S. Eris, Z. Daşdelen, Y. Yıldız, F. Sen, Nanostructured polyaniline-rGO decorated platinum catalyst with enhanced activity and durability for methanol oxidation, Int. J. Hydrogen Energy 43 (2018) 1337–1343, https://doi.org/10.1016/j.ijhydene.2017.11.051.

[11] S. Eris, Z. Daşdelen, F. Sen, Enhanced electrocatalytic activity and stability of monodisperse Pt nanocomposites for direct methanol fuel cells, J. Colloid Interface Sci. 513 (2018) 767–773, https://doi.org/10.1016/j.jcis.2017.11.085.

[12] F. Şen, G. Gökağaç, S. Şen, High performance Pt nanoparticles prepared by new surfactants for C1 to C3 alcohol oxidation reactions, J. Nanopart. Res. 15 (2013) 1979, https://doi.org/10.1007/s11051-013-1979-5.

[13] F. Şen, G. Gökağaç, Pt nanoparticles synthesized with new surfactants: improvement in C1–C3 alcohol oxidation catalytic activity, J. Appl. Electrochem. 44 (2014) 199–207, https://doi.org/10.1007/s10800-013-0631-5.

[14] E. Erken, Y. Yıldız, B. Kilbaş, F. Şen, Synthesis and characterization of nearly monodisperse Pt nanoparticles for C 1 to C 3 alcohol oxidation and dehydrogenation of dimethylamine-borane (DMAB), J. Nanosci. Nanotechnol. 16 (2016) 5944–5950, https://doi.org/10.1166/jnn.2016.11683.

[15] S. Ertan, F. Şen, S. Şen, G. Gökağaç, Platinum nanocatalysts prepared with different surfactants for C1–C3 alcohol oxidations and their surface morphologies by AFM, J. Nanopart. Res. 14 (2012) 922, https://doi.org/10.1007/s11051-012-0922-5.

[16] Y. Yıldız, S. Kuzu, B. Sen, A. Savk, S. Akocak, F. Şen, Different ligand based monodispersed Pt nanoparticles decorated with rGO as highly active and reusable catalysts for the methanol oxidation, Int. J. Hydrogen Energy 42 (2017) 13061–13069, https://doi.org/10.1016/j.ijhydene.2017.03.230.

[17] Z. Ozturk, F. Sen, S. Sen, G. Gokagac, The preparation and characterization of nano-sized Pt–Pd/C catalysts and comparison of their superior catalytic activities for methanol and ethanol oxidation, J. Mater. Sci. 47 (2012) 8134–8144, https://doi.org/10.1007/s10853-012-6709-3.

[18] Ö. Karatepe, Y. Yildiz, H. Pamuk, S. Eris, Z. Dasdelen, F. Sen, Enhanced electrocatalytic activity and durability of highly monodisperse Pt@PPy–PANI nanocomposites as a novel catalyst for the electro-oxidation of methanol, RSC Adv. 6 (2016) 50851–50857, https://doi.org/10.1039/c6ra06210e.

[19] B. Çelik, S. Kuzu, E. Erken, H. Sert, Y. Koşkun, F. Şen, Nearly monodisperse carbon nanotube furnished nanocatalysts as highly efficient and reusable catalyst for dehydrocoupling of DMAB and C1 to C3 alcohol oxidation, Int. J. Hydrogen Energy 41 (2016) 3093–3101, https://doi.org/10.1016/j.ijhydene.2015.12.138.

[20] Z. Daşdelen, Y. Yıldız, S. Eriş, F. Şen, Enhanced electrocatalytic activity and durability of Pt nanoparticles decorated on GO-PVP hybride material for methanol oxidation reaction, Appl. Catal. Environ. 219 (2017) 511–516, https://doi.org/10.1016/j.apcatb.2017.08.014.

[21] B. Çelik, G. Başkaya, H. Sert, Ö. Karatepe, E. Erken, F. Şen, Monodisperse Pt(0)/DPA@GO nanoparticles as highly active catalysts for alcohol oxidation and

dehydrogenation of DMAB, Int. J. Hydrogen Energy 41 (2016) 5661–5669, https://doi.org/10.1016/j.ijhydene.2016.02.061.

[22] S. Şen, F. Şen, G. Gökağaç, Preparation and characterization of nano-sized Pt-Ru/C catalysts and their superior catalytic activities for methanol and ethanol oxidation, Phys. Chem. Chem. Phys. 13 (2011) 6784–6792, https://doi.org/10.1039/c1cp20064j.

[23] Use of Carbon-Nanotube Based Materials in Microbial Fuel Cells, 2019, pp. 151–176, https://doi.org/10.21741/9781644900116-7.

[24] F. Şen, Mesoporous materials in biofuel cells, Mater. Res. Found. (2019) 157–172, https://doi.org/10.21741/9781644900079-7.

[25] K. Mazloomi, C. Gomes, Hydrogen as an energy carrier: prospects and challenges, Renew. Sustain. Energy Rev. 16 (2012) 3024–3033, https://doi.org/10.1016/j.rser.2012.02.028.

[26] W. Vielstich, A. Lamm, H.A. Gasteiger, H. Yokokawa, Handbook of Fuel Cells: Fundamentals, Technology, and Applications, Wiley, 2003.

[27] B.C. Ong, S.K. Kamarudin, S. Basri, Direct liquid fuel cells: a review, Int. J. Hydrogen Energy 42 (2017) 10142–10157, https://doi.org/10.1016/j.ijhydene.2017.01.117.

[28] U.B. Demirci, How green are the chemicals used as liquid fuels in direct liquid-feed fuel cells? Environ. Int. 35 (2009) 626–631, https://doi.org/10.1016/J.ENVINT.2008.09.007.

[29] C. Lamy, A. Lima, V. LeRhun, F. Delime, C. Coutanceau, J.-M. Léger, Recent advances in the development of direct alcohol fuel cells (DAFC), J. Power Sources 105 (2002) 283–296, https://doi.org/10.1016/S0378-7753(01)00954-5.

[30] A.M. Zainoodin, S.K. Kamarudin, M.S. Masdar, W.R.W. Daud, A.B. Mohamad, J. Sahari, Investigation of MEA degradation in a passive direct methanol fuel cell under different modes of operation, Appl. Energy 135 (2014) 364–372, https://doi.org/10.1016/J.APENERGY.2014.08.036.

[31] F. Sen, G. Gokagac, Activity of carbon-supported platinum nanoparticles toward methanol oxidation reaction: role of metal precursor and a new surfactant, *tert*-octanethiol, J. Phys. Chem. C 111 (3) (2007) 1467–1473.

[32] W. Qian, D.P. Wilkinson, J. Shen, H. Wang, J. Zhang, Architecture for portable direct liquid fuel cells, J. Power Sources 154 (2006) 202–213, https://doi.org/10.1016/J.JPOWSOUR.2005.12.019.

[33] N. Fujiwara, Z. Siroma, S. Yamazaki, T. Ioroi, H. Senoh, K. Yasuda, Direct ethanol fuel cells using an anion exchange membrane, J. Power Sources 185 (2008) 621–626, https://doi.org/10.1016/J.JPOWSOUR.2008.09.024.

[34] C. Yang, P. Costamagna, S. Srinivasan, J. Benziger, A.B. Bocarsly, Approaches and technical challenges to high temperature operation of proton exchange membrane fuel cells, J. Power Sources 103 (2001) 1–9, https://doi.org/10.1016/S0378-7753(01)00812-6.

[35] S. Song, W. Zhou, Z. Liang, R. Cai, G. Sun, Q. Xin, V. Stergiopoulos, P. Tsiakaras, The effect of methanol and ethanol cross-over on the performance of PtRu/C-based anode DAFCs, Appl. Catal. Environ. 55 (2005) 65–72, https://doi.org/10.1016/J.APCATB.2004.05.017.

[36] A. Jablonski, A. Lewera, Electrocatalytic oxidation of ethanol on Pt, Pt-Ru and Pt-Sn nanoparticles in polymer electrolyte membrane fuel cell—role of oxygen permeation, Appl. Catal. Environ. 115–116 (2012) 25–30, https://doi.org/10.1016/J.APCATB.2011.12.021.

[37] J. Garche, C.K. Dyer, Encyclopedia of Electrochemical Power Sources, Academic Press, 2009.

[38] R. Christoph, B. Schmidt, U. Steinberner, W. Dilla, R. Karinen, Glycerol, in: Ullmann's Encycl. Ind. Chem., Wiley-VCH Verlag Gmb H & Co. KGaA, Weinheim, Germany, 2006, https://doi.org/10.1002/14356007.a12_477.pub2.

Chapter 24 The synthesis and characterization of size-controlled monometallic nanoparticles **461**

[39] J. Van Gerpen, Biodiesel processing and production, Fuel Process. Technol. 86 (2005) 1097–1107, https://doi.org/10.1016/J.FUPROC.2004.11.005.

[40] Z. Zhang, L. Xin, J. Qi, D.J. Chadderdon, W. Li, Supported Pt, Pd and Au nanoparticle anode catalysts for anion-exchange membrane fuel cells with glycerol and crude glycerol fuels, Appl. Catal. Environ. 136–137 (2013) 29–39, https://doi.org/10.1016/J.APCATB.2013.01.045.

[41] Z. Zhang, L. Xin, W. Li, Supported gold nanoparticles as anode catalyst for anion-exchange membrane-direct glycerol fuel cell (AEM-DGFC), Int. J. Hydrogen Energy 37 (2012) 9393–9401, https://doi.org/10.1016/J.IJHYDENE.2012.03.019.

[42] X. Han, D.J. Chadderdon, J. Qi, L. Xin, W. Li, W. Zhou, Numerical analysis of anion-exchange membrane direct glycerol fuel cells under steady state and dynamic operations, Int. J. Hydrogen Energy 39 (2014) 19767–19779, https://doi.org/10.1016/J.IJHYDENE.2014.08.144.

[43] E. Antolini, E.R. Gonzalez, Alkaline direct alcohol fuel cells, J. Power Sources 195 (2010) 3431–3450, https://doi.org/10.1016/j.jpowsour.2009.11.145.

[44] A.M. Zainoodin, S.K. Kamarudin, W.R.W. Daud, Electrode in direct methanol fuel cells, Int. J. Hydrogen Energy 35 (2010) 4606–4621, https://doi.org/10.1016/J.IJHYDENE.2010.02.036.

[45] M.Z.F. Kamarudin, S.K. Kamarudin, M.S. Masdar, W.R.W. Daud, Review: direct ethanol fuel cells, Int. J. Hydrogen Energy 38 (2013) 9438–9453, https://doi.org/10.1016/j.ijhydene.2012.07.059.

[46] J.N. Tiwari, R.N. Tiwari, G. Singh, K.S. Kim, Recent progress in the development of anode and cathode catalysts for direct methanol fuel cells, Nano Energy 2 (2013) 553–578, https://doi.org/10.1016/J.NANOEN.2013.06.009.

[47] M.A.F. Akhairi, S.K. Kamarudin, Catalysts in direct ethanol fuel cell (DEFC): an overview, Int. J. Hydrogen Energy 41 (2016) 4214–4228, https://doi.org/10.1016/j.ijhydene.2015.12.145.

[48] S. Minteer, Alcoholic Fuels, CRC Press, 2006, https://doi.org/10.1201/9781420020700.

[49] A. Więckowski, J.K. Nørskov, Fuel Cell Science: Theory, Fundamentals, and Biocatalysis, Wiley, 2010.

[50] S.X. Liu, L.W. Liao, Q. Tao, Y.X. Chen, S. Ye, The kinetics of CO pathway in methanol oxidation at Pt electrodes, a quantitative study by ATR-FTIR spectroscopy, Phys. Chem. Chem. Phys. (2011), https://doi.org/10.1039/c0cp01728k.

[51] M.T.M. Koper, Fuel Cell Catalysis: A Surface Science approach, Wiley, 2009.

[52] S. Basri, S.K. Kamarudin, W.R.W. Daud, Z. Yaakub, Nanocatalyst for direct methanol fuel cell (DMFC), Int. J. Hydrogen Energy 35 (2010) 7957–7970, https://doi.org/10.1016/j.ijhydene.2010.05.111.

[53] A. Ghumman, C. Vink, O. Yepez, P.G. Pickup, Continuous monitoring of CO_2 yields from electrochemical oxidation of ethanol: catalyst, current density and temperature effects, J. Power Sources 177 (2008) 71–76, https://doi.org/10.1016/J.JPOWSOUR.2007.11.009.

[54] S.C. Zignani, V. Baglio, J.J. Linares, G. Monforte, E.R. Gonzalez, A.S. Aricò, Endurance study of a solid polymer electrolyte direct ethanol fuel cell based on a Pt–Sn anode catalyst, Int. J. Hydrogen Energy 38 (2013) 11576–11582, https://doi.org/10.1016/J.IJHYDENE.2013.04.162.

[55] Z. Zhang, L. Xin, K. Sun, W. Li, Pd–Ni electrocatalysts for efficient ethanol oxidation reaction in alkaline electrolyte, Int. J. Hydrogen Energy 36 (2011) 12686–12697, https://doi.org/10.1016/j.ijhydene.2011.06.141.

[56] P.-C. Su, H.-S. Chen, T.-Y. Chen, C.-W. Liu, C.-H. Lee, J.-F. Lee, T.-S. Chan, K.-W. Wang, Enhancement of electrochemical properties of Pd/C catalysts toward ethanol oxidation reaction in alkaline solution through Ni and Au alloying,

Int. J. Hydrogen Energy 38 (2013) 4474–4482, https://doi.org/10.1016/J. IJHYDENE.2013.01.173.

[57] A.N. Geraldes, D.F. da Silva, E.S. Pino, J.C.M. da Silva, R.F.B. de Souza, P. Hammer, E.V. Spinacé, A.O. Neto, M. Linardi, M.C. dos Santos, Ethanol electro-oxidation in an alkaline medium using Pd/C, Au/C and PdAu/C electrocatalysts prepared by electron beam irradiation, Electrochim. Acta 111 (2013) 455–465, https://doi.org/10.1016/j.electacta.2013.08.021.

[58] Y.S. Li, T.S. Zhao, A high-performance integrated electrode for anion-exchange membrane direct ethanol fuel cells, Int. J. Hydrogen Energy 36 (2011) 7707–7713, https://doi.org/10.1016/J.IJHYDENE.2011.03.090.

[59] S.Y. Shen, T.S. Zhao, J.B. Xu, Y.S. Li, Synthesis of PdNi catalysts for the oxidation of ethanol in alkaline direct ethanol fuel cells, J. Power Sources 195 (2010) 1001–1006, https://doi.org/10.1016/J.JPOWSOUR.2009.08.079.

[60] Y. Chen, L. Zhuang, J. Lu, Non-Pt anode catalysts for alkaline direct alcohol fuel cells, Chinese J. Catal. 28 (2007) 870–874, https://doi.org/10.1016/S1872-2067(07)60073-4.

[61] H. Gao, S. Liao, Z. Liang, H. Liang, F. Luo, Anodic oxidation of ethanol on core-shell structured Ru@PtPd/C catalyst in alkaline media, J. Power Sources 196 (2011) 6138–6143, https://doi.org/10.1016/J.JPOWSOUR.2011.03.031.

[62] J. Cai, Y. Huang, Y. Guo, Bi-modified Pd/C catalyst via irreversible adsorption and its catalytic activity for ethanol oxidation in alkaline medium, Electrochim. Acta 99 (2013) 22–29, https://doi.org/10.1016/J.ELECTACTA.2013.03.059.

[63] N. Li, Y.-X. Zeng, S. Chen, C.-W. Xu, P.-K. Shen, Ethanol oxidation on Pd/C enhanced by MgO in alkaline medium, Int. J. Hydrogen Energy 39 (2014) 16015–16019, https://doi.org/10.1016/J.IJHYDENE.2013.12.122.

[64] L. Cao, G. Sun, H. Li, Q. Xin, Carbon-supported IrSn catalysts for a direct ethanol fuel cell, Electrochem. Commun. 9 (2007) 2541–2546, https://doi.org/10.1016/J.ELECOM.2007.07.031.

[65] M. Zhiani, H.A. Gasteiger, M. Piana, S. Catanorchi, Comparative study between carbon supported platinum and non-noble metal cathode catalyst in alkaline direct ethanol fuel cell (ADEFC), Int. J. Hydrogen Energy 36 (2011) 5110–5116, https://doi.org/10.1016/J.IJHYDENE.2011.01.079.

[66] N.A. Karim, S.K. Kamarudin, L.K. Shyuan, Z. Yaakob, W.R.W. Daud, A.A.H. Khadum, Novel cathode catalyst for DMFC: study of the density of states of oxygen adsorption using density functional theory, Int. J. Hydrogen Energy 39 (2014) 17295–17305, https://doi.org/10.1016/J.IJHYDENE.2014.06.110.

[67] M. Winter, R.J. Brodd, What are batteries, fuel cells, and supercapacitors? Chem. Rev. 104 (2004) 4245–4270, https://doi.org/10.1021/cr020730k.

[68] J. Zhang, PEM Fuel Cell Electrocatalysts and Catalyst Layers, Springer London, London, 2008, https://doi.org/10.1007/978-1-84800-936-3.

[69] T. Lopes, E. Antolini, E.R. Gonzalez, Carbon supported Pt–Pd alloy as an ethanol tolerant oxygen reduction electrocatalyst for direct ethanol fuel cells, Int. J. Hydrogen Energy 33 (2008) 5563–5570, https://doi.org/10.1016/J.IJHYDENE.2008.05.030.

[70] T. Lopes, E. Antolini, F. Colmati, E.R. Gonzalez, Carbon supported Pt–Co (3, 1) alloy as improved cathode electrocatalyst for direct ethanol fuel cells, J. Power Sources 164 (2007) 111–114, https://doi.org/10.1016/J.JPOWSOUR.2006.10.052.

[71] S. Beyhan, N.E. Şahin, S. Pronier, J.-M. Léger, F. Kadırgan, Comparison of oxygen reduction reaction on Pt/C, Pt-Sn/C, Pt-Ni/C, and Pt-Sn-Ni/C catalysts prepared by Bönnemann method: a rotating ring disk electrode study, Electrochim. Acta 151 (2015) 565–573, https://doi.org/10.1016/J.ELECTACTA.2014.11.053.

Chapter 24 The synthesis and characterization of size-controlled monometallic nanoparticles **463**

[72] J.B. Xu, T.S. Zhao, Y.S. Li, W.W. Yang, Synthesis and characterization of the Au-modified Pd cathode catalyst for alkaline direct ethanol fuel cells, Int. J. Hydrogen Energy 35 (2010) 9693–9700, https://doi.org/10.1016/J. IJHYDENE.2010.06.074.

[73] Y.-H. Qin, Y. Li, R.-L. Lv, T.-L. Wang, W.-G. Wang, C.-W. Wang, Pd-Au/C catalysts with different alloying degrees for ethanol oxidation in alkaline media, Electrochim. Acta 144 (2014) 50–55, https://doi.org/10.1016/J. ELECTACTA.2014.08.078.

[74] F. Şen, G. Gökagaç, Activity of carbon-supported platinum nanoparticles toward methanol oxidation reaction: role of metal precursor and a new surfactant, tert-octanethiol, J. Phys. Chem. C 111 (2007) 1467–1473, https://doi.org/10.1021/jp065809y.

[75] F. Sen, Y. Karatas, M. Gulcan, M. Zahmakiran, Amylamine stabilized platinum (0) nanoparticles: active and reusable nanocatalyst in the room temperature dehydrogenation of dimethylamine-borane, RSC Adv. 4 (2014) 1526–1531, https://doi.org/10.1039/c3ra43701a.

[76] B. Sen, B. Demirkan, B. Şimşek, A. Savk, F. Sen, Monodisperse palladium nanocatalysts for dehydrocoupling of dimethylamineborane, Nano-Struct. Nano-Objects 16 (2018) 209–214, https://doi.org/10.1016/j.nanoso.2018.07.008.

[77] B. Sen, E. Kuyuldar, A. Şavk, H. Calimli, S. Duman, F. Sen, Monodisperse ruthenium–copper alloy nanoparticles decorated on reduced graphene oxide for dehydrogenation of DMAB, Int. J. Hydrogen Energy 44 (2019) 10744–10751, https://doi.org/10.1016/j.ijhydene.2019.02.176.

[78] B. Şen, B. Demirkan, M. Levent, A. Şavk, F. Şen, Silica-based monodisperse PdCo nanohybrids as highly efficient and stable nanocatalyst for hydrogen evolution reaction, Int. J. Hydrogen Energy 43 (2018) 20234–20242, https://doi.org/10.1016/j.ijhydene.2018.07.080.

[79] B. Sen, S. Kuzu, E. Demir, S. Akocak, F. Sen, Highly monodisperse RuCo nanoparticles decorated on functionalized multiwalled carbon nanotube with the highest observed catalytic activity in the dehydrogenation of dimethylamine–borane, Int. J. Hydrogen Energy 42 (2017) 23292–23298, https://doi.org/10.1016/j.ijhydene.2017.06.032.

[80] B. Şen, B. Demirkan, A. Savk, R. Kartop, M.S. Nas, M.H. Alma, S. Sürdem, F. Şen, High-performance graphite-supported ruthenium nanocatalyst for hydrogen evolution reaction, J. Mol. Liq. 268 (2018) 807–812, https://doi.org/10.1016/j.molliq.2018.07.117.

[81] Y. Yildiz, E. Erken, H. Pamuk, H. Sert, F. Şen, Monodisperse Pt nanoparticles assembled on reduced graphene oxide: highly efficient and reusable catalyst for methanol oxidation and dehydrocoupling of dimethylamine-borane (DMAB), J. Nanosci. Nanotechnol. 16 (2016) 5951–5958, https://doi.org/10.1166/jnn.2016.11710.

[82] F. Şen, S. Şen, G. Gökağaç, Efficiency enhancement of methanol/ethanol oxidation reactions on Pt nanoparticles prepared using a new surfactant, 1,1-dimethyl heptanethiol, Phys. Chem. Chem. Phys. 13 (2011) 1676–1684, https://doi.org/10.1039/c0cp01212b.

[83] F. Şen, G. Gökagaç, Different sized platinum nanoparticles supported on carbon: an XPS study on these methanol oxidation catalysts, J. Phys. Chem. C (2007), https://doi.org/10.1021/JP068381B.

25

Topics on the fundamentals of the alcohol oxidation reactions in acid and alkaline electrolytes

Vildan Erduran[a,b], Merve Akin[a,b], Hakan Burhan[a], Iskender Isik[b], and Fatih Şen[a]

[a]Şen Research Group, Department of Biochemistry, Dumlupinar University, Kütahya, Turkey. [b]Department of Materials Science and Engineering, Faculty of Engineering, Dumlupinar University, Kütahya, Turkey

1 Introduction

The interest in alcohol oxidation reactions, which started with research in the field of energy and fuel that began in the early 1900s, has gained interest with studies on meeting the world's increasing energy demands and finding and storing alternative energy sources [1].

Electrochemical oxidation of alcohols is especially important for direct alcohol fuel cells (DAFCs). In the studies carried out in this field, this issue has developed and methanol and ethanol are the most commonly used alcohols in studies on alcohol oxidation reactions [2–4]. Because the relative reactivity of these alcohols for their electrooxidation reactions is very high, they have a high energy density, can be easily obtained from renewable sources, and most importantly, they are ideal fuels because they can be converted directly into electrical energy [5–9]. Apart from these alcohols, studies involving isopropanol and butanol have been performed [1, 5, 10–17]. In contrast to all these advantages, the oxidation reactions of these alcohols on the platinum (Pt) and Pt-containing anode electrode the surface cannot be completed and take place slowly [18]. Pt has also some disadvantageous due to its limited reserves and high cost. Therefore, the search for alternative electrodes continues. As a catalyst, pure metals can be used directly in most systems, as well as in combination

---

Nanomaterials for Direct Alcohol Fuel Cells. https://doi.org/10.1016/B978-0-12-821713-9.00003-2
Copyright © 2021 Elsevier Inc. All rights reserved.

with various metals. In addition, most electrode materials are used with the carbon support surface. The desired properties of catalysts to be used in alcohol oxidation reactions include the following [1, 6, 14, 16, 19, 20]:

- Large surface area
- High electrical conductivity
- Good support and nanoparticle interaction
- Appropriate porosity
- Corrosion resistance
- Purity and inertness
- Thermal conductivity and resistance
- Ability to work with anion and cation exchange membranes
- Adsorption of gas or liquid on the surface

Alcohol oxidation reactions are complex reactions involving multielectron transfer and many intermediate steps. The mechanisms of these reactions involve the formation of many adsorbed intermediates, products, and by-products. It is very difficult to explain the mechanisms of these complex reactions of alcohols because the oxidation of alcohols occurs in many different ways, catalytic surfaces, and environments [21, 22].

Oxidation reactions of short-chain alcohol molecules are explained by a double-path mechanism [23]. Alcohol molecules contain two active centers: the hydroxide group and the carbon atom to which this group is attached. During the adsorption process, these active centers interact with the metal catalyst and can be adsorbed on the surface of the metal catalyst, forming intermediate products that break down the oxygen-hydrogen bond or the carbon-hydrogen bond before the carbon-carbon bond and the carbon-oxygen bond are split. This phenomenon is the first step of alcohol oxidation.

Both carbon-containing and oxygen-containing groups are turned into aldehyde by removing protons. According to the dual-path mechanism of primary alcohols, the reaction follows two paths [14, 23, 24]:

- Adsorption of alcohol on the catalyst surface and the formation of carbon monoxide (CO) adsorbed by direct dehydrogenation will occur. This molecule will then be oxidized to carbon dioxide (CO_2).
- Nonadsorbed intermediates will turn into oxidation products after a series of reactions.

Due to the complex reaction path and a large number of intermediates and products that form, alcohol oxidation reactions wear down the Pt surface, and catalyst reactivity decreases. With the increasing length of the carbon chain (methanol > ethanol > propanol > butanol), the oxidation reaction reactivity on the Pt surface decreases [14, 25]. Another problem that reduces reactivity is the adsorption of the carbon monoxide (CO) gas formed on the catalyst surface, which causes poisoning of the catalyst. The removal of this problem depends on the oxidizing species in the environment. These species are formed as a result of the ionization of water at high potentials.

Alcohols contain three types of bonds. These bonds include a carbon-hydrogen bond, which is very easy to break; a carbon-carbon bond, which is more difficult to break; and a carbon-oxygen bond, which does not break even if a catalyst such as Pt is used. A good understanding of the oxidation reactions of alcohols can be done using methanol, the simplest alcohol molecule. Methanol is carbon-carbon-bond-free alcohol that provides six electrons with its oxidation to CO_2, and it is the most used alcohol in alcohol oxidation reactions. Even in this state, it is a molecule that is completely oxidized by a multistep sequence of reactions and creates many intermediate products such as formaldehyde and formic acid. Therefore, it is clear that other alcohol molecules containing carbon-carbon bonds will form more complex reaction sequences, more steps, and more intermediates and products to be adsorbed to oxidize to CO_2. Therefore, the most commonly used alcohols in alcohol oxidation reactions are methanol and ethanol, although 2-propanol, glycerol, ethylene glycol, allyl alcohol, and butanol are also used [1, 16, 26].

Another issue for alcohol oxidation reactions is the environment where the reaction will be carried out. As most reactions that have been explored in research have been in an acidic environment, there are many publications that prefer studies on basic media. These two environments, the required potential for ion adsorption, the appropriate operating range, and pH, provide superiority due to the activity of the catalyst used.

In electrochemical processes that take place in an aqueous environment, the range studied is limited to the potential range of water. According to the Nernst equation, this range shifts to -59 mV for every pH value that increases. The working potential range shifts nominally 0.83 V as a result of a change from a 1-N strong acid solution to a 1-N strong base solution when measured with a standard hydrogen electrode. This shift changes the electrical double-layer structure formed in electrochemical processes and the electric field at the electrode-electrolyte interface. This change is very important because the adsorption force of neutral species changes. Due to this shift, the adsorption force of a precious ion can be weakened by roughly 0.83 eV. As a result, the adsorption of anions in alkaline solutions becomes weaker than acidic solutions. Weak anion adsorption in the alkaline environment can make most electrocatalytic processes easier in the alkaline environment than in the acidic environment [1, 16, 19, 26–28].

2 Alcohol oxidation reactions in acidic media

2.1 Methanol and ethanol oxidation in acidic media

The methanol and ethanol oxidation reactions taking place in the acidic medium are complex reactions involving successive parallel sequences. Although the detailed description of the reaction mechanism remains uncertain, some intermediate products and products have been identified by spectroscopic methods such as Fourier transform infrared spectroscopy (FTIR) and online differential electrochemical mass [29].

If we look at the methanol oxidation reaction,

$$CH_3OH + H_2O \rightarrow CO_2 + 6H^+ + 6e^-$$

Methanol theoretically has a low oxidation potential (0.02 V). This value is comparable to hydrogen. It also means that it can be used as an effective fuel for methanol at low temperatures. However, during this oxidation reaction that takes place on the surface of the Pt electrode, species poisoning the Pt electrode surface are released, leading to a decrease in the reactivity of the electrode. Absorbent CO and CO_{ads} are the leading species in these studies [14, 29, 30], as shown in Fig. 1. It is necessary to oxidize to remove CO from the catalyst surface. For this purpose, there is a need for oxygen created by the decomposition of water. However, Pt

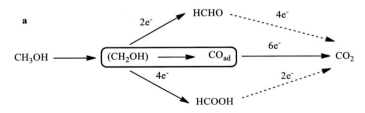

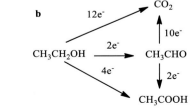

Fig. 1 Schematic representation of (A) methanol and (B) ethanol on the electrode surface of the Pt electrode surface.

Chapter 25 Topics on the fundamentals of the alcohol oxidation reactions **469**

poisons the CO catalyst on the surface, as it cannot chemically absorb water at potentials below 0.7 V [14, 23, 25]. Therefore, pure Pt is a weak catalyst at low potentials. In this case, a high oxidation rate in the oxidation of alcohol occurs only at high potentials. Therefore, the alcohol oxidation reaction at the anode is delayed. This is an unfavorable situation for all systems where alcohols are used to obtain energy by electrooxidation reactions.

The oxidation reaction of ethanol on the Pt electrode surface is shown below [24]:

$$CH_3CH_2OH + 3H_2O \rightarrow 2CO_2 + 12H^+ + 12e^-$$

The ethanol oxidation reaction has a potential of 0.08 V. This reaction on the Pt surface is more difficult to achieve than the oxidation reaction of methanol. Because the number of changing electrons is twice and the activation of CC bonds is stronger than the activation of C—H bonds, it is more difficult— [23–25, 31].

The main products formed by the oxidation of ethanol in an acidic medium are CO_2, acetaldehyde (CH_3CHO), and acetic acid (CH_3COOH) [1, 7, 32, 33]. Similar to the methanol oxidation reaction, ethanol oxidation produces species poisoning on the Pt surface. Due to these absorbed intermediates, the electrode is rapidly poisoned, and electrode reactivity decreases.

In ethanol oxidation, breaking the C—C bonds for the total oxidation of CO_2 is a huge problem. The conversion of acetaldehyde, which is the partial oxidation path of ethanol to acetic acid, is the oxidation step that prevails in the reaction on Pt. This causes a decrease in the number of electrons earned from ethanol, which results in a serious decrease in fuel efficiency [16, 32].

2.2 2-Propanol oxidation in acidic media

Although the studies on the oxidation of 2-propanol started with the discovery of the Nafion membrane, it gained more interest with the study by Wang et al. on fuel cells in 1995 [34]. It has been proven by researchers that the electrochemical activity decreases because the carbon-carbon bond in alcohols is difficult to break on the Pt surface. However, some catalysts gave better results with the increasing number of carbon atoms. Therefore, studies with high carbon alcohols have increased during the last decade [10, 11].

The smallest secondary alcohol is 2-propanol. Therefore, its catalytic oxidation is very interesting in terms of its molecular structure. The products formed as a result of total oxidation of alcohol are CO_2 and water, and those formed as a result of oxidation of 3-carbon primary alcohols on Pt in the acidic environment are CO_2, aldehyde, and carboxylic acid. Unlike these alcohols, ketone and very little CO_2 are formed when 2-propanol is oxidized. Therefore, the explanation of the mechanism of the oxidation reaction of 2-propanol is very important, in that it differs from other alcohols and directs other secondary alcohols to be studied.

The total oxidation reaction of 2-propanol to CO_2 in an acidic medium is given here:

$$C_3H_7OH + 5H_2O \rightarrow 3CO_2 + 18H^+ + 18e^-$$

While 2 electrons are released from the oxidation reaction of hydrogen, 6 electrons from the oxidation reaction of methanol, and 12 electrons from the oxidation of ethanol, 18 electrons are released from the reaction of 2-propanol. This is an important advantage of high-carbon alcohols over low-carbon alcohols.

3 Alcohol oxidation reaction in alkaline media

3.1 Methanol and ethanol oxidation in alkaline media

Methanol oxidation reactions taking place in an alkaline medium are especially important for the development of direct methanol fuel cells. However, these reactions in the alkaline environment face some problems, the most serious of which is the carbonization of the alkali electrolyte. To deal with this issue, various solutions have been derived by adapting the experimental environment design to the carbonate electrolyte using recycled electrolytes or anion-exchange membrane electrolytes [5, 6, 12, 26].

The methanol oxidation reaction that takes place in the alkaline medium is given here. As seen in both half-reactions, carbonate, and formations are formed in the medium.

$$CH_3OH + 8OH^- \rightarrow CO_3^{2-} + 6H_2O + 6e^-$$

$$CH_3OH + 5OH^- \rightarrow HCOO^- + 4H_2O + 4e^-$$

In both reactions, this resulted in many adsorbable intermediates. CO is produced first:

$$CH_3OH + 4OH^- \rightarrow CO_{ads} + 4H_2O + 4e^-$$

Adsorbed CO is removed from the medium according to the following equation:

$$CO_{ads} + 2OH_{ads} \rightarrow CO_2 + H_2O$$

Adsorbed OH is formed as a result of the conversion of hydroxide ions in the environment according to the following equation:

$$OH^- \rightarrow OH_{ads} + e^-$$

If a reaction is carried out in a strong alkali environment such as sodium hydroxide, the CO_2 gas released can also be converted to carbonate ion:

$$CO_2 + 2OH^- \rightarrow CO_3^{2-} + H_2O$$

The methanol oxidation reaction rate in an alkaline medium is higher than the acidic medium when pure Pt, Pt-supported carbon electrode or Pt-based electrodes are used. The reason for this is that the ion absorption property of the alkaline medium and the maximum conversion of the adsorbed hydroxide required for methanol oxidation at low potential [20].

Ethanol oxidation reactions taking place in an alkaline medium on the surface of Pt electrode occur faster than reactions in an acidic medium.

In an alkaline medium, ethanol is oxidized to CO_2 on the anode electrode surface. The total reaction is given here:

$$CH_3CH_2OH + 12OH^- \rightarrow 2CO_2 + 9H_2O + 12e^-$$

It is known that the oxidation reaction of ethanol is carried out via a double-path mechanism. According to this mechanism, ethanol is oxidized to acetaldehyde and acetic acid, and as a result of this reaction sequence, only four electrons are transferred. The acetic acid formation is a complete dead end for this reaction because acetic acid is not more oxidized under these ambient conditions. In these reactions, while the carbon-carbon bond in ethanol is broken, acetaldehyde, CO_{ad}, and $CH_{x,ad}$ ($x = 1$ in acidic media) are the only carbon types that are adsorbed [19, 26–28, 30, 35–38]. These species are oxidized to CO_2 after a series of reactions and a total of 12 electrons are released:

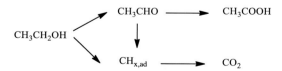

The oxidation reaction of ethanol with Pt, palladium (Pd), or gold (Au) catalyst in an alkaline medium and intermediate

472 Chapter 25 Topics on the fundamentals of the alcohol oxidation reactions

products and products resulting from these reactions are given in the following reaction sequence:

$$M + OH^- \rightarrow M - OH^-_{ad}$$
$$M + CH_3CH_2OH \rightarrow M - [CH_3CH_2OH]_{ad}$$
$$+ 3OH^- \rightarrow M - [CH_2CO]_{ad} + 3H_2O + 3e^-$$
$$M - OH^-_{ad} + M - [CH_2CO]_{ad} \rightarrow M - CH_3COOH + M$$
$$M - CH_3COOH + OH^- \rightarrow M - CH_3COO^- + H_2O$$
$$M : Pt, \ Pd \ or \ Au$$

The rate-determining step in this reaction is the penultimate one, in which the adsorbed ethoxy ion and the hydroxide ions that are adsorbed move away from the environment as a result of the reaction [17]. During the oxidation of ethanol in an alkaline environment, the electrolyte is prone to progressive carbonation, while CO_2 is retained at this time and the electrolyte is disabled over time. The researchers have used the solid alkali electrolyte to create systems that allow electrolyte recirculation by using carbonate electrolytes that dissolve CO_2 to overcome these undesirable situations and make the alkaline environment superior to the acidic environment.

3.2 2-Propanol oxidation in alkaline media

Oxidation reactions of alcohols started to be carried out in an alkaline environment by clarifying the kinetics of oxidation reactions. The fact that the fuel cells operating in the alkaline environment have made significant progress and achieved good results paved the way for the development of vehicles that can work in this environment. Ensuring the connection with hydroxide ions contributed to the circulation of the ionic current with the proton-connected membranes in the opposite direction.

Other metals in alkaline solutions are as effective as Pt. In addition, electrode poisoning caused by the adsorption of reaction intermediates is less apparent in the alkaline environment. The hydroxide ion is known to suppress poisoning intermediates. It offers a wide range of electrode types that can be studied in a wide range of cathodic and anodic alkaline media compared to the acidic medium.

Although the number of carbons in 2-propanol is higher than the other alcohols discussed here, better results have been obtained with the alkaline environment.

The oxidation reaction of 2-propanol in the alkaline environment is given here:

$$C_3H_7OH + 2OH^- \rightarrow CH_3COCH_{3ad} + 2H_2O + 2e^-$$
$$CH_3COCH_{3ad} + OH^- \rightarrow CH_2COCH_3^- ad + H_2O$$

The products resulting from oxidation, and especially intermediate products, cannot be revealed by electrochemical methods. Intermediate products can be determined by applying spectroscopic methods to the system during the reaction. As a result of the oxidation of methanol in an alkaline medium, the only product is CO_2. However, the product formed by the oxidation reaction of 2-propanol in an alkaline medium is acetone, independent of the catalyst at a low overvoltage. However, acetone is converted to enolate ion by catalytic oxidation. Neither CO_2 nor carbonate species have been found in the oxidation of 2-propanol [10, 11, 14].

4   Catalysts for alcohol oxidation reactions

The catalyst to be used in alcohol oxidation reactions differs according to anodic or cathodic properties. Cathodic catalysts are generally preferred as Pt or non-Pt catalysts. However, catalysts to be used as anode materials are classified according to acidic or basic environmental conditions. While catalysts to be used as anode materials in acidic environments are mostly classified as Pt and Pt-based catalysts, the catalysts to be used in alkaline environments are classified as Pt catalysts, Pt-based catalysts, and non-Pt catalysts. This classification is given in Fig. 2.

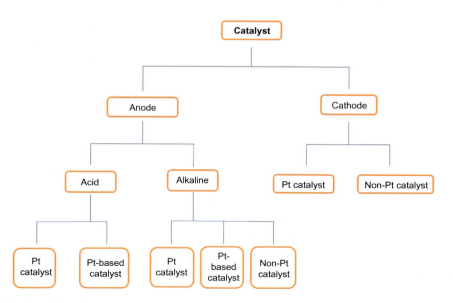

Fig. 2 Classification of the catalysts used in alcohol oxidation.

The Pt electrode for alcohol oxidation remains a unique catalyst. However, due to some negative properties of the Pt electrode, it is considered an inadequate catalyst in alcohol oxidation, and the search for alternative catalysts is ongoing. In recent years, experimental studies are carried out by adding cocatalyst to improve the catalytic performance of Pt.

Studies in this area have involved Pt-Pd [33], Pt-Ni [36], Pt-Mo and Pt-Mo/C [15], and Pt-Sn and Pt-Sn/C [7, 39, 40], Pt-Ru-In and Pt-Ru-In/C [29] oxidation reactions are examples of Pt-supported cocatalysts. Apart from these catalysts, Pd [25], Pd/C, Au/C and PdAu/C [41], PdNi [35], ReS_2 [42], Mo_2C and Mo_2N [43], Ru, RuO_2, Ir and IrO_2 [44], CoSe [45], Fe-N/C [46], and Ni [47] catalysts are other examples of non-Pt catalysts that are used in methanol and ethanol oxidation.

Among the catalysts formed with Pt, the most commonly used catalysts in alcohol oxidation reactions are Pt-Pd, Pt-Ru, and Pt-Sn catalysts. These perform very well at high potentials, especially when the Pt content is high, while they perform very well at low potentials when the other metal content is high [48]. They are used mostly in DAFCs at a 1:1 M ratio.

Pt-Ru-based catalyst is the most active binary catalyst for methanol, ethanol, and 2-propanol oxidation. This catalyst performs very well in acidic environments, especially in oxidation reactions using methanol. The Pt-Ru catalyst was used in ethanol oxidation by Camara et al. [49]. While pure ruthenium (Ru) is ineffective in ethanol oxidation, the best performance is achieved when using a Pt-Ru catalyst.

In some studies, a second or third metal was added to the Pt-Ru catalyst and new catalyst systems were developed and used in alcohol oxidation reactions. These metals are osmium (Os), nickel (Ni), tungsten (W), molybdenum (Mo), lead (Pb), iridium (Ir), and their oxidized components. In these studies, the best performance was achieved with the Pt-Ru-Os-Ir catalyst [5, 13, 15, 31, 36, 44, 47, 49].

Pt-Sn electrocatalysts are among the catalysts that researchers emphasize most, especially because they increase the performance of methanol oxidation reaction and contribute to the oxidation of the released CO [50]. Pt-Sn catalysts performed much better in methanol oxidation than Pt-Ru catalysts. Pt-Sn catalysts attached to the carbon support also performed much better under the same conditions than Pt-Ru/C catalysts [51]. Tin (Sn) contributes better than Ru in the oxidation of ethanol. Moreover, studies have shown that the Pt-Sn catalyst (without support) is more effective than Pt-Ru catalyst [52].

Pt-tin electrodes are also combined with different metals to design triple- and quadruple-electrode systems. The metals most commonly used in these processes and studies can be given as Ru, Ir, Ni, cerium (Ce), W, Pd, rhodium (Rh), rhenium (Re), Mo, and titanium (Ti) [1, 15, 17, 19, 53]. It is also used by mixing electrodes containing Pt and Sn with some metal oxides. Here, WO_2, TiO_2, and CeO_2 compounds are used [12, 14, 19, 28, 30, 39, 42, 44].

5 Conclusion

In this chapter, alcohol oxidation reactions in acid and alkaline media were explained; it also discussed the alcohols and catalysts used, the effect of the medium in question on these reactions, and alternative catalyst studies of Pt and Pd. Reactions on the catalyst surface are described according to the alcohol used in both acidic and alkaline conditions.

The most preferred alcohols in such reactions are methanol and ethanol, but 2-propanol and its isomers and butanol are also used. However, the reactions that take place as the chain gets longer in the alcohol molecule become more complex. While Pt is the most used and most effective catalyst in acidic media, Pt has a better effect than Pd in alkaline media. However, these catalysts have some disadvantages, such as limited reserves and high costs. No new catalyst has been found that can perform as well as these catalysts. However, new types of catalysts, where they are used at a minimum rate, have similar reactivity and durability. It has been proven in experiments that they have excellent effects. It can be seen from the results in the research literature that when using catalysts designed with good architectural intelligence by creating dual, triple, and even quadruple catalyst systems with Pt and Pd, the performance obtained is better than that obtained only by using Pt and Pd. In addition to the Ir and Ru used in these systems, Sn, Os, and Mo-like metals have interesting results.

The aim of the next studies on the new catalysts will not only be to find alternatives to Pt and Pd, but also to achieve better results and help better understand the reactions by increasing the type of alcohol that can be used in such reactions.

References

[1] M. Shao, Electrocatalysis in Fuel Cells—A Non- and Low- Platinum Approach, Springer London, London, 2013, https://doi.org/10.1007/978-1-4471-4911-8.

[2] Ö. Karatepe, Y. Yildiz, H. Pamuk, S. Eris, Z. Dasdelen, F. Sen, Enhanced electrocatalytic activity and durability of highly monodisperse Pt@PPy-PANI

nanocomposites as a novel catalyst for the electro-oxidation of methanol, RSC Adv. 6 (2016) 50851–50857, https://doi.org/10.1039/c6ra06210e.

[3] S. Eris, Z. Daşdelen, Y. Yıldız, F. Sen, Nanostructured polyaniline-rGO decorated platinum catalyst with enhanced activity and durability for methanol oxidation, Int. J. Hydrog. Energy 43 (2018) 1337–1343, https://doi.org/10.1016/j.ijhydene.2017.11.051.

[4] F. Şen, S. Şen, G. Gökağaç, Efficiency enhancement of methanol/ethanol oxidation reactions on Pt nanoparticles prepared using a new surfactant, 1,1-dimethyl heptanethiol, Phys. Chem. Chem. Phys. 13 (2011) 1676–1684, https://doi.org/10.1039/c0cp01212b.

[5] C. Lamy, E.M. Belgsir, J.M. Léger, Electrocatalytic oxidation of aliphatic alcohols: application to the direct alcohol fuel cell (DAFC), J. Appl. Electrochem. 31 (2001) 799–809, https://doi.org/10.1023/A:1017587310150.

[6] K.I. Ozoemena, RSC advances alkaline direct alcohol fuel cells: catalyst design, principles and applications, RSC Adv. 6 (2016) 89523–89550, https://doi.org/10.1039/C6RA15057H.

[7] P. Boolchand, A.P. Angelopoulos, Improved Electrocatalytic Ethanol Oxidation Activity in Acidic and Alkaline Electrolytes Using Size-Controlled Pt−Sn Nanoparticles, 2013, https://doi.org/10.1021/la403704w.

[8] Y. Yıldız, S. Kuzu, B. Sen, A. Savk, S. Akocak, F. Şen, Different ligand based monodispersed Pt nanoparticles decorated with rGO as highly active and reusable catalysts for the methanol oxidation, Int. J. Hydrog. Energy 42 (2017) 13061–13069, https://doi.org/10.1016/j.ijhydene.2017.03.230.

[9] Y. Yildiz, E. Erken, H. Pamuk, H. Sert, F. Şen, Monodisperse Pt nanoparticles assembled on reduced graphene oxide: highly efficient and reusable catalyst for methanol oxidation and dehydrocoupling of dimethylamine-borane (DMAB), J. Nanosci. Nanotechnol. 16 (2016) 5951–5958, https://doi.org/10.1166/jnn.2016.11710.

[10] J. Ye, J. Liu, C. Xu, S. Jiang, Y. Tong, Electrooxidation of 2-propanol on Pt, Pd and Au in alkaline medium, Electrochem. Commun. 9 (2007) 2760–2763, https://doi.org/10.1016/j.elecom.2007.09.016.

[11] M.E.P. Markiewicz, D.M. Hebert, S.H. Bergens, Electro-oxidation of 2-propanol on platinum in alkaline electrolytes, J. Power Sources 161 (2006) 761–767, https://doi.org/10.1016/j.jpowsour.2006.05.002.

[12] H.R. Corti, E.R. Gonzalez, Direct Alcohol Fuel Cells, Springer Netherlands, Dordrecht, 2014, https://doi.org/10.1007/978-94-007-7708-8.

[13] C.A. Ottoni, Glycerol and ethanol oxidation in alkaline medium using PtCu/C electrocatalysts, Int. J. Electrochem. Sci. 13 (2018) 1893–1904, https://doi.org/10.20964/2018.02.58.

[14] Y. Wang, Nanomaterials for Direct Alcohol Fuel Cell, Pan Stanford Publishing, Singapore, 2016, https://doi.org/10.1201/9781315364902.

[15] W.J. Pech-Rodríguez, C. Calles-Arriaga, D. González-Quijano, G. Vargas-Gutiérrez, C. Morais, T.W. Napporn, F.J. Rodríguez-Varela, Electrocatalysis of the ethylene glycol oxidation reaction and in situ Fourier-transform infared study on PtMo/C electrocatalysts in alkaline and acid media, J. Power Sources 375 (2018) 335–344, https://doi.org/10.1016/j.jpowsour.2017.07.081.

[16] S. Du, B. Pollet, Applications of nanomaterials in fuel cells, in: D. Rickerby (Ed.), Nanotechnol. Sustain. Manuf, CRC Press, 2014, pp. 113–152, https://doi.org/10.1201/b17046-8.

[17] S.L. Suib, New and Future Developments in Catalysis, Elsevier, 2013, https://doi.org/10.1016/C2010-0-68687-1.

[18] S. Eris, Z. Daşdelen, F. Sen, Enhanced electrocatalytic activity and stability of monodisperse Pt nanocomposites for direct methanol fuel cells, J. Colloid Interface Sci. 513 (2018) 767–773, https://doi.org/10.1016/j.jcis.2017.11.085.

Chapter 25 Topics on the fundamentals of the alcohol oxidation reactions **477**

[19] N. Ramaswamy, Q. He, D. Abbott, M. Bates, Electrocatalysis in Alkaline Electrolytes—Research Overview, 2011.

[20] J.S. Spendelow, A. Wieckowski, Electrocatalysis of oxygen reduction and small alcohol oxidation in alkaline media, Phys. Chem. Chem. Phys. 9 (2007) 2654, https://doi.org/10.1039/b703315j.

[21] S. Eris, Z. Daşdelen, F. Sen, Investigation of electrocatalytic activity and stability of Pt@f-VC catalyst prepared by in-situ synthesis for methanol electrooxidation, Int. J. Hydrog. Energy 43 (2018) 385–390, https://doi.org/10.1016/j.ijhydene.2017.11.063.

[22] F. Şen, G. Gökağaç, Different sized platinum nanoparticles supported on carbon: an XPS study on these methanol oxidation catalysts, J. Phys. Chem. C 111 (2007) 5715–5720, https://doi.org/10.1021/jp068381b.

[23] D. Bayer, S. Berenger, M. Joos, C. Cremers, J. Tübke, Electrochemical oxidation of C2 alcohols at platinum electrodes in acidic and alkaline environment, Int. J. Hydrog. Energy 35 (2010) 12660–12667, https://doi.org/10.1016/j.ijhydene.2010.07.102.

[24] Y. Wang, S. Zou, W.-B. Cai, Recent advances on electro-oxidation of ethanol on Pt- and Pd-based catalysts: from reaction mechanisms to catalytic materials, Catalysts 5 (2015) 1507–1534, https://doi.org/10.3390/catal5031507.

[25] B. Wang, L. Tao, Y. Cheng, F. Yang, Y. Jin, C. Zhou, H. Yu, Y. Yang, Electrocatalytic oxidation of small molecule alcohols over Pt, Pd, and Au catalysts: the effect of alcohol's hydrogen bond donation ability and molecular structure properties, Catalysts 9 (2019) 387, https://doi.org/10.3390/catal9040387.

[26] E. Antolini, E.R. Gonzalez, Alkaline direct alcohol fuel cells, J. Power Sources 195 (2010) 3431–3450, https://doi.org/10.1016/j.jpowsour.2009.11.145.

[27] E.A. Monyoncho, T.K. Woo, E.A. Baranova, Ethanol electrooxidation reaction in alkaline media for direct ethanol fuel cells, SPR Electrochem. 15 (2019) 1–57, https://doi.org/10.1039/9781788013895-00001.

[28] S.C.S. Lai, M.T.M. Koper, Ethanol electro-oxidation on platinum in alkaline media, Phys. Chem. Chem. Phys. 11 (2009) 10446, https://doi.org/10.1039/b913170a.

[29] M.C.L. Santos, J. Nandenha, J.M.S. Ayoub, M.H.M.T. Assumpção, A.O. Neto, Methanol oxidation in acidic and alkaline electrolytes using PtRuIn/C electrocatalysts prepared by borohydride reduction process, J. Fuel Chem. Technol. 46 (2018) 1462–1471, https://doi.org/10.1016/S1872-5813(18)30060-4.

[30] E.H. Yu, U. Krewer, K. Scott, Principles and materials aspects of direct alkaline alcohol fuel cells, Energies 3 (2010) 1499–1528, https://doi.org/10.3390/en3081499.

[31] N. Ramaswamy, S. Mukerjee, Fundamental mechanistic understanding of electrocatalysis of oxygen reduction on Pt and non-Pt surfaces: acid versus alkaline media, Adv. Phys. Chem. 2012 (2012) 1–17, https://doi.org/10.1155/2012/491604.

[32] A. Santasalo-Aarnio, S. Tuomi, K. Jalkanen, K. Kontturi, T. Kallio, The correlation of electrochemical and fuel cell results for alcohol oxidation in acidic and alkaline media, Electrochim. Acta 87 (2013) 730–738, https://doi.org/10.1016/j.electacta.2012.09.100.

[33] L.L. Carvalho, A.A. Tanaka, F. Colmati, Palladium-platinum electrocatalysts for the ethanol oxidation reaction: comparison of electrochemical activities in acid and alkaline media, J. Solid State Electrochem. 22 (2018) 1471–1481, https://doi.org/10.1007/s10008-017-3856-0.

[34] J. Wang, Evaluation of ethanol, 1-propanol, and 2-propanol in a direct oxidation polymer-electrolyte fuel cell, J. Electrochem. Soc. 142 (1995) 4218, https://doi.org/10.1149/1.2048487.

[35] Z. Qi, H. Geng, X. Wang, C. Zhao, H. Ji, C. Zhang, J. Xu, Z. Zhang, Novel nanocrystalline PdNi alloy catalyst for methanol and ethanol electro-oxidation in alkaline media, J. Power Sources 196 (2011) 5823–5828, https://doi.org/10.1016/j.jpowsour.2011.02.083.

[36] L.S.R. Silva, I.G. Melo, C.T. Meneses, F.E. Lopez-Suarez, K.I.B. Eguiluz, G.R. Salazar-Banda, Effect of temperature on the ethanol electrooxidation at PtNirich@PtrichNi/C catalyst in acidic and alkaline media, J. Electroanal. Chem. 857 (2020) 113754, https://doi.org/10.1016/j.jelechem.2019.113754.

[37] L.L. Wang, Q.X. Li, T.Y. Zhan, Q.J. Xu, A review of Pd-based electrocatalyst for the ethanol oxidation reaction in alkaline medium, Adv. Mater. Res. 860–863 (2013) 826–830, https://doi.org/10.4028/www.scientific.net/AMR.860-863.826.

[38] R. Jiang, D.T. Tran, J.P. McClure, D. Chu, A class of (Pd–Ni–P) electrocatalysts for the ethanol oxidation reaction in alkaline media, ACS Catal. 4 (2014) 2577–2586, https://doi.org/10.1021/cs500462z.

[39] L. Jiang, A. Hsu, D. Chu, R. Chen, Ethanol electro-oxidation on Pt/C and PtSn/C catalysts in alkaline and acid solutions, Int. J. Hydrog. Energy 35 (2010) 365–372, https://doi.org/10.1016/j.ijhydene.2009.10.058.

[40] R. Rizo, D. Sebastián, M.J. Lázaro, E. Pastor, On the design of Pt-Sn efficient catalyst for carbon monoxide and ethanol oxidation in acid and alkaline media, Appl. Catal. B Environ. 200 (2017) 246–254, https://doi.org/10.1016/j.apcatb.2016.07.011.

[41] A.N. Geraldes, D.F. da Silva, E.S. Pino, J.C.M. da Silva, R.F.B. de Souza, P. Hammer, E.V. Spinacé, A.O. Neto, M. Linardi, M.C. dos Santos, Ethanol electro-oxidation in an alkaline medium using Pd/C, Au/C and PdAu/C electrocatalysts prepared by electron beam irradiation, Electrochim. Acta 111 (2013) 455–465, https://doi.org/10.1016/j.electacta.2013.08.021.

[42] M.B. Askari, P. Salarizadeh, Ultra-small ReS2 nanoparticles hybridized with rGO as cathode and anode catalysts towards hydrogen evolution reaction and methanol electro-oxidation for DMFC in acidic and alkaline media, Synth. Met. 256 (2019) 116131, https://doi.org/10.1016/j.synthmet.2019.116131.

[43] X. Chen, J. Qi, P. Wang, C. Li, X. Chen, C. Liang, Polyvinyl alcohol protected Mo2C/Mo2N multicomponent electrocatalysts with controlled morphology for hydrogen evolution reaction in acid and alkaline medium, Electrochim. Acta 273 (2018) 239–247, https://doi.org/10.1016/j.electacta.2018.04.033.

[44] S. Cherevko, S. Geiger, O. Kasian, N. Kulyk, J.-P. Grote, A. Savan, B.R. Shrestha, S. Merzlikin, B. Breitbach, A. Ludwig, K.J.J. Mayrhofer, Oxygen and hydrogen evolution reactions on Ru, RuO_2, Ir, and IrO_2 thin film electrodes in acidic and alkaline electrolytes: a comparative study on activity and stability, Catal. Today 262 (2016) 170–180, https://doi.org/10.1016/j.cattod.2015.08.014.

[45] H. Wang, X. Wang, D. Yang, B. Zheng, Y. Chen, Co0.85Se hollow nanospheres anchored on N-doped graphene nanosheets as highly efficient, nonprecious electrocatalyst for hydrogen evolution reaction in both acid and alkaline media, J. Power Sources 400 (2018) 232–241, https://doi.org/10.1016/j.jpowsour.2018.08.027.

[46] L. Osmieri, R. Escudero-Cid, M. Armandi, A.H.A. Monteverde Videla, J.L. García Fierro, P. Ocón, S. Specchia, Fe-N/C catalysts for oxygen reduction reaction supported on different carbonaceous materials. Performance in acidic and alkaline direct alcohol fuel cells, Appl. Catal. B Environ. 205 (2017) 637–653, https://doi.org/10.1016/j.apcatb.2017.01.003.

[47] M. Sunitha, N. Durgadevi, A. Sathish, T. Ramachandran, Performance evaluation of nickel as anode catalyst for DMFC in acidic and alkaline medium, J. Fuel Chem. Technol. 46 (2018) 592–599, https://doi.org/10.1016/S1872-5813(18)30026-4.

Chapter 25 Topics on the fundamentals of the alcohol oxidation reactions **479**

[48] J.-M. Leger, C. Lamy, The direct oxidation of methanol at platinum based catalytic electrodes: what is new since ten years? Ber. Bunsenges. Phys. Chem. 94 (1990) 1021–1025, https://doi.org/10.1002/bbpc.19900940928.

[49] G. Camara, R. de Lima, T. Iwasita, Catalysis of ethanol electrooxidation by PtRu: the influence of catalyst composition, Electrochem. Commun. 6 (2004) 812–815, https://doi.org/10.1016/j.elecom.2004.06.001.

[50] K. Wang, H.A. Gasteiger, N.M. Markovic, P.N. Ross, On the reaction pathway for methanol and carbon monoxide electrooxidation on Pt-Sn alloy versus Pt-Ru alloy surfaces, Electrochim. Acta 41 (1996) 2587–2593, https://doi.org/10.1016/0013-4686(96)00079-5.

[51] V.R. Stamenković, M. Arenz, C.A. Lucas, M.E. Gallagher, P.N. Ross, N.M. Marković, Surface chemistry on bimetallic alloy surfaces: adsorption of anions and oxidation of CO on Pt_3 Sn(111), J. Am. Chem. Soc. 125 (2003) 2736–2745, https://doi.org/10.1021/ja028771l.

[52] L. Jiang, G. Sun, S. Sun, J. Liu, S. Tang, H. Li, B. Zhou, Q. Xin, Structure and chemical composition of supported Pt–Sn electrocatalysts for ethanol oxidation, Electrochim. Acta 50 (2005) 5384–5389, https://doi.org/10.1016/j.electacta.2005.03.018.

[53] N. Ramaswamy, Electrocatalysis of Oxygen Reduction in Alkaline Media and a Study of Perfluorinated Ionomer Membrane Degradation, Thesis, 2011, pp. 1–197.

26

Direct alcohol-fed solid oxide fuel cells

Hakan Burhan[a]**, Kubilay Arıkan**[a]**, Sadin Ozdemir**[c]**, Iskender Isik**[b]**, and Fatih Şen**[a]

[a]*Şen Research Group, Department of Biochemistry, Dumlupinar University, Kütahya, Turkey.* [b]*Department of Materials Science and Engineering, Faculty of Engineering, Dumlupinar University, Kütahya, Turkey.* [c]*Mersin University, Food Processing Programme, Technical Science Vocational School, Mersin, Turkey*

1 Introduction

Solid oxide fuel cells (SOFCs) are remarkable due to their great energy conversion efficiency and flexible fuel, as well as their environmental friendliness. SOFCs provide high energy conversion, with a performance of up to 80%, with combined heat and power. In addition to H_2, hydrocarbon fuels are used in this system. In addition, it is seen that carbon dioxide (CO_2) emission does not occur in studies with H_2 [1–4]. SOFCs require high temperatures according to their working principles. However, due to the high temperature of this system, thermal expansion between cell components, phase transition, corrosion of components, and consequently a decrease in cell performance occur. Due to these problems, SOFCs have been successful only in stationary applications [5, 6]. The problems experienced with high-temperature values have enabled advanced research to be carried out for the development of SOFCs that can operate in the low-temperature range [4, 5, 7–24]. The ability to operate at low temperatures in SOFCs increases their efficiency in cell performance and allows the use of sealing materials and low-cost interconnections. These positive developments bring opportunities that increase the application areas of SOFCs and directly affect their affordability [16].

It is also very important to choose suitable fuel to be used in SOFCs. The storage cost of hydrogen and limitations in

---

Nanomaterials for Direct Alcohol Fuel Cells. https://doi.org/10.1016/B978-0-12-821713-9.00016-0
Copyright © 2021 Elsevier Inc. All rights reserved.

Table 1 Properties of methanol, ethanol, and hydrogen in fuel cells [25].

Methanol	Ethanol	Hydrogen	Property
CH_3OH	C_2H_5OH	H_2	Formula
702	1325	237	$-\Delta G°$ (kJ/mol)
726	1367	286	$-\Delta H$ (kJ/mol)
6.09	8.00	33	Energy density, LHV (kWh/kg)
4.80	6.32	2.96×10^{-3}	Energy density, LHV (kWh/L)
1.21	1.14	1.23	E^0 cell (V)
3350	2330	26.802	Energy stored (Ah/kg)
2653	1841	2.40	Energy stored (Ah/L)

distribution infrastructure increase the desirability of liquid alcohols. Here, the distribution problems and storage costs of alcohols such as methanol and ethanol are extremely low, increasing their availability for fuel in SOFCs [25]. The comparative properties of the alcohols and hydrogen mentioned here are given in Table 1. Methanol and ethanol can be used directly as fuel, and they are also a source of hydrogen. Because SOFCs are designed for fixed use, a complete study has not been done on fuels such as methanol and ethanol [26]. In addition, alcohols such as methanol and ethanol are seen as a more attractive fuel source than gasoline, diesel, or liquefied petroleum gas (LPG) in portable applications [27, 28]. Many studies have been conducted using methanol as fuel in SOFCs operating at low temperatures. In addition, it has recently been used as a fuel in SOFCs in ethanol. While ethanol is portable and clean and has high energy density similar to methanol, it is more difficult to oxidize than are methanol and hydrogen. The usability of ethanol for SOFCs at low temperatures has also been reported in most studies [29, 30]. In addition, the usability of glycerol has also been tested for SOFCs at low temperatures [4]. One of the problems with alcohol-fired, low-temperature SOFCs is poor cell performance due to carbon monoxide (CO) poisoning at the anode. This situation does not occur in systems based on hydrogen fuel. In addition to this problem, there is still a need for improvement in terms of commercialization of SOFCs, as the service life and power density of alcohol-based systems are far below expectations [25]. Therefore, the anode, cathode, electrolyte, structure, and components need to be developed to expand the application areas of SOFCs and to achieve commercial success.

2 Direct alcohol-fed, low-temperature solid oxide fuel cells

2.1 SOFC structure

The SOCF equation here gives the anode, cathode, open circuit voltage (OCV), concentration loss, ohmic loss, and cell voltage [31]:

$$V = V_{oc} - \eta_{act,\, anode} - \eta_{act,\, cathode} - \eta_{ohmic} - \eta_{conc}$$

Hydrogen fuel can undergo a loss of concentration of oxidant or fuel molecules in mass transportation processes, and this situation can be neglected in SOFCs. However, for direct alcohol-fed SOFCs, this loss of concentration can result in cell performance that is not negligible given the high molecular weight of the ethanol fuel. To develop SOFCs with high power density and cell performance, it is necessary to choose a fuel where the concentration loss is negligible. The ohmic activation loss is very small in high-temperature SOFCs. However, losses due to electrochemical processes, ionic transport via electrolyte, charge transfer reaction rate, and exponential dependence of ionic conductivity in relation to temperature were observed to increase in low-temperature SOFCs [16]. Activation loss at the cathode is associated with slow oxygen depletion for low-temperature SOFCs and is more dominant at the anode. Therefore, the electrolyte and the design of cathode components in hydrogen-fed, low-temperature SOFCs is very important due to this activation loss [16, 31].

Another loss of anode activation shows that alcohol molecules are followed by a more complex oxidation process by indirect feeding in low-temperature SOFCs compared to hydrogen fuel, and therefore the activation loss is negligible [25]. Carbon formation as a result of alcohol oxidation creates an instability in the anode reaction [26]. The direct alcohol-fed working scheme of low-temperature SOFCs is shown in Fig. 1.

Graphite is more advantageous in SOFCs operating at lower temperatures compared to conventional SOFCs. This means that alcohols have a lower equilibrium potential in pyrolysis, as shown in Fig. 2 [33].

2.2 Fuels and anode reactions

A schematic representation of the electrochemical reactions of low-temperature SOFCs with direct alcohol feed on the anode, the cathode, various electrolyte types, and methanol and ethanol fuels are as shown in Fig. 1. Detailed individual reaction paths given

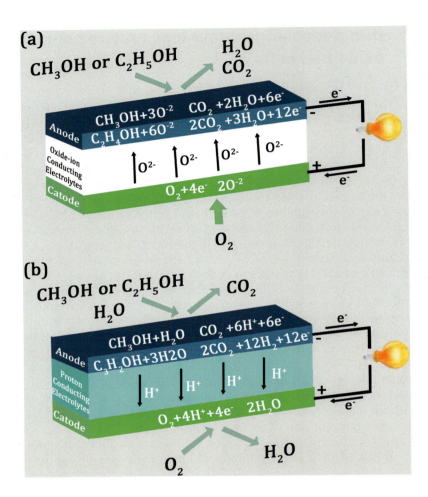

Fig. 1 Direct alcohol-fed, low-temperature SOFCs: (A) oxide ion-conducting electrolytes, (B) proton-conducting electrolytes [32].

in many studies [26, 34–47] show that there is a change in net water (H_2O) formation depending on the electrolyte type. While there are oxide-conducting electrolytes at the anode, the presence of proton-conducting electrolytes at the cathode can be mentioned. Finally, the produced water molecules are included in the reaction. Water molecules can participate in anodic reactions in cells [25]. The most commonly preferred anodic reaction when using oxide-ion conducting electrolyte for SOFCs will be discussed next.

In the first step, the formation of products such as hydrogen and CO occurs after the pyrolysis of the fuel. The combination of the products formed as a result of pyrolysis and fuel molecules causes catalytic degradation. In the next step, the fuel molecules, catalytic degradation products, and pyrolysis products react with oxygen ions transferred via the electrolyte. As a result of this new reaction, CO_2 and H_2O molecules are formed. Then another

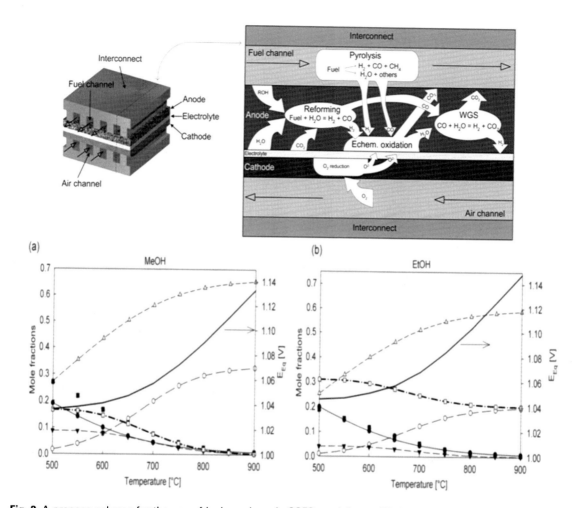

Fig. 2 A process scheme for the use of hydrocarbons in SOFCs and the equilibrium and potential change of temperature as a function for (A) methanol and (B) ethanol [26].

reaction takes place, allowing the fuel, CO_2 and H_2O molecules to form CO and H_2. Due to the decomposition and reforming of CO and H_2O, a reaction takes place that releases CO_2 and H_2. In addition, carbon formation and removal occurs along with it. All of these situations occur depending on conditions such as the catalytic properties of the anode, flow state, pressure, temperature, and current density. The anodic reaction path for methanol proceeds as follows [34]:

$$CH_3OH + H_2O = CO_2 + 3H_2 \text{ (MeOH steam reforming)}$$

$$CH_3OH = CO + 2H_2 \text{ (MeOH decomposition)}$$

$$CH_3OH + 1/2O_2 = CO_2 + 2H_2 \text{ (partial oxidation)}$$

The occurrence reaction of hydrogen production occurs as follows:

$$CO + H_2O = CO_2 + H_2$$

Methanation reaction occurrence is as follows:

$$CO + 3H_2 = CH_4 + H_2O$$

$$CO_2 + 4H_2 = CH_4 + 2H_2O$$

Carbon formation occurs by CO disproportion:

$$2CO = C + CO_2$$

Direct electrochemical oxidation of methanol with oxide ions can occur at the anode-electrolyte interface:

$$CH_3OH + 3O_2^- = CO_2 + 2H_2O + 6e^-$$

The intermediate reaction formations are as follows [40]:

$$CH_3OH + 2O_2^- = HCOOH + H_2O + 4e^-$$

$$HCOOH + O_2^- = CO_2 + H_2O + 2e^-$$

More complex reaction paths in the anode section can occur with steam reforming and thermal decomposition of ethanol. This formation can occur at 400–600°C as $C_2H_5OH + H_2O = 2CO + 4H_2$. Electrochemical oxidation of low-temperature SOFCs creates CO-rich particles that are used as fuel [41, 44–46]:

$$H_2 + O_2^- = H_2O + 2e^-$$

$$CO + O_2^- = CO_2 + 2e^-$$

The water formed as a result of electrochemical oxidation at the anode is used in the reforming of ethanol [35–38]: then direct electrochemical oxidation of ethanol takes place:

$$C_2H_5OH + 3H_2O = 2CO_2 + 6H_2$$

$$H_2 + O_2^- = H_2O + 2e^-$$

Influence of oxygen ions [39] occurs as follows:

$$C_2H_5OH + 3O_2^- = 3H_2 + 2CO_2 + 6e^-$$

3 Membranes

The membrane design in low-temperature SOFCs consists of three components: the anode, cathode, and electrolyte. The anode and cathode electrodes are fabricated with porous structures to reduce the activation loss and maximize the reaction area. For alcohol oxidation, noble and nonnoble metals suitable for catalytic activity should be selected in the anode section. Noble metals such as; While alloys such as platinum (Pt), palladium (Pd), and Pt-Ru are preferred, nickel (Ni), copper (Cu), and their alloys, which are nonnoble and mixed with electrolyte materials, are also used. In the cathode part, it is seen that the oxygen reduction reaction takes place and that appropriate active materials are preferred. Noble metals such as Pt and gold (Ag) can be selected in this section, while $Ba_{1-x}Sr_xCo_yFe_{1-y}O_{3-\delta}$ (BSCF) and $La_{1-x}Sr_xCoO_{3-\delta}$ (LSC) materials are chosen as oxide-based. Electrolytes such as proton conductivity (e.g., oxide-ion conductive or yttria-doped barium zirconate (BYZ)) ceramics (e.g., samaria-doped ceria (SDC), gadolinia-doped ceria (GDC) or yttria-stabilized zirconia (YSZ)) consists of conductive oxides [32]. Membrane structures in direct alcohol-fed low-temperature SOFCs are briefly described in Table 2.

3.1 Anode

3.1.1 Nonnoble metal-based

Preferred materials for the anode are categorized as noble or nonnoble. In addition, attention should be paid to factors such as chemical mechanical stability, alcohol oxidation kinetics, and carbon coking, which affect anode performance (as described in Table 2). Nickel (Ni) is among the most suitable nonnoble metals for direct alcohol-fed low-temperature SOFCs. Ni is a metal that gives successful results for hydrogen electrocatalytic oxidation and high electrical conductivity. Compared to Pt group metals, it has been observed that Ni, alcohols, and hydrocarbons give close results at temperatures up to 500°C for oxidation [49, 50, 67, 68]. Ni is susceptible to coke formation in the anode section because it activates the thermal decomposition of hydrocarbons. This situation adversely affects the performance of direct alcohol-fed SOFCs [50]. To overcome this coking, many studies have been carried out that use Ni with Cu. As a result of using Cu as an anode catalyst, coking did not occur, because of the catalytic cracking property of Cu for hydrocarbons [51, 69]. Azimova et al. studied low-temperature SOFCs with direct alcohol-fed Ni, Ni-Cu, and Cu-based anode materials. They obtained the anode material via the infiltration of Cu acetate and Ni acetate solutions. While the hydrocarbon-reforming reaction of Cu decreases the anode performance, it also decreases the coking that occurs in the anode with the addition of high amounts of Cu, reducing the decrease in performance [48].

Table 2 Summary of performance data of direct alcohol-fed, low-temperature SOFCs.

Main points	Anode	Cathode	Electrolyte	Maximum power density (mW/cm²)	Fuel	Ref.
Alloying Ni with Cu to limit carbon formation at the anode	Ni (150 μm)	$BaCe_{0.48}Zr_{0.4}Yb_{0.1}Co_{0.02}O_{3-\delta}$ (BCZYbCo)/$La_{1-x}Sr_xCoO_{3-\delta}$ (LSC) (25 μm)	$BaCe_{0.48}Zr_{0.4}Yb_{0.1}Co_{0.02}O_{3-\delta}$ (BCZYbCo) (30–35 μm)	78 at 500°C	2:1 methanol/water	[48]
				88 at 500°C	4:1 methanol/water	
				98 at 500°C	4:1 ethanol/water	
	Cu-based (150 μm)			31 at 500°C	2:1 methanol/water	
				40 at 500°C	4:1 methanol/water	
				51 at 500°C	4:1 ethanol/water	
	Ni-Cu (150 μm)			22 at 500°C	2:1 methanol/water	
				25 at 500°C	4:1 methanol/water	
				17 at 500°C	4:1 ethanol/water	
Increased performance due to the small dipole moment response of bioethanol	$Li_{0.2}Ni_{0.7}Cu_{0.1}O$-Samarium-doped ceria-sodium carbonate	$Li_{0.2}Ni_{0.7}Cu_{0.1}O$-Samarium doped ceria-sodium carbonate	Samarium-doped ceria-sodium carbonate	150 at 580°C 215 at 580°C	Bioethanol Glycerol	[41]
Functional nanocomposite electrodes that are effective for liquid-based fuel cells	Ni-Cu-Zn (50%) composite with electrolyte (50%) (0.4 mm)	Ni-Cu-ZnO composite with electrolyte (0.3 mm)	Nanocomposite of samaria-doped ceria (SDC) and Na_2CO_3 (0.3 mm)	400 at 480°C 514 at 520°C 584 at 570°C	Bioethanol	[39]

Addition of pyridine to ethanol fuel gas suppressing coke formation and improving stability	$NiO + (Y_2O_3)_{0.1}(ZrO_2)_{0.9}$ (YSZ) with $NiAl_2O_3$ catalyst (800 μm)	$Ba_{1-x}Sr_xCo_yFe_{1-y}O_{3-\delta}$ (BSCF)/$Sm_{0.5}Sr_{0.25}CoO_{3-\delta}$ (SSC) + $Sm_{0.2}Ce_{0.8}O_2$ (SDC) (25 μm)	$(Y_2O_3)_{0.1}(ZrO_2)_{0.9}$(YSZ) (10 μm)	300 at 600°C	Ethanol/ pyridine	[49]
ALD Ru islands improving surface kinetics for methanol oxidation and thermal stability with less Ni coarsening and reducing the content of adsorbed carbon on the anode	Ru-coated Ni (45–53 nm)	Pt (~150 nm)	$Gd_{0.1}Ce_{0.9}O_{2-\delta}$ (GDC) (350 μm)	0.5 at 300°C	Methanol	[50]
Low activation and concentration polarization for methanol with small molecular weight	Ni+Yttria stabilized zirconia (YSZ) (anode support: 1 mm) (anode interlayer: ~20 μm)	$La_{0.8}Sr_{0.2}MnO_{3-\delta}$ (LSM) (~80 μm)	Yttria-stabilized zirconia (YSZ) (~10 μm)	200 at 550°C	Methanol	[51]
Catalytic activity of methanol decomposition lost due to sintering at 600°C (Cu-ceria)	Cu-$Ce_{0.9}Gd_{0.1}O_{1.95}$ (GDC) (30–50 μm) Cu-ceria (30–40 μm)	$La_{0.6}Sr_{0.4}Co_{0.2}Fe_{0.8}O_{3-\delta}$ (LSCF) (15 μm)	$Ce_{0.9}Gd_{0.1}O_{1.95}$ (GDC) (0.7 mm)	– 9 at 600°C	Methanol/ water steam	[34]
Direct methanol-fueled SOFC showing long-term stability without coking	NiO/samaria-doped ceria (SDC)	$Sm_{0.5}Sr_{0.25}CoO_3$ (SSC)/ samaria-doped ceria (SDC)	Samaria-doped ceria (SDC) (24 μm)	223 at 550°C 430 at 600°C	Methanol	[52]
No carbon deposition with methanol flame vs severe carbon deposition with ethanol flame	$Ni + Sm_{0.2}Ce_{0.8}O_{1.9}$ (SDC) (50 μm)	$Ba_{1-x}Sr_xCo_yFe_{1-y}O_{3-\delta}$ (BSCF) + $Sm_{0.2}Ce_{0.8}O_{1.9}$ (SDC) (30 μm)	$Sm_{0.2}Ce_{0.8}O_{1.9}$ (SDC) (360 μm)	500 at 600°C	Methanol flame	[53]

Continued

Table 2 Summary of performance data of direct alcohol-fed, low-temperature SOFCs.—cont'd

Main points	Anode	Cathode	Electrolyte	Maximum power density (mW/cm^2)	Fuel	Ref.
More active sites for fuel oxidation reaction due to thin electrolyte with uniform pores	Ni-Ce$_{0.85}$Sm$_{0.15}$O$_{1.925}$ (SDC) (anode support: 0.6 mm) (active anode layer: ~12 μm)	Ce$_{0.85}$Sm$_{0.15}$O$_{1.925}$ (SDC)— La$_{0.6}$Sr$_{0.4}$Co$_{0.2}$Fe$_{0.8}$O$_{3-\delta}$ (LSCF)	Ce$_{0.85}$Sm$_{0.15}$O$_{1.925}$ (SDC) (~6 μm)	110 at 450°C 260 at 500°C 520 at 550°C 820 at 600°C	Methanol	[54]
Improvement of electronic conductivity of anode by adding carbon having high electronic conductivity and oxidation-resistance	C-MO-SDC (C = activation carbon/carbon black, M = Cu, Ni, and Co, SDC = Ce$_{0.9}$Sm$_{0.1}$O$_{1.95}$)	–	Ceria-salt composite(CSC) (1 mm)	250 at 560°C	Methanol	[55]
Pd facilitating the internal reforming/decomposition of methanol at the anode	Pd-added NiO—Yttria stabilized zirconia (YSZ) (1 mm)	La$_{0.6}$Sr$_{0.4}$Co$_{0.2}$Fe$_{0.8}$O$_{3-\delta}$ (LSCF)	Ce$_{0.9}$Gd$_{0.1}$O$_{1.95}$ (GDC) (5 μm)	32 at 550°C 65 at 600°C 15 at 550°C 40 at 600°C	Methanol (reformed CH$_3$OH) methanol/H$_2$O	[56]
BZCYYb has good oxygen ion conductivity, small particle size, high water storage capability, and good coking resistance	NiO + BaZr$_{0.1}$Ce$_{0.7}$Y$_{0.1}$Yb$_{0.2}$O$_{3-\delta}$ (BZCYYb)	Ba$_{1-x}$Sr$_x$Co$_y$Fe$_{1-y}$O$_{3-\delta}$ (BSCF) (15 μm)	Sm$_{0.2}$Ce$_{0.8}$O$_{1.9}$ (SDC) (−20 μm)	519 at 600°C	Ethanol	[57]

Improvement of hydrogen selectivity and coking resistance due to Ni $+Ce_{0.8}Zr_{0.2}O_2$ catalyst layer	$NiO+(Y_2O_3)_{0.1}(ZrO_2)_{0.9}$ (YSZ) with $Ni+Ce_{0.8}Zr_{0.2}O_2$ catalyst later	$Ba_{1-x}Sr_xCo_yFe_{1-y}O_{3-\delta}$ (BSCF) + $Sm_{0.5}Sr_{0.25}CoO_{3-\delta}$ (SSC)	$(Y_2O_3)_{0.1}(ZrO_2)_{0.9}$ (YSZ) (10 μm)	179 at 550°C 324 at 600°C	Ethanol/O_2	[58]
Good coking resistance due to the strong interaction between Ni and $Ce_{0.8}Zr_{0.2}O_2$; also has good catalytic activity for EtOH conversion to hydrogen	$NiO+(Y_2O_3)0.1(ZrO_2)0.9$ (YSZ) with $Ni-Ce_{0.8}Zr_{0.2}O_2$ catalyst layer	$Ba_{1-x}Sr_xCo_yFe_{1-y}O_{3-\delta}$ (BSCF)	$(Y_2O_3)_{0.1}(ZrO2)_{0.9}$ (YSZ)	162 at 600°C 74 at 550°C	Ethanol/water steam gas	[59]
Lower catalytic activity of Pt-Ru alloy anode compared to Pt anode	$Pt_{0.4}Ru_{0.6}$ (80 nm)	Pt (60 nm)	8 mol% yttria stabilized zirconia (YSZ) (100 μm)	0.0015 at 250°C 0.0073 at 300°C 0.071 at 350°C	Methanol	[60]
Prevention of coarsening of Pt cluster due to passivated layer by Ru oxidation	Ru-coated (<20 nm) Pt (150 nm)	Pt (150 nm)	$Gd_{0.1}Ce_{0.9}O_{2-\delta}$ (350 μm)	5.5 at 400°C 8.5 at 450°C	Methanol	[61]
Increased surface kinetics and reduced anode impedance by ALD Ru coating without coking	Ru-coated (<10 nm) Pt (150 nm)	Pt	$Gd_{0.1}Ce_{0.9}O_{2-\delta}$ (350 μm)	13 at 500°C	Ethanol	[62]

Continued

Table 2 Summary of performance data of direct alcohol-fed, low-temperature SOFCs.—cont'd

Main points	Anode	Cathode	Electrolyte	Maximum power density (mW/cm^2)	Fuel	Ref.
Ag anode shows inferior performance compared to Pt anode	Pt	Pt	Yttria stabilized zirconia (YSZ)	4 at 550°C	3.5 kPa: 7.5 kPa ethanol/water	[62]
				3 at 550°C	5 kPa: 7.5 kPa ethanol/water	
	Ag	Ag		2 at 550°C	5 kPa: 7.5 kPa ethanol/water	
Increase of bond cleavage energy due to the use of ethanol fuel	Pd	Pt	Y-BaZrO$_3$ (BYZ) (130 nm)	15.3 at 400°C	Ethanol vapor	[63]
Ceria-salt composite ceramic electrolytes with high ionic conductivity and good chemical stability	NiO (40%)—Ge$_{0.1}$Ce$_{0.9}$O$_{1.95}$ (GDC) (40%) (1–2 mm)	La$_{0.6}$Sr$_{0.4}$Co$_{0.2}$Fe$_{0.8}$O$_{3-\delta}$ (LSCF)	Gd$_{0.1}$Ce$_{0.9}$O$_{1.95}$ (GDC + MOH) (M = Li, Na), MXi (M = Li, Na, Ca, Sr, Ba; X = C1, F; i = 1, 2)	330 at 600°C 300 at 600°C	2 M methanol 1 M ethanol	[64]
Improvement of cell performance by superionic conduction of the composite electrolyte	Ni-Cu-ZnO composite Ce$_{0.8}$Sm$_{0.2}$O$_{1.5}$ (SDC) (1 mm)	Ni-Cu-ZnO + Ce$_{0.8}$Sm$_{0.2}$O$_{1.5}$ (SDC)	Carbon 10 wt% Ce$_{0.8}$Sm$_{0.2}$O$_{1.5}$ (SDC)	213 at 500°C 390 at 600°C	Methanol/water steam	[65]
			Carbon 20 wt% Ce$_{0.8}$Sm$_{0.2}$O$_{1.5}$ (SDC)	295 at 500°C 516 at 600°C		
			Carbon 25 wt% Ce$_{0.8}$Sm$_{0.2}$O$_{1.5}$ (SDC)	431 at 500°C 603 at 600°C		
			Carbon 30 wt% Ce$_{0.8}$Sm$_{0.2}$O$_{1.5}$ (SDC)	376 at 500°C 577 at 600°C		

Description	Anode	Cathode	Electrolyte	Performance	Fuel	Ref.
Used highly ion-conductive BYZ and improved OCV by controlling the thickness and the pore size of BYZ layers	Pt(300 nm)	Pt (200 nm)	Yttria-doped barium zirconate (BYZ) (900 nm)	5.6 at 250°C	Methanol/water vapor	[66]
Trimetal oxide materials with good catalytic activity in direct operation of alcohol	Cu-Ni, Cu-Ni-C (1 mm)	$Ba_{1-x}Sr_xCo_yFe_{1-y}O_{3-\delta}$ (BSCF)	Ceria carbon composite	200 at 500°C	Methanol	[45]
				180 at 500°C	Ethanol	
		$LeFe_{0.8}Ni_{0.2}O_3$ (LFN) (LaFeO-based)		200 at 500°C	Methanol	
				180 at 500°C	Ethanol	
		Trimetal oxide ($CuNiO_x$-ZnO)		300 at 500°C	Methanol	
				380 at 550°C		
				500 at 580°C		

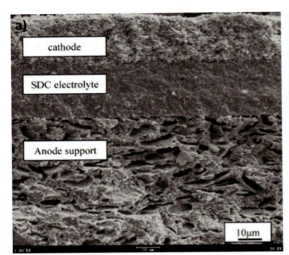 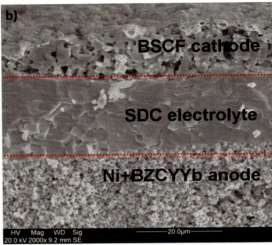

Fig. 3 (A) SEM image of SSC/SDC, SDC, and Ni/SDC; (B) SEM image of Ni+BZCYYb [52, 57].

Metal oxide electrolyte composites that prevent coking and maximize density at the interface where the electrochemical reaction occurs and the electrode, gas, and electrolyte are together have been extensively studied to improve anode performance. Composites such as Ni and proton-conductive Ni-BaZr$_{0.1}$Ce$_{0.7}$Y$_{0.2}$O$_{3-y}$ (BZCY7) and Ni-BaZr$_{0.1}$Ce$_{0.7}$Y$_{0.1}$Yb$_{0.2}$O$_{3-y}$ (BZCYYb) are reported as oxides. Composites such as Ni-Cu and doped ceria (e.g., SDC and GDC) are used as oxide ion conductors in the anode section [34, 41, 42, 48, 49, 51–59, 64, 65, 70, 71]. Liu et al., working on methanol-fed SOFC, studied the NiO-SDC anode composite as a result of preliminary calculations and principalization of some powders (Fig. 3A) [52]. In this study, they demonstrated that methanol-fed SOFCs are stable against carbon coking. The superior kinetic performance of the anode section is explained by the Ni-ceria composite anode. In another study, the Ni-SDC anode composite was produced by ball milling by Meng et al. This composite has shown a more active uniform porous cell quality for fuel oxidation [54]. Catalytic cell performance values were measured as 820, 520, 260, and 110 mW/cm^2 at 600°C, 550°C, 500°C, and 450°C in direct alcohol-fed SOFCs, which use methanol as fuel. In a study of methanol and ethanol fuel-fed SOFCs, cell performance at 600°C was measured as 330 and 300 mW/cm^2 by Zhu et al. In this study, Ni (40%)—Gd$_{0.1}$Ce$_{0.9}$O$_{1.95}$ (GDC) (40%) anode composite was obtained by sintering. Ni-SDC (Sm$_{0.2}$Ce$_{0.8}$O$_{1.9}$), an anode composite for ethanol-fed SOFC, causes low cell performance due to carbon coking [64].

Carbonate, carbon, or other layers are used to develop anode activity. In one study, it was determined that the C-MO-SDC composite provides high conductivity as an anode catalyst [55]. Imran et al. synthesized Ni-Cu-ZnO and SDC-Na_2CO_3 anode composites with high catalytic performance (584 mW/Cm^2—570°C) by the solid-state reaction method. It has been proved to be a very efficient nanocomposite for water-based fuel cells [39]. It was determined that adding Pd to the NiO-YSZ anode affects catalytic cell performance and facilitates the oxidation of methanol [56]. In the Ni-BZCYYb cermet anode, features of BZCYYb such as high water storage capacity, small particle size, good coking resistance, and high oxide-ion conductivity provide high catalytic performance (Fig. 3B) [57]. The Ni-Al_2O_3 catalyst, which gives high performance and strong working stability, improved the oxidation of ethanol (>100 h) (Fig. 4C) [49]. An anode composite was obtained by depositing Ru on Ni via atomic layer deposition (ALD) (Fig. 4A and B) [50].

3.1.2 Noble metal-based

Direct alcohol-fed SOFCs are also preferred as noble metals such as Pt, Pd, Ag, or Pt-Ru in the anode catalyst. The Pt catalyst is seen as most suitable for alcohol oxidation [72–76]. Several studies have used Pt catalysts in the anode section for direct alcohol-fed low-temperature SOFCs. In SOFCs, alcohol-fueled Pt catalysts have CO accumulation. In addition, since complex reactions occur in SOFCs where alcohol is used instead of hydrogen fuel, the anode performance is lower. Anode catalysts developed with noble metals are produced by the spraying technique, thus ensuring that they have a porous structure. The aim is to increase anode performance by integrating with other cell components to improve anode morphology. Ha et al. have conducted studies on the use of Pt catalysts with aluminum oxide support in thin-film SOFCs (Fig. 5A) [66].

Thanks to Ru added to the Pt anode catalyst, the newly formed nanocatalyst increases the thermal stability of the anode and at the same time provides a precaution against carbon coking. To demonstrate this situation, Komadina and Jeong compared Pt and Pt-Ru alloys [60–62]. The addition of Ru to the anode catalyst results in the formation of hydroxyl groups at the catalyst surface. This situation combats carbon coking and causes CO to oxidize. A low cell performance measurement was taken at 250–450°C (<1 mW cm^{-2}). In this study, where methanol was used as fuel, they prepared the Pt-Ru anode catalyst by spraying and showed significantly lower catalytic activity compared to the Pt catalyst [60, 77]. The Pt-Ru catalyst was synthesized by combining Pt and Ru via ALD (Fig. 5B and C) [61, 62]. The Ru-coated Pt anode catalyst used in a methanol and ethanol fuel cell performance test provided a 5- to 10-fold increase in cell performance compared to

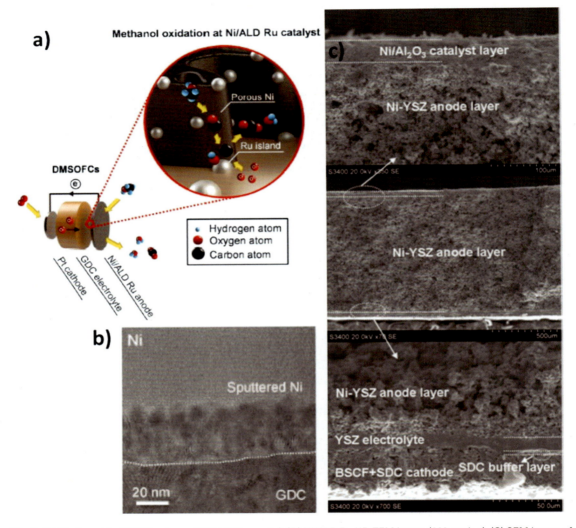

Fig. 4 (A) Methanol-fed SOFC scheme (Ni/ALD Ru anode). (B) Ni/ALD Ru HR-TEM image (300 cycles). (C) SEM image of Ni/Al$_2$O$_3$ catalyst [49, 50]. (A, B) Reprinted (adapted) with permission from H. Jeong, J.W. Kim, J. Park, J. An, T. Lee, F.B. Prinz, et al., Bimetallic nickel/ruthenium catalysts synthesized by atomic layer deposition for low-temperature direct methanol solid oxide fuel cells, ACS Appl. Mater. Interfaces 8(44) (2016) 30090–8. Available from: https://pubs.acs.org/doi/10.1021/acsami.6b08972. Copyright (2016) American Chemical Society.

a pure Pt catalyst. One of the main reasons for this increase is the coating of the surface of the Pt catalyst with Ru by ALD. While Ru increases cell performance and surface kinetics, it ensures that the CO absorbed at the Pt surface is oxidized. In addition, the presence of Ru prevents coarsening of the Pt cluster by creating a depressing oxidation layer on the Pt surface. In this way, a long-term stability in the catalyst is achieved. It is also seen that

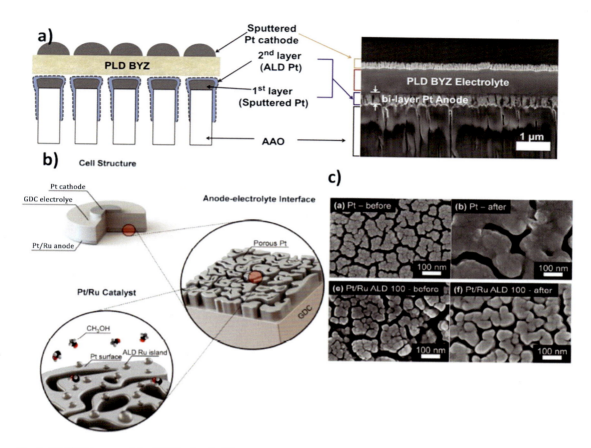

Fig. 5 (A) SEM image of the BYZ fuel cell; (B) schematic image of GDC-based SOFC. (C) Pt/Ru ALD FESEM image [61, 66]. (A–C) Reprinted (adapted) with permission from S. Ha, P.-C. Su, S. Ji, S.W. Cha, Low temperature solid oxide fuel cells with proton-conducting Y:BaZrO₃ electrolyte on porous anodic aluminum oxide substrate, Thin Solid Films 544 (2013) 125–8. Available from: https://linkinghub.elsevier.com/retrieve/pii/S0040609013006949. Copyright (2015) American Chemical Society.

Ag anode catalysts are preferred to avoid the high cost of using Pt catalysts. Poulianitis et al., who investigated the cell performances of Pt and Ag catalysts in an ethanol-fueled cell, stated that the realization rate of the alcohol oxidation reaction mechanism on the Ag surface was slower than that of Pt [78]. In a study conducted by Li et al. with Pd, which could be an alternative to Pt anode catalysts, they looked at cell performance results using ethanol fuel. They obtained the Pd catalyst in a nanoporous structure using the spraying method [63]. In ethanol-fired Pd anode catalyst work, it is thought that the spread of Pd particles and agglomeration is possible due to the complex oxidation of the ethanol fuel. In addition, there appears to be no carbon coking on the Pd surface, and due to this anticoking property, Pd was considered to be a superior catalyst over Pt.

3.2 Electrolytes

As the conducting electrolytes of oxide ions in low-temperature SOFCs, fluorite oxides such as doped seria containing oxygen ion spaces or doped zirconia are preferred. YSZ electrolyte material is used as an oxide-ion conductor in direct alcohol-fed low-temperature SOFCs due to its thermal and mechanical properties and high ionic conductivity, as it is preferred in hydrogen fueled low-temperature SOFCs. In addition, this material is chemically stable under both oxidizing and reducing conditions. Doped ceria materials such as SDC or GDC, which have approximately onefold higher ionic conductivity than YSZ, are among the preferable electrolyte candidates in direct alcohol-fed low-temperature SOFCs due to their lower activation energies than YSZ (at 400–600°C) [52, 54, 56]. To minimize omic resistance, the thickness of the electrolytes should be produced as thin films [49, 51, 58]. This can be done using techniques such as die-pressing [53], coprecipitation [79], tape-casting [56] cofiring [54], particle suspension coating, and wet powder spraying [49, 51, 52, 58]. Today, as new electrolyte materials for direct alcohol-fed, low-temperature SOFCs, composites produced in ceria and salt phases have gained importance. While these electrolytes prevent low cell performance caused by slow anode kinetics, they also prevent ohmic loss and provide superior ionic conductivity. Using composite SDC electrolyte, Gao et al. presented the results when they mixed Na_2CO_3 and pure $Sm_{0.2}Ce_{0.8}O_{1.5}$ (SDC) powder with a carbonate such as Li_2CO_3 [65, 79] (see Fig. 6A and B).

Among the electrolyte materials for SOFCs, proton-conducting electrolytes are being investigated for persistence of high ionic conductivity at low temperatures. These materials are responsible for the conduction of protons (H^+). Oxide ions are considerably larger than proton ions, so the mobility of protons is higher. This situation leads to higher ionic conductivity [80–85]. Generally, ABO_3-structured perovskite materials are used as proton conductor materials. Among these materials, doped barium greenhouse ($BaCeO_3$) and doped barium zirconate ($BaZrO_3$) electrolytes have been investigated for use in low-temperature SOFCs. $BaZrO_3$, when investigated for direct alcohol-fed low-temperature SOFCs, appeared to be more stable in CO_2 production than $BaCeO_3$. However, it has less proton conductivity than $BaCeO_3$. More stability can be achieved for $BaZrO_3$ by adding yttria (Y_2O_3) [86]. In studies conducted by Li [63] and Ha [66], its usability for thin-film, yttria-doped

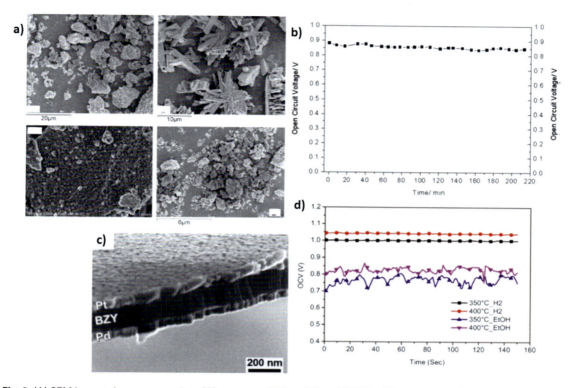

Fig. 6 (A) SEM image of nanocomposites; (B) long-term OCV stability of SDC/Na$_2$CO$_3$ composite; (C) Pt-BZY-Pd μ-SOFC cross-section image; (D) OCV development for ethanol and hydrogen fuel testing [63, 79].

barium zirconate (BYZ, BaZr$_{0.8}$Y$_{0.2}$O$_3$-use) SOFCs developed via pulse laser deposition (PLD) was tested (Fig. 6C and D).

3.3 Cathodes

Many studies for direct alcohol-fed, low-temperature SOFCs have been carried out on anode catalysts. However, the important parameter for cathode catalysts is to provide thermal stability and a reliable power output in the cell. The oxygen reduction reaction (ORR), which occurs in the quota section and is a speed-determining step in temperature processes, is very important for low-temperature SOFCs. In this section, cathode polarization loss is seen as an important situation in the operation of SOFCs [4, 87]. While the most common cathode material used in SOFCs

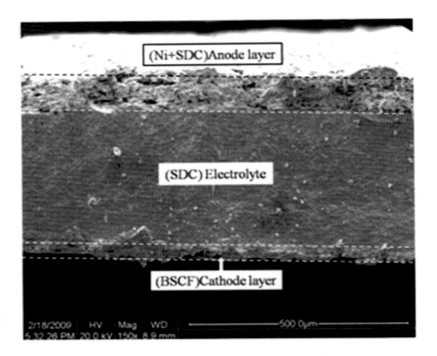

Fig. 7 SEM image of the BSCF cathode layer in SDS [53, 62].

working with hydrogen fuel was perovskite ceramic with additives, this material was also used in direct alcohol-fed SOFCs. BSCF [53, 58], LSC [48] and LSCF [34] cathodes are widely preferred (Fig. 7). Nowadays, new materials that can be alternatives among cathode materials are being tested for direct alcohol-fed, low-temperature SOFCs. Some noble metal composites have shown high catalytic performance for ORRs in low-temperature perovskite oxides. In some studies, Pt or Ag catalysts have been used as both anode and cathode material in direct alcohol-fed, low-temperature SOFCs [78]. The Pt cathode catalyst showed a higher peak density at 550°C compared to Ag. Noble metal catalysts, which are among the cathode materials made so far, have been less preferable than oxide-based materials due to their low thermal stability and high costs [88]. To increase the thermal stability of noble metal catalysts, it is necessary to work with transition metals such as Ni or Co as a Pt-based alloy [89–91]. In addition, it is thought that this thermal stability problem can be solved with oxides such as doped zirconia and doped seria [92–97]. In addition, the suitability of new composite materials (nanospheres or core-shell nanofibers) that are successful in hydrogen fuel-fed, low-temperature SOFCs can be tested for direct alcohol-fed, low-temperature SOFCs [98, 99].

4 Electrochemical performance

Studies of direct alcohol-fed low-temperature SOFCs at temperatures of 600°C and below with a maximum power density exceeding 200 mW/cm^2 for are summarized as in Fig. 8. By comparing the hydrogen-fed, low-temperature SOFCs to these results, the resulting power density differences are segmented. Direct alcohol-fed SOFCs have lower densities than hydrogen-fed SOFCs due to anodic activation loss. In addition, in the comparison of methanol- and ethanol-fed SOFCs, the concentration loss of ethanol was higher than that of methanol [51]. It is necessary to summarize the high-performance membrane electrode assembly (MEA) designs of direct alcohol-fed, low-temperature SOFCs. Many of these high-performance systems are seen to be in anode-supported cell structures. Such anodes are Ni or Ni-Cu alloyed composites known to develop alcohol oxidation kinetics and have anticoking properties [39, 45, 54, 65].

Doped seria materials are preferred as the most common electrolytes, but there appears to be an increasing interest in new materials that can operate at low temperatures, such as ceria plus carbonate [39, 65] or ceria plus carbon [45]. The preferred cathode materials LSCF [54] and BSCG [45, 57] for direct alcohol-fed low-temperature SOFCs are widely used in

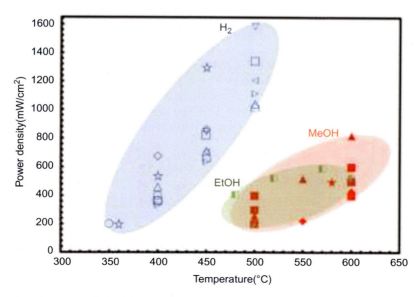

Fig. 8 Summary of performance of alcohol- and hydrogen-fed SOFCs [32].

hydrogen-fed SOFCs. However, trimetal oxide and $LaFe_{0.8}Ni_{0.2}O_3$ (LFN) cathode materials that give high performance at low temperatures for direct alcohol-fed SOFCs are also being investigated [45]. Direct alcohol-fed SOFCs show low catalytic performance compared to hydrogen-fueled low-temperature SOFCs ($\sim$1300 mW/cm^2 at 500°C) compared to low-temperature SOFCs ($\sim$430 mW/cm^2 at 500°C) using methanol or ethanol [19, 65]. To overcome these problems, it is necessary to develop new materials and designs and to investigate direct alcohol-fed SOFCs that provide high performance.

5 Conclusion

In this chapter, results of studies on anode, cathode, and membrane designs and materials for direct alcohol-fed, low-temperature SOFCs are reported. Performance values of hydrogen-fueled SOFCs and SOFCs, where alcohols such as methanol and ethanol are preferred, are given comparatively. Because of the costly problems of accessing hydrogen fuel, direct alcohol-fed, low-temperature SOFCs seem promising, especially as portable power sources, thanks to the ease of access and high energy density of liquid alcohols. However, due to the poor performance of direct alcohol-fed SOFCs, this is one of the main challenges that the energy system must overcome.

Great efforts have been made to ensure that SOFCs operate at low temperatures, and it has been observed that in particular, hydrogen-fueled SOFCs create a power density of approximately 1.6 W/cm^2 at temperatures as low as 500°C. However, direct alcohol-fed, low-temperature SOFCs still need to be investigated in this context [4, 19]. In a study summarizing this situation, Meng et al. [54] measured the catalytic performance of 820 mW/cm^2 for methanol-fed SOFCs. To demonstrate the applicability of these results, it is necessary to reach at least 1 W/cm^2 power density.

References

[1] B.C.H. Steele, A. Heinzel, Materials for fuel-cell technologies, in: Materials for Sustainable Energy, Co-Published with Macmillan Publishers Ltd, UK, 2010, pp. 224–231. Available from: http://www.worldscientific.com/doi/abs/10.1142/9789814317665_0031.
[2] O. Yamamoto, Solid oxide fuel cells: fundamental aspects and prospects, Electrochim. Acta 45 (15–16) (2000) 2423–2435. Available from: https://linkinghub.elsevier.com/retrieve/pii/S0013468600003303.

Chapter 26 Direct alcohol-fed solid oxide fuel cells **503**

[3] S. Park, J.M. Vohs, R.J. Gorte, Direct oxidation of hydrocarbons in a solid-oxide fuel cell, Nature 404 (6775) (2000) 265–267. Available from: http://www.nature.com/articles/35005040.

[4] J. An, Y.-B. Kim, J. Park, T.M. Gür, F.B. Prinz, Three-dimensional nanostructured bilayer solid oxide fuel cell with 1.3 W/cm 2 at 450 °C, Nano Lett. 13 (9) (2013) 4551–4555. Available from: https://pubs.acs.org/doi/10.1021/nl402661p.

[5] E.D. Wachsman, K.T. Lee, Lowering the temperature of solid oxide fuel cells, Science 334 (6058) (2011) 935–939. Available from: https://www.sciencemag.org/lookup/doi/10.1126/science.1204090.

[6] N.P. Brandon, S. Skinner, B.C.H. Steele, Recent advances in materials for fuel cells, Annu. Rev. Mater. Res. 33 (1) (2003) 183–213. Available from: http://www.annualreviews.org/doi/10.1146/annurev.matsci.33.022802.094122.

[7] A. Evans, A. Bieberle-Hütter, J.L.M. Rupp, L.J. Gauckler, Review on microfabricated micro-solid oxide fuel cell membranes, J. Power Sources 194 (1) (2009) 119–129. Available from: https://linkinghub.elsevier.com/retrieve/pii/S0378775309005606.

[8] S. Ji, J. Ha, T. Park, Y. Kim, B. Koo, Y.B. Kim, et al., Substrate-dependent growth of nanothin film solid oxide fuel cells toward cost-effective nanostructuring, Int. J. Precis. Eng. Manuf. Technol. 3 (1) (2016) 35–39. Available from: http://link.springer.com/10.1007/s40684-016-0005-7.

[9] T. Park, G.Y. Cho, Y.H. Lee, W.H. Tanveer, W. Yu, Y. Lee, et al., Effect of anode morphology on the performance of thin film solid oxide fuel cell with PEALD YSZ electrolyte, Int. J. Hydrog. Energy 41 (22) (2016) 9638–9643. Available from: https://linkinghub.elsevier.com/retrieve/pii/S0360319915319650.

[10] W. Yu, S. Ji, G.Y. Cho, S. Noh, W.H. Tanveer, J. An, et al., Atomic layer deposition of ultrathin blocking layer for low-temperature solid oxide fuel cell on nanoporous substrate, J. Vac. Sci. Technol. A 33 (1) (2015), 01A145. Available from: http://avs.scitation.org/doi/10.1116/1.4904206.

[11] M. Li, M. Zhao, F. Li, W. Zhou, V.K. Peterson, X. Xu, et al., A niobium and tantalum co-doped perovskite cathode for solid oxide fuel cells operating below 500°C, Nat. Commun. 8 (1) (2017) 13990. Available from: http://www.nature.com/articles/ncomms13990.

[12] W. Zhou, J. Sunarso, M. Zhao, F. Liang, T. Klande, A. Feldhoff, A highly active perovskite electrode for the oxygen reduction reaction below 600°C, Angew. Chem. Int. Ed. 52 (52) (2013) 14036–14040. Available from: http://doi.wiley.com/10.1002/anie.201307305.

[13] Y. Zhu, Z.-G. Chen, W. Zhou, S. Jiang, J. Zou, Z. Shao, An A-site-deficient perovskite offers high activity and stability for low-temperature solid-oxide fuel cells, ChemSusChem 6 (12) (2013) 2249–2254. Available from: http://doi.wiley.com/10.1002/cssc.201300694.

[14] Y. Zhu, W. Zhou, R. Ran, Y. Chen, Z. Shao, M. Liu, Promotion of oxygen reduction by exsolved silver nanoparticles on a perovskite scaffold for low-temperature solid oxide fuel cells, Nano Lett. 16 (1) (2016) 512–518. Available from: https://pubs.acs.org/doi/10.1021/acs.nanolett.5b04160.

[15] S. Choi, S. Yoo, J. Kim, S. Park, A. Jun, S. Sengodan, et al., Highly efficient and robust cathode materials for low-temperature solid oxide fuel cells: $PrBa_{0.5}Sr_{0.5}Co_{2-x}Fe_xO_{5+\delta}$, Sci. Rep. 3 (1) (2013) 2426. Available from: http://www.nature.com/articles/srep02426.

[16] S. Yoo, A. Jun, Y.-W. Ju, D. Odkhuu, J. Hyodo, H.Y. Jeong, et al., Development of double-perovskite compounds as cathode materials for low-temperature solid oxide fuel cells, Angew. Chem. 126 (48) (2014) 13280–13283. Available from: http://doi.wiley.com/10.1002/ange.201407006.

[17] H. Huang, M. Nakamura, P. Su, R. Fasching, Y. Saito, F.B. Prinz, High-performance ultrathin solid oxide fuel cells for low-temperature operation, J. Electrochem. Soc. 154 (1) (2007) B20. Available from: https://iopscience.iop.org/article/10.1149/1.2372592.

[18] Y.B. Kim, T.M. Gür, S. Kang, H.-J. Jung, R. Sinclair, F.B. Prinz, Crater patterned 3-D proton conducting ceramic fuel cell architecture with ultra thin Y:BaZrO3 electrolyte, Electrochem. Commun. 13 (5) (2011) 403–406. Available from: https://linkinghub.elsevier.com/retrieve/pii/S138824811100049X.

[19] C.-C. Chao, C.-M. Hsu, Y. Cui, F.B. Prinz, Improved solid oxide fuel cell performance with nanostructured electrolytes, ACS Nano 5 (7) (2011) 5692–5696. Available from: https://pubs.acs.org/doi/10.1021/nn201354p.

[20] Z. Fan, J. An, A. Iancu, F.B. Prinz, Thickness effects of yttria-doped ceria interlayers on solid oxide fuel cells, J. Power Sources 218 (2012) 187–191. Available from: https://linkinghub.elsevier.com/retrieve/pii/S0378775312011135.

[21] P.-C. Su, C.-C. Chao, J.H. Shim, R. Fasching, F.B. Prinz, Solid oxide fuel cell with corrugated thin film electrolyte, Nano Lett. 8 (8) (2008) 2289–2292. Available from: https://pubs.acs.org/doi/10.1021/nl800977z.

[22] M. Tsuchiya, B.-K. Lai, S. Ramanathan, Scalable nanostructured membranes for solid-oxide fuel cells, Nat. Nanotechnol. 6 (5) (2011) 282–286. Available from: http://www.nature.com/articles/nnano.2011.43.

[23] J.D. Baek, C.-C. Yu, P.-C. Su, A silicon-based nanothin film solid oxide fuel cell array with edge reinforced support for enhanced thermal mechanical stability, Nano Lett. 16 (4) (2016) 2413–2417. Available from: https://pubs.acs.org/doi/10.1021/acs.nanolett.5b05221.

[24] Y. Zhang, R. Knibbe, J. Sunarso, Y. Zhong, W. Zhou, Z. Shao, et al., Solid-oxide fuel cells: recent progress on advanced materials for solid-oxide fuel cells operating below 500°C, Adv. Mater. 29 (48) (2017), 1770345. Available from: http://doi.wiley.com/10.1002/adma.201770345.

[25] S.P.S. Badwal, S. Giddey, A. Kulkarni, J. Goel, S. Basu, Direct ethanol fuel cells for transport and stationary applications—a comprehensive review, Appl. Energy 145 (2015) 80–103.

[26] M. Cimenti, J. Hill, Direct utilization of liquid fuels in SOFC for portable applications: challenges for the selection of alternative anodes, Energies 2 (2) (2009) 377–410. Available from: http://www.mdpi.com/1996-1073/2/2/377.

[27] D. Yue, S. Pandya, F. You, Integrating hybrid life cycle assessment with multiobjective optimization: a modeling framework, Environ. Sci. Technol. 50 (3) (2016) 1501–1509. Available from: https://pubs.acs.org/doi/10.1021/acs.est.5b04279.

[28] F.G. Üçtuğ, S.M. Holmes, Characterization and fuel cell performance analysis of polyvinylalcohol–mordenite mixed-matrix membranes for direct methanol fuel cell use, Electrochim. Acta 56 (24) (2011) 8446–8456. Available from: https://linkinghub.elsevier.com/retrieve/pii/S0013468611010334.

[29] F. Vigier, C. Coutanceau, A. Perrard, E.M. Belgsir, C. Lamy, Development of anode catalysts for a direct ethanol fuel cell, J. Appl. Electrochem. 34 (4) (2004) 439–446. Available from: http://link.springer.com/10.1023/B:JACH.0000016629.98535.ad.

[30] M. Ni, D.Y.C. Leung, M.K.H. Leung, A review on reforming bio-ethanol for hydrogen production, Int. J. Hydrog. Energy 32 (15) (2007) 3238–3247. Available from: https://linkinghub.elsevier.com/retrieve/pii/S0360319907002479.

[31] J. An, J.H. Shim, Y.-B. Kim, J.S. Park, W. Lee, T.M. Gür, et al., MEMS-based thin-film solid-oxide fuel cells, MRS Bull. 39 (9) (2014) 798–804. Available from: https://www.cambridge.org/core/product/identifier/S0883769414001717/type/journal_article.

Chapter 26 Direct alcohol-fed solid oxide fuel cells **505**

[32] B.C. Yang, J. Koo, J.W. Shin, D. Go, J.H. Shim, J. An, Direct alcohol-fueled low-temperature solid oxide fuel cells: a review, Energy Technol. 7 (1) (2019) 5–19. Available from: http://doi.wiley.com/10.1002/ente.201700777.

[33] M. Cimenti, J.M. Hill, Thermodynamic analysis of solid oxide fuel cells operated with methanol and ethanol under direct utilization, steam reforming, dry reforming or partial oxidation conditions, J. Power Sources 186 (2) (2009) 377–384. Available from: https://linkinghub.elsevier.com/retrieve/pii/S0378775308019575.

[34] D.J.L. Brett, A. Atkinson, D. Cumming, E. Ramírez-Cabrera, R. Rudkin, N.P. Brandon, Methanol as a direct fuel in intermediate temperature (500–600) solid oxide fuel cells with copper based anodes, Chem. Eng. Sci. 60 (21) (2005) 5649–5662. Available from: https://linkinghub.elsevier.com/retrieve/pii/S0009250905004641.

[35] P. Vernoux, J. Guindet, M. Kleitz, Gradual internal methane reforming in intermediate-temperature solid-oxide fuel cells, J. Electrochem. Soc. 145 (10) (1998) 3487–3492. Available from: https://iopscience.iop.org/article/10.1149/1.1838832.

[36] J.-M. Klein, S. Georges, Y. Bultel, Modeling of a SOFC fueled by methane:anode barrier to allow gradual internal reforming without coking, J. Electrochem. Soc. 155 (4) (2008) B333. Available from: https://iopscience.iop.org/article/10.1149/1.2838139.

[37] J.-M. Klein, M. Hénault, P. Gélin, Y. Bultel, S. Georges, A solid oxide fuel cell operating in gradual internal reforming conditions under pure dry methane, Electrochem. Solid-State Lett. 11 (8) (2008) B144. Available from: https://iopscience.iop.org/article/10.1149/1.2936228.

[38] J.-M. Klein, M. Hénault, C. Roux, Y. Bultel, S. Georges, Direct methane solid oxide fuel cell working by gradual internal steam reforming: analysis of operation, J. Power Sources 193 (1) (2009) 331–337. Available from: https://linkinghub.elsevier.com/retrieve/pii/S0378775308022921.

[39] S.K. Imran, R. Raza, G. Abbas, B. Zhu, Characterization and development of bio-ethanol solid oxide fuel cell, J. Fuel Cell Sci. Technol. 8 (6) (2011). Available from: https://asmedigitalcollection.asme.org/electrochemical/article/doi/10.1115/1.4004475/467535/Characterization-and-Development-of-BioEthanol.

[40] M. Mogensen, K. Kammer, Conversion of hydrocarbons in solid oxide fuel cells, Annu. Rev. Mater. Res. 33 (1) (2003) 321–331. Available from: http://www.annualreviews.org/doi/10.1146/annurev.matsci.33.022802.092713.

[41] H. Qin, Z. Zhu, Q. Liu, Y. Jing, R. Raza, S. Imran, et al., Direct biofuel low-temperature solid oxide fuel cells, Energy Environ. Sci. 4 (4) (2011) 1273. Available from: http://xlink.rsc.org/?DOI=c0ee00420k.

[42] C. Rossi, C. Alonso, O. Antunes, R. Guirardello, L. Cardozofilho, Thermodynamic analysis of steam reforming of ethanol and glycerine for hydrogen production, Int. J. Hydrog. Energy 34 (1) (2009) 323–332. Available from: https://linkinghub.elsevier.com/retrieve/pii/S0360319908012779.

[43] T. Valliyappan, N.N. Bakhshi, A.K. Dalai, Pyrolysis of glycerol for the production of hydrogen or syn gas, Bioresour. Technol. 99 (10) (2008) 4476–4483. Available from: https://linkinghub.elsevier.com/retrieve/pii/S096085240700733X.

[44] A. Lima da Silva, I.L. Müller, Thermodynamic study on glycerol-fuelled intermediate-temperature solid oxide fuel cells (IT-SOFCs) with different electrolytes, Int. J. Hydrog. Energy 35 (11) (2010) 5580–5593. Available from: https://linkinghub.elsevier.com/retrieve/pii/S0360319910004386.

506 Chapter 26 Direct alcohol-fed solid oxide fuel cells

[45] M. Mat, X. Liu, Z. Zhu, B. Zhu, Development of cathodes for methanol and ethanol fuelled low temperature (300–600°C) solid oxide fuel cells, Int. J. Hydrog. Energy 32 (7) (2007) 796–801. Available from: https://linkinghub.elsevier.com/retrieve/pii/S0360319906006227.

[46] B. Zhu, X.Y. Bai, G.X. Chen, W.M. Yi, M. Bursell, Fundamental study on biomass-fuelled ceramic fuel cell, Int. J. Energy Res. 26 (1) (2002) 57–66. Available from: http://doi.wiley.com/10.1002/er.765.

[47] S.D. Nobrega, M.V. Galesco, K. Girona, D.Z. de Florio, M.C. Steil, S. Georges, et al., Direct ethanol solid oxide fuel cell operating in gradual internal reforming, J. Power Sources 213 (2012 Sep) 156–159. Available from: https://linkinghub.elsevier.com/retrieve/pii/S0378775312007331.

[48] M.A. Azimova, S. McIntosh, On the choice of anode electrocatalyst for alcohol fuelled proton conducting solid oxide fuel cells, J. Electrochem. Soc. 158 (12) (2011) B1532. Available from: https://iopscience.iop.org/article/10.1149/2.101112jes.

[49] W. Wang, F. Wang, R. Ran, H.J. Park, D.W. Jung, C. Kwak, et al., Coking suppression in solid oxide fuel cells operating on ethanol by applying pyridine as fuel additive, J. Power Sources 265 (2014) 20–29. Available from: https://linkinghub.elsevier.com/retrieve/pii/S0378775314006120.

[50] H. Jeong, J.W. Kim, J. Park, J. An, T. Lee, F.B. Prinz, et al., Bimetallic nickel/ruthenium catalysts synthesized by atomic layer deposition for low-temperature direct methanol solid oxide fuel cells, ACS Appl. Mater. Interfaces 8 (44) (2016) 30090–30098. Available from: https://pubs.acs.org/doi/10.1021/acsami.6b08972.

[51] Y. Jiang, A.V. Virkar, A high performance, anode-supported solid oxide fuel cell operating on direct alcohol, J. Electrochem. Soc. 148 (7) (2001) A706. Available from: https://iopscience.iop.org/article/10.1149/1.1375166.

[52] M. Liu, R. Peng, D. Dong, J. Gao, X. Liu, G. Meng, Direct liquid methanol-fueled solid oxide fuel cell, J. Power Sources 185 (1) (2008) 188–192. Available from: https://linkinghub.elsevier.com/retrieve/pii/S0378775308013529.

[53] L. Sun, Y. Hao, C. Zhang, R. Ran, Z. Shao, Coking-free direct-methanol-flame fuel cell with traditional nickel–cermet anode, Int. J. Hydrog. Energy 35 (15) (2010) 7971–7981. Available from: https://linkinghub.elsevier.com/retrieve/pii/S0360319910009833.

[54] X. Meng, Z. Zhan, X. Liu, H. Wu, S. Wang, T. Wen, Low-temperature ceria-electrolyte solid oxide fuel cells for efficient methanol oxidation, J. Power Sources 196 (23) (2011) 9961–9964. Available from: https://linkinghub.elsevier.com/retrieve/pii/S037877531101500X.

[55] B. Feng, C.Y. Wang, B. Zhu, Catalysts and performances for direct methanol low-temperature (300 to 600 C) solid oxide fuel cells, Electrochem. Solid-State Lett. 9 (2) (2006) A80–A81. Available from: https://iopscience.iop.org/article/10.1149/1.2151128.

[56] M. Sahibzada, B.C.H. Steele, K. Hellgardt, D. Barth, A. Effendi, D. Mantzavinos, et al., Intermediate temperature solid oxide fuel cells operated with methanol fuels, Chem. Eng. Sci. 55 (16) (2000) 3077–3083. Available from: https://linkinghub.elsevier.com/retrieve/pii/S0009250999005692.

[57] W. Wang, Y. Chen, F. Wang, M.O. Tade, Z. Shao, Enhanced electrochemical performance, water storage capability and coking resistance of a Ni $+BaZr_{0.1}Ce_{0.7}Y_{0.1}Yb_{0.1}O_{3-\delta}$ anode for solid oxide fuel cells operating on ethanol, Chem. Eng. Sci. 126 (2015) 22–31. Available from: https://linkinghub.elsevier.com/retrieve/pii/S0009250914007258.

[58] W. Wang, C. Su, T. Zheng, M. Liao, Z. Shao, Nickel zirconia cerate cermet for catalytic partial oxidation of ethanol in a solid oxide fuel cell system, Int. J.

Hydrog. Energy 37 (10) (2012) 8603–8612. Available from: https://linkinghub.elsevier.com/retrieve/pii/S0360319912005022.

[59] M. Liao, W. Wang, R. Ran, Z. Shao, Development of a Ni–Ce$_{0.8}$Zr$_{0.2}$O$_2$ catalyst for solid oxide fuel cells operating on ethanol through internal reforming, J. Power Sources 196 (15) (2011) 6177–6185. Available from: https://linkinghub.elsevier.com/retrieve/pii/S0378775311006008.

[60] J. Komadina, Y.B. Kim, J.S. Park, T.M. Gür, S. Kang, F.B. Prinz, Low temperature direct methanol fuel cell with YSZ electrolyte, ECS Trans. 35 (1) (2019) 2855–2866. Available from: https://iopscience.iop.org/article/10.1149/1.3570285.

[61] H.J. Jeong, J.W. Kim, K. Bae, H. Jung, J.H. Shim, Platinum–ruthenium heterogeneous catalytic anodes prepared by atomic layer deposition for use in direct methanol solid oxide fuel cells, ACS Catal. 5 (3) (2015) 1914–1921. Available from: https://pubs.acs.org/doi/10.1021/cs502041d.

[62] H.J. Jeong, J.W. Kim, D.Y. Jang, J.H. Shim, Atomic layer deposition of ruthenium surface-coating on porous platinum catalysts for high-performance direct ethanol solid oxide fuel cells, J. Power Sources 291 (2015) 239–245. Available from: https://linkinghub.elsevier.com/retrieve/pii/S0378775315008666.

[63] Y. Li, L.M. Wong, H. Xie, S. Wang, P.-C. Su, Nanoporous palladium anode for direct ethanol solid oxide fuel cells with nanoscale proton-conducting ceramic electrolyte, J. Power Sources 340 (2017) 98–103. Available from: https://linkinghub.elsevier.com/retrieve/pii/S0378775316315610.

[64] B. Zhu, Advantages of intermediate temperature solid oxide fuel cells for tractionary applications, J. Power Sources 93 (1–2) (2001) 82–86. Available from: https://linkinghub.elsevier.com/retrieve/pii/S0378775300005644.

[65] Z. Gao, R. Raza, B. Zhu, Z. Mao, Development of methanol-fueled low-temperature solid oxide fuel cells, Int. J. Energy Res. 35 (8) (2011) 690–696. Available from: http://doi.wiley.com/10.1002/er.1718.

[66] S. Ha, P.-C. Su, S. Ji, S.W. Cha, Low temperature solid oxide fuel cells with proton-conducting Y:BaZrO$_3$ electrolyte on porous anodic aluminum oxide substrate, Thin Solid Films 544 (2013) 125–128. Available from: https://linkinghub.elsevier.com/retrieve/pii/S0040609013006949.

[67] J.R. Rostrupnielsen, J.H.B. Hansen, CO$_2$-reforming of methane over transition metals, J. Catal. 144 (1) (1993) 38–49. Available from: https://linkinghub.elsevier.com/retrieve/pii/S0021951783713126.

[68] M.C. Sanchez-Sanchez, R.M. Navarro Yerga, D.I. Kondarides, X.E. Verykios, J.L.G. Fierro, Mechanistic aspects of the ethanol steam reforming reaction for hydrogen production on Pt, Ni, and PtNi catalysts supported on γ-Al$_2$O$_3$, J. Phys. Chem. A 114 (11) (2010) 3873–3882. Available from: https://pubs.acs.org/doi/10.1021/jp906531x.

[69] S. Park, R. Craciun, J.M. Vohs, R.J. Gorte, Direct oxidation of hydrocarbons in a solid oxide fuel cell: I. Methane oxidation, J. Electrochem. Soc. 146 (10) (1999) 3603–3605. Available from: https://iopscience.iop.org/article/10.1149/1.1392521.

[70] J.Y. Won, H.J. Sohn, R.H. Song, S.I. Woo, Glycerol as a bioderived sustainable fuel for solid-oxide fuel cells with internal reforming, ChemSusChem 2 (11) (2009) 1028–1031. Available from: http://doi.wiley.com/10.1002/cssc.200900170.

[71] N. Laosiripojana, S. Assabumrungrat, Catalytic steam reforming of methane, methanol, and ethanol over Ni/YSZ: the possible use of these fuels in internal reforming SOFC, J. Power Sources 163 (2) (2007) 943–951. Available from: https://linkinghub.elsevier.com/retrieve/pii/S0378775306020581.

508 Chapter 26 Direct alcohol-fed solid oxide fuel cells

[72] C. Lamy, A. Lima, V. LeRhun, F. Delime, C. Coutanceau, J.-M. Léger, Recent advances in the development of direct alcohol fuel cells (DAFC), J. Power Sources 105 (2) (2002) 283–296.

[73] E. Antolini, Catalysts for direct ethanol fuel cells, J. Power Sources 170 (1) (2007) 1–12. Available from: https://linkinghub.elsevier.com/retrieve/pii/S0378775307007161.

[74] E. Kuyuldar, S.S. Polat, H. Burhan, S.D. Mustafov, A. Iyidogan, F. Sen, Monodisperse thiourea functionalized graphene oxide-based PtRu nanocatalysts for alcohol oxidation, Sci. Rep. 10 (1) (2020) 7811. Available from: http://www.nature.com/articles/s41598-020-64885-6.

[75] H. Burhan, H. Ay, E. Kuyuldar, F. Sen, Monodisperse Pt-Co/GO anodes with varying Pt: Co ratios as highly active and stable electrocatalysts for methanol electrooxidation reaction, Sci. Rep. 10 (1) (2020) 6114. Available from: http://www.nature.com/articles/s41598-020-63247-6.

[76] E. Kuyuldar, H. Burhan, A. Şavk, B. Güven, C. Özdemir, S. Şahin, et al., Enhanced electrocatalytic activity and durability of PtRu nanoparticles decorated on rGO material for ethanol oxidation reaction, in: Graphene Functionalization Strategies, 2019, pp. 389–398. Available from: http://link.springer.com/10.1007/978-981-32-9057-0_16.

[77] D.A. Stevens, J.M. Rouleau, R.E. Mar, A. Bonakdarpour, R.T. Atanasoski, A.K. Schmoeckel, et al., Characterization and PEMFC testing of $Pt_{1-x} M_x$ (M = Ru, Mo, Co, Ta, Au, Sn) anode electrocatalyst composition spreads, J. Electrochem. Soc. 154 (6) (2007) B566. Available from: https://iopscience.iop.org/article/10.1149/1.2724591.

[78] C. Poulianitis, V. Maragou, A. Yan, S. Song, P. Tsiakaras, Investigation of the reaction of ethanol-steam mixtures in a YSZ electrochemical reactor operated in a fuel cell mode, J. Fuel Cell Sci. Technol. 3 (4) (2006) 459–463. Available from: https://asmedigitalcollection.asme.org/electrochemical/article/3/4/459/451162/Investigation-of-the-Reaction-of-EthanolSteam.

[79] Z. Gao, R. Raza, B. Zhu, Z. Mao, C. Wang, Z. Liu, Preparation and characterization of $Sm_{0.2}Ce_{0.8}O_{1.9}/Na_2CO_3$ nanocomposite electrolyte for low-temperature solid oxide fuel cells, Int. J. Hydrog. Energy 36 (6) (2011) 3984–3988. Available from: https://linkinghub.elsevier.com/retrieve/pii/S0360319910024031.

[80] K.D. Kreuer, Proton-conducting oxides, Annu. Rev. Mater. Res. 33 (1) (2003) 333–359. Available from: http://www.annualreviews.org/doi/10.1146/annurev.matsci.33.022802.091825.

[81] K. Bae, D.Y. Jang, H.J. Choi, D. Kim, J. Hong, B.-K. Kim, et al., Demonstrating the potential of yttrium-doped barium zirconate electrolyte for high-performance fuel cells, Nat. Commun. 8 (1) (2017) 14553. Available from: http://www.nature.com/articles/ncomms14553.

[82] H. Iwahara, H. Uchida, K. Ono, K. Ogaki, Proton conduction in sintered oxides based on $BaCeO_3$, J. Electrochem. Soc. 135 (2) (1988) 529–533. Available from: https://iopscience.iop.org/article/10.1149/1.2095649.

[83] W. Münch, G. Seifert, K. Kreuer, J. Maier, A quantum molecular dynamics study of the cubic phase of $BaTiO_3$ and $BaZrO_3$, Solid State Ionics 97 (1–4) (1997) 39–44. Available from: https://linkinghub.elsevier.com/retrieve/pii/S0167273897000854.

[84] T. Hibino, A. Hashimoto, M. Suzuki, M. Sano, A solid oxide fuel cell using Y-doped $BaCeO_3$ with Pd-loaded FeO anode and $Ba_{0.5}Pr_{0.5}CoO_3$ cathode at low temperatures, J. Electrochem. Soc. 149 (11) (2002) A1503. Available from: https://iopscience.iop.org/article/10.1149/1.1513983.

[85] J.H. Shim, J.S. Park, J. An, T.M. Gür, S. Kang, F.B. Prinz, Intermediate-temperature ceramic fuel cells with thin film yttrium-doped barium zirconate electrolytes, Chem. Mater. 21 (14) (2009) 3290–3296. Available from: https://pubs.acs.org/doi/10.1021/cm900820p.

[86] E. Fabbri, L. Bi, D. Pergolesi, E. Traversa, Towards the next generation of solid oxide fuel cells operating below 600°C with chemically stable proton-conducting electrolytes, Adv. Mater. 24 (2) (2012) 195–208. Available from: http://doi.wiley.com/10.1002/adma.201103102.

[87] H.-S. Noh, K.J. Yoon, B.-K. Kim, H.-J. Je, H.-W. Lee, J.-H. Lee, et al., The potential and challenges of thin-film electrolyte and nanostructured electrode for yttria-stabilized zirconia-base anode-supported solid oxide fuel cells, J. Power Sources 247 (2014) 105–111. Available from: https://linkinghub.elsevier.com/retrieve/pii/S0378775313014225.

[88] J. An, Y.-B. Kim, F.B. Prinz, Ultra-thin platinum catalytic electrodes fabricated by atomic layer deposition, Phys. Chem. Chem. Phys. 15 (20) (2013) 7520. Available from: http://xlink.rsc.org/?DOI=c3cp50996f.

[89] N.M. Marković, T.J. Schmidt, V. Stamenković, P.N. Ross, Oxygen reduction reaction on Pt and Pt bimetallic surfaces: a selective review, Fuel Cells 1 (2) (2001) 105–116. Available from: https://onlinelibrary.wiley.com/doi/10.1002/1615-6854(200107)1:2%3C105::AID-FUCE105%3E3.0.CO;2-9.

[90] L. Xiong, A. Manthiram, Effect of atomic ordering on the catalytic activity of carbon supported PtM (M = Fe, Co, Ni, and Cu) alloys for oxygen reduction in PEMFCs, J. Electrochem. Soc. 152 (4) (2005) A697. Available from: https://iopscience.iop.org/article/10.1149/1.1862256.

[91] H.A. Gasteiger, S.S. Kocha, B. Sompalli, F.T. Wagner, Activity benchmarks and requirements for Pt, Pt-alloy, and non-Pt oxygen reduction catalysts for PEMFCs, Appl. Catal. B Environ. 56 (1–2) (2005) 9–35. Available from: https://linkinghub.elsevier.com/retrieve/pii/S0926337304004941.

[92] H.J. Kim, J.-G. Yu, S. Hong, C.H. Park, Y.-B. Kim, J. An, Ridge-Valley nanostructured Samaria-doped ceria interlayer for thermally stable cathode Interface in low-temperature solid oxide fuel cell, Phys. Status Solidi 214 (11) (2017) 1700465. Available from: http://doi.wiley.com/10.1002/pssa.201700465.

[93] S. Oh, S. Hong, H.J. Kim, Y.-B. Kim, J. An, Enhancing thermal-stability of metal electrodes with a sputtered gadolinia-doped ceria over-layer for low-temperature solid oxide fuel cells, Ceram. Int. 43 (7) (2017) 5781–5788. Available from: https://linkinghub.elsevier.com/retrieve/pii/S0272884217301505.

[94] K.-Y. Liu, L. Fan, C.-C. Yu, P.-C. Su, Thermal stability and performance enhancement of nano-porous platinum cathode in solid oxide fuel cells by nanoscale ZrO_2 capping, Electrochem. Commun. 56 (2015) 65–69. Available from: https://linkinghub.elsevier.com/retrieve/pii/S1388248115001058.

[95] Y.H. Lee, G.Y. Cho, I. Chang, S. Ji, Y.B. Kim, S.W. Cha, Platinum-based nanocomposite electrodes for low-temperature solid oxide fuel cells with extended lifetime, J. Power Sources 307 (2016) 289–296. Available from: https://linkinghub.elsevier.com/retrieve/pii/S0378775315307011.

[96] Y.K. Li, H.J. Choi, H.K. Kim, N.K. Chean, M. Kim, J. Koo, et al., Nanoporous silver cathodes surface-treated by atomic layer deposition of $Y:ZrO_2$ for high-performance low-temperature solid oxide fuel cells, J. Power Sources 295 (2015) 175–181. Available from: https://linkinghub.elsevier.com/retrieve/pii/S0378775315300264.

[97] I. Chang, S. Ji, J. Park, M.H. Lee, S.W. Cha, Ultrathin YSZ coating on Pt cathode for high thermal stability and enhanced oxygen reduction reaction activity, Adv. Energy Mater. 5 (10) (2015) 1402251. Available from: http://doi.wiley.com/10.1002/aenm.201402251.

[98] V. D'Innocenzo, G. Grancini, M.J.P. Alcocer, A.R.S. Kandada, S.D. Stranks, M.M. Lee, et al., Excitons versus free charges in organo-lead tri-halide perovskites, Nat. Commun. 5 (1) (2014) 3586. Available from: http://www.nature.com/articles/ncomms4586.

[99] F. Liang, W. Zhou, Z. Zhu, A highly stable and active hybrid cathode for low-temperature solid oxide fuel cells, ChemElectroChem. 1 (10) (2014) 1627–1631. Available from: http://doi.wiley.com/10.1002/celc.201402143.

27

Commercial aspects of direct alcohol fuel cells

Elif Esra Altuner[a], Kubilay Arıkan[a], Hakan Burhan[a], Sadin Ozdemir[b], and Fatih Şen[a]

[a]Şen Research Group, Department of Biochemistry, Dumlupinar University, Kütahya, Turkey, [b]Mersin University, Food Processing Programme, Technical Science Vocational School, Mersin, Turkey

1 Introduction

For the last 10 years, direct alcohol fuel cells (DAFCs) have been in great demand. Especially in electrical devices, direct fuel cells or proton exchange membrane cells are widely used [1, 2]. Fuel cells and membranes are required to enable the power output of the catalysts in DAFC systems using different alcohols, membranes, and catalysts [1]. DAFCs are preferred because they provide high efficiency and are environmentally friendly and economical. The most preferred alcohol in DAFC is methanol. Besides, other alcohols such as ethanol, propanol, ethylene glycol, and glycerol are also used in these cells [1, 3]. Methanol is the most commonly used alcohol in direct alcohol fuel cells, but the low reaction mechanism of most alcohols in direct alcohol fuel cells confuses. Methanol contains only one carbon and no C–C bond, and electrooxidation of methanol is faster and acidic compared to other alcohols and [1]. in addition, carbon monoxide (CO) poisoning is one of the big problem of platinum (Pt) catalysts.

This chapter is a review of the commercial position of DAFC systems in the industrial field. Direct alcohol fuel cells are very important devices for researchers due to their environmental friendliness. Several direct alcohol fuel cell studies started with methanol, the smallest member of alcohols. The direct methanol fuel cell is the most prone to commercialization within direct alcohol fuel cells and in addition, alcohols such as ethanol, ethylene glycol, etc. follow this series. Methanol and ethanol fuel

---

Nanomaterials for Direct Alcohol Fuel Cells. https://doi.org/10.1016/B978-0-12-821713-9.00012-3
Copyright © 2021 Elsevier Inc. All rights reserved.

cells are most popular for their practical, high density, and energy production. Hydrogen-derived fuel cells provide energy above 100°C however there is no such obstacle in direct alcohol fuel cells [4, 5]. Due to the advantages of DAFCs, small direct methanol fuel cells are widely used for various portable devices. These devices are in use, such as cell phones, notebooks, wireless systematic devices, laptops, tablets, etc. [6–10]. The environment-friendly, noiseless, and high-quality energy production of DAFCs has made their use in the commercial field even more attractive, and especially portable devices (mobile phones, laptops, notebooks, etc.) are fed thanks to the performance of direct fuel cells. Commercial approaches of direct alcohol fuel cells in this chapter and their use in the commercial field are detailed with reference to the compilations.

2 Fuel cells

Fuel cells are used as a renewable energy source. Besides, they are **special energy converter devices** and therefore do not cause pollution. Fuel cells have high energy efficiency. The structure of the electrolytes in which the fuel cells are used affects the operating characteristics of the fuel cells. Fossil energy sources are in a structure that can harm the environment and the use of fuel cells has gained great importance due to their environmentally friendly, and renewable properties Generally, fuel cell systems convert chemical energy into electrical energy [11–14]. Fuel cells systematically consist of an electrolyte, membrane, an anode and a cathode. In direct alcohol fuel cells, the concentration of alcohol affects the power efficiency of the cell. Fuel cells are cells created by various fuels such as hydrogen, methanol, ethanol and natural gas. A fuel cell contains anode and cathode poles and these two poles are separated from each other. The systematic shape of direct alcohol fuel cells is shown in Fig. 1. Fuel cells have great advantages as they are small, portable, efficient, quiet, environmentally friendly devices that work with little power.

The types of fuel cells include proton exchange membrane fuel cells (PEMFCs), DAFCs, phosphoric acid fuel cells, alkaline fuel cells, molten carbonate fuel cells (MCFCs), and solid oxide fuel cells (SOFCs) [16–18]. These fuel cells are categorized according to their temperatures according to their working principles. MCFCs and SOFCs are called high-grade fuel cells, which operate above 600°C. The others are called low-grade fuel cells, which operate below 250°C [16,19,20]. PEMFCs are a collection of systems that involve electrolytic reactions that occur based on

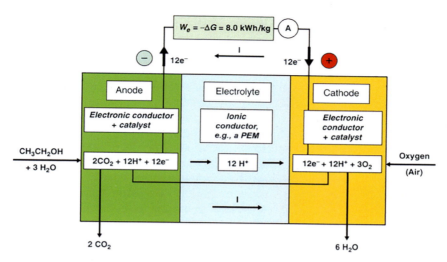

Fig. 1 The schematic of direct ethanol fuel cells [15].

hydrogen ion transfer. Hydrogen ions rush from the anode to the cathode. This system consists of an anode, a cathode, and electrodes. They are in the low-temperature fuel cell class because they operate at low temperatures. PEMFCs are supported by solid polymer membranes [11,21,22]. For the working principle of MCFCs, a method that involves blends of active metals in carbonate form, especially a mixture of lithium (Li) carbonate and potassium carbonate, is followed. Because carbonate is an abrasive material, attention should be paid to the possibility of abrasion in places where it exists in the system. MCFCs are in the group of fuel cells operating at high temperatures, as they operate in the range of 600–700°C [23]. SOFCs consist of a solid oxidized ceramic material used for electrolytes. They also are among the high-temperature fuel cells, as they operate between 750°C and 1000°C. Air oxygen acts as an anode, as in other fuel cells [24]. In phosphoric acid fuel cells, phosphoric acid is used for electrolytes, and it carries protons. The protons are moved from the anode to the cathode by the driving force of phosphoric acid [25]. Negatively charged OH$^-$ ions are produced in an alkaline fuel cell. These negatively charged OH$^-$ ions rush toward the anode [26]. The aspect that this chapter will focus on is the evaluation of alcohol fuel cells from a commercial point of view. DAFCs are used in industry and portable devices because they give fewer side reactions, work better in portable devices, and are more economical than the previously mentioned fuel cells [27,28].

2.1 Overview of direct alcohol fuel cells

Fossil fuels have a structure that can harm the environment. Therefore, in recent years, there has been a great interest in DAFCs as they are environmentally friendly and renewable. One of the most important materials used for the oxidation reaction in fuel cells is platinum as a catalyst. Pure platinum can sometimes give a carbon monoxide poisoning reaction on its surface in reactions, and therefore it can also be used as platinum alloys. There are studies that have explored increasing the efficiency of pure Pt without the need for additional materials and alloys [29]. Gokagac and Sen supported Pt with carbon instead of pure Pt and characterized their results in various ways [30]. By using Pt with reduced graphene oxide, the researchers managed to record oxidation reactions efficiently [31]. Likewise, important results were obtained by supporting Pt with polyaniline and polypyrrole [32]. In addition, studies have been carried out on increasing the oxidation of the electrocatalytic reaction with non-Pt alloys and support materials. Some of these studies have looked at palladium (Pd) that provides catalytic activity [33], ruthenium-copper (Ru-Cu) alloys [34], and Pd-cobalt (Co) alloys [35]. In particular, the use of DAFCs in portable devices (such as mobile phones, laptops, and notebooks) and in cars are examples of their important place in technology today. Among the alcohols, methanol has been the most used because it is the lowest-chain alcohol, easily dissolves in the electrolyte, and is cheap. Other alcohols that have been tried in DAFCs include ethanol, ethylene glycol, propanol, and glycerol [3,36].

In general, the reactions in alcohol fuel cells at the anode and cathode are as follows [4]:

$$\text{Anode}: C_nH_{2n+1}OH + 2n - 1\,H_2O \rightarrow nCO_2 + 6n\,H^+ + 6n\,e^-$$

$$\text{Cathode}: 3n/2\,O_2 + 6n\,e^- + 6n\,H^+ \rightarrow 3n\,H_2O$$

$$\text{Total reaction}: C_nH_{2n+1}OH + 3n/2\,O_2 \rightarrow n\,CO_2 + n + 1\,H_2O$$

2.1.1 Direct methanol fuel cells

Generally, it is difficult to compress and store hydrogen in liquid and gaseous form in PEMFCs, for catalytic reactions [37,38]. The catalytic oxidation of methanol in an acid medium and the electro-oxidation of methanol are given in Eqs. (1), (2), respectively [4]. Methanol is oxidized to carbon dioxide and an oxidation reaction is seen in these equations. As a result of the oxidation reaction, a carbon monoxide (CO) radical is formed on the surface of the platinum catalyst. The CO radical formed is single-electron, has a very high energy, is unstable and has a very high desire to react. Therefore, an easy reaction mechanism will develop and as a result of the intermediate CO radical, the reaction chain continues and thus the reaction is concluded and carbon dioxide and water is formed [38].

$$CH_3OH + H_2O \rightarrow 3/2\,CO_2 + 6\,H^+ + 6\,e^- \qquad (1)$$

$$CH_3OH + 3/2\,O_2 \rightarrow CO_2 + 2H_2O \qquad (2)$$

Previously, pure oxygen was used instead of air in DMFC devices, and articles have always been reported about it. Today, quality efficiency can be obtained in DMFC devices operating under high pressure and low-temperature air, and today most DMFC devices are single-celled [38–40]. Solid polymer membranes were used in the first direct alcohol fuel cell [41]. Besides, direct methanol fuel cells are the most prone to commercialization.

2.1.2 Direct ethanol fuel cells

Ethanol is an edible biofuel obtained from biomass fermentation and agricultural products. Direct ethanol fuel cells are also highly preferred energy devices and have an electrochemical activity comparable to direct methanol fuel cells [42] and due to the some drawbacks that occur in direct methanol fuel cells, researchers have focused on direct ethanol fuel cells , which is less toxic compared to direct methanol fuel cells because ethanol is made from agricultural products by fermentation. Thus, researchers started to investigate direct ethanol fuel cells [42,43]. Direct ethanol fuel cells operate by catalytic reactions based on the oxidation of ethanol [42]. The total oxidation system of ethyl alcohol in acidic media is as follows [42–45]:

$$CH_3CH_2OH \rightarrow [CH_3CH_2OH]ad \rightarrow C_1 ad, C_2 ad \rightarrow CO_2$$

The advantages of direct ethanol fuel cells are as follows; ethanol is not toxic. Ethanol is a high-energy fuel and has an enormous energy density, so direct ethanol fuel cells are highly used fuel cell-like direct methanol fuel cells [42].

2.1.3 Direct ethylene glycol fuel cells

Direct ethylene glycol fuel cells are created like direct alcohol fuel cells of other alcohol types. Oxidation reaction takes place at the anode, and electron accumulation occurs in the environment with hydroxyl ions. Oxidation occurs in the anode as in other fuel cells, and it is used in Pt alloys [46–52]. The electrocatalytic total reaction is as follows [46]:

$$(CH_2OH)_2 + 2H_2O \rightarrow 2CO_2 + 10H^+ + 10e^-$$

Ethylene glycol is less toxic, efficient, safe, and more energy quality than some of the alcohols, and therefore researchers have sought a more reliable fuel. In terms of energy quality, direct ethylene glycol fuel cells have 17% more energy than direct methanol fuel cells [48].

3 Development of direct alcohol fuel cells

Numerous studies have been carried out on DAFCs [15,53–82]. This line of research began on methanol. Due to the some disadvantages of methyl alcohol, ethanol fuel cells were performed directly. In view of some undesirable side reactions in direct ethyl alcohol fuel cells, studies were continued with ethylene glycol, propanol, and other alcohols in fuel cells. As a result of these studies, many academic and scientific studies/patents were obtained. Most patents were taken directly from methanol fuel cell studies. The main components of a direct methanol fuel cell are given in Fig. 2 [81].

While direct methanol fuel cells generate more energy than Li-based power supplies, they are also larger than other power supplies, which unfortunately is also undesirable. Therefore, various studies have been done to solve this problem. First, efforts have been made to increase the durability of devices (e.g., laptops, notebooks) for this situation. General DAFC image is given in Fig. 2. For example, a notebook that can run for 6 h with an ordinary Li-based power source should be able to last 15 h with a direct methyl alcohol fuel cell. Therefore, various academic reviews have examined them [4,54]. Another study looked at making alcohol more efficient in portable, miniaturized devices [4, 8].

4 Process of direct alcohol fuel cells

The systematic process of DAFCs is shown below (see Fig. 3). A modeling scheme is created by making the device on network more comprehensive in order to define the I-V curve.

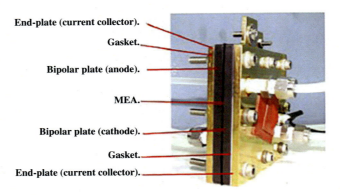

Fig. 2 The main components of direct methanol fuel cell [81].

Thus, interconnected data in the system is analyzed with an artificial modeling network [55]. Recently, according to a number of studies, it was determined that DAFCs perform better in an alkaline environment than an acidic environment. It has been observed that when polymer-based membranes are used in direct fuel cells operating in an alkaline environment, both conductivity and activity increase [56–70]. The alcohol content in a normal DAFC should be in the range of 1 mol/L or 5 mol/L.

Gwak et al. examined the concentration of methanol in methanol fuel cells, the flow process, and every detail required for this process [70–75]. The principles of the working mechanism of direct alcohol fuel cells are shown in Fig. 4.

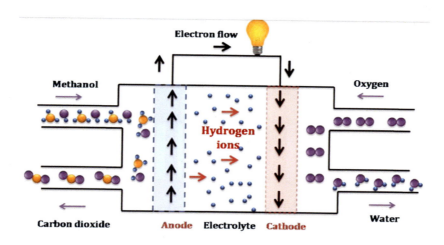

Fig. 3 The systematic process of DAFCs [15].

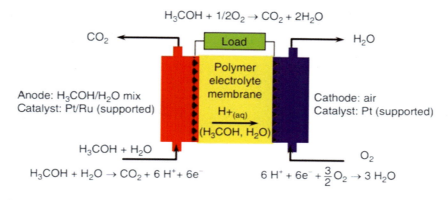

Fig. 4 The principles of DMFC [79].

5 The commercial market of direct alcohol fuel cells

Notebooks using DAFCs were first put on the market in 1986. This resulted in 2% notebook use worldwide. In 2008, notebooks provided approximately 150 million buyers all over the world, while in 2009 this ratio reached approximately 180 million [4]. Looking directly at alcohol fuel cells, portable devices that work directly with methyl alcohol reached a commercial market of about $190 million in 2020 [76]. Initially, the number of notebooks and mobile phones using DAFCs was very low because at that time, procurement of the relevant materials and systems was expensive and time-consuming. Today, with the development of technology, the use of DAFCs has become much more economical in commercial applications, and hence more prevalent. A portable device that use DAFCs are given in Fig. 5.

Methanol is the most studied DAFC and some companies have been worked directly on the methanol fuel cell. These companies are Direct Methanol Fuel Cell Corporation, Toshiba, Motorola, Samsung, Fideris (formerly Lynntech Industries), Manhattan Scientifics, NEC, Giner Electrochemical Systems, Yuasa Corporation, MTI Micro Fuel Cells, and Smart Fuel Cell [77].

5.1 Motorola

Motorola was an American telecommunications company founded in 1928; it went out of business in 2011. It has a methanol evaporator and a vapor converter due to its direct alcohol fuel cell system content [77].

Fig. 5 (A) A DMFC device, (B) a laptop computer with DMFC [80].

(A) (B)

5.2 Toshiba

Toshiba is a Japanese company selling mobile phones that directly utilize methanol fuel cells. Their devices produce approximately 30 mW/cm^2 of power. In addition, the company announced that it will use fuel cells directly in its laptops, notebooks, and tablets as well. In newly purchased Toshiba devices, the direct alcohol fuel cell needs to be charged for about 12 h before it is more active [77]. The portable chargers manufactured by Toshiba has a 400 mA current and 5-volt potential. Toshiba's portable chargers are 150 × 21 × 74.5 mm [54].

5.3 NEC

NEC is a Japanese commercial firm whose devices have approximately 10 times more robust power density than Li-ion based power supplies. Since 2005, it has been trading in notebooks and mobile phones that feature a direct methanol fuel cell system [77].

5.4 Yuasa

Yuasa is a Japanese telecommunications firm competing with NEC, Toshiba, and other firms. Its fuel cell devices produce 100–300 W of power [77].

5.5 Samsung

Samsung is a Korean company that produces notebooks that can last 6 h without recharging. In addition, this company plans to reduce its mobile phones with direct methanol fuel cells to the size of automated teller machine (ATM) cards [77]. For portable devices (phones, cameras, etc.) Samsung produced 2.6 W devices with a lifespan of more than 400 hours through replaceable methanol cartridges in 2006 and in these devices, methanol is used as fuel. The devices produced as Passive chargers are 5 mm thick, 180 g in weight [54].

5.6 Lynntech Industries/Fideris

Lynntech Industries/Fideris has developed devices with four prototype DAFCs that offer a 15 W/12 V power supply, which it sells to the US army [77].

6 Conclusion

The deterioration of the ecological system in the world and the destruction of nature with various chemical fuels have led human beings to seek alternative energy sources, and as a result fuel cells have been discovered. Fuel cells are devices that work by converting chemical energy into electrical energy, and fuel cells are environmentally friendly and provide quality energy production. Direct alcohol fuel cells can be applied to even the smallest electronic components, and therefore their commercial use in the industry of direct alcohol fuel cells is of great importance. In this chapter, information about direct fuel cells and their types is explained with reference and their production in the commercial field is discussed. Some companies have entered into a fierce competition using these cells. Companies such as Toshiba, Motorola, NEC, Yuasa, and Samsung have developed DAFCs that are more durable and more powerful. At the same time, some companies have not been able to keep up with current technology.

To sum up, DAFCs are used in portable devices such as mobile phones, laptops, and notebooks because they give very strong signals and are durable, long-term, and economical. These features make these devices very successful commercially.

References

[1] E.H. Yu, U. Krewer, K. Scott, Principles and materials aspects of direct alkaline alcohol fuel cells, Energies 3 (8) (2010) 1499–1528.

[2] K. Dircks, Recent Advances in Fuel Cells for Transportation Applications, 1999.

[3] C. Lamy, J.-M. Léger, S. Srinivasan, Direct methanol fuel cells: from a twentieth century electrochemist's dream to a twenty-first century emerging technology, in: Modern Aspects of Electrochemistry, Springer, 2002, pp. 53–118.

[4] H.R. Corti, E.R. Gonzalez, Direct alcohol fuel cells, in: H.R. Corti, E.R. Gonzalez (Eds.), Materials, Performance, Durability and Applications Introduction to Direct Alcohol Fuel Cells, vol. 1, Springer, New York, 2014.

[5] V. Antonucci, Direct methanol fuel cells for mobile applications: a strategy for the future, Fuel Cells Bull. 2 (7) (1999) 6–8.

[6] A. Arico, S. Srinivasan, V. Antonucci, DMFCs: from fundamental aspects to technology development, Fuel Cells 1 (2) (2001) 133–161.

[7] R. Dillon, et al., International activities in DMFC R&D: status of technologies and potential applications, J. Power Sources 127 (1–2) (2004) 112–126.

[8] E.A. Franceschini, H.R. Corti, Applications and durability of direct methanol fuel cells, in: Direct Alcohol Fuel Cells, Springer, 2014, pp. 321–355.

[9] Y.R. Thomas, M.M. Bruno, H.R. Corti, Characterization of a monolithic mesoporous carbon as diffusion layer for micro fuel cells application, Microporous Mesoporous Mater. 155 (2012) 47–55.

[10] K.A. Mauritz, R.B. Moore, State of understanding of Nafion, Chem. Rev. 104 (10) (2004) 4535–4586.

[11] N. Esmaeili, E.M. Gray, C.J. Webb, Non-fluorinated polymer composite proton exchange membranes for fuel cell applications—a review, ChemPhysChem 20 (16) (2019) 2016–2053.

[12] N.H. Kim, et al., Synthesis of sulfonated poly (ether ether ketone)/layered double hydroxide nanocomposite membranes for fuel cell applications, Chem. Eng. J. 272 (2015) 119–127.

[13] A.K. Mishra, et al., Silicate-based polymer-nanocomposite membranes for polymer electrolyte membrane fuel cells, Prog. Polym. Sci. 37 (6) (2012) 842–869.

[14] K. Rabaey, W. Verstraete, Microbial fuel cells: novel biotechnology for energy generation, Trends Biotechnol. 23 (6) (2005) 291–298.

[15] C. Lamy, Direct alcohol fuel cells (DAFCs), Encyclopedia of Applied Electrochemistry, Springer, 2014, https://doi.org/10.1007/978-1-4419-6996-5_188.

[16] A. Olabi, T. Wilberforce, M.A. Abdelkareem, Fuel cell application in the automotive industry and future perspective, Energy 214 (2020) 118955.

[17] A.M. Nassef, et al., Maximizing SOFC performance through optimal parameters identification by modern optimization algorithms, Renew. Energy 138 (2019) 458–464.

[18] W.H. Tanveer, et al., Improving fuel cell performance via optimal parameters identification through fuzzy logic based-modeling and optimization, Energy 204 (2020) 117976.

[19] T. Wilberforce, et al., Effect of bipolar plate materials on performance of fuel cells, in: Reference Module in Materials Science and Materials Engineering, Elsevier Inc, 2018, pp. 1–15.

[20] M.A. Abdelkareem, et al., Elimination of toxic products formation in vapor-feed passive DMFC operated by absolute methanol using air cathode filter, Chem. Eng. J. 240 (2014) 38–44.

[21] I. Hanif, et al., Fossil fuels, foreign direct investment, and economic growth have triggered CO_2 emissions in emerging Asian economies: some empirical evidence, Energy 171 (2019) 493–501.

[22] E. Ogungbemi, et al., Fuel cell membranes–Pros and cons, Energy 172 (2019) 155–172.

[23] A.L. Dicks, Molten carbonate fuel cells, Curr. Opinion Solid State Mater. Sci. 8 (5) (2004) 379–383.

[24] R.M. Ormerod, Solid oxide fuel cells, Chem. Soc. Rev. 32 (1) (2003) 17–28.

[25] N. Sammes, R. Bove, K. Stahl, Phosphoric acid fuel cells: fundamentals and applications, Curr. Opinion Solid State Mater. Sci. 8 (5) (2004) 372–378.

[26] F. Bidault, et al., Review of gas diffusion cathodes for alkaline fuel cells, J. Power Sources 187 (1) (2009) 39–48.

[27] F. Vigier, et al., Development of anode catalysts for a direct ethanol fuel cell, J. Appl. Electrochem. 34 (4) (2004) 439–446.

[28] C. Wen, et al., Improving the electrocatalytic properties of Pd-based catalyst for direct alcohol fuel cells: effect of solid solution, Sci. Rep. 7 (1) (2017) 1–11.

[29] S. Eris, Z. Daşdelen, F. Sen, Enhanced electrocatalytic activity and stability of monodisperse Pt nanocomposites for direct methanol fuel cells, J. Colloid Interface Sci. 513 (2018) 767–773.

[30] F. Şen, G. Gökağaç, Different sized platinum nanoparticles supported on carbon: an XPS study on these methanol oxidation catalysts, J. Phys. Chem. C 111 (15) (2007) 5715–5720.

[31] Y. Yıldız, et al., Monodisperse Pt nanoparticles assembled on reduced graphene oxide: highly efficient and reusable catalyst for methanol oxidation and dehydrocoupling of dimethylamine-borane (DMAB), J. Nanosci. Nanotechnol. 16 (6) (2016) 5951–5958.

522 Chapter 27 Commercial aspects of direct alcohol fuel cells

[32] Ö. Karatepe, et al., Enhanced electrocatalytic activity and durability of highly monodisperse Pt@ PPy–PANI nanocomposites as a novel catalyst for the electro-oxidation of methanol, RSC Adv. 6 (56) (2016) 50851–50857.

[33] B. Sen, et al., Monodisperse palladium nanocatalysts for dehydrocoupling of dimethylamineborane, Nano-Struct. Nano-Obj. 16 (2018) 209–214.

[34] B. Sen, et al., Monodisperse rutheniumcopper alloy nanoparticles decorated on reduced graphene oxide for dehydrogenation of DMAB, Int. J. Hydrog. Energy 44 (21) (2019) 10744–10751.

[35] B. Çelik, et al., Monodispersed palladium–cobalt alloy nanoparticles assembled on poly (N-vinyl-pyrrolidone)(PVP) as a highly effective catalyst for dimethylamine borane (DMAB) dehydrocoupling, RSC Adv. 6 (29) (2016) 24097–24102.

[36] C. Lamy, et al., Recent advances in the development of direct alcohol fuel cells (DAFC), J. Power Sources 105 (2) (2002) 283–296.

[37] N. Edwards, et al., On-board hydrogen generation for transport applications: the HotSpot™ methanol processor, J. Power Sources 71 (1–2) (1998) 123–128.

[38] M. Baldauf, W. Preidel, Status of the development of a direct methanol fuel cell, J. Power Sources 84 (2) (1999) 161–166.

[39] J.O.M. Brockris, S. Srinivasan, Fuel Cells: Their Electrochemistry, 1969.

[40] A. Demirbas, Direct use of methanol in fuel cells, Energy Sources Part A 30 (6) (2008) 529–535.

[41] H.F. Hunger, The mechanism of oscillatory behavior during the anodic oxidation of formaldehyde, J. Electrochem. Soc. 115 (5) (1968) 492.

[42] E. Antolini, Catalysts for direct ethanol fuel cells, J. Power Sources 170 (1) (2007) 1–12.

[43] J. Wang, S. Wasmus, R. Savinell, Evaluation of ethanol, 1-propanol, and 2-propanol in a direct oxidation polymer-electrolyte fuel cell: a real-time mass spectrometry study, J. Electrochem. Soc. 142 (12) (1995) 4218.

[44] L. Jiang, et al., Structure and chemical composition of supported Pt–Sn electrocatalysts for ethanol oxidation, Electrochim. Acta 50 (27) (2005) 5384–5389.

[45] E. Antolini, E.R. Gonzalez, A simple model to assess the contribution of alloyed and non-alloyed platinum and tin to the ethanol oxidation reaction on Pt–Sn/C catalysts: application to direct ethanol fuel cell performance, Electrochim. Acta 55 (22) (2010) 6485–6490.

[46] A. Serov, C. Kwak, Recent achievements in direct ethylene glycol fuel cells (DEGFC), Appl. Catal. B Environ. 97 (1–2) (2010) 1–12.

[47] E. Peled, V. Livshits, T. Duvdevani, High-power direct ethylene glycol fuel cell (DEGFC) based on nanoporous proton-conducting membrane (NP-PCM), J. Power Sources 106 (1–2) (2002) 245–248.

[48] R. De Lima, et al., On the electrocatalysis of ethylene glycol oxidation, Electrochim. Acta 49 (1) (2003) 85–91.

[49] A. Neto, et al., Electro-oxidation of ethylene glycol on PtRu/C and PtSn/C electrocatalysts prepared by alcohol-reduction process, J. Appl. Electrochem. 35 (2) (2005) 193–198.

[50] R. Chetty, K. Scott, Catalysed titanium mesh electrodes for ethylene glycol fuel cells, J. Appl. Electrochem. 37 (9) (2007) 1077–1084.

[51] V. Selvaraj, M. Vinoba, M. Alagar, Electrocatalytic oxidation of ethylene glycol on Pt and Pt–Ru nanoparticles modified multi-walled carbon nanotubes, J. Colloid Interface Sci. 322 (2) (2008) 537–544.

[52] N.W. Maxakato, C.J. Arendse, K.I. Ozoemena, Insights into the electro-oxidation of ethylene glycol at Pt/Ru nanocatalysts supported on MWCNTs: adsorption-controlled electrode kinetics, Electrochem. Commun. 11 (3) (2009) 534–537.

Chapter 27 Commercial aspects of direct alcohol fuel cells **523**

[53] S. Badwal, et al., Direct ethanol fuel cells for transport and stationary applications—a comprehensive review, Appl. Energy 145 (2015) 80–103.

[54] K. Kleiner, Assault on batteries, Nature 441 (7097) (2006) 1046–1047.

[55] N.A. Tapan, M.E. Günay, R. Yildirim, Constructing global models from past publications to improve design and operating conditions for direct alcohol fuel cells, Chem. Eng. Res. Des. 105 (2016) 162–170.

[56] E. Antolini, E. Gonzalez, Alkaline direct alcohol fuel cells, J. Power Sources 195 (11) (2010) 3431–3450.

[57] K. Matsuoka, et al., Alkaline direct alcohol fuel cells using an anion exchange membrane, J. Power Sources 150 (2005) 27–31.

[58] E.H. Yu, K. Scott, Development of direct methanol alkaline fuel cells using anion exchange membranes, J. Power Sources 137 (2) (2004) 248–256.

[59] C. Coutanceau, et al., Development of electrocatalysts for solid alkaline fuel cell (SAFC), J. Power Sources 156 (1) (2006) 14–19.

[60] L. Demarconnay, et al., Ethylene glycol electrooxidation in alkaline medium at multi-metallic Pt based catalysts, J. Electroanal. Chem. 601 (1–2) (2007) 169–180.

[61] L. Demarconnay, C. Coutanceau, J.-M. Léger, Study of the oxygen electroreduction at nanostructured PtBi catalysts in alkaline medium, Electrochim. Acta 53 (8) (2008) 3232–3241.

[62] K. Miyazaki, et al., Perovskite-type oxides $La_{1-x}Sr_xMnO_3$ for cathode catalysts in direct ethylene glycol alkaline fuel cells, J. Power Sources 178 (2) (2008) 683–686.

[63] H. Hou, et al., Alkali doped polybenzimidazole membrane for high performance alkaline direct ethanol fuel cell, J. Power Sources 182 (1) (2008) 95–99.

[64] H. Hou, et al., Alkali doped polybenzimidazole membrane for alkaline direct methanol fuel cell, Int. J. Hydrog. Energy 33 (23) (2008) 7172–7176.

[65] C.-C. Yang, S.-J. Chiu, W.-C. Chien, Development of alkaline direct methanol fuel cells based on crosslinked PVA polymer membranes, J. Power Sources 162 (1) (2006) 21–29.

[66] C.-C. Yang, C.-T. Lin, S.-J. Chiu, Preparation of the PVA/HAP composite polymer membrane for alkaline DMFC application, Desalination 233 (1–3) (2008) 137–146.

[67] C.-C. Yang, et al., Study of poly (vinyl alcohol)/titanium oxide composite polymer membranes and their application on alkaline direct alcohol fuel cell, J. Power Sources 184 (1) (2008) 44–51.

[68] J.R. Varcoe, et al., Investigations into the ex situ methanol, ethanol and ethylene glycol permeabilities of alkaline polymer electrolyte membranes, J. Power Sources 173 (1) (2007) 194–199.

[69] Z. OGUMI, et al., Preliminary study on direct alcohol fuel cells employing anion exchange membrane, Electrochemistry 70 (12) (2002) 980–983.

[70] Y. Li, T. Zhao, Z. Liang, Effect of polymer binders in anode catalyst layer on performance of alkaline direct ethanol fuel cells, J. Power Sources 190 (2) (2009) 223–229.

[71] D. Fadzillah, et al., Critical challenges in the system development of direct alcohol fuel cells as portable power supplies: an overview, Int. J. Hydrog. Energy 44 (5) (2019) 3031–3054.

[72] J. Kondoh, et al., Development of a shear horizontal surface acoustic wave sensor system for liquids with a floating electrode unidirectional transducer, Jpn. J. Appl. Phys. 47 (5S) (2008) 4065.

[73] Q. Mao, U. Krewer, Sensing methanol concentration in direct methanol fuel cell with total harmonic distortion: theory and application, Electrochim. Acta 68 (2012) 60–68.

524 Chapter 27 Commercial aspects of direct alcohol fuel cells

[74] J. Geng, et al., An alternating pulse electrochemical methanol concentration sensor for direct methanol fuel cells, Sensors Actuators B Chem. 147 (2) (2010) 612–617.

[75] J.S. Yang, et al., I–V characteristics of a methanol concentration sensor for direct methanol fuel cell (DMFC) by using catalyst electrode of Pt dots, Curr. Appl. Phys. 10 (2) (2010) 370–372.

[76] A.M. Pinto, V.S. Oliveira, D.S.C. Falcão, Direct Alcohol Fuel Cells for Portable Applications: Fundamentals, Engineering and Advances, Academic Press, 2018.

[77] G. Apanel, E. Johnson, Direct methanol fuel cells–ready to go commercial? Fuel Cells Bull. 2004 (11) (2004) 12–17.

[78] T. Chen, et al., Applications of lithium-ion batteries in grid-scale energy storage systems, Trans. Tianjin Univ. 26 (2020) 208–217.

[79] F. Vigier, et al., Electrocatalysis for the direct alcohol fuel cell, Top. Catal. 40 (2006) 1–4, https://doi.org/10.1007/s11244-006-0113-7.

[80] P. Joghee, et al., A review on direct methanol fuel cells—in the perspective of energy and sustainability, MRS Energy Sustain. 3 (2015), https://doi.org/10.1557/mre.2015.4.

[81] S.K. Kamarudin, et al., Overview on the application of direct methanol fuel cell (DMFC) for portable electronic devices, Int. J. Hydr. Energy 34 (2009) 6902–6916, https://doi.org/10.1016/j.ijhydene.2009.06.013.

[82] A.M.F.R. Pinto, et al., ., Direct Alcohol Fuel Cells for Portable Applications, Elsevier, 2018, pp. 1–15.

Index

Note: Page numbers followed by *f* indicate figures, *t* indicate tables, and *s* indicate schemes.

A

Acidic media, oxidation
ethanol, 468–469, 468*f*
methanol, 468–469, 468*f*
propanol, 469–470
Acidification, 288
AEM-based DEGFCs, 47
AgAu nanoparticles, 149
AgCu nanoparticles, 149
Al_2O_3-ZrO_2 nanoceramics, 27
Alcohol fuel cells (AFCs), 174*f*, 357
advantages, 78–79
carbon-based materials, 373–374
carbon-based nanomaterials (CBNs) (*see* Carbon-based nanomaterials (CBNs))
carbon monoxide, 374
carbon-polymer hybrid nanomaterials, 372
cathode and anode, 373–374
climate change, 371
conductive polymers, 372
cost-effective electrocatalysts, 89
dendrimer-based nanocomposites, 346–347
direct ethanol fuel cells (DEFCs), 324–325, 340–341
direct ethylene glycol fuel cells (DEGFCs), 342
direct methanol fuel cells (DMFCs), 325–326, 339–340
electrical appliances, 372
electrical energy, 372–373
electrochemical devices, 372
electrochemical storage systems, 372
electrolysis, 373
electronic devices, 372
energy consumption, 371–372
environmental damage, 371
hydrogen, 373–374
methanol, 337–338, 374, 374*f*
operating temperature/pressure., 373–374
oxidizer, 373
planet Earth, 371
platinum (Pt), 374
polymer materials (*see* Polymer materials)
portable applications, 75
power sources, 372–373
properties, 82*t*
pure hydrogen, 323
renewable energy sources, 371–372
ruthenium (Ru), 374
slow oxidation kinetics, 75
thermodynamics
ethanol oxidation, temperatures, 83*f*
first law, 79–80
gibbs free energy, 81
second law, 80–81
third law, 81
zeroth law, 79
working principle, 323–324, 324*f*
Alcohol-fueled polymer electrolyte membranes, 176–177
Alcohol-fueled proton-exchange membrane fuel cell, 176
Alcohol oxidation kinetics, 39
Alcohol oxidation reactions
acidic media (*see* Acidic media)
alkaline media (*see* Alkaline media)
bonds, 467
carbon-carbon bond, 466
carbon chain, 466–467
carbon-hydrogen bond, 466
carbon monoxide (CO) gas, 466–467
catalysts, 465–466, 473–475, 473*f*
double-path mechanism, 466
dual-path mechanism, 466
electrical energy, 465–466
electric field, 467–468
electrochemical
oxidation, 465–466
processes, 467–468
ethanol, 465–466
formaldehyde, 467
formic acid, 467
ion adsorption, 467
methanol, 465–467
Nernst equation, 467–468
oxygen-hydrogen bond, 466
platinum (Pt), 465–466
Alcohols, 36–37, 77–79
properties, 174–175, 175*t*
reduction technique, 118
Alkaline direct ethanol fuel cells (ADEFCs), 131
Alkaline fuel cells (AFCs), 12, 35–36, 59–61, 357, 512–513
Alkaline media, oxidation
ethanol, 470–472
methanol, 470–472
2-propanol, 472–473
Alloy electrocatalysts

525

526 Index

Alloy electrocatalysts
(*Continued*)
  adsorption and
    decomposition, ethanol,
    214
  alcohol oxidation reactions,
    211, 214–216
  base-metal oxide, 211
  bifunctional mechanism, 211,
    213–214
  catalytic activity, 211, 214
  direct alcohol fuel cells
    (DAFCs), 211
  particle sizes, 214, 215*f*
  platinum-ruthenium
    nanoparticles, 211–212
  Pt-based alloys, 212–213
  Pt-Pd catalysts, 212
  Pt-Ru-based ternary and
    quaternary alloys, 211
  Pt-Ru catalyst, 211–212
  PtRu nanocatalysts, 211–212
  PtSn/C electrocatalysts, 212
  RuO_2, 214
  Ru-Pt nanoparticles, 211–212
Anhydrous proton-conducting
  polymers, 252
Anion exchange membrane fuel
  cells (AEMFCs), 100
Anion exchange membranes
  (AEMs), 41–45, 96
Anode-charged hydrogen, 59–60
Anode electrocatalyst, 238
Anode fuel, 84
Anodizing protection, 8
Asymmetric acrylic membranes,
  296
Atomic force microscopy (AFM),
  276, 277*f*
Au nanoparticles, 150
AuPd nanoparticle synthesis,
  148*f*

B
Bacon fuel cell, 59
Batteries, 9–12
Bimetallic catalysts, 120
Bimetallic nanomaterials

  alcohol oxidation catalyzed,
    150–152
  applications, 152
  AuPd nanoparticle synthesis,
    148*f*
  bimetallic catalytic device,
    147–148
  catalytic performance, 147
  design, 147
  nanoparticles, 149–150
  reactions, 152
  structures, 147
  synthesis, 147–150
  two-alloy metals, 147, 152
Bimetallic nanoparticles
  bottom-down method,
    434–435
  effect of nickel, 441–442
  electrical current
    (*see* Electrical current)
  fabrication/synthesis,
    434–435
  ionic metals, 435–436, 436*f*
  iron-copper, 441
  iron-nickel, 441
  metals, 434–435
  microemulsion, 437
  monometallic, 434–435, 443
  nanocatalysts, 433–434
  nanoscale, 433–434
  palladium (Pd), 442
  physicochemical properties,
    433
  platinum (Pt), 442
  porous form and surface area,
    439–440
  reduction agent, 434–435
  solid nanoparticles, 433
  solid particle, 433
  synergic effect, 434
  synthesis, 435
  thermal method, 436
  top-down method, 434–435
Binary nanomaterials, 165–166
Blending method, 297
Boron-nitrogen doped graphene
  (BNG), 232
Buckminsterfullerene, 310–311

C
Calcium hydroxyapatite
  (CaHAP), 27
Carbonaceous materials,
  221–222
Carbonaceous NMs, 23
Carbonates, 29
Carbon-based nanocomposite
  materials, 122, 344*f*
Carbon-based nanomaterials
  (CBNs), 23–25, 24*f*,
    190–191, 327*f*
  carbon black, 222–223, 329
  carbon nanofibers (CNFs),
    227–229, 308–310
  carbon nanotubes (CNTs),
    223–227, 306–308, 328
  nanosizes, 305
  catalysts, 327–328, 330
  chemical and physical
    properties, 327–328
  electrical conductivity,
    319–320
  electrode material, 375
  energy storage, 375
  fuel cells, 319–320
  fullerenes, 310–311, 310*f*,
    329–330, 377–378
  graphene, 231–233, 328–329,
    375–376
  nanocarbons, 375
  nanodiamonds (ND), 329
  nanosizes, 305
  nanotubes, 376–377
Carbon-based supporting
  materials
  carbon blacks, 222–223
  carbon nanofibers (CNFs),
    227–229
  carbon nanotubes (CNTs),
    223–227
  graphene, 231–233
  mesoporous carbon, 229–231
Carbon blacks, 221–223, 329
Carbon dioxide, 395
Carbon fibers (CFs), 308
Carbon formation, 483
Carbon materials, 151–152

Carbon nanofibers (CNFs), 66, 163–164, 227–229, 308–310, 309*f*

Carbon nanotubes (CNTs), 24, 163–164, 199, 328, 376–377
- carbonaceous nanomaterials, 223–225
- catalysts, 224–225, 306*f*, 307, 312
- chemical structure, 225–226
- conventional carbon powders, fuel cells, 307
- covalent modification, 225–226
- cylinder-shaped macromolecules, 306
- direct alcohol fuel cells (DAFCs), 224–225
- direct methanol fuel cells (DMFCs), 307–308
- fuel cell applications, 224–225
- fullerenes, 308
- hydrophilic characteristics, 306–307
- metal nanoparticles, 224–226
- multiwalled nanotubes (MWCNTs), 306–307
- nanofibers, 308
- Ni and Fe materials, 307
- powder metal catalysts, 307
- Pt/MWCNT, 226–227, 226*f*
- Pt-Ru/MWCNT, 226–227, 226*f*
- single-celled DMFC tests, 308
- single-walled carbon nanotubes (SWCNTs), 306–307
- structures
  - multi-walled carbon nanotube (MWCNT), 224*f*, 226–227
  - single-walled carbon nanotube (SWCNT), 224*f*
- supercritical fluid (SCfs), 307
- 3D nanostructures, 308
- 2D nanostructures, 306
- types, 223–224
- ultrasonic treatment, 306–307

Carbon-supported catalysts, 196*f*, 325–326

Catalyst layer (CL), 134–137

Catalyst rear-blade, 190–191

Catalysts, 131–133, 138, 152
- alcohol fuel cell process, 177
- alkaline fuel cells (AFC), 63
- Au, 63–65
- catalytic electrodes, 187
- characterization (*see* Characterization methods of catalysts)
- chromium-based, 180
- direct alcohol fuel cells (DAFCs), 62–65, 178
- electrocatalytic activity, 70
- electro-oxidation reaction, 322
- ethanol fuel cells, 178–179
- ethanol SOFC, 180
- fuel cells, 62, 77, 304, 319–323
- fuel electro-oxidation kinetic performance, 322–323
- high-temperature alcohol fuel cells, 175–177, 180
- materials, 177
- methanol, 178
- methanol-fueled SOFC, 179
- Ni-based catalysts, 177–179
- nickel (Ni), 63–65, 178
- oxygen reduction reaction (ORR), 322
- platinum (Pt), 62–65, 319–320
- polybenzimidazole-based membrane, 180
- precious-metal-based, 180
- process features, 177
- samarium-doped ceria (SDC), 178
- support material, 305
- synthesis, 65–66
- ZnO, 179

Catalytic reactivity, 133–134

Catalytic vapor deposition (CVD), 308

Cathode electrocatalyst, 137

Cathodes, 499–500

Ceramics nanomaterials, 27

Cerium oxide, 25–26

Characterization and synthesis, polybenzimidazole-base membranes.. *See also* Polybenzimidazole-base membranes, synthesis and characterization
- atomic force microscopy (AFM), 276, 277*f*
- Fourier transform infrared (FTIR), 272, 274*f*
- high-temperature proton exchange membrane fuel cell (HT-PEMFCs), 269
- nafion-based polymer systems, 270
- nanocomposite membranes
  - energy distributions, 276–277, 278*f*
  - mechanics, 276–277, 278*f*
  - storage modules, 276–277, 278*f*
- nanocomposite proton conductive membranes, 270
- poly[2,20-(*m*-phenylene)-5,50-(bibenzimidazole), 269*f*
- Raman spectroscopy, 272–275, 274*f*
- scanning electron microscopy (SEM), 272, 273*f*
- thermal analysis, 275–276, 275*f*
- thermal gravimetric analysis, 270–272, 271*f*
- X-ray diffraction (XRD) analysis, 270, 271*f*

Characterization methods of catalysts
- calcium oxide (CaO) phases, 68
- carbon fuels, 68
- material properties, 66
- new preparation technique, 66
- particle size, 66
- Pt nanocatalysts, 66
- Pt NPs, 67*f*

528 Index

Characterization methods of catalysts *(Continued)*
silver (Ag) NPs, 67
size distribution, 66
structure/mineral phase, fuel samples, 67, 68f
thermogravimetric (TGA) analysis, 69f
x-ray photoelectron spectroscopy, 69
Chemical activation, 222–223
Chitosan, 285
Chlorine alkali process, 288
Chromium-based catalyst, 180
Classification of nanomaterials (NMs)
chemical composition
carbonates, 29
composite, 29
inorganic-based NMs (*see* Inorganic-based NMs)
nitrides, 29
organic-based NMs, 29
silicates, 29
criteria, 19f
dimensionality
one-dimensional (1D), 20–21
three-dimensional (3D), 22
two-dimensional (2D), 21–22
types, 18
zero dimensional (0D), 19–20
morphology, 22–23
Clay (magnesium aluminum silicate), 29
Clean energy technology, 99–100
Coal-derived fuels, 109–110
Composite-based NMs, 29
Composite materials, 379
Composite membranes
acid-base structure
Nafion@PPy (polypyrrole) membranes, 293
Pall IonClad membranes, 293
inorganic organic structure

molybophosphoric acid modification, 293
Nafion, 293
Nafion-Zr membranes, 292
silica, 293
properties, 253–254
Compound gas diffusion layer (GDL), 195–199
Conducting oxide materials, 236–237
Conducting polymers (CPs), 233–235
Conductivity measurements, 266, 267f
Coprecipitation, 419
Core-shell nanoparticles, 28, 149
Corrosion, 7–8
Crown-Joy structure, 148
Crystal MOFs, 359–360
Cyclic voltammetry (CV), 114–115

D
Dehydrogenation, 152
Dendrimer-based nanocomposites
alcohol fuel cells, 346–347
cancer drugs, 345
direct alcohol fuel cells (DAFCs), 337–338, 345
first generation, 345
formation, 346f
macromolecules, 347–348
molecular structure, 337–338
pharmaceutical chemistry, 346–348
platinum-coated support materials, 347–348
polymerization cycle, 345
structure, 346, 346f
textiles, 345
wastewater cleaners, 345
zeroth generation, 345
Dimethyl carbonate (DMC), 120–121
Direct alcohol fuel cells (DAFCs), 304, 374, 407, 465–466
acidic and alkaline electrolytes, 450–451

acidic proton exchange membranes (PEMs), 39
advantages, 137, 410–411, 511–512
air transport, 511–512
alcohol oxidation
kinetics, 39
reaction, 454–456
anode
alcohols, 161f
catalysts, 454–456
electrocatalysts, 210
reactions, 159–160, 168
carbonate, 512–513
carbon dioxide (CO_2), 411
carbon monoxide (CO), 511
catalytic effects, 159
catalysts (*see* Characterization methods of catalysts)
catalyzation, anion exchange membrane, 160f
cathode, 514
alcohols, 161f
catalysts, 456–458
reactions, 159–160, 168
chemical reaction, 410–411
commercial industrial market, 511–512
commercial market
direct methanol fuel cell corporation, 519
Lynntech Industries/ Fideris, 519
methyl alcohol, 518
Motorola, 519
NEC, 519
portable devices, 518, 518f
Samsung, 519
Toshiba, 518
Yuasa, 519
cost and reaction kinetics, 157–158
development, 516, 516f
disadvantages, 137
ethanol fuel cells, 413–414, 515
ethylene glycol fuel cells, 159, 413–414, 515–516
electrical devices, 511
electrical energy, 512

Index **529**

electrochemical oxidation
reaction mechanisms,
39–40
electronic devices, 451
energy
density, 451t
sources, 514
ethylene glycol fuel cells, 159
features, 38
fossil fuels, 514
fuel cells, 512, 513f
high-grade fuel cells,
512–513
hydrogen ions, 450–451,
512–513
hydrogen oxidation reaction
(HOR), 450–451
industry and portable devices,
512–513
liquid and renewable alcohol
fuels, 53–54
low-grade fuel cells, 512–513
material research and
development, 54
materials (*see* Materials, direct
alcohol fuel cells
(DAFCs))
membrane electrode assembly
(MEA), 453–454
methanol, 159, 337–338
methanol fuel cells, 159,
411–412, 514–515
methyl alcohol, 511, 520
nanocomposýtes
(*see* Nanocomposýtes)
nanoelectrocatalysts, 238
operation, 136f
oxygen reduction rate (ORR),
453–454
phosphoric acid fuel cells,
512–513
platinum (Pt), 159–161,
337–338
polyaniline, 514
polypyrrole, 514
portable power applications,
78–79
process, 516–517, 517f
propanol fuel cells, 414

proton exchange membrane
(PEM), 209
proton exchange membrane
fuel cells (PEMFCs),
512–513
Pt and Pt group metals, 137
reactions, 137, 178t
renewable energy sources, 512
temperature, 159
ternary/quaternary
nanomaterials, 165–168
types, 452–453, 452t
uses, 168
Direct borohydride fuel cells
(DBFCs), 99–100, 357
Direct carbon fuel cells (DCFCs),
357
Direct ethanol fuel cells (DEFCs),
96, 324–325, 451
acidic media, 43–44, 44f
advantages, 88, 112
alkaline media, 44–45, 45f
anode-cathode reaction
process, 342f
carbon-supported platinum,
341
disadvantages, 88
electro-oxidation, 341
liquid fuel ethanol, 86–87
nafion membrane, 341
performance, 88–89
platinum (Pt), 113–114, 341
platinum electrocatalyst, 341
power source, mobile devices,
38–39
reactions, 340–341
working principles, 87–88, 87f
Direct ethylene glycol fuel cells
(DEGFCs), 159
acidic media, 45–47
alkaline media, 47–48
anode-cathode reactions,
343f
disadvantages, 342
electrocatalytic reactions, 342
oxidation reaction, 342
process, 343f
Direct glucose fuel cells
(DGFCs), 357

Direct liquid fuel cells (DLFCs),
451
benefits, 38
direct ethanol fuel cells
(DEFCs), 96
hydrogen fuel cells, 48–49
sustainable energy systems, 96
Direct methanol fuel cells
(DMFCs), 12, 57–58,
325–326, 451
acidic media, 40–41
advantages, 85, 286
alkaline media, 41–43
anode catalysts, 84–85
anode-cathode reactions, 340f
catalysts, 286–287
companies, 283–284
direct alcohol fuel cells
(DAFCs), 38
direct ethanol fuel cells
(DEFCs), 209–210
direct hydrogen/reforming
systems, 285–286
disadvantages, 85
electrochemical process,
209–210
energy sources, 339
liquid methanol, 83
methanol transition, 84–85
operating principles, 287
oxygen, 285–286
performance, 85–86, 86f
portable applications, 83
power devices to energy ratios,
340f
process, 340f
Pt-based electrocatalysts, 210
reduce costs, 85–86
requirements, 287
structure, 286f
technological devices, 339,
339f
working principles, 58f, 84–85,
84f
Direct polymer electrolyte fuel
cells, energy densities of,
284t
Disordered mesoporous carbon
(DOMC), 230

Doped ceria materials, 498–499
Doped seria, 499–500
Doped zirconia, 499–500
Double layer current, 115
Drect carbon fuel cell (DCFC), 77
Dry cells, 9
DuPont Nafion membranes, 290–291

E

Electrical current
  highly energetic radiations, 437–438
  metal distributions, 438–439
  sol-gel form, 438
  tetra-alkyl ammonium, 437
Electrical energy, 129–130
Electricity, 2, 319, 321
  generation, 325
Electroactive conjugated polymers, 233–234
Electrocatalysts, 62, 118, 120, 134, 166, 195, 305, 325–326
Electrocatalytic properties, metals, 134–135, 138
Electrochemical cells, 354
  battery and classifications
    fossil fuels, 9
    fuel cells, 11–12
    lithium ion, 10f
    primary, 9
    reserve batteries, 10–11
    secondary/rechargeable cells, 9–10
  chemical and biological fuel cells, 4
  compact Helmholtz layer, 4
  diffuse layer, 4
  electrochemical systems, 4
  electrode solution systems, 5
  electrolytic cell, 3f
  electron transfer, 4–5
  galvanic cells, 3–4, 3f
  types, 2–4
Electrochemical impedance spectroscopy, 258–259
Electrochemical techniques, fuel cells, 218

cyclic voltammetry (CV), 114–115
hydrodynamic voltammetric methods, 115–116
Electrochemistry
  cell potentials, 6–7
  chemical structures, 1
  chemical *vs.* electrical effects, 1
  corrosion, 7–8
  electrical energy, 1
  electrochemical cells, 2–5
  electrochemical-energy generation process, 13
  electrode, 6–7
  history, 1
  oxidation–reduction redox reactions, 2
Electrode and cell potentials
  cell response to stability, 6
  Faraday constant, 6
  half-cell potentials, 6
  Nernst equation, 7
  reference purposes, 6
  thermodynamics, 6–7
Electrode solution potential, 4
Electrolysis, 355
Electrolytes, 321, 498–499
  membrane, 41–43, 289–290
Electrolytic cell, 3f, 4
Electrons, 37, 59–60, 84, 321
Electro-oxidation, 453
Electrospun nanofiber, 146
Energy, 35, 138, 251, 303
  conversion, 175
  production
    fuel cells, 303–304
    renewable energy sources, 303–304
    turbine generators, 303–304
  requirements, 95
  storage, 95
Entropy, 80
Environmental pollution, 319
Enzymatic fuel cells (EFCs), 12
ETFE-SA membranes, 295
Ethanol, 38–39, 86–87, 96, 323–324, 453, 475

Ethanol fuel cells, operation of, 161f
Ethanol oxidation reactions (EORs)
  alkaline electrolyzers (AELs), 97–98
  alkaline environment, 97–98
  anion exchange membrane fuel cells (AEMFCs), 100
  carbon-based materials, 100–101
  carbon nanoparticles, 100–101
  chemical conversion and storage, energy, 99
  clean energy technology, 99–100
  direct borohydride fuel cells (DBHFCs), 99–100
  direct ethanol fuel cells (DEFCs), 96–97, 98f
  electrocatalyst systems, 97–98
  fuel cells, 96, 100
  2,5-furandicarboxylic acid (FDCA) electrocatalytic oxidation, 98–99
  hydroxide exchange membrane electrolyzer (HEMEL), 97–98
  5-hydroxymethylfurfural (HMF), 98–99
  metal-based materials, 100–101
  nanomaterials, 98, 101–102
  NF-supported NiO, 100
  platinum group metals (PGMs), 100
  polymer electrolyte membrane fuel cells (PEMFCs), 99–100
  proton exchange membrane electrolyzers (PEMELs), 97–98
  Pt-Ru nanosized materials, 101
  quantum mechanics, 99
  reaction mechanism, 97f
  techniques, 96–97
  water electrolysis cell, 99
Ethanol transition rate, 88

Ethylene glycol, 39, 46–47
  fuel cells, 159
  oxidation reaction (EGOR),
    46–48
Ex situ conductivity test,
    264–266, 265f
Extracellular matrix (ECM), 146

F
Fabrication methods,
    nanocomposite
    membranes
  blending, 297
  infiltration, 299
  inorganic fillers, 297
  sol-gel method, 298
  types, 298f
Faraday constant, 6
FeCo-EDA-600R catalyst, 203
Fenton tests, 259–261
Fermi energy, 134
Ferromagnetic transition metals,
    146
Fiber supporting materials,
    227–228
Filled porous polymeric
    membranes, 191
First law of thermodynamics,
    79–80
Fossil fuels, 138, 209
  direct liquid fuel cells (DLFCs),
    96
  electrochemical devices,
    101–102
  energy, 53, 138
  energy demand, 129–130
  ethanol oxidation reactions
    (EORs), 96–101
  global warming, 95
  liquid alkaline systems, 96
  materials, 96
  proton exchange membrane
    (PEM), 96
Fourier transform infrared
    (FTIR), 259, 272, 274f, 468
Free enthalpy, 81
Fuel and oxygen oxidizers,
    353–354
Fuel cell components

bipolar (BP) plates, 322
catalysts, 322–323
gas diffusion layers (GDLs),
    322
membranes, 321–322
Fuel cells
  advantages, 110, 130
  alcohols, 36–37, 77–79, 130,
    174–175
  alkaline fuel cells (AFCs),
    35–36, 59–61
  anode, 11
  applications, 355
  bipolar plates, 76
  catalysts319–320
    (see Catalysts)
  categories, 55
  cathode, 11
  chemical energy, fuels,
    129–130
  conversion productivities, 36t
  direct alcohol fuel cells
    (DAFCs), 159–162
  direct methanol fuel cells
    (DMFCs), 57–58
  disadvantages, 110
  ectrochemical devices, 35
  electrical efficiency, 303–304,
    312
  electrical energy, 129–130
  electricity, 53–54, 129–130
  electrochemical
    cells, 75, 354
    device, 53–54, 319
    energy conversion/storage,
      320
    energy generation process,
      13
    reactions, 110
    tools, 54–55
  electrode, 76
  electrolyte, 11, 320
    energy
      conversion, 35, 54
      generating tools, 354
      production, 53, 129–130
  environmentally friendly
    device, 320
  features, 36t, 76

fuel energy, 353–354
galvanic cell-type devices, 11
gas diffusion layers, 76
general scheme, 304f
history, 77, 78f
hydrogen, 173, 283, 319–320
liquid electroyte, 158–159
membrane fuel cells
    (PEMFCs), 35–36
membranes, 76
  filled porous polymeric
    membranes preparation,
    191
  hybrid organic-inorganic
    membranes, 192–194
  proton-conducting
    composite membranes,
    191–192
  zirconium phosphonate
    particles, 192f
metal anodes, 11
methanol, 283–284
molten carbon, 157–158
nanohybrid materials,
    158–159
operating principle, 356
phosphoric acid fuel cells
    (PAFCs), 58–59
potable water applications,
    176f
principles, 354
properties, 76, 131t
proton exchange membrane
    (PEM), 109–110
proton exchange membrane
    fuel cells (PEMFCs),
    55–57, 157–158
renewable potential energy
    techniques, energy
    security, 158
research and development,
    109–110
science and technology, 157
solid electrolytes, 158–159
solid oxide fuel cells (SOFCs),
    61–62
space studies, 109–110
substances, 353–354
temperature, 158–159, 304

532 Index

Fuel cells *(Continued)*
  types, 11–12, 54–55, 131t, 157–159, 158f, 320, 356–357
  usage, 130, 320
  working principle, 321, 321f
Fullerene C_{60}, 310–311
Fullerenes, 310–311, 310f, 329–330, 376–378
Furfuryl alcohol, 122

G
Gallium nitride, 27
Galvanic cells, 3–4, 3f
Gas diffusion layer (GDL), polymeric electrolyte membrane fuel cells, 76
  carbon-supported catalysts, 196f, 203
  catalyst without carbon support, 199–203
  compound, 195–199
  flow area designs, fuel cells, 196f
  improvement, 194–195
  MEA preparation, 203
  metallic materials, 194
  nonprecious catalysts, 203
  polymeric electrolyte membrane fuel cells (PEMFCs), 194
  PTFE blade, 194
  scanning electron microscopy (SEM), foam with pressed channel, 197, 198f
  water flow, 194
  X-ray computed tomography (CT), metallic foam, 197, 198f
  XY-plane cross section, hybrid flow channel, 197, 198f
  XY-plane porosity analysis, hybrid flow channel, 197, 198f
Gibbs free energy, 37, 81
Glycerol, 453
Graphene, 164–165, 231–233, 328–329, 375–376

Graphene oxides (GOs), 164, 164f, 232–233, 344–345
Graphite, 483
Green energy resources, 75

H
Heteroatoms, 232
Heterogeneous distribution, 438–439
High-temperature alcohol fuel cells, 175–177
High-temperature proton exchange membrane fuel cell (HT-PEMFCs), 56–57, 59
  challenges, 253
  opportunities, 253
  PFSA membranes, 252–253
History of fuel cells, 77, 78f
Hollow nanostructures, 149
Hybrid organic-inorganic membranes
  fillers, 192–193
  inorganic and inorganic-organic proton conductive particles, 192–193
  nanocomposite PEMs
    miscellaneous fillers, 193–194
    MOFs, 193
    TiO_2, 193
Hybrid supporting materials, 235–237
Hydrodynamic voltammetric methods, 115–116
Hydrogen, 95, 173, 408–410
  fuel cells, 173, 483
Hydrogen oxidation reactions (HORs), 135, 450–451
Hydrogen-oxygen fuel cell model, 199, 202f
Hydrolysis, 288
Hydrothermal method, 420
Hydroxide exchange membrane electrolyzer (HEMEL), 97–98

I
Indirect alcohol fuel cells (IAFCs), 450–451
Indirect fuel cells, 173
Infiltration method, 299
Inorganic-based NMs
  carbon, 23–25, 24f
  ceramic, 27
  core-shell, 28
  metallic, 25–26
  polymer, 27–28
  quantum dots (QD), 26
  semiconductor, 27
Ion exchange capacity (IEC), 257–258, 260
  measurements, 263

L
Ligand exchange, 364
Liquid alkaline systems, 96
Liquid fuel ethanol, 86–87
Lithium ion battery, 10f
Low-temperature fuel cells, 304
Low-temperature PEMFCs (LT-PEMFCs), 56–57, 59

M
Magnetic Pd-Co bimetallic nanoparticles, 151–152
Materials, direct alcohol fuel cells (DAFCs)
  carbon, 162
  cobalt-polymer multiwalled carbon nanotube (MWCNTs), 162
  electricity, 162
  electron and proton transport, 162
  nanomaterials
    carbon nanofibers (CNFs), 163–164
    carbon nanotubes (CNTs), 163
    graphene, 164–165
    mesoporous nanocarbon, 165
    nanostructured carbon, 163
  oxidation, 162
  Pt-Ru catalyst, 162

Index **533**

MAu/CB/P nanoparticles, 150–151
Membrane electrode assembly (MEA), 44–45, 96, 188, 285–286, 322, 453–454
Membrane fuel cells, 35–36
Membranes, 76, 321–322
Mesoporous carbon, 229–231
  nanocarbon, 165
  supported Pt catalysts, 193
Metal catalysts, 63
Metal-H bond energy, 134–135
Metallic/carbon nanotubes (CNTs)
  electrical characteristics, fuel cell, 202*f*
  nanometer scale, 201*f*
  Ni nanoparticles, 201*f*
  particles, 200*f*
  structure, 200*f*
Metallic nanomaterials
  metal oxides, 25–26
  ultrasonic vibration, 25
Metallic nanoparticles, 132
Metal nanoparticles, 132, 133*f*
Metal organic frameworks (MOFs)
  applications, 361–362
  biomedical use, biological species, 360–361
  catalysis, 360–361, 365
  catalytic chemistry, 360–361
  crystal, 359–360
  crystalline materials, 360
  dimensional structures, 359
  fuel cells
    components, 363–364
    industrial manufacturing, 363
    ligand exchange, 364
    metal exchange, 364
    stratified synthesis, 364
  functionalization, 362–363
  functional sites, 360
  metal-organic structures, 360
  nanocomposites, 365
  organic ligands, 359, 359*f*
  photocatalyst processes, 360–361

  self-assembly, 359–360
  structural classification, 361
  structural features, 362
  synthesis, 361
Metal oxides, 26
  nanomaterials, 25–26
Metals, 22–23
Methanol, 178, 323, 475
  electrooxidation, 117
  oxidation kinetics, 85
Methanol oxidation reaction (MOR), 42*f*
Microbial fuel cells (MFCs), 12
Microemulsion, 418–419, 437
Mobile devices, 35
Molten carbonate, 393–395, 394*f*
Molten carbonate fuel cells (MCFCs), 11, 304, 357, 512–513
Monomers, 233–234
Monometallic nanomaterials, direct alcohol fuel cells (DAFCs), 146
  catalyst layer (CL), 134–137
  nanostructured materials, 131–134, 138
Monometallic nanoparticles, 133, 138
  anode and cathode catalysts, 449–450
  direct alcohol fuel cells (DAFCs), 449
  electricity, 449
  electrocatalyst, 449–450
  electrolyte, 449
  hydrogen (H_2), 449
  hydrogen economy, 449
  membrane electrode assembly (MEA), 449
  nanostructuredmaterials, 449–450
Multi-walled carbon nanotube (MWCNT), 223–224, 224*f*, 235–236, 306, 328, 375, 436

N
Nafion, 41–42, 58–59, 96, 190–191, 254–255, 260–261, 289–290, 290*f*
  composite membranes, 193
  membranes, 176, 259–263, 267–268
  membrane-supported direct ethanol fuel cells, 341
  nanocomposite polymers, 343–344
  platinum nanoparticles, 118
  PTA membranes, 260–261
Nafion@PPy (polypyrrole) membranes, 293
Nafion-ZrO_2 sol-gel nanocomposite, 264–266
Nanoalumina, 25–26
Nanocatalyst synthesis
  ammonia-borane, 415–416
  carbon nanomaterials, 416
  colloid method, 418
  coprecipitation, 419
  graphene, 416
  heterogeneous, 415
  homogeneous, 415
  hydrothermal method, 420
  impregnation method, 416–417
  macromolecules, 416
  metal, 415–416
  microemulsion, 418–419
  nanomaterial synthesis (*see* Nanomaterial synthesis)
  nanoscale, 415
  nanostructures, 416
  polyol method, 420
  sol-gel method, 419–420
Nanocomposite PEMs
  metal organic frameworks (MOFs), 193
  miscellaneous fillers, 193–194
  TiO_2, 193
Nanocomposites
  advantages, 358–359
  carbon nanocomposite materials, 344–345

534 Index

Nanocomposites *(Continued)*
  dendrimers (*see* dendrimer-based nanocomposites)
  graphene oxide, 344–345
  materials, 22, 344–345, 358, 379
  nafion polymers, DAFCs, 343–344
  nanometer-sized particles, 358–359
  nanoparticles, 358
  nanotechnological studies, 337–338
  platinum (Pt), 343–344
  platinum-induced CO toxicity, 345
  polyhydroxy polymers, 344–345
  polymerization, 337–338, 343–344, 358
  properties, 358–359
  quaternized polyvinyl alcohol/fumed silica (QPVA/FS), 343–344
  research development, 343–344, 358
  tructural applications, 358
Nanodiamonds (ND), 329
Nanofibers, 21
Nanofillers, 22
Nanohybrid materials, 158–159
Nanomaterials, 162–165
  alcohol fuel cells, 157
  alcohols, 406
  artificial and natural forms, 326
  carbon nanofibers (CNFs), 163–164
  carbon nanotubes (CNTs), 163
  catalysts, 406
  chemical components, 17
  classification (*see* Classification of nanomaterials)
  definition, 17, 145
  diesel/gasoline-powered vehicles, 408

direct alcohol fuel cells (*see* Direct alcohol fuel cells)
electrical energy, 408
electrolysis, 405–406
electrospun nanofiber, 146
energy, 405
engineered nanomaterials, 146
extracellular matrix (ECM), 146
fossil fuels, 405
graphene, 164–165
Grove's theory, 406–407
hydrogen, 407–410
large-scale energy production, 408
medical industry, 146
mesoporous nanocarbon, 165
microprocessor fuel cells, 408
nanocatalyst synthesis (*see* Nanocatalyst synthesis)
nanometer scale, 17
nanoscale, 326
nanostructured carbon, 163
one-dimensional (1D), 326–327
operating principle, fuel cell, 406–407, 407*f*
oxygen, 407
particles, 17
physical chemistry features, 29–30
polytetrafluoroethylene, 407
production, 146
solid oxide electrolytes, 406–407
terminology, 18*t*
tractors and trucks, 408
two-dimensional (2D), 326–327
uses, 326–327
zero-dimensional (0D), 326–327
Nanomaterial synthesis
  alcohol
    fuel cells, 423, 425
    oxidation, 421–422

catalyst poisoning, 422
Co-MOF, 424
direct methanol fuel cells, 421
economical electrocatalysts, 422
electrochemical active surface, 424
ethanol electrooxidation, 423
ethanol fuel cells, 423
fuel cells, 420–421
hydrothermal method, 424
nanocatalysts, 423
nanocrystalline oxides, 423
nanoparticle materials, 421
Ni-based nanocatalysts, 422
nitrogen gas, 424
Pt and Ru alloys, 420–421
Pt nanoparticles, 423
Pt-Ru catalysts, 422
Pt/TiO2-C catalysts, 424
research and development (R&D) studies, 420–421
titanium isopropoxide, 424
triplemetal nanocatalyst, 424
X-ray diffraction (XRD), 424
Nanoparticles, 131–134
Nano-Pd-HCHO-EDTA, 218
Nanoplates, 21–22
Nanoreactor, 346–347
Nanorods, 21
Nanoscale, 17, 21–22, 29, 337–338
Nanoscience, 131–132
Nanosilicon, 27
Nanosized materials, 65
Nanostructures, 65–66
  carbon, 163
  conducting polymers, 233–235
  fullerenes, 329–330
  materials, 131–134, 138, 162–163
Nanotechnological devices, 168
Nanotechnology, 26, 65, 131–132, 157
Nanotubes, 21
Nanowires, 21
National Aeronautics and Space Administration (NASA), 391–392, 407

Nernst equation, 7
Next-generation fuel cell technology, 253
Ni-based catalysts, 177
Nickel/samarium-doped ceria (Ni/SDC) catalyst, 178
Nitrides, 29
Noble metals, 227
Non-noble metals, 218–220
Nonprecious catalysts, 203

O

Octa decylamine (ODA), 439
Ohmicloss field, 56
One-dimensional (1D) nanomaterials, 20–21, 326–327
Ordered mesoporous carbon (OMC), 230
Organic-based NMs, 29
Oxidation of alcohol, 150–151s, 151–152
Oxidation–reduction redox reactions, 2
Oxidized carbon nanofibers (OCNFs), 310
Oxygen evolution reactions (OERs), 353–354
Oxygen reduction rate (ORR), 453–454
Oxygen reduction reactions (ORRs), 135, 203, 322, 499–500

P

Paint coating, 8
Pall IonClad membranes, 293
Patina, 7–8
Pd-based electrocatalysts, 217–218
PdCu nanomaterial, 149
Pd-In$_2$O$_3$/MWCNT, 237
PDMYH, 355
Pd nanorods, 219f
Pd nanostructures, 218
Pd/PPy-graphene, 235–236
Perfluorinated polymer materials, 289

Perfluorosulfonic acid (PFSA), 251–254
  membranes, 289
Perimeter, 79–80
Phosphoric acid fuel cells (PAFCs), 12, 58–59, 357, 392–393, 394f, 512–513
Photochemical technique, 436
Platinum (Pt), 319–320, 337–338, 341
  catalysts, 62–63
Platinum-ruthenium materials, 86
  nanoparticles, 211–212
Polyalcohols, 420
Polyamidoamine dendrimers, 346–347
Polyaniline (PANI), 233–234, 236–237, 379
Polyarylene ether sulfone-based membranes, 296
Polybenzimidazole-base membranes, synthesis and characterization, 294
  cell temperatures, 254–255
  disulfonated poly(arylene ether sulfonate), 255
  electrochemical impedance spectroscopy, 258–259
  Fenton tests, 259–261
  Fourier transform infrared (FTIR) analyses, 259
  high-temperature, 255
  high-temperature electrolyte membrane, 254
  hybrid membrane systems, 256
  ion exchange capacity (IEC), 257–258
  membrane production, 256
  nafion, 254–255
  oxygen, 254
  poly(2,2′-(p-phenylene)-5,5′-bibenzimidazole), 254
  polymer chain, 256
  polyphosphoric acid, 254
  temperature and conductivity, 254–255
  thermal stability, 255

  water uptake capacities, 256–257
  zirconium phenyl phosphonate, 255
Poly(3,4-ethylenedioxythiophene) (PEDOT), 379
Polyethylene-sub-tetrafluoroethylene (ETFE) polymer, 295
Polyethylene terephthalate (PET), 453
Polyfuel polycarbon membranes, 297
Polymer based nanocatalyts
  aerospace industry, 390
  alkaline fuel cells, 395
  batteries, 390
  carbonate cell, 391–392
  carbon-containing substances, 397–398
  carbon materials, 397
  chemical energy, 390
  CNT- and molten carbonate, 398
  electrical energy, 389–391
  electrocatalysts, 397
  electrochemical converters, 389–390
    reaction, 390–391
  electrodes, 392
  electrolyte, 392, 393t
  electronic conductivity, 398
  energy systems, 389–390
  fossil fuels, 389
  fuel cell
    industry, 392
    technology, 391–392
  hydrocarbon membranes, 397
  hydrogen, 389–391, 398
  liquid petroleum gas (LPG), 390
  metal nanoparticles, 397
  molten carbonate fuel cells, 393–395, 394f, 398
  natural gas, 390
  phosphoric acid fuel cells (PAFCs), 392–393, 394f

536 Index

Polymer based nanocatalyts
 (Continued)
 physiochemical approaches,
 397
 polymer electrolyte fuel cells
 (PEFCs), 396–397, 396*f*
 proton exchange membrane
 fuel cells (PEMFCs), 392
 solar cells, 389
 solid oxide
 electrolytes, 392
 fuel cells, 395–396
 traditional power systems, 399
 transformation, 391, 391*f*
Polymer-based nanocomposites,
 337–338
Polymer electrolyte fuel cells
 (PEFCs), 396–397, 396*f*
Polymer electrolyte membrane
 (PEM)
 characteristics, 286
 chitosan, 285
 composite membranes used
 with fluorine
 acid-base structure, 293
 inorganic organic structure,
 292–293
 contact with fluorine
 chemical DOW-XUS
 membranes, 291
 3P energy membranes, 291
 direct methanol fuel cells
 (DMFCs), 283–286, 297,
 299
 DuPont Nafion membranes,
 290–291
 electrochemical devices,
 284–285
 materials, 287–289
 membrane electrode assembly
 (MEA), 285–286
 methanol, 283–284
 methoxide oxidation,
 285–286
 Nafion, 283–284
 polyvinyl alcohol (PVA), 285
 properties, 283–284, 287–297
 unmodified composite
 membranes with fluorine

composite membranes with
 inorganic/organic
 structure, 293–297
Polymeric electrolyte membrane
 fuel cells (PEMFCs),
 99–100, 411
 carbon-based nanomaterials,
 190–191
 catalysts, 190–191
 particles, 190–191
 rear-blade, 190–191
 electrons, 189, 189*f*
 fossil fuels, 251
 fuel cells, 188, 191–194
 gas diffusion layer (GDL)
 (*see* Gas diffusion layer
 (GDL))
 gas transport, 189, 189*f*
 high-temperature, 252
 membrane electrode assembly
 system (MEA), 188
 nafion, 190–191
 parts, 187
 perfluorinated composite
 membrane, 190–191
 processes, 251–252
 proton-exchange membranes,
 188
 protons, 189, 189*f*
 Pt-based catalysts, 190–191
 temperatures, 188
 transportation of protons, 189
 view, 188*f*
 working principle, 251
Polymeric nanocomposites, 28
Polymeric nanomaterials, 27–28
Polymerization, 343–344
Polymers, 233–234
 materials
 carbon-polymer hybrid-
 supported, 379–380
 conductive polymers (CPs),
 379
 synthetic polymers, 378–379
 membranes, 354
 nanocomposites, 358
Polymer-stabilized bimetallic
 AuPd nanoparticles, 151
Poly-3-methyl thiophene, 379

Polyol method, 420
Polyvinyl alcohol (PVA), 285
Porous carbon, 165, 229–230
Porous polytetrafluoroethylene
 (PTFE), 190–191
Portable power devices, 75
Precious-metal-based catalysts,
 180
Primary batteries, 9
1-Propanol, 414
2-Propanol, 414
Proton-conducting composite
 membranes, 191–192
Proton exchange membrane
 (PEM), 96, 188, 209, 278,
 289–290
Proton exchange membrane
 electrolyzers (PEMELs),
 97–98
Proton exchange membrane fuel
 cells (PEMFCs), 12, 55–57,
 55*f*, 109–110, 157–158,
 304, 356, 512–513
 characteristic curve, 56*f*
 electrocatalytic performance,
 121–122
 hydrogen oxidation reaction,
 122
 operation, 112*f*
 Pt-loaded DMCs, 121–122
 Pt-loaded mesoporous
 carbons, 121–122
 reactions, 113*f*
Protons, 84, 321
Pt-based alloys, 62–63, 212–213
Pt-based catalysts, alcohol
 oxidation
 electrochemical techniques,
 fuel cells, 114–116
 electrodes, 122–123
 fuel cell performance,
 122–123
 plasma effects, 111–114
 single-cell tests (*see* Single-cell
 tests, Pt-based catalysts)
 solvent environmental effects,
 111
Pt-based electrocatalysts, 84–85,
 210

alloy (*see* Alloy electrocatalysts)
Pt-oxide electrocatalysts, 216–217
Pt-based nanosheets (S-PtPdPt$_{BN}$), methanol oxidation reactions, 152
Pt-free electrocatalysts, 217–220
Pt group metals (PGMs), 135
Pt-LiCoO$_2$ catalyst, 117
Pt NW arrayed electrode, 216–217
Pt-oxide electrocatalysts, 216–217
Pt/PIn-MWCNT, 235–236
Pt-Ru alloys, 63
Pt-Ru nanocatalysts, 234f
Pt-Ru-based catalyst, 474
Pt-Sn electrocatalysts, 474
Pt-Sn nanocatalysts, 234f
Pt-tin electrodes, 475
Pulse laser deposition (PLD), 498–499
PVDF-based membranes, 296

Q

Quantum dots (QD), 26
  nanotechnology, 26
Quantum mechanics, 99
Quarternary phosphonium polymers, 167–168

R

Radioisotope thermal generator (RTG), 408
Raman spectroscopy, 272–275, 274f
Reduced carbon nanofibers (RCNFs), 310
Reformed alcohol fuel cells (RAFCs), 450–451
Reinforcement, 358–359
Renewable energy, 303, 319–320
Reserve batteries, 10–11

S

Scanning electron microscopy (SEM), 272, 273f

Secondary/rechargeable cells, 9–10
Second law of thermodynamics, 80–81
Semiconductor nanomaterials, 27
Silicates, 29
Single-cell tests, Pt-based catalysts
  alcohol reduction technique, 118
  bimetallic catalysts, 120
  carbon-based materials, 122
  density function theory (DFT) study, 119
  dimethyl carbonate (DMC), 120–121
  FeNi and AgCu alloys, 117
  LiCoO$_2$, 117–118
  materials, 117
  metal-adsorbent bond, 116
  metal alloys, 116
  nafion-platinum nanoparticles, 118
  ORR mechanisms, 120–121
  oxygen-metal adsorption bond, 119
  particle size, 120
  Pd-Co catalysts, 118–119
  PEM fuel cell performance, 121–122
  powder silica particles, 121
  Pt-CeO$_2$, 118
  PtFoM/C catalysts, 119
  Pt-LiCoO$_2$ catalyst, 117
  Pt metals, 117
  Pt-Ru electrocatalyst, 120
  PtSnNi and PtSnRh triple alloys, 117
  quaternary catalysts, 119
  Tafel equation, 116
  transition metals, 119
Single-walled carbon nanotubes (SWCNTs), 223–224, 224f, 306–307
Sodium borohydride, 439
Sol-gel method, 298, 419–420
Solid electrochemical devices, 61

Solid oxide fuel cells (SOFCs), 11, 61–62, 175–176, 304, 357, 395–396, 512–513
  carbon dioxide (CO$_2$) emission, 481
  carbon monoxide (CO), 481–482
  cathodes, 499–500
  components, 486–500
  electrochemical performance, 501–502, 501f
  electrolytes, 498–499
    materials, 486–487
  energy conversion efficiency, 481
  ethanol, 481–482, 482t
  flexible fuel, 481
  fuels and anode reactions, 483–486
  glycerol, 481–482
  high-temperature values, 481
  hydrogen, 481–482, 482t
  liquid alcohols, 481–482
  metals, 486–487
  methanol, 481–482, 482t
  noble metal-based, 495–497, 497f
  nonnoble metal-based anode, 487
    catalytic cell performance values, 494
    high electrical conductivity, 487
    hydrocarbons, 487
    hydrogen electrocatalytic oxidation, 487
    metal oxide electrolyte, 494
    Ni-BZCYYb cermet anode, 494–495
    performance data, 487, 488–493t
    solid-state reaction method, 494–495
  structure, 483
Sulfonated polyether ether ketone (SPEEK)-based membranes, 294–295, 297
Sulfonated polyphosphate (sPPZ), 294–295

Supercritical fluid (SCf), 307
Supporting materials
  alcohol oxidation, 220–221
  carbonaceous materials,
    221–222
  carbon-based (*see* Carbon-
    based supporting
    materials)
  carbon blacks, 221–222
  challenges, 237–238
  conducting polymers (CPs),
    233–235
  DAFCs, 221–222
  electrocatalysts, 238
  future perspectives, 237–238
  hybrid, 235–237
  loading catalysts, 220–221
Surfactants, 418
Synthesis and characterization,
    nanocomposite
    membranes
  conductivity measurements,
    266, 267*f*
  ex situ conductivity test,
    264–266, 265*f*
  ion-exchange capacity
    measurements, 263
  in situ sol-gel synthesis
    methods, 261–262, 262*f*
  thermomechanical
    characterization, 267–268,
    268–269*f*
  water uptake measurements,
    262–263, 263–264*f*
Synthesis of catalysts
  chemical composition, 65
  CO, fuel cells, 65

nanostructures, 65–66
properties, 65

T
Teflon, 289
Ternary/quaternary
    nanomaterials, direct
    alcohol fuel cells (DAFCs)
  binary nanomaterials, 165–166
  electrochemical applications,
    167–168
  electrochemical oxidation
    bismuth, 167
    nickel, 167
  energy generation, 165
  metal dissolution, 166
  metal ternary alloys, 167–168
  methyl alcohol, 166
  platinum (Pt), 165–166
  platinum-palladium-cobalt
    triple system, 166
  PtPdCo/C catalysts, 166
  quarternary phosphonium
    polymers, 167–168
  Sn ions, 167–168
  synthesis, Pt-Pt-Ni ternary
    metals, 166
Thermal analysis, 275–276, 275*f*
Thermodynamics, 6–7
Thermogravimetric (TGA)
    analysis, 68, 270–272
Thermomechanical
    characterization, 267–268,
    268–269*f*
Third law of thermodynamics, 81
Three-dimensional (3D)
    nanomaterials, 22

Transmission electron
    microscopy (TEM)
    images, 415–416, 415*f*
Two-dimensional (2D)
    nanomaterials, 21–22,
    326–327

U
Ultrasonic treatment, 306–307
Ultraviolet (UV) spectra, 441

V
Volcano curve, metal electrode,
    134–135, 135*f*
Vulcan XC72R, 223

W
Water uptake
  capacities, 256–257
  measurements, 262–263

X
X-ray diffraction (XRD) analysis,
    270
X-ray photoelectron
    spectroscopy, 69

Z
Zeolite, 29
Zero-dimensional (0D)
    nanomaterials, 19–20,
    326–327
Zeroth law of thermodynamics,
    79
Zinc oxide, 25–26

Printed in the United States
by Baker & Taylor Publisher Services